# Progress in
# Inorganic Chemistry

## Volume 46

# Advisory Board

# PROGRESS IN
# INORGANIC CHEMISTRY

*Edited by*

KENNETH D. KARLIN

DEPARTMENT OF CHEMISTRY
JOHNS HOPKINS UNIVERSITY
BALTIMORE, MARYLAND

## VOLUME 46

AN INTERSCIENCE® PUBLICATION
**JOHN WILEY & SONS, INC.**
**New York · Chichester · Weinheim · Brisbane · Singapore · Toronto**

Cover Illustration of ''a molecular ferric wheel'' was adapted from Taft, K. L. and Lippard, S. J., *J. Am. Chem. Soc.*, **1990,** 112, 9629.

This text is printed on acid-free paper.

An Interscience® Publication

Library of Congress Catalog Card Number 59-13035
ISBN 0-471-17992-2

Printed in the United States of America

10 9 8 7 6 5 4 3 2 1

# Contents

# Progress in
# Inorganic Chemistry

## Volume 46

# Anion Binding and Recognition by Inorganic Based Receptors

**PAUL D. BEER** *and* **DAVID K. SMITH**

*Inorganic Chemistry Laboratory, University of Oxford*
*Oxford, UK*

## CONTENTS

*Progress in Inorganic Chemistry, Vol. 46*, Edited by Kenneth D. Karlin.
ISBN 0-471-17992-2 © 1997 John Wiley & Sons, Inc.

## I.  INTRODUCTION

In comparison to cation coordination chemistry, anion coordination is a recent development. The birth of this field occurred in the late 1960s with the synthesis of the first artificial host molecules (1–3). It is perhaps surprising that anion recognition was so slow to begin, bearing in mind the importance of anions in many chemical and biological processes. Specific receptors capable of binding anionic guests are dependent on effectively addressing the characteristic features of anions, such as their negative charge, their size (larger than analogous cations), their wide variety of shapes, and their pH dependence.

Activity in this field during the 1970s and early 1980s was reported in a range of review articles (4–6). These articles outlined the basic principles involved in anion binding, which helped to delineate the arena for further experimental investigations. In the ensuing years, the investigation of anion recognition systems has continued apace (7–15), and involved the development of a whole new range of receptors, many incorporating metal centers. To date, however, there has not been any comprehensive review of this type of anion receptor based on inorganic systems, an omission this chapter intends to rectify.

Anions are of key importance across many fields of scientific life; making their selective binding and sensing a critical research target, as the following illustrative examples indicate:

*Chemically*, anions have various roles as catalysts, bases and redox mediators. The use of receptors to coordinate anions can alter their chemical reativity (4), and may also be helpful for mixture separation or stabilization of unstable species.

*Environmentally*, anions pose a considerable pollution problem. In particular, the nitrate anion (used in fertilizers on agricultural land) often pollutes river water to unacceptable levels. This pollution leads to eutrophocation and consequent disruption of aquatic life cycles (16). Radioactive pertechnetate anions also cause a pollution problem in the nuclear fuel cycle. Selective binding and sensing of environmentally sensitive anions is, therefore, an important goal.

*Biologically*, adenosine triphosphate (ATP), the free energy of life processes, is itself an anion, bound by enzymes in order to perform its many metabolic functions. Deoxyribonucleic acid (DNA) is also a polyanion, its binding by proteins being of great importance in transcription and translation processes. Anion-binding biomimicry could therefore yield much information about fundamental biological processes.

*Medically*, anions are of great importance in many disease pathways. Cystic fibrosis, a genetic illness affecting a significant proportion of society, is caused by misregulation of chloride channels (17). There is, therefore, a real need for selective halide detection, as established methods of chloride analysis are unsuitable for biological applications (18). Cancer is caused by the uncontrolled replication of polyanionic DNA. Anion-binding proteins have also been implicated in the mechanism of Alzheimer's disease (19).

Many enzymes bind anions very successfully in biological systems; the majority of substrates and cofactors bound by enzymes are anionic (20). It is clearly worth considering the ways in which enzymes carry out this function before going on to discuss synthetic chemical approaches to the problem of anion binding.

## II.  BIOLOGICAL APPROACHES TO ANION BINDING

### A.  Binding through Hydrogen Bonding

A recurrent theme in biological chemistry is hydrogen bonding, and the action of receptors for anions is no exception. The crystal structure of the sulfate-binding protein Salmonella typhimurium provides a remarkable example of the coordinative strength of hydrogen bonds for anions (21). The sulfate anion is buried in a cleft 7 Å below the surface of the enzyme. The only stabilization for its dinegative charge is provided by neutral hydrogen-bond donors; peptide groups on the protein backbone are particularly important.

More commonly, there is salt bridge hydrogen-bond formation with a protonated amino group (lysine, arginine, or the N-terminus of the protein) which provides additional anion stabilization through electrostatic attraction (22). Extensive surveys of the active sites of solely organic, hydrogen-bonding anion-binding proteins have been made (23, 24). Chakrabarti (23) showed that, on average, oxoanions are held by $7(\pm 3)$ hydrogen bonds, of which the protein contributes $5(\pm 3)$ with the rest provided by water molecules.

However, nature is not solely organic as the elegant philosophy of Williams and Frausto da Silva (25) stresses. Frequent use is made of the metal ions available in natural systems to provide additional anion-binding interactions.

## B. Metal Ion Based Anion Binding

The guanosine diphosphate (GDP)-bound Ran protein crystal structure (Fig. 1) provides a perfect illustration of how a metal ion can be combined with hydrogen-bonding groups to augment the strength of anion binding and form the focus for an anion-binding pocket (26). (The Ran proteins are located primarily in the nucleus of eukaryotic cells and are involved in protein nuclear import and DNA synthetic control.) The magnesium ion is held in place by four water molecules and two protein residues. This ion binds the anionic GDP substrate in an "end-on" manner. The remaining functionalities of the bound anionic substrate are then satisfied by hydrogen bonding with preorganized sections of the enzyme superstructure.

Lactoferrin, an important iron-binding protein (Fig. 2), synergistically binds carbonate anions and iron(III) (27). The carbonate is bound through a combination of metal–anion coordinate bonds, hydrogen bonds, and electrostatic interactions. It is argued that anion-binding assists iron(III) binding by causing a buildup of negative charge at the iron-binding site.

Many structurally refined enzymes have been found to contain more than one metal ion at the active site. These metal ions are often bridged by an anionic substrate.

Superoxide dismutase (an enzyme for destroying biotoxic superoxide anions) contains copper(II) and zinc(II) ions at its active site, bridged by an imidazolate anion from a histidine residue (28). Phospholipase C has a number of zinc(II) ions at its active site, with two of them being bridged by hydroxide and aspartate (from a protein residue) anions (29). Phospholipase C evolved to bind and hydrolyze negatively charged phosphate esters, illustrating an important principle: the link between anion binding and functioning metal centers. This important concept of function will be returned to at a later stage (Section IV).

Of course, there are many other enzymes that bind anions and many more for which the binding mode is not as yet elucidated. The examples chosen here are merely intended to be illustrative.

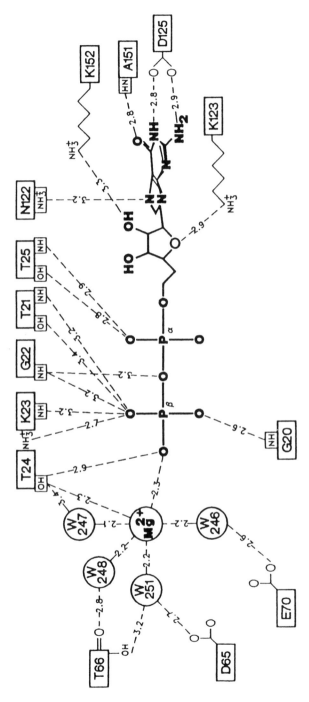

Figure 1. The binding site of Mg·GDP to Ran protein. A schematic drawing of the nucleotide-binding site with selected interactions (dashed lines) between Mg²⁺, GDP, and the protein, with the corresponding distances (in angstrom units). [Reprinted with permission from *Nature* (K. Scheffzek, C. Klebe, K. Fritz-Wolf, A. Wittinghofer, *374* 378 (1995). Copyright © Macmillan Magazines Limited.]

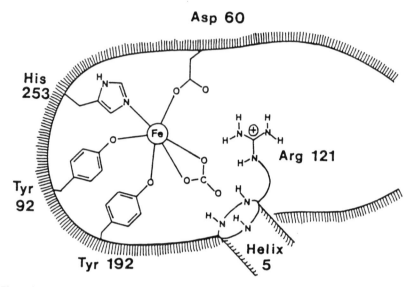

Figure 2. Schematic diagram of the iron- and anion-binding site in human lactoferrin. [Reprinted with permission of (27).]

## III. ORGANIC RECEPTORS FOR ANIONS

Dietrich's reviews have provided detailed accounts of early approaches to this type of anion receptor (4, 13). This chapter will, therefore, only briefly outline the basic types of organic receptors for anions as they will prove of relevance in understanding some of the recent inorganic systems that form the main body of discussion.

The first synthetic anion receptor (**1**) was based on a protonated nitrogen system (1). Compound **1** was shown to bind anions in the cavity in a "katapinate" manner (with the hydrogen atoms pointed inward toward the anion) (2). The source of interaction with anions in such a system is therefore twofold: electrostatic attraction and hydrogen bonding. Many protonated polyammonium

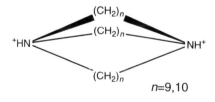

**1**

macrocycles were subsequently studied for their ability to bind anions, and indeed strong interactions were observed in aqueous solution, especially with carboxylates and phosphates (30–35). Lehn and co-workers (36, 37) in particular, made outstanding contributions in this field of work, synthesizing, for example, compound **2**, which was selective for dicarboxylates of specific chain length (as structure (**2**) illustrated). Compound **3** (BISTREN), when hexaprotonated, selectively bound azide anions due to the elliptical nature of the binding site (38, 39).

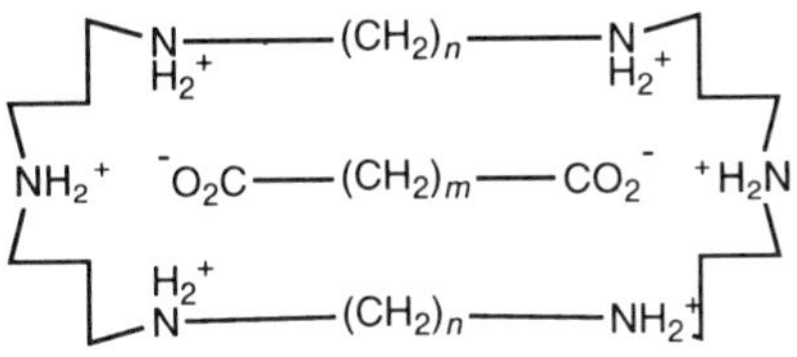

$$n=7 \text{ or } 10$$
$$m=2,3 \text{ or } 5,6$$

**2**

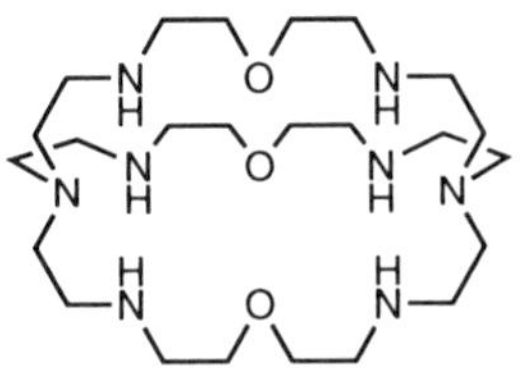

**3**

Unfortunately, polyammonium hosts are limited by the pH range over which they are protonated. This pH range is the same as that at which anions (such as phosphate and carboxylate) also begin to protonate. Consequently, the utility of this class of receptor was limited, thus causing the development of guanidinium based hosts.

Guanidinium (**4**) is protonated across a wider pH range than polyammonium systems and consequently avoids many of their pH limitations. It is extensively

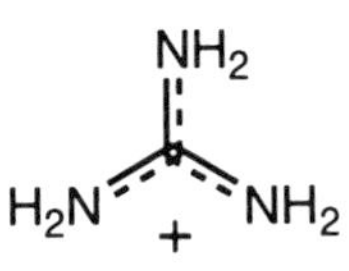

**4**

used in enzymatic anion-binding systems [such as Staphylococcal nuclease (40, 41)] in the form of arginine residues. It was therefore a natural choice to be incorporated within organic anion receptors, its action relying, once again, on a combination of electrostatic and hydrogen-bonding interactions. Lehn and co-workers (42) were the first to propose the use of polyguanidinium systems as synthetic anion complexones. The strength of anion binding, however, was less than for analogous polyammonium systems, probably due to the greater charge delocalization across guanidinium.

Nonetheless, much excellent use has been made of this system. Receptor 5, for example, extracts $p$-nitrobenzoate quantitatively from water into chloroform (43), and the chirality of the receptor allows the possibility of chiral anion recognition (44). Guanidinium has also been incorporated into devices, such as a hydrogen sulfite selective electrode (45). Recently, Mendoza and co-workers (46) reported a chiral double helical array of polyguanidinium strands assembled around sulfate templating anions, the first anion centered helical structure.

5

Another method enabling protonation of nitrogenous host molecules at accessible pH values has been developed by Sessler et al. (47), which was based on the ease of protonation of expanded porphyrins. Studies of sapphyrin [an expanded porphyrin (6)] yielded a crystal structure of diprotonated host with a bound anion, fluoride, found to be in the plane of the macrocyclic ring. These readily protonated expanded porphyrin systems are now well established as receptors for anions. Compound 7, for example, elegantly illustrates two-point binding, with the expanded porphyrin binding the negatively charged phosphate group while the nucleic acid base is complementary to the base of the bound substrate (48).

A different approach to the pH problem involved forming permanently positive nitrogen centers by quaternization. This methodology is exemplified by

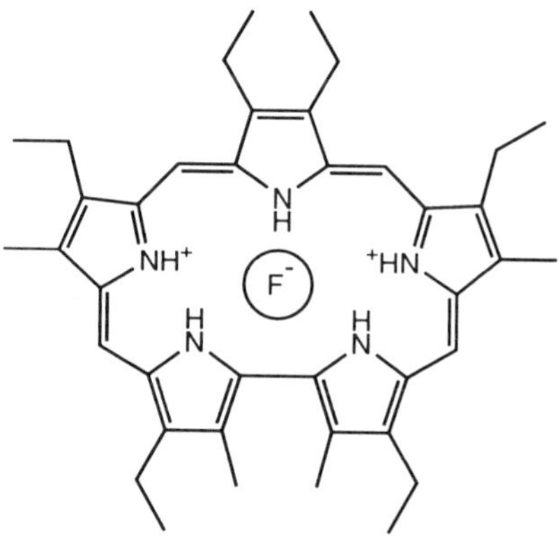

6

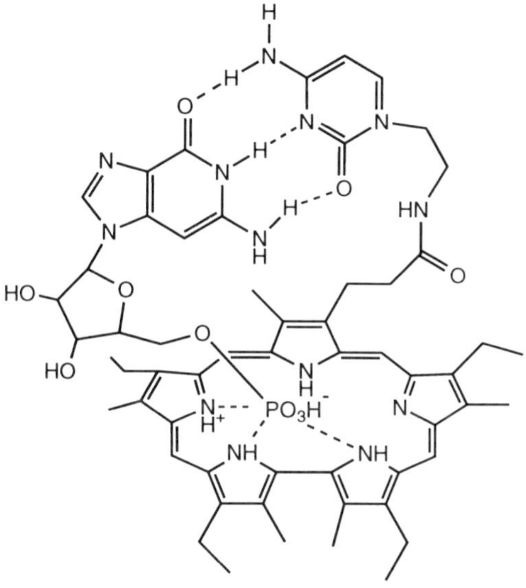

7

Compound **8**, which was synthesized by Schmidtchen (49). This receptor provides a fixed-binding site for anions that operates through a combination of electrostatic and hydrophobic forces (50). The absence of donor protons prevents any opportunity for hydrogen bonding, but the crystal structure of the iodide complex still indicated that the anion-binding site was in the center of the cavity (51). Zwitterionic hosts such as **9**, have also been reported (52, 53). These net neutral hosts prevent the need for the substrate to compete against a

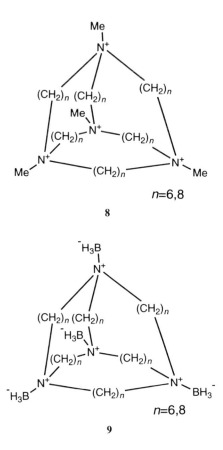

**8**

**9**

counteranion in the binding process, hence enhancing the strength of anion binding.

Dipolar electrostatic interactions have also been manipulated for the purposes of anion binding. Macrocyclic receptor **10** was shown to be capable of binding halide anions through interactions with the positive ends of the S = O and P = O dipoles (54). Evidence was also provided for the simultaneous binding of primary alkyl ammonium cations (to the oxygen atoms) and halide anions (to the dipoles). This topic of simultaneous cation and anion recognition is of considerable current interest. Further examples will be encountered during this chapter.

Another way of avoiding the problems of pH range and counterion competition is to use a neutral hydrogen-bonding receptor based on amide functionalities. Peptide groups from the protein backbone are, of course, well known to be involved in enzyme anion binding as discussed earlier. Amide involve-

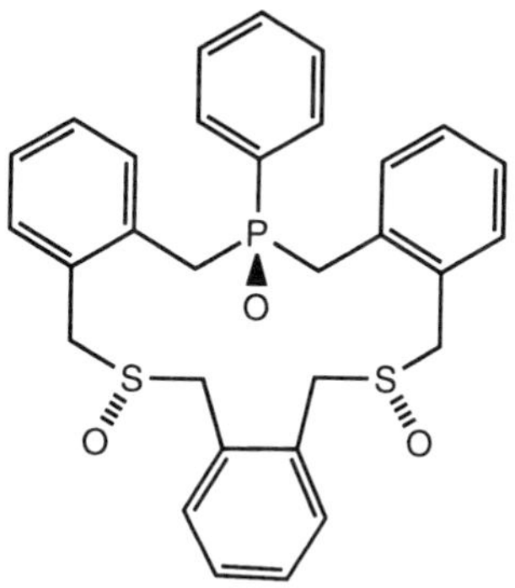

**10**

ment in the binding of anions by synthetic hosts was first suggested by Kimura et al. (32) for the protonated receptor (**11**). This involvement was later crystallographically proven for azide ion binding (55). In 1986, Pascal et al. (56) prepared **12**, the first purely amide based receptor, which had three amide protons pointed into the cyclophane cavity and showed evidence of fluoride binding in dimethyl sulfoxide $d_6$ (DMSO-$d_6$) solution. Since this report, considerable use has been made of amides and, in particular, ureas for the construction of neutral anion receptors (57–61).

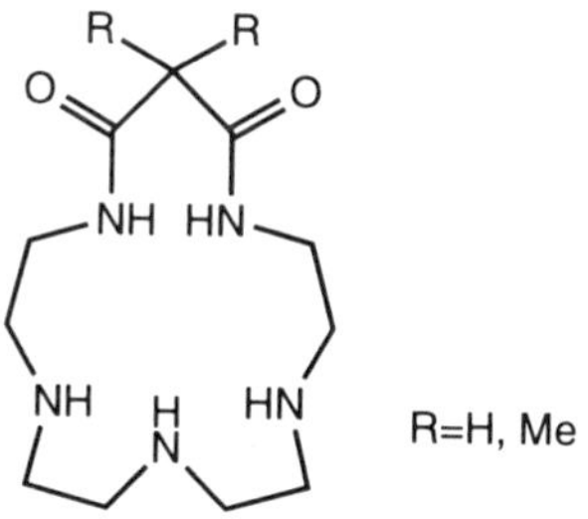

R=H, Me

**11** 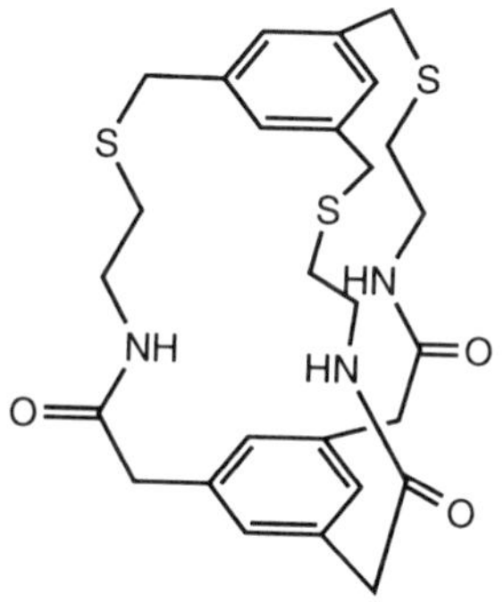                         **12**

All the receptors discussed so far, however, are based on traditional organic chemistry and while often effective, there are several good reasons for approaching the problem of anion binding from a slightly different, inorganic viewpoint.

## IV.  INTRODUCTION TO INORGANIC BASED ANION
## RECEPTORS

The organic receptors illustrated above take their inspiration from the first class of biological receptor to be discussed, utilizing solely organic hydrogen bonds, electrostatic attraction, and hydrophobic forces for anion bindnig. Nature, however, as shown earlier, casts its net wider in search of effective means of binding anions, thus incorporating metal centers into many of its anion receptors. There are several excellent reasons other than pure biomimicry, however, for attempting to incorporate metal centers:

1. *Source of Interactions with Anions.* Metals are usually either positively charged or formally electron deficient. This knowledge leads to either an enhanced electrostatic interaction with negatively charged substrates or the chance for orbital overlap and formation of bonding interactions, thus increasing the stability of any complex species formed.

2. *Structural Factors.* Metal compounds often have precisely defined geometries [e.g., Cu(I)-tetrahedral, Cu(II)-distorted octahedral]. These geometries can be manipulated by the inorganically minded anion recognition chemist to create receptor molecules with well-controlled and interesting relative geometries of ligating groups. This knowledge can be used to enhance selectivity for specifically shaped anions, or create unusual switching and conformational effects on binding.

3. *Incorporation of Functionality.* Metal ions possess a huge range of function, and this is perhaps one of the most compelling reasons for their incorporation into receptor structures. Redox activity, ultraviolet–visible (UV–vis) spectroscopic properties (color), catalytic ability, fluorescent and energy-transfer properties, and radioactivity could all form the basis of potentially useful molecular machines dependent on the recognition of anions. This mechanism could in turn lead to advances in sensor technology (62), anion transport, drug delivery, and catalysis (63); naming only a few applications.

In this chapter, we will try to illustrate the reasons in each case for the incorporation of inorganic centers into the receptors and the advantages conferred to each host by doing so.

In a simplistic treatment, of course, an isolated metal ion could itself be viewed as a receptor for anions as it fulfills the basic criterion of reversible anion binding. The concept of an anion receptor, however, requires a reversal of this metal-centered viewpoint. For the purposes of this chapter, a receptor will be considered as a molecule designed for binding anions, which does so through a combination of bonding interactions (rather than through a single coordinate bond). The receptors we will discuss are often designed to incorporate anion selectivity and frequently manipulate a variety of noncovalent interactions.

As will be illustrated, the use of inorganic and organometallic chemistry has, in the past 10 years, enabled the development of a rich and exciting range of novel, functional anion receptors. Inorganic anion receptors can be usefully subdivided into four classes dependent on the bonding interactions responsible for anion binding:

1. Neutral receptors based on multiple Lewis acid–anion orbital overlap interactions.
2. Positively charged receptors based on multiple coordination interactions from transition metals.
3. Charged receptors based primarily on intermolecular electrostatic attraction from positively charged metal centers.
4. Receptors incorporating hydrogen-bonding interactions.

## V.  INORGANIC APPROACHES TO ANION BINDING

### A.  Neutral, Lewis Acidic Receptors

Lewis acidic centers are, due to their electron deficiency, capable of interacting with anions through an orbital overlap, causing a bonding interaction. Many novel, neutral receptors incorporating multiple numbers of this kind of anion-binding interaction have been developed.

#### 1.  Tin Based Receptors

Organotin compounds have been used as neutral carriers for selected anions in membrane electrodes since the late 1960s (64). The compounds generally used, however, were mononuclear tin species such as trioctyl tin chloride. The mechanism of interaction was elucidated to be the formation of a single-bonding interaction between the four coordinate neutral tin center and the anionic guest (65, 66). As such, these molecules do not fall under our criterion of multiple anion-binding interactions for a designed anion receptor.

The first attempts to marshal a multiple number of Lewis acidic tin centers in order to create a receptor specifically for the purpose of binding anions were made by Newcomb and co-workers (67). In 1984 they reported the synthesis of several tin based macrocycles (e.g., **13** and **14**), the first macrocycles to contain Lewis acidic acceptor groups rather than lone-pair donors. Receptor **13** was available in gram quantities making it an ideal receptor for anion-binding in-

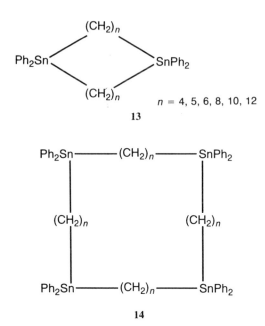

13

$n = 4, 5, 6, 8, 10, 12$

14

vestigations. The first coordination study results were reported in 1987 (68). This kind of host was shown to form both $1:1$ and $1:2$ stoichiometric complexes with chloride ions in acetonitrile solution. Stability constants ranged from 400–850 $M^{-1}$ with little difference between first and second anion coordination (although some uncertainty was expressed over the reliability of these values). This casts some doubt over whether the tin atoms act cooperatively or independently in this receptor. A small size selective effect was observed ($n = 8$ binding more strongly than $n = 10$) and a small macrocyclic effect was observed on comparison with an acyclic analogue.

Macrobicyclic receptors (**15**) were also reported (69). Binding studies indicated kinetically slower binding than with their macrocyclic analogues (probably due to a more enclosed binding site) The stoichiometry of halide ion binding was exclusively $1:1$ and encapsulation of the guest anion was postulated (70). Nuclear magnetic resonance (NMR) $^{119}$Sn studies were used to illustrate that for receptor **15**, with $n = 6$, the fluoride ion ($K \approx 200\ M^{-1}$) was bound five orders of magnitude more strongly than chloride ions ($K \leq 0.01\ M^{-1}$) in chloroform (71). Crystallography later showed that the fluoride ion was encapsulated within the cavity between the two tin atoms ($r_{Sn-F} = 2.12/2.28\ \text{Å}$), whereas in the chloride complex, the ion was strongly bound to one tin center but only weakly interacting with the second (72). For fluoride ions, therefore, the enhancement in binding energy is caused by cooperative interactions with both the tin atoms in the host molecule, making **15** a selective fluoride receptor.

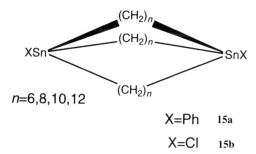

X=Ph  15a

X=Cl  15b

Receptor **16** was reported to form a 1:1 complex with chloride ions in chloroform ($K = 500\ M^{-1}$), exhibiting fast exchange on the NMR time scale (73). Such hosts, containing four tin binding sites, were shown to be considerably more effective than mononuclear organotin compounds for chloride binding. In 1991, Newcomb and co-worker (74) published modeling studies for these tin based hosts as well as a crystal structure of **17**.

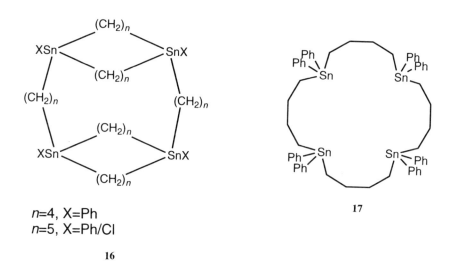

The trinuclear receptor **18** and its propyl linked analog **19** have been reported by Jurkschat et al. (75, 76). Receptor **18** was shown to transport chloride and bromide ions from water through dichloromethane, but the process was slow. Crystallography of **19a** indicated chloride binding between a pair of tin atoms, while NMR studies indicated that **19b** involved all three tin atoms in the binding process. No quantitative evaluation of binding affinities was attempted.

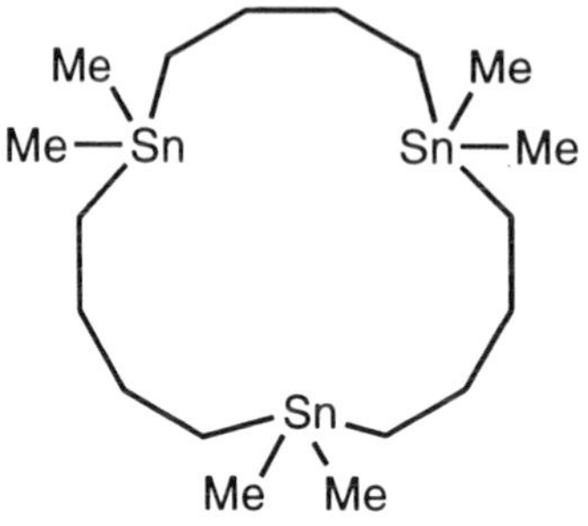

18

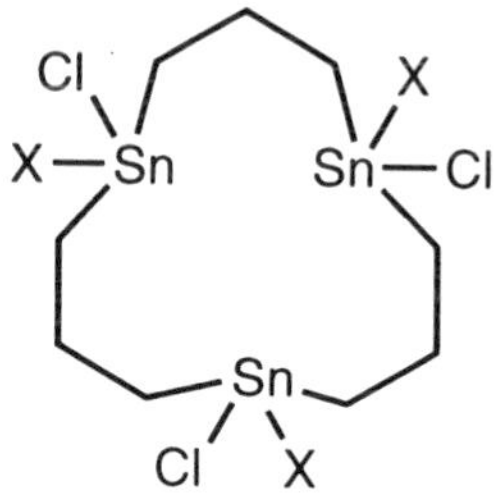

X = Cl (**19a**)
X = Me (**19a**)

Compounds such as **20** have recently been reported and crystallographically proven to form Sn–X–Sn intramolecular bridging interactions (77). These bis-coordinated bromide and iodide ions are observed to be held with bonds of intermediate bonding length, indicative of cooperative tin multicentered anion binding. Multidentate acyclic organyl tin species have recently been used as phosphate selective carriers in polymer based liquid membranes (78). This result indicates the potential for practical application of organic multitin systems in the field of anion recognition.

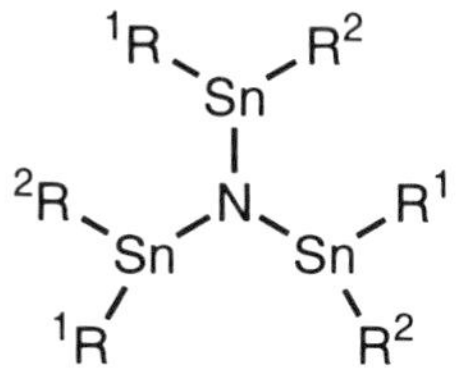

20

## 2. Boron Based Receptors

Although technically a heteroelement rather than a metal, some interesting Lewis acidic receptors for anions have been based on boron-containing systems. The first evidence for this type of anion receptor was published in 1967, one of the earliest examples of anion binding. Compound **21**, when compared with **22**, exhibited a chelate effect in the binding of methoxide anions (3).

$F_2B$     $BF_2$

21

$BF_3$

22

Katz (79, 80) used **23** as a receptor for anions, studying its interaction with hydride, fluoride, and hydroxide ions. Comparison with **24** once again indicated a chelate effect: compound **23** abstracting hydride or fluoride from complexed **24**. The crystal structure of the hydride sponge **23**, showed the hydride ion bound between the pair of boron atoms with short strong bonds. A crystal structure of the chloride complex was also elucidated and showed the same bridged structure (81).

**23**                    **24**

Compound **25** has been synthesized by Reetz et al. (82, 83). It has been used to complex ionic pair species such as KF. The potassium ion is complexed by the crown ether and the fluoride ion is held by a combination of orbital overlap with the Lewis acidic boron atom and electrostatic attraction to the positively charged potassium. This receptor provides an elegant example of the use of a combination of intermolecular forces.

**25**

A study has also been made of the theoretical propensity of organoboron macrocyclic hosts for anion binding (84). This study indicated that for the hypothetical hosts studied, anion inclusion occurred with cavity shrinkage and partial boron rehybridization. This effect would be expected for these anion receptors, which are dependent on orbital overlap interactions for their mode of action.

Recently, Shinkai and co-workers (85) reported a ferroceneboronic acid receptor (**26**). This receptor showed selective electrochemical recognition of fluoride ions over other halides, the redox potential for the ferrocene unit being perturbed on the addition of $F^-$. There is, however, only one orbital overlap interaction provided by this host and it is likely the selectivity arises solely due to the "hardness" of the fluoride anion.

Compound **27**, a mixed boron–silicon system, was synthesized to investigate the influence of organosilicon on the anion-binding process (86). A fluoride complex was isolated and NMR and crystallographic studies showed that the silicon was involved in binding the anion, but only weakly.

## 3. Silicon and Germanium Based Receptors

In light of the work carried out on boron systems as described above, a silacrown (**28**) was synthesized in fair yield via nine steps (87). This transported bromide ions more effectively than chloride, but no evidence for the binding mechanism or evaluation of binding affinity was provided. In recent studies, Compound **29** showed, as would be expected, a chelate effect with fluoride ions, exhibiting a high-binding constant ($\log K > 9$ in acetone-$d_6$), which had to be determined via a stepwise procedure (88).

Germanium based macrocycles (**30a–b**) have also recently been synthesized (89, 90). Receptor **30b** has been shown to transport chloride ions in preference to bromide, although the degree of transport was only 20% in 35L ($H_2O/CH_2Cl_2$) (91). The expanded hosts (**31**) have also been synthesized and show similar effects to **30**, indicating that ring size is not the only controlling factor in the anion complexation process (92).

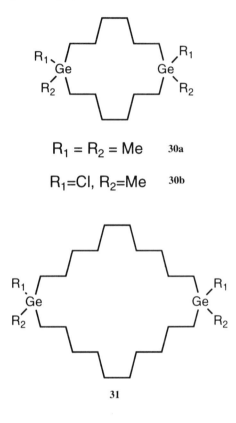

$$R_1 = R_2 = Me \qquad \text{30a}$$

$$R_1=Cl, R_2=Me \qquad \text{30b}$$

31

## 4.  Mercury Based Receptors

Perhaps the most eye catching of the receptors utilizing Lewis acidic centers are those based on mercury. Mercury is *sp* (linear) hybridized which has two consequences.

1. Mercury has two empty *p* orbitals available for orbital overlap with guests and consequently binds anions with practically no energetically unfavorable geometric reorganization (unlike Sn, Si, B, and Ge, which have to partially rehybridize on anion binding).

2. Large macrocyclic receptors can be synthesized due to the linear geometry (93).

The first literature example of a mercury based receptor (32) for anions was analogous to the chelating boron receptor discussed earlier. The crystal structure indicated that two molecules of Compound 32 asociate with one chloride ion, which sits in a four-coordinate binding site. Solution studies, however, gave results indicative of 1 : 1 binding for halide anions (94, 95). This simple

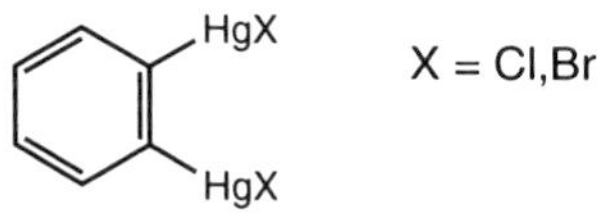

32

receptor unit was subsequently incorporated into macrocyclic structures (33 and 34) and investigations indicated interactions with electron-donating molecules such as tetrahydrofuran (THF) (96). Unfortunately, there was no marked enhancement in binding attributable to a macrocyclic effect. Receptor 34 has, however, been further functionalized and built into a polymeric membrane electrode that shows selective response to $SCN^-$ and $Cl^-$, in the presence of $NO_3^-$ and $ClO_4^-$ (97).

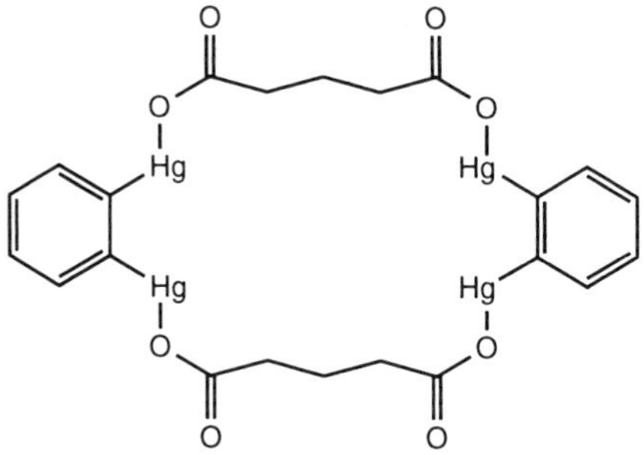

33

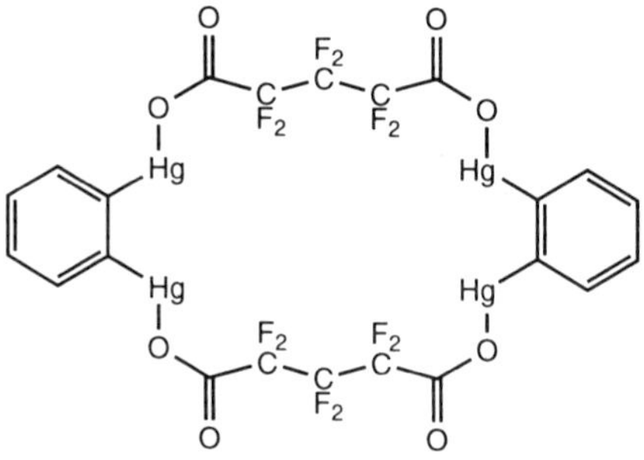

34

Macrocyclic mercury compounds have also been investigated by Shur et al. (98). Receptor **35** was proven to complex halide ions by $^{199}$Hg NMR studies but complex isolation failed. On fluorination of the benzene rings yielding receptor **36**, complex isolation became possible (99). As seen from the crystal structure of **36** (Fig. 3), each bromide ion is coordinated to six mercury atoms and is sandwiched between two mercury macrocycles. The Hg—Br distances are 3.07–3.39 Å, shorter than the van der Waals contact but longer than a covalent bond. Receptor **37** was also crystallographically proven to coordinate chloride ions (100). The crystal structure indicated one chloride ion bound above the plane of the macrocycle and one below it, both coordinated by all five mercury atoms ($r_{Hg-Cl} = 3.09–3.39$ Å). The chloride–chloride distance was short (3.25 Å), probably only tolerated due to the presence of the mercury "sandwich filling."

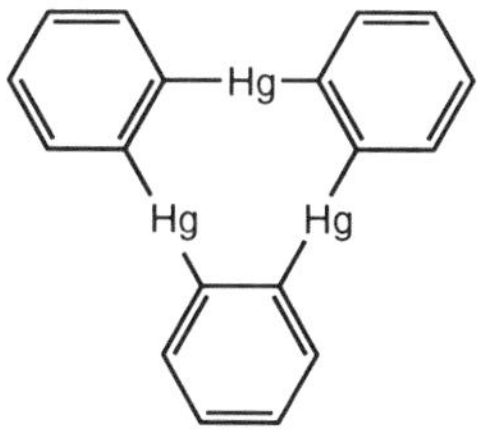

35

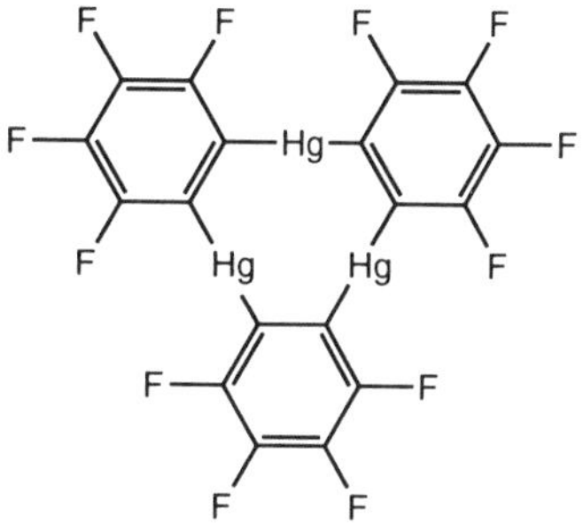

36

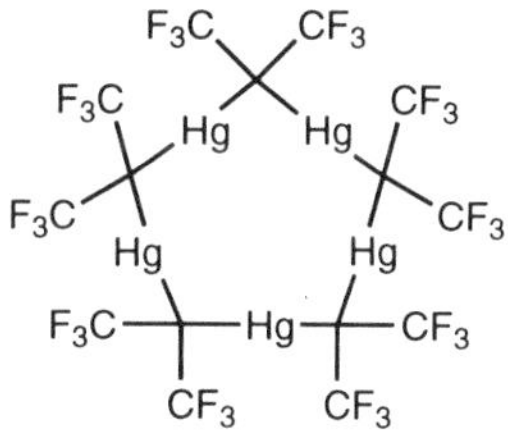

37

Of particular note are the beautiful mercuracaboranes of Hawthorne and co-workers (101), which synthetically link the carbon atoms of carborane cages with mercury atom bridges. Receptor **38** was reported along with the crystal structure of its chloride complex [Fig. 4(a)]. The Hg—Cl bond distances were 2.94 Å shorter than those for Shur's hosts listed above. The square planar coordination of a chloride ion was unprecedented, and analogy was drawn with the lithium selective cation receptor 12-crown-4, the receptor being named [12]mercuracarborand-4 and classed as an "anticrown." It was proposed that the presence of chloride ion during the synthesis templated the formation of

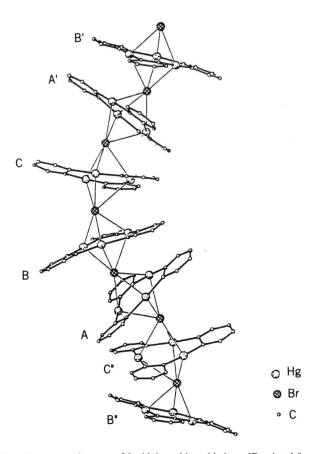

Figure 3. Crystal structure of receptor **36** with bound bromide ions. [Reprinted from *J. Organomet. Chem.*, *418*, C29, V. B. Shur, A. Tikhonova, A. I. Yanovsky, Y. T. Struchkov, P. V. Petrovskii, S. Y. Panov, G. G. Furin, and M. H. Vol'pin. Crown compounds for anions. Unusual complex of tremeric perfluoro-*o*-phenylene mercury with the bromide anion having a polydecker sandwich structure C29, 1991, with kind permission of Elsevier Science SA, Lausanne, Switzerland.]

products avoiding the formation of oligomeric species. When mercuric acetate (incapable of templating) was used in the synthesis, the yield was diminished, providing further evidence for this templated process. X-ray structures have also shown both one and two iodide anions [Fig. 4(b)] bound in the cavity (102, 103). The host has been shown, crystallographically to form a supramolecular aggregate with $[B_{10}H_{10}]^{2-}$, the anion fitting snugly within the binding cavity.

A smaller analogue, [9]mercuracarborand-3 (**39**), has also been synthesized, and evidence has been provided for its interaction with chloride ions (104).

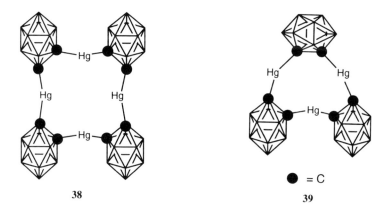

38

39

● = C

Recently, a hexamethyl[9]mercuracarborand-3 derivative has been reported (105). This receptor showed smaller shifts in its $^{199}$Hg NMR peaks on anion complexation, probably due to the methyl groups making the mercury centers less electron deficient. Chloride ions formed 1:1 complexes, while bromide and iodide formed complexes of 2:1 stoichiometry.

A tetraphenyl substituted derivative of [12]mercuracarborand-4 (**40**), which binds one iodide ion in its cavity due to steric hinderance, has been reported. The stereochemistry of the phenyl groups was found to depend on the mercury counteranion used during the synthesis (106). This observation provided yet further evidence for a direct anion templating effect in mercuracarborane syntheses. Most recently, the C—Hg—C link in this type of host has been replaced with a B—Hg—B link, which alters the electron demands of the mercury centers (reducing their electron deficiency) and apparently switches off anion complexation (107).

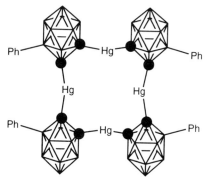

40

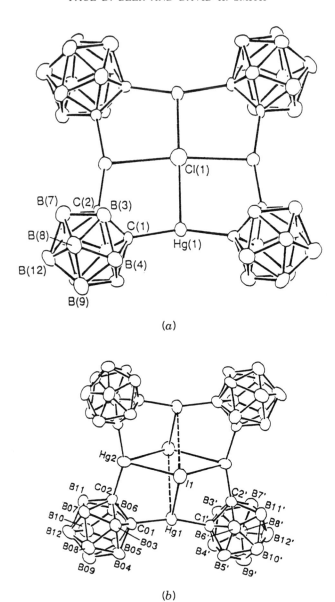

(a)

(b)

Figure 4.  (a) Crystal structure of receptor **38** with bound chloride anion. [Reprinted with permission of (101).] (b) Crystal structure of receptor **38** with bound iodide anion. The iodide anion is too large to bind in the plane of the mercury crown. [Reprinted with permission from X. Yang, C. B. Knobler, and M. F. Hawthorne, *J. Am. Chem. Soc.*, *114*, 380 (1992). Copyright © 1992 American Chemical Society.]

## 5.  Summary

The hosts outlined above generally utilize their metal centers to provide interactions with anions. The orbital overlap between unfilled heteroelement orbitals and filled anion orbitals yields bonding interactions. In the case of the mercury based receptors, the role of the metal atoms is partly structural, with their *sp* hybridization linear geometry specifically yielding large macrocyclic cavities of suitable diameters for anion encapsulation. The metal atoms have often been functionally used as an NMR "handle" on the anion coordination process.

In summary, various Lewis acidic host molecules have been used for the complexation of anions, particularly halide ions, in organic solvents. Stability constants, when elucidated, are often low in magnitude ($K \leq 1000\ M^{-1}$). The exceptions to this rule are those cases where boron or silicon are coordinating to an electron dense "hard" anion such as fluoride, hydroxide, or hydride. In these cases anion binding is strong, but reservations have to be expressed as to the degree of control that can be exerted over the selectivity of the binding process. Lewis acid centers are also well known for their moisture and air sensitivity, leading to a lack of physical robustness. None the less, multidenate Lewis acidic hosts have shown novel anion-binding modes and selectivities and have, as has been illustrated, considerable potential for application into chemical sensor technology.

## B.  Multiple Positively Charged Metal Ion Based Coordinative Receptors

The receptors considered so far have been based on neutral Lewis acidic centers, covalently built into the host framework with organic $\sigma$-bonds. Combinations of positively charged metal ions held within one host, however, can also fulfill the function of anion binding by forming multiple coordination interactions with a single cobound anionic substrate. In some ways, these hosts are analogous to the Lewis acidic hosts discussed above, the interactions being based on orbital overlap between the electron-rich anion and empty metal centered orbitals. However, the metal centers are positively charged and are coordinated into the receptor framework rather than held in place by covalent bonds.

The past 25 years have spawned literally hundreds of examples of this type of multimetal center anion coordination. Therefore, rather than being exhaustive, this chapter will attempt to summarize the main types of receptor investigated and analyze their anion-binding potential. A recent review highlighted the general importance of binuclear metal complexes (and concomitant anion binding) in the modeling of biological catalytic processes, and some of the examples contained therein will be returned to later (108).

### 1. "Robson-Type" Receptors

The first macrocyclic receptors (**41–43**) capable of completely circumscribing two metal ions were reported in 1970 (109–111). Receptors **41** and **43** bound two Ni$^{II}$ ions while **42** bound two Cu$^{II}$ ions. No evidence was found of anionic species bridging the metal ions because in these examples the metal ions were either too closely held or were coordinatively saturated. If multimetal anion coordination interactions are to be observed, it is critical that the metal ions involved have vacant coordination sites.

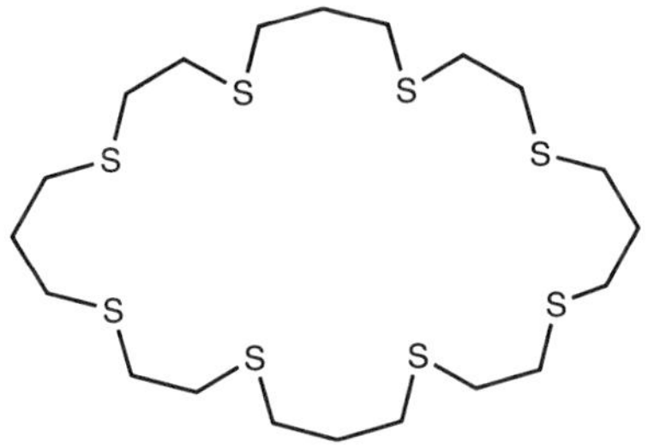

**41**

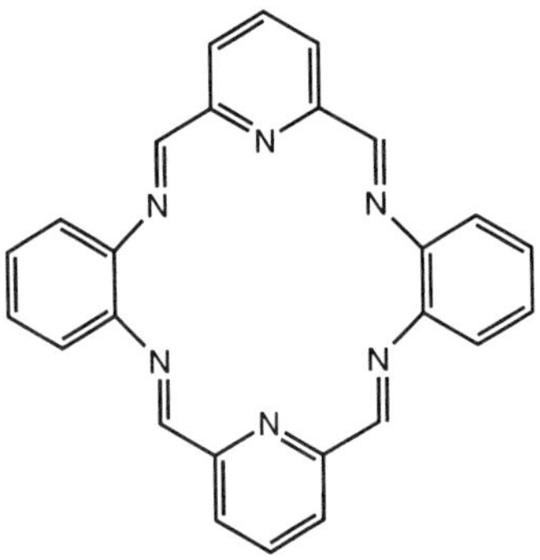

**42**

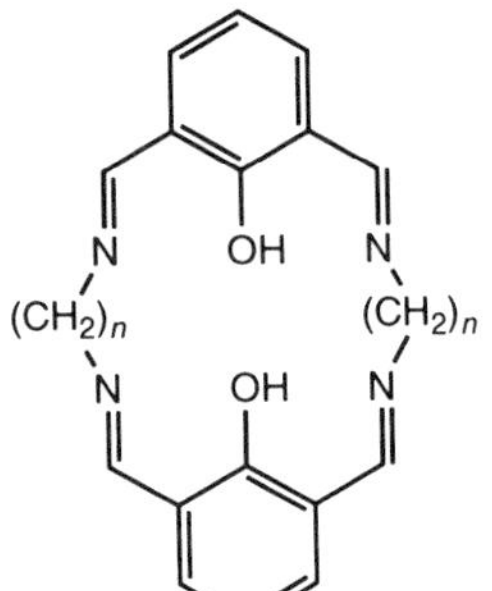

**43**

To this end, Robson et al. (112, 113) investigated a series of acyclic analogues (**44**) of receptor **41**. As illustrated, these receptors do not coordinatively saturate the two metal ions. Consequently, there is room for a bridging anion X. A variety of alkoxide anions were observed to bridge the pair of metal ions.

This area of chemistry was particularly fertile in the early 1970s with a whole range of analogous binucleated anion receptors (e.g., **45** and **46**) being synthesized and analyzed either spectroscopically or crystallographically (114–118). A good overview of early binucleated systems, including those with bridging anions, was provided by Groh (119).

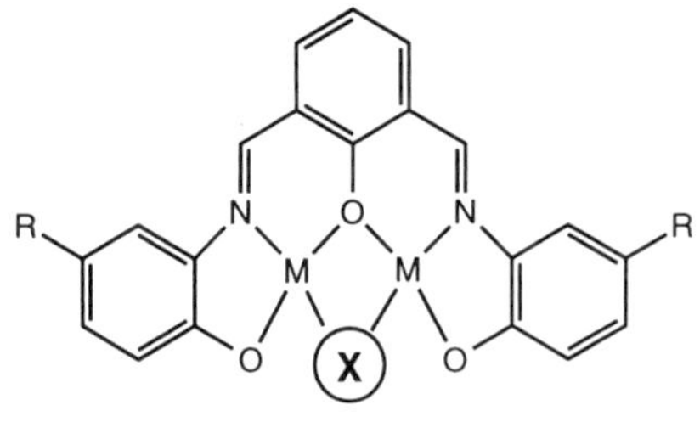

$X = EtO^- / PhCH_2O^- / MeO(CH_2)_2O^-$

$R = H / Me_3 / Cl / NO_2$

**44**

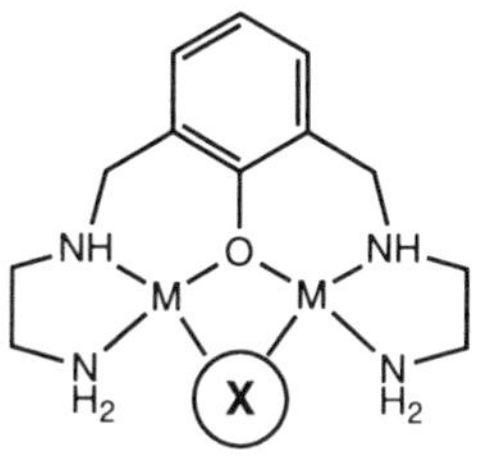

$X = OH^- / Cl^- / Br^- / NCS^-$

**45**

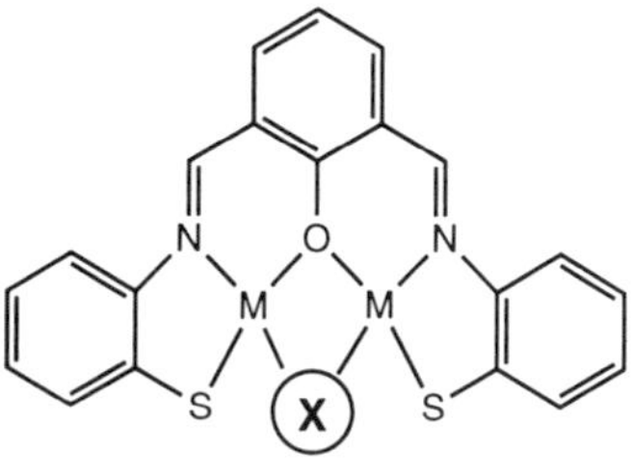

$X = NH_2^- / MeO^- / PhCH_2O^-$

**46**

These acyclic ''Robson'' systems provided initial evidence for this mode of binding. Recently, a dicopper(II) complex of this type (**47**) was proven to bind phosphate esters and increase their rate of hydrolytic cleavage (120), which

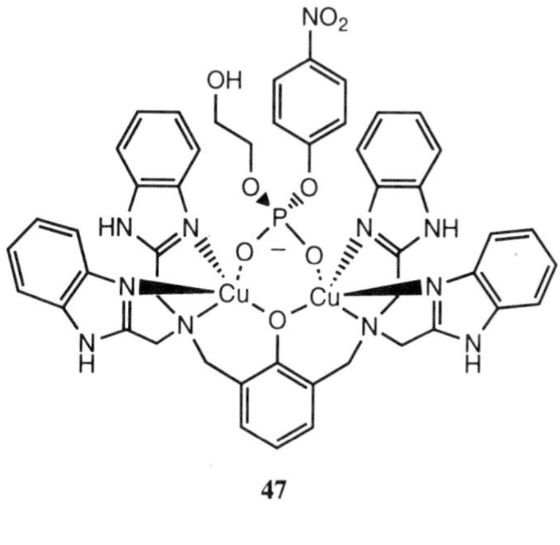

**47**

indicates the potential of these systems for the incorporation of catalytic function as well as anion recognition properties.

### 2. Macrocyclic Receptors

In three erudite review articles, Lehn (121–123) expounded his early philosophies of supramolecular chemistry. In these articles, he considered the development of generalized multisite receptors such as **48**; analogous with the anion-bridging receptors discussed above. He argued that such receptors would provide an entry into higher forms of molecular behavior normally associated with enzymes, such as cooperativity, allostery, and regulation; describing the bound substrate as being ''cascaded.'' With this assimilation of functional concepts and shift in emphasis from coordination chemistry to ''supramolecular'' substrate binding, much new attention was focused on these anion-binding systems.

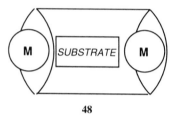

**48**

Receptors such as OBISDIEN (**49**) when protonated, had already been studied by Lehn and co-workers (30, 31) as organic hosts for anions. OBISDEN

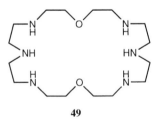

**49**

was now used in its unprotonated state to bind two metal ions in a coordinatively unsaturated manner and, subsequently, cascade an anion in between them. Receptor **50** was also used to the same end. Along with Lippard and co-workers (124), Lehn found that the biologically important imidazolate anion bridged the bis-copper(II) complex of **49**. Many studies were subsequently made of the potential of this system to act as an effective enzyme model (125–127). The crystal structure of the bis-copper(II)azide complex, however, indicated that cascade complexes were not always formed by such dinucleated receptors (128).

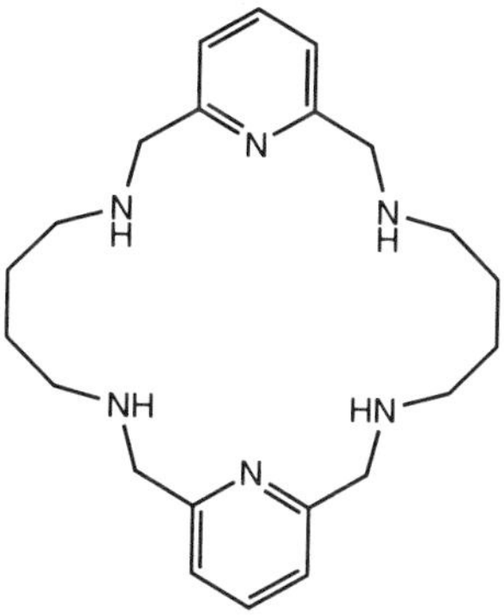

**50**

A structural study of receptors **51**, **52**, and **49**, showed three different modes of azide complexation to the binuclear copper(II) host (123); 1, 1′ cascaded, 1,3 cascaded, and noncascaded, respectively. These structures indicate that the nature of the macrocyclic framework of the receptor is important in determining the mode of anion coordination. Figure 5(*a* and *b*) shows the crystal structures of **52**, 2 Cu(II)·azide, and **52**, 2Cu(II)·chloride, respectively, for comparison (129). As can be seen, it is the length of the azide bridge that makes cascade complexation possible, whereas for the smaller mononuclear chloride anion this obviously cannot occur. Receptor **49** has also been shown to cascade bind pyrophosphate [as its bis-copper(II) complex] (130) and sulfate [as its bis-iron(II) complex](131). At about the same time, Nelson and co-workers (132) published a similar bis-copper(II) complex structure that cascaded an azide anion.

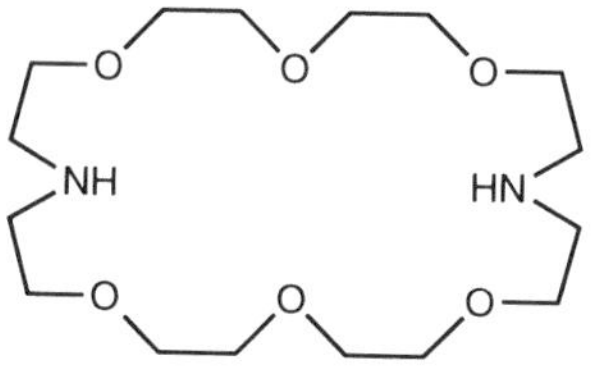

**51** 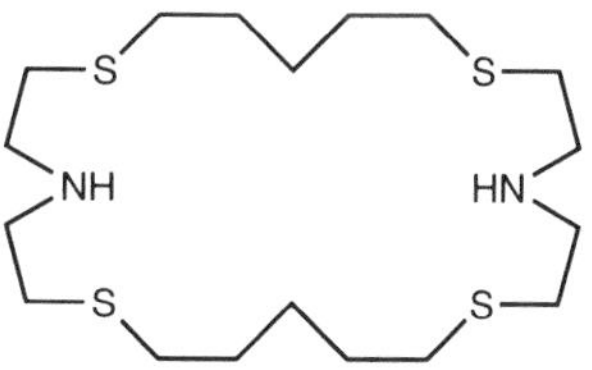          **52**

In the examples discussed thus far, the coordinating metal sites have had a twofold role. They have (obviously) been the source of interaction with the anion and also played a structural role; their degree of separation influencing the type of anion that can be cascade bound. Martell and Motekaitis (134, 135) have, however, also succeeded in introducing a functional role for the metal centers.

In 1983, it was shown that receptor **49** was capable of binding dioxygen

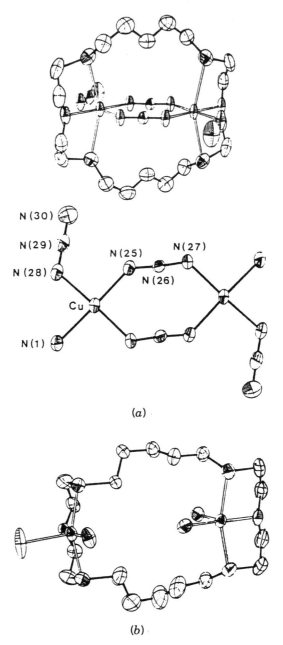

(a)

(b)

Figure 5.   (a) Crystal structure of bis-copper(II) **52** with cascade-bound azide anions. (b) Crystal structure of bis-copper(II) **52** with bound chloride anions. [Reprinted with permission from Y. Agnus, R. Louis, and R. Weiss, *J. Am. Chem. Soc.*, *101*, 3381 (1979). Copyright © 1979 American Chemical Society.]

between a pair of cobalt ions (133). Reports in 1988 then proved the capacity of this system to perform oxidation reactions (134, 135). The dioxygen bound receptor **53** still has one vacant coordination site on each cobalt(II) ion. Oxalate dianions were shown to be able to bridge the two metal centers. On warming to 45°C rapid oxidative degradation of the oxalate dianion to carbon dioxide was observed ($k = 4 \times 10^{-4}$ s$^{-1}$). Activation of the dioxygen toward oxalate oxidation was considered to be due to the simultaneous coordination of oxidant and reductant to the two cobalt centers. Similarly, the properties of the complex with phosphite anions were investigated (136). The oxygen free receptor bound phosphite (log $K = 4.0$), while the oxygenated version (**53**) bound the anion even more strongly (log $K = 8.96$). This result indicates the presence of a degree of preorganization for binding in the dioxygen complex. On heating under argon, phosphite was oxidized to phosphate and dioxygen was reduced to water, the cobalt centers becoming permanently Co(III) in nature. These are good examples of the marriage of a cascade anion recognition with a catalytic function and indicate the potential application of this type of anion-binding system.

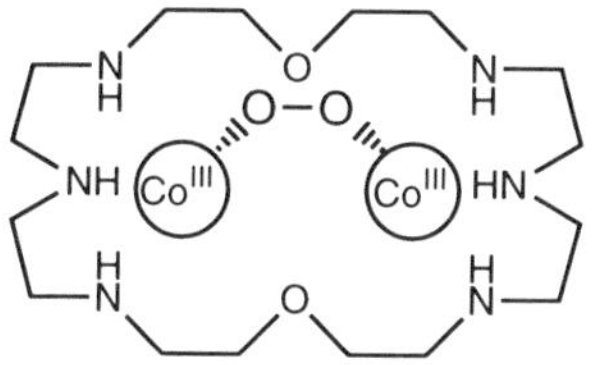

53

## 3. Linked Macrocyclic Receptors

Another approach to forming receptors capable of cascade binding is to link pairs of macrocycles with a variety of spacer units. Each macrocycle is capable of binding a single metal ion. An example of this has recently been reported in a catalytic system with an anthracene unit rigidly holding two macrocyclic-binding functions (137). Receptor **54** with two bound Co$^{III}$ ions catalyzes the

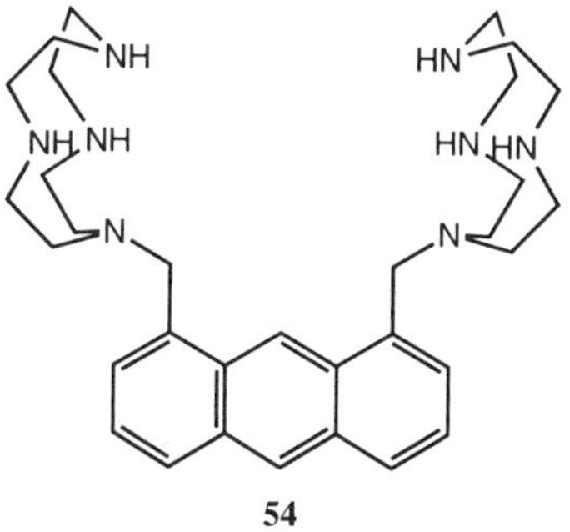

54

hydrolysis of phosphate esters. In particular, the rate of p-nitrophenylphosphate hydrolysis is 10 times greater ($k = 1.3 \times 10^{-2}$ s$^{-1}$) than observed with a mononuclear cobalt based analogue ($k = 1.2 \times 10^{-3}$ s$^{-1}$). The end product of the catalytic pathway contains a phosphate anion cascaded between the two cobalt metal centers. It is believed to be important for the catalytic rate enhancement that phosphate fits snugly in the rigidly held internuclear pocket.

### 4. Macrobicyclic Receptors

The macrobicyclic cryptand BISTREN (**3**), as discussed earlier, binds anions when hexaprotonated. This compound is also capable of binding pairs of transition metal ions [e.g. the bis-copper(II) Complex (**55**)] when unprotonated (39, 138). The bis-copper(II) based receptor (**55**) was observed to bind a chloride ion cascaded between the two metal ions. In fact, receptor **55** bound chloride more strongly (log $K = 3.55$) than the hexaprotonated form of **3** (log $K = 2.36$), and this was attributed to the formation of strong coordinate bonds. Hydroxide was also cascade bound forming a thermodynamically very stable complex (log $K = 11.56$), the resulting complex being of a different type from that formed by non-cryptate complexes of the Cu(II) ion.

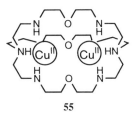

55

In 1987, the synthesis of a variety of macrobicyclic systems using tris(2-aminoethyl)amine (tren) as a head group was made more general (**56, 57**) (139, 140). These hosts were used as dinucleating metal ion receptors, and their cascade coordination potential was investigated particularly by Nelson and coworkers (141). They used spectroscopic evidence to show the capacity of receptor **57a** to bind two copper(II) ions and then cascade coordinate an azide anion (141). This highlights the ability of transition metal ions to *respond* to the presence of cascaded substrates through their spectroscopic behavior. Martell and co-workers (142) reported the crystal structure of such a cryptand (**58**) with two Cu$^{II}$ ions bridged by a carbonate anion. Fabbrizzi et al. (143) also recently reported the anion-binding properties of this receptor. They found that the bis-copper(II) cryptate recognizes neither anion size nor coordinating tendencies, but that the key factor in the recognition process was the anion bite length (the distance between the two extreme atoms of the complexed anion)

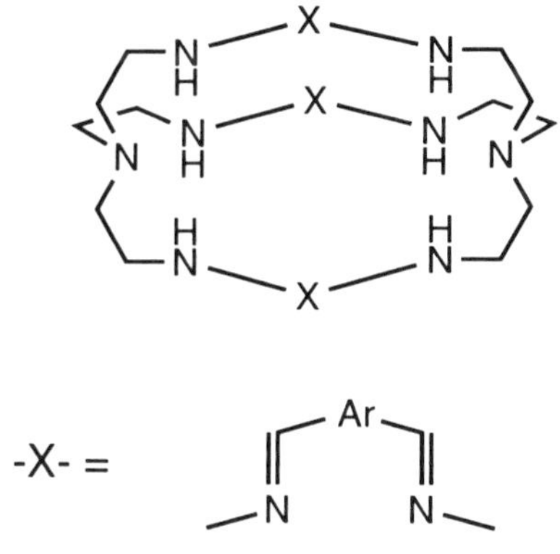

**56**

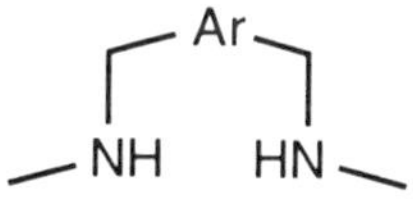

**57**

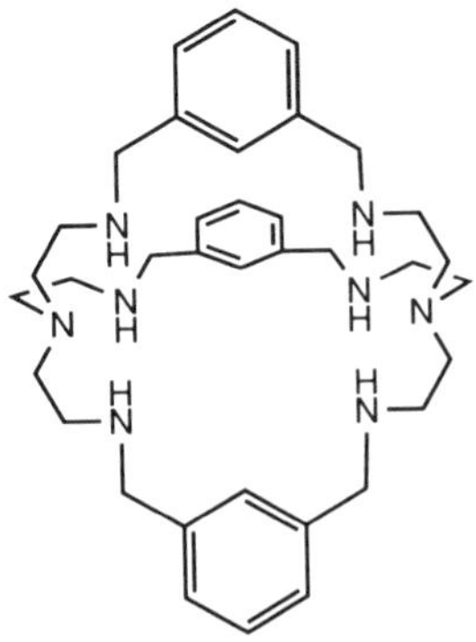

**58**

(143). It remains to be seen whether this mode of anion selectivity is generally true for other cascade complexes.

We have also reported a novel redox-active macrobicycle (**59**) (144). This receptor has been shown to form dinuclear complexes with $Cu^{II}$ and $Zn^{II}$ ions.

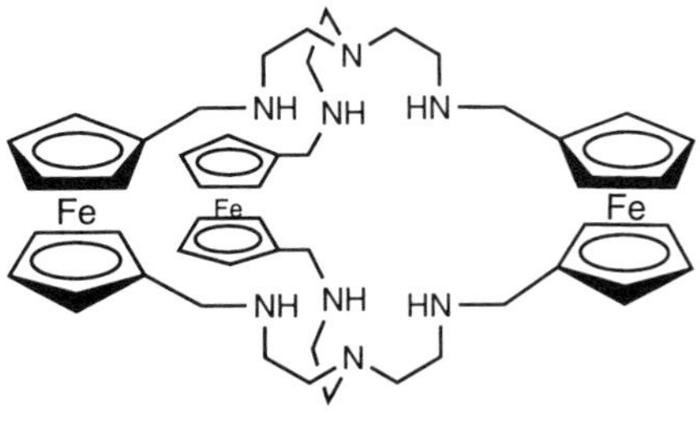

**59**

On addition of sodium azide to the zinc(II) cryptate, a solid cascade complex was isolated. Evidence of cascade formation was provided by an investigation of azide ion infrared (IR) stretching frequencies (145).

## 5.  Receptors Based on a Calixarene Framework

The nonplanar framework of the calixarenes has been known for many years (146, 147). Syntheses to modify the upper and lower rims of these "basket-like" structures have been developed through the efforts of many research groups (148–151). This research has allowed the calixarenes to be used as a three-dimensional framework upon which ligating groups can be affixed for the coordination of guest species.

We have reported the synthesis of functionalized calixarene **60**. The crystal structure of the nickel azide complex of this dinuclear host indicates two nickel ions bound (one centered in each of the appended lower rim macrocycles), with three unique azide 1,1' end on bridging ligands cascaded in between (Fig. 6) (152).

60

Puddephatt and co-workers (153, 154) reported a novel use of a calixresorcinarene for anion binding. Receptor **61** was synthesized and tetracopper(I) and tetrasilver(I) complexes were isolated. In each case, an electron deficient cavity is formed with an array of four metal ions around its upper rim. These metal ions are capable of coordinating an anion between them. Crystal structures [Fig. 7(a and b)] of two such anion inclusion complexes were resolved and a discussion of highest occupied molecular orbital/lowest unoccupied molecular orbital (HOMO/LUMO) overlap was provided. The nature of the cavity was shown to

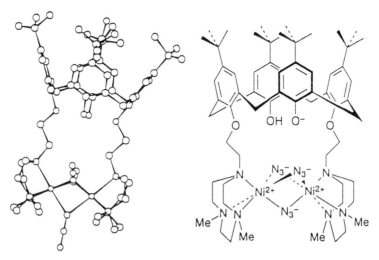

Figure 6. Crystal structure and schematic diagram of bis-nickel(II) (**60**) with three 1,1'-cascade bound azide anions.

bring about size selective halide binding. The nucleophilic properties of the imprisoned anion were also shown to be dramatically changed compared with the free anion (e.g., an aliphatic substitution reactions). This result suggests potential applications of such anion complexes in chemical processes. The work in this section illustrates how, by increasing the number of metal ions and utilizing a preformed binding cavity, multiple coordination arrays of metal ions may prove to be a means of introducing strong and selective anion binding.

**61**

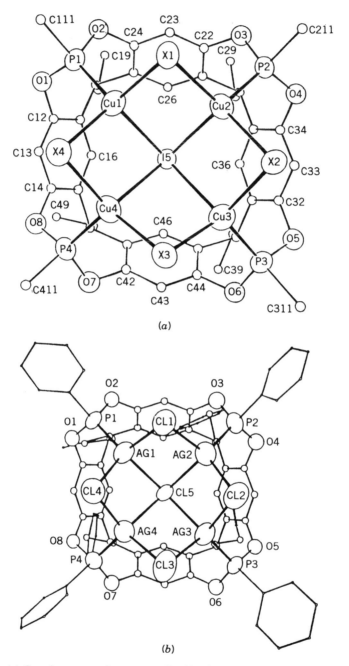

Figure 7. (a) Crystal structure of tetra-copper(I) (**61**) with included iodide anion. (b) Crystal structure of tetra-silver(I) (**61**) with included chloride anion. [Reprinted with permission from W. Xu, J. J. Vittal, R. J. Puddephatt, *J. Am. Chem. Soc., 117*, 8362 (1995). Copyright © 1995 American Chemical Society.]

## 6. Receptors Combining Small Ligating Groups

Rather than building a macrocyclic ligand or using an existing three-dimensional framework, some receptors of this cascading type bring together small ligands around arrays of metal ions. Receptor **62**, for example, forms a binuclear nickel(II) complex. This complex was shown to form 1 : 1 cascade complexes with chloride and bromide ions, but not with thiocyanate, which is too large to bridge the metal centers (155).

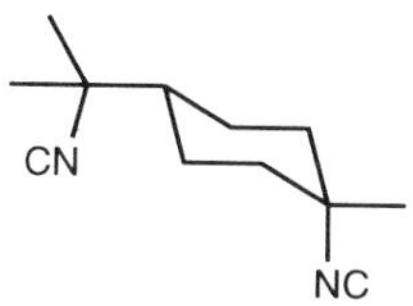

**62**

In 1990, Pecoraro and co-worker (156) formulated the metallocrown analogy. In this analogy, they visualized the synthesis of cyclic compounds, which were analogous to the crown ethers, only they contained metal ions built into the cyclic framework. One of the earliest reports of this type of compound (**63**) consisted of a trimetallic cyclic framework with the electron rich oxygen atoms pointing into the center of the cavity (157). This receptor contained a central bound metal ion. These four metal ions were bridged by carboxylate anions. Many reports followed this initial communication, illustrating the extension of the principle yield tetra- and pentametallic systems (158,159). In particular, recent studies have been concerned with the solution integrity of these complex

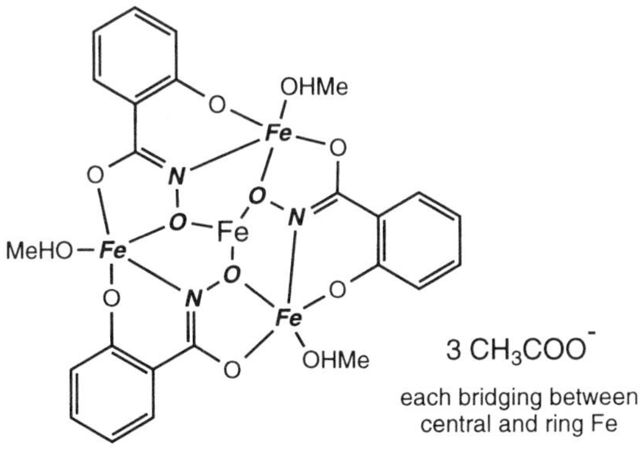

each bridging between
central and ring Fe

**63**

species (160, 161). It has been observed that in acetonitrile or $N$, $N$-dimethyl-formamide (DMF) solution, these structures apparently hold together and be-have as indicated by solid crystallographic determinations. In the case of the metallocrowns, the central cavity is electron rich and suitable for cation bind-ing, the anions then being bound in bridging modes. However, Pecoraro and co-workers (162) recently reported an analogous inverse 12-metallocrown-4 (64) based on bringing the association of discrete zinc(II) bound fragments to form a cyclic tetrametallic crown. In this case, however, the metal ions are closest to the center of the cavity, and the electron requirements of the macrocycle are reversed. Consequently, the crystal structure indicated two bound hydroxide anions situated within the metallocrown ring. This tunable class of receptor could prove of considerable future importance in the field of anion binding and selective recognition.

64

Harvey and co-workers (163) reported novel arrays of palladium centers, held together in trimeric arrays by phenylphosphine and phenylarsine ligands. Such receptors have been shown to strongly bind halide anions, the binding mechanism being a mixture of interaction with the metal centers and hydropho-bic interactions within the ''pocket'' formed by the phenyl groups (164). The use of arsine rather than phosphine ligands was shown to tune the size of the cavity (expanding it), thus enhancing the stability constants for anion binding.

A recent macrocyclic host (65) also incorporates the metal ions into the walls of the macrocycle (165). Various small tetrahedral or Y-shaped anions were observed to selectively bridge the two metal centers and a novel host–guest chemistry was proposed. However, the system acts analogously to many of the systems already discussed.

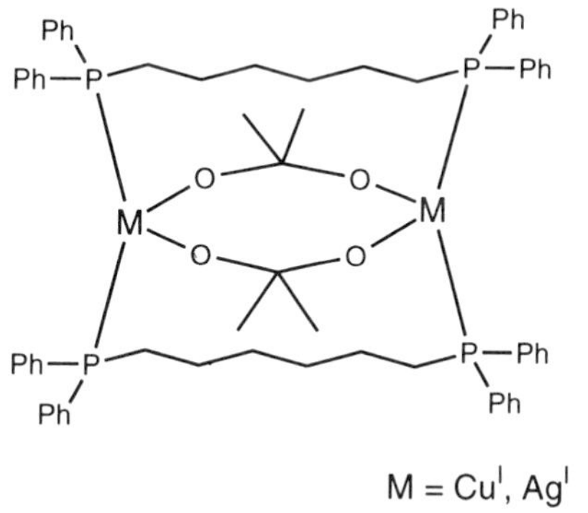

M = Cu$^I$, Ag$^I$

**65**

Fish et al. (166, 167) produced novel bioorganometallic trimers (**66**) which are themselves based on an anion recognition process. Cyclopentadienyl (Cp) rhodium fragments are used to coordinate deprotonated anionic aminopurines. These self-assembled bioorganometallic arrays show remarkable stability in aqueous solution at physiological pH. Coordination of aromatic amino acids into these preformed cavities has also been performed in aqueous media at pH 7, when the amino acids are zwitterionic in nature (168). The complexation is believed to occur through largely hydrophobic interactions. Recently, anionic deprotonated methyl thymine has been coordinated by rhodium centers forming

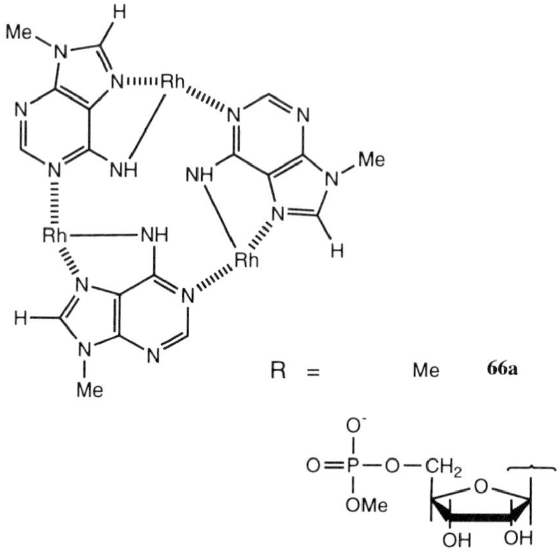

67

Compound **67** (169). Crystallographic analysis of this compound showed that the anionic methylthymine is bound into the structure by a combination of coordination [to the linear Rh(I) center] and $\pi$–$\pi$ stacking (with the Cp attached to rhodium). It was proposed that the unusual linear Rh(I) center is stablized by the four adjacent C=O groups providing steric and electronic shielding.

## 7. Receptors that Supplement a Single Coordination Interaction

Some anion receptors base themselves on one coordination-type interaction coupled with additional noncovalent interactions. Receptor **68**, for example, is prepared by modification of a cyclodextrin with flexible caps (170). This receptor was shown to bind hydrophobic carboxylate anions, such as **69**, 330 times more strongly than unfunctionalized $\beta$-cyclodextrin. Both coordination and hydrophobic interactions were of importance in the binding process, their relative contributions being $\Delta G = 3.4$ kcal mol$^{-1}$ and $-4.0$ kcal mol$^{-1}$ respectively (171).

Schwabacher et al. (1972) prepared a cyclophane-type structure with metal ions coordinated into the walls of the macrocycle (**70**). This system was first reported in 1992 and shown to transport neutral aromatic hydrocarbons through an aqueous membrane. Even though this receptor has a net anionic charge, it has been illustrated to bind indole and naphthalene units functionalized with

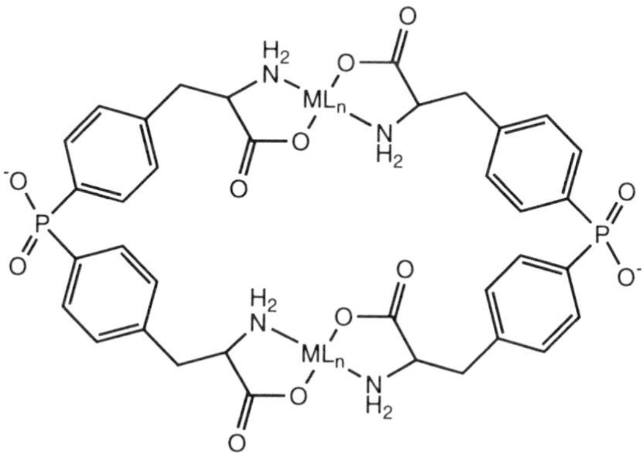

**70**

carboxylates (173). By optimal choice of the distance of the carboxylate from the aromatic group, the binding constant for the guest could be enhanced by a factor of up to 10 compared with the unsubstituted system. Dicarboxylates bound yet more strongly than monocarboxylates. The proposed binding mechanism has a twofold nature. The carboxylate group becomes ligated to the metal center and the aromatic indole or naphthalene unit is hydrophobically encapsulated within the cyclophane. In this manner, the cyclophane nature of the receptor endows a degree of additional anion selectivity. The metal therefore has both structural and functional roles in this receptor.

Receptor **71** binds deoxythymidine in preference to the other deoxy nucleic acid bases (174). Crystallographic study with analogous guests showed that this is due to a combination of anion–Zn$^{II}$ coordination and $\pi$–$\pi$ stacking. As such, this study is an elegant example of two-point binding. The imide functionality of *dT* was also found to be of importance in the selectivity of the recognition process. The complex formed is very stable and does not dissociate even in aqueous solution at physiological pH.

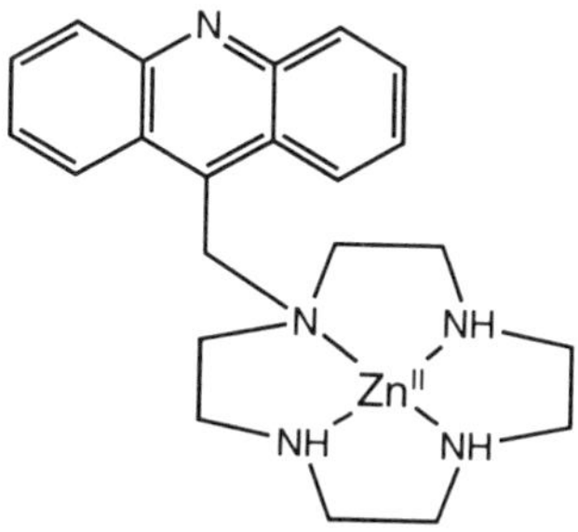

**71**

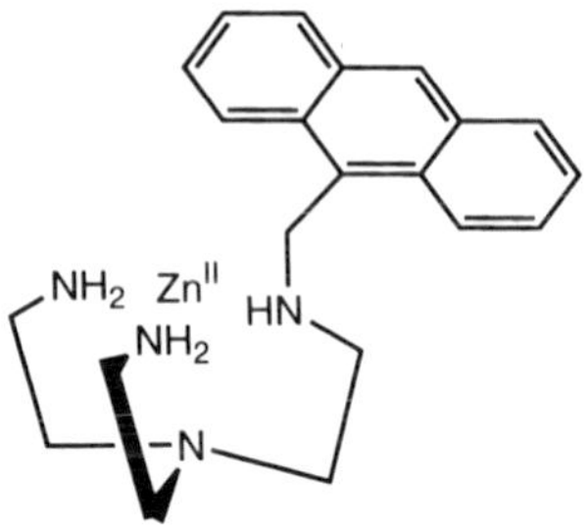

72

Fabbrizzi and co-workers (175) recently reported receptor **72**. This host also binds anions through a combination of coordination interaction and $\pi$-stacking. It also shows a fluorescence quenching sensing response on the binding of anions capable of undergoing electron transfer with the excited anthracene subunit. Both electron-donor carboxylates (such as ferrocene–monocarboxylate) and electron-accepting carboxylates (such as 4-nitrobenzoate) were shown to elicit a quenching response from the receptor. These examples illustrate the importance of noncovalent intermolecular forces in anion-binding processes.

This enhancement of a metal coordination binding site with the potential for molecular recognition via other intermolecular forces allows the synthesis of anion receptors with novel anion selectivities.

Metalloporphyrin receptors (e.g., **73**) based solely on a single metal ion–anion coordinative interaction have seen considerable application as receptor species for incorporation in electrode systems (176–180). The main goal in

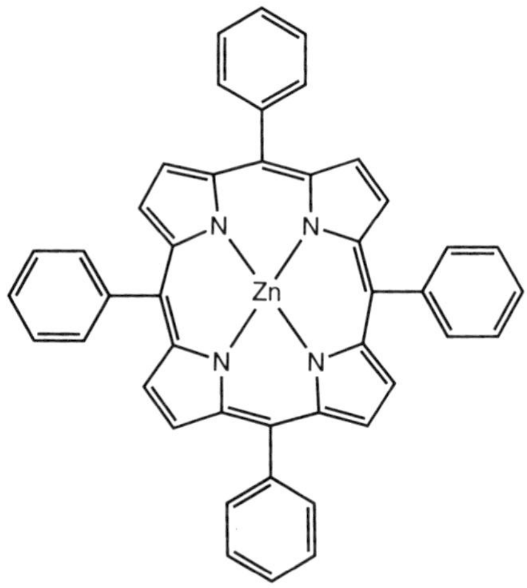

73

the development of these receptors is to overcome the Hofmeister series for response to anions (181). This selectivity trend is dependent mainly on anion lipophilicity and is usually observed in electrode systems. Several metallo-porphyrin (and phthalocyanine) based anion receptors have been developed that subvert this order of anion selection, notably being nitrite and thiocyanate selective (182–185). These molecules, however, although of practical utility, depend on a single coordinative interaction and are not receptors as defined at the outset of this review. Instead, they are traditional coordination compounds. Metalloporphyrins have, however, also been functionalized with secondary hydrogen-bonding ligating groups. These receptors will be discussed in Section V.D.

### 8.  Summary

Multiple metal ion coordination of anions, or cascade binding, has been an area of intense recent investigation for coordination chemists. The role of the metals in these receptors is threefold. They provide the source of interaction with anions through the formation of coordinate bonds. They fulfill a structural role, the separation of the metal ions being used to impart anion selectivity or different modes of anion binding, and in some cases they have taken a functional role. In particular, the use of this type of receptor for biocatalytic mimicking and rate enhancement has already been established and could prove particularly fruitful.

The use of intermolecular forces in concert with coordination interactions as discussed in Section V.B.7 has led to receptors with novel anion selectivities. In the next sections, we examine receptors that function primarily through non-covalent intermolecular interactions.

### C.  Charged Receptors Based Primarily on Electrostatic Attraction

A different approach to that engendered by the hosts discussed thus far, which function through orbital overlap interactions, has been to design positively charged inorganic receptors capable of binding anions primarily via through-space electrostatic attraction. This kind of host is, in its mode of binding at least, analogous to the organic quaternary ammonium systems discussed earlier.

### 1.  Cobaltocenium Based Receptor

We reported the first transition metal centered anion receptor to operate solely through electrostatic attraction in 1989 [9, 186]. Receptor **74** contains two positively charged, 18-electron, air stable, redox active cobaltocenium moieties. The reversible reduction potential of these redox active centers was observed to shift cathodically (up to 45 mV) on the addition of excess bromide ions. This

**74**

finding is indicative of bromide interaction with the receptor through electro-static attraction to the positively charged, organometallic fragments. Conse-quently, receptor **74** was the first redox active sensor for anions, the role of the metal being twofold:

1. Interaction with the anion through electrostatic attraction.
2. Functional anion sensing properties.

Subsequently we have made much use of the cobaltocenium fragment in com-bination with hydrogen-bonding amide substituents for the production of func-tional anion sensors. These hosts will be discussed in Section V.D.

## 2.  Metal Ion Cornered Macrocyclic Receptors

Fujita, et al. (187) first reported a novel class of tetrameric metal cornered macrocycles in 1990. These octapositive macrocycles (**75**) have potential anion-binding properties due to their high charge and, indeed, evidence for the binding of electron-rich 1,3,5-trimethoxybenzene was presented, although no anion binding was discussed. The slow self-assembly of **75a** was studied by NMR techniques (188), and then in 1991 a series of molecular recognition results were presented (189). Stoichiometric 1 : 1 complexes were observed with elec-tron-rich aromatic guests and stability constants evaluated in $D_2O$, the strongest complex being formed with 1,3,5-trimethoxybenzene ($K = 750\ M^{-1}$). Recog-nition of aryl carboxylate anionic guests was also indicated but the stoichiom-etry of binding was unclear and stability constants for the process could not be determined. The NMR chemical shifts of proton resonances on the bound car-boxylates were dramatically affected (up to 2.8 ppm) by complexation, and the presence of an aromatic ring on the guest anion appeared to be essential for binding. This indicates that complexation probably occurs due to a combination of electrostatic and hydrophobic intermolecular forces. The role of the plati-num/palladium in these receptors is therefore twofold, being both the source of interaction with anions and structurally ensuring the formation of the four-cor-nered "molecular-box" due to square planar $d^8$ metal ion coordination.

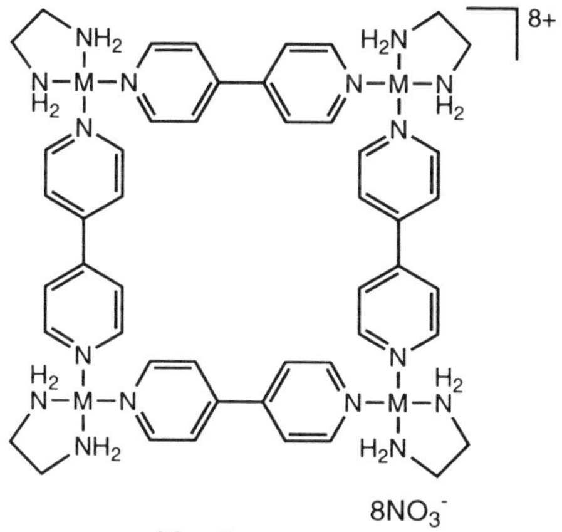

M = Pt    75a
M= Pd    75b

Stang and co-workers (190–192) also prepared an analogous series of platinum and palladium cornered macrocycles (**76–78** are typical), but as yet, only weak recognition of electron-rich 1,5-dihydroxynaphthalene has been reported. Anion-binding studies are yet to be published, although the obvious potential of these systems is often stressed. An elegant iodonium macrocycle (**79**) with anion-binding potential has also been reported. Once again, binding studies are awaited (193, 194). Additionally, the synthesis of the macrocycle **80** has been reported (195).

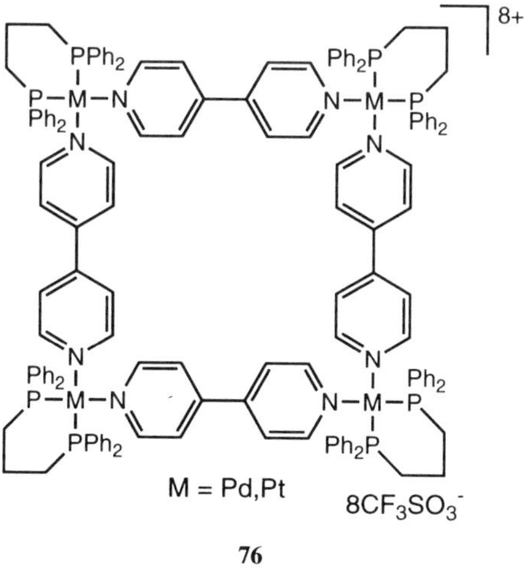

M = Pd,Pt    $8CF_3SO_3^-$

**76**

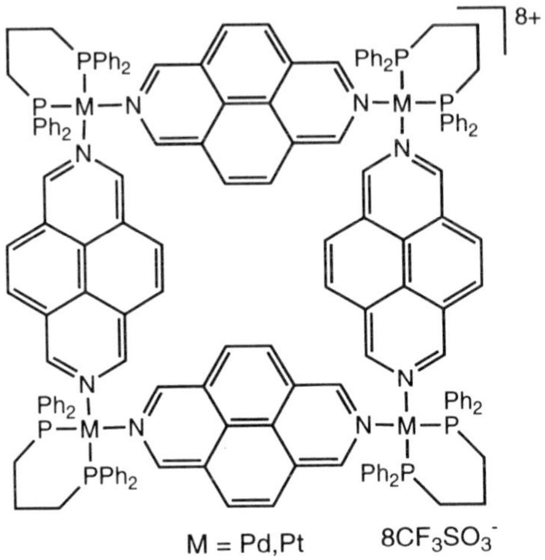
8+
PPh2
P—M—N
Ph2
N
Ph2P
N—M—P
Ph2
N
N
Ph2
P—M—N
PPh2
N
Ph2
N—M—P
Ph2P
M = Pd,Pt
8CF3SO3-

**77**

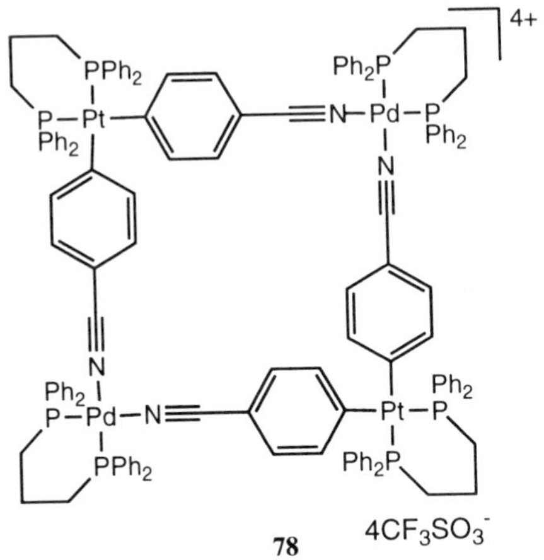
4+
PPh2
P—Pt
Ph2
Ph2P
N—Pd—P
Ph2
N
N
Ph2
P—Pd—N
PPh2
Ph2
Pt—P
Ph2P
4CF3SO3-

**78**

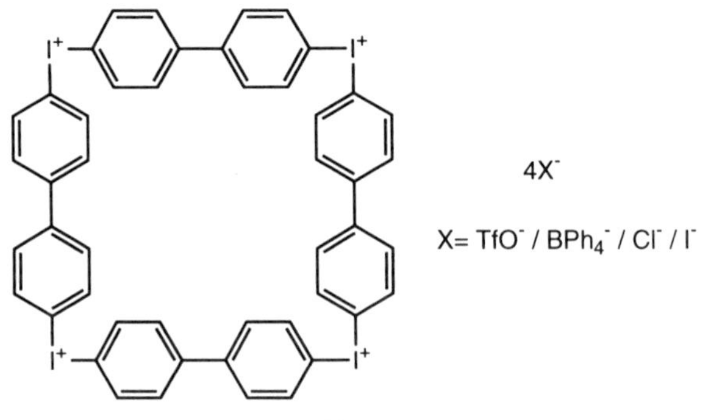
I+
I+
I+
I+
4X-
X= TfO- / BPh4- / Cl- / I-

**79**

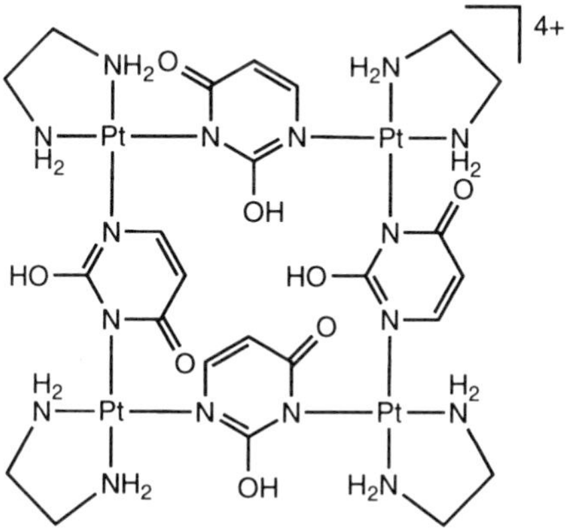

**80**

A recent report by Hupp et al. (196) provides the first evidence for anion sensing by this type of host molecule (**81**). Time-resolved luminescence studies indicate that **81** has a shorter excited-state lifetime ($\tau = 17$ ns) than the rhenium only model (**82**) ($\tau = 645$ ns). This shorter lifetime causes a lower intensity of emission and is presumably due to palladium quenching effects. On addition of tetraethylammonium perchlorate to a solution of **81**, however, luminescent intensity increases, as does the excited-state lifetime ($\tau = 21$ ns). It is argued that perchlorate binding to the receptor affects the energetics of the quenching process, consequently causing luminescent anion sensing. The stability constant for the binding process was elucidated as 900 $M^{-1}$, the solvent being acetone.

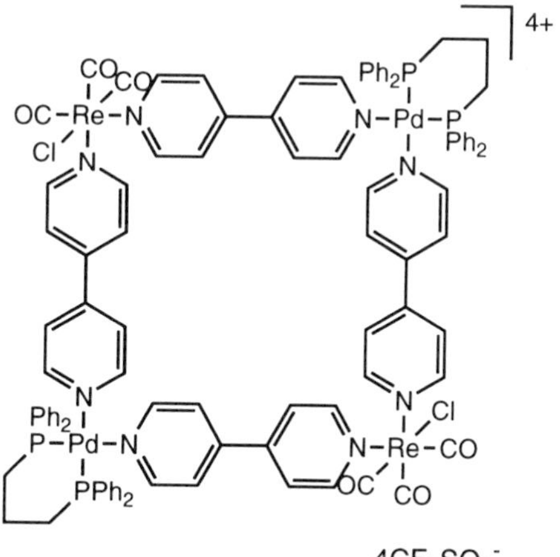

**81**

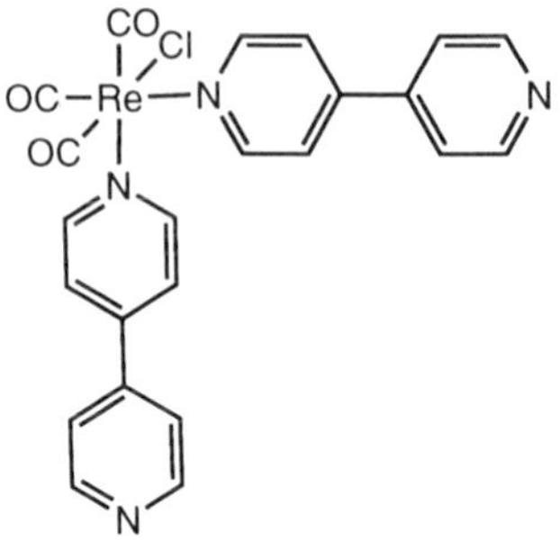

**82**

Unfortunately, no mention is made of what effect, if any, other anions have on the luminescent process, making selectivity difficult to assess. The role of the metal ions in this system is threefold:

1. Electrostatic interaction (due to positive charge).
2. Structural (coordination geometries define macrocyclic structure).
3. Functional (luminescent sensing properties of the receptor).

Perhaps most spectacularly, in 1995, Fujita et al. (197) reported the synthesis of a three-dimensional, palladium based molecular cage (**83**). This receptor was formed most effectively in the presence of an organic carboxylate anion. In the absence of such a guest, a considerable quantity of oligomeric and polymeric material was obtained (receptor yield: 60%). Figure 8 provides the receptor yields in the presence of each templating carboxylate anion employed in the study and the induced NMR chemical shift (ppm) of the guest anion's pro-

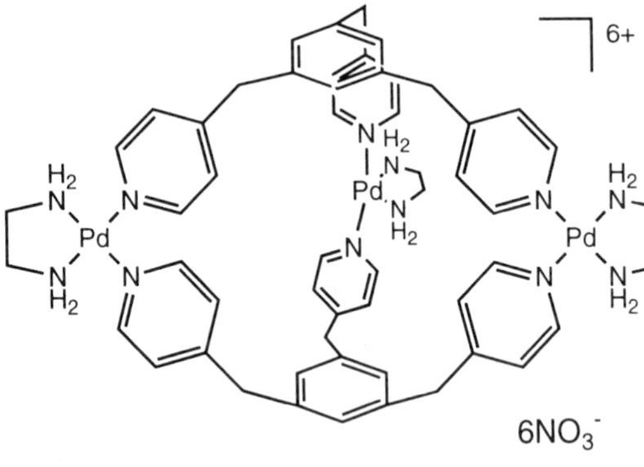

**83**

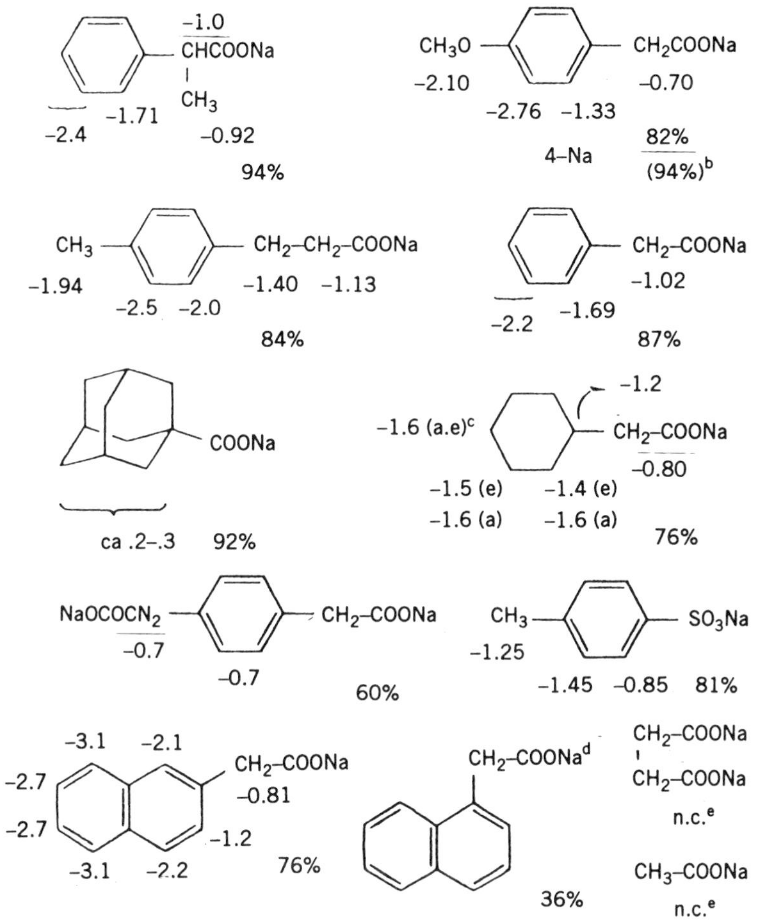

−1.0
CHCOONa
|
CH₃
−1.71
−2.4        −0.92
94%

CH₃O —⟨⟩— CH₂COONa
−2.10                  −0.70
−2.76  −1.33
4–Na
82%
(94%)[b]

CH₃ —⟨⟩— CH₂–CH₂–COONa
−1.94        −1.40  −1.13
−2.5  −2.0
84%

⟨⟩— CH₂–COONa
−1.02
−2.2   −1.69
87%

COONa
ca .2–.3   92%

−1.6 (a.e)[c]  ⟨⟩— CH₂–COONa   −1.2
−1.5 (e)   −1.4 (e)        −0.80
−1.6 (a)   −1.6 (a)
76%

NaOCOCN₂ —⟨⟩— CH₂–COONa
−0.7
−0.7
60%

CH₃ —⟨⟩— SO₃Na
−1.25
−1.45  −0.85   81%

−3.1  −2.1
−2.7              CH₂–COONa
−2.7              −0.81
−3.1  −2.2   −1.2
76%

CH₂–COONa[d]

CH₂–COONa
|
CH₂–COONa
n.c.[e]

CH₃–COONa
36%              n.c.[e]

Figure 8.   Templating anions used in the synthesis of macrobicyclic receptor **83**. [a]Receptor yields are quoted, as are shifts of the NMR peaks for the included anion. [b]Receptor yield quoted under conditions of more concentrated guest. [c]a = axial proton, e = equatorial proton. [d]Δδ could not be analyzed due to signal overlap. [e]Not complexed (no upfield shift of guest protons). [Reprinted with permission from M. Fujita, S. Nagas, and K. Ogura, *J. Am. Chem. Soc.*, **117**, 1649 (1995). Copyright © 1995 American Chemical Society.]

tons. The effectiveness of the carboxylate template can be summarized:

bulky hydrophobic monocarboxylate > phenyl or cyclohexyl
monocarboxylate (or tosylate) > dicarboxylate or acetate

(1-naphthyl) acetate actually decreased the yield of the receptor obtained (36%), probably due to its steric bulk. These results are, once again, in accord with binding through a mixture of hydrophobic and electrostatic forces.

### 3.  Metalated Calixarene and CTV Based Receptors

As above, the judicious choice of metal ions and coordinating ligands can lead to the formation of "molecular boxes" capable of complexing anions. An alternative approach, however, is to take an existing cavity, and functionalize it with positively charged metal sites, which can subsequently act as a source of electrostatic attraction. Calix[4]arene is, as discussed earlier, a bowl-shaped receptor with aromatic walls. These aromatic rings have been metalated by Atwood and co-workers (198) utilizing organometallic synthetic strategies. The crystal structure of **84** was obtained, and the host was shown to include a molecule of diethyl ether. The tetraruthenated calix[4]arene (**85**), was also synthesized (199). This water soluble, positively charged receptor was crystallographically proven to include tetrafluoroborate anions deeply embedded within its cavity (Fig. 9). The bound anion showed an unusually low degree of thermal motion, indicative of its tight fit within the binding site.

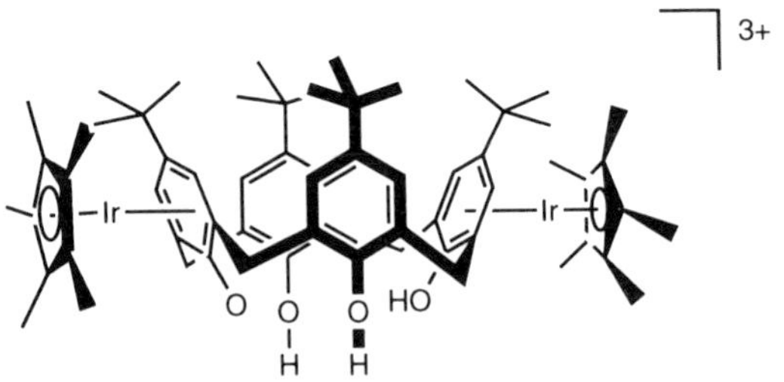

**84**

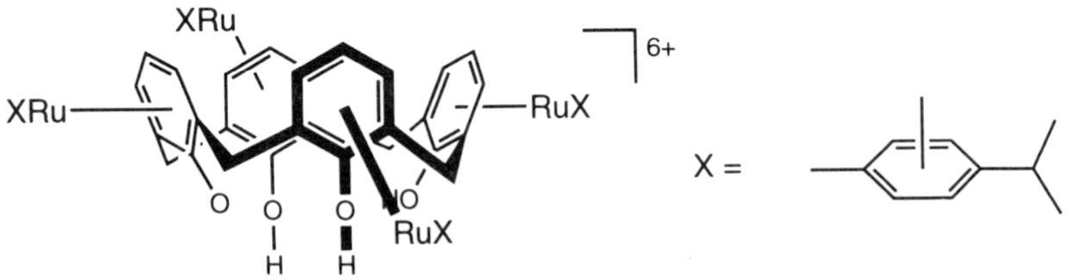

**85**

Cyclotriveratrylene (CTV) has also been used as a bowl-shaped macrocycle containing aromatic rings, the cavity being wider and shallower than that of calix[4]arene. Atwood and co-worker (200) attached transition metal centers to the outer surface of CTV yielding potential anion-binding hosts, and preventing CTV stacking, which is normally observed in the solid state. Initially, mono-metallic receptors such as **86** were synthesized and illustrated the general principle that CTV metalation did indeed prevent stacking and modify the electronic

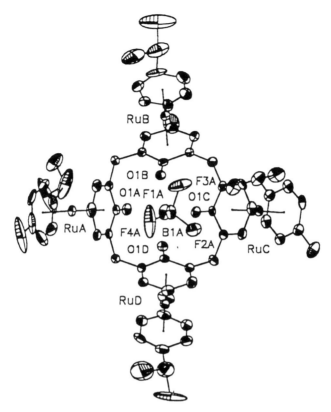

Figure 9. Crystal structure of the tetrametallic receptor **85** showing the deeply embedded tetra-fluoroborate anion. [Reprinted with permission of (199).]

**86**

**87**

properties to potentially allow anion binding. More recently, receptor **87** has been shown to bind the perrhenate anion within the cavity, the anion being slightly displaced toward the two metalated rings (201). This binding is of particular importance due to the close relationship between perrhenate and the isoelectronic, environmentally polluting, radioactive pertechnetate anion. Promising radiochemical two-phase extraction experiments (saline/nitromethane) with $^{188}$Re and $^{99}$Tc have indicated marked selectivity for these anions in the presence of up to 10-fold excesses of chloride, nitrate, and sulfate anions. The receptor was also observed to show a small redox response (15 mV) to the presence of the perrhenate anion. The source of binding energy in these metalated organic receptors is primarily thought to be electrostatic attraction to the appended metal centers. Solvation effects may also be of importance.

### 4.  Metallacrown Based Receptor

Lehn et al. (202) produced an elegant macrotricyclic chiral receptor (**88**) capable of anion recognition through electrostatic attraction to alkali metal cations. The bis(naphthyl) spacer was used to confer chirality onto the host molecule. The receptor was complexed to 2 equiv of alkali metal cation salt, and then the coordination of racemic mandelic acid salts was investigated, the interaction occurring between the mandelate anion and the positively charged alkali metal cations bound in the crown ether rings. Receptor/anionic substrate ratios of approximately 1:1 stoichiometry were observed (by $^1$H NMR experiments) and a small enantiomeric excess in the recognition process was detected (15% with $Cs^+$ bound in crown ether rings). The receptor was also found to be capable of transporting the anion at rates of up to 1.5 mmol $h^{-1}$ across a phase

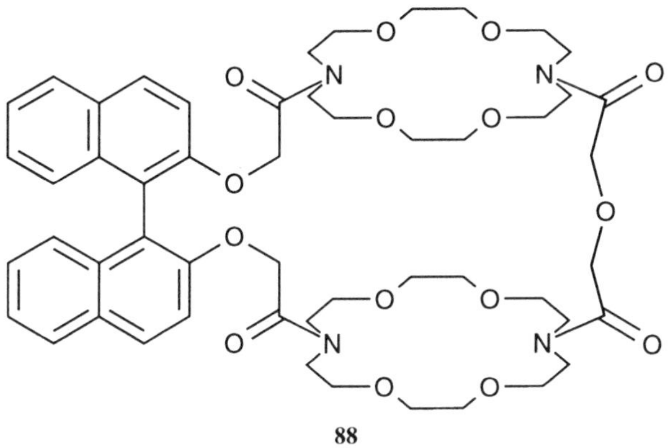

**88**

boundary (with small enantioselection effects). This receptor is consequently capable of simultaneous anion and cation recognition, an important current area of research. Crown ether bound alkali metal cations have also been used for anion binding in combination with hydrogen-bonding functionalities, and these receptors will be further discussed in Section V.D.

## 5. Vanadate Based Receptors

For many years zeolites were well-known to incorporate small guest species in the solid state. The first soluble oxide inclusion complex, however, was based on an isopolyoxavanadate cluster, which was shown to bind acetonitrile (203). These polyoxavanadates (e.g., **89–90**) were also shown to be capable of incorporating anions (204, 205). Structurally different vanadates were observed to form in the presence of different anions (Fig. 10), the anion exhibiting a templating effect. The vanadate host molecules are themselves, heteropolyanions, and it may therefore seem surprising on electrostatic grounds that anion encapsulation occurs at all. The receptors contain vanadium(IV) and (V) centers (in almost any desired ratio) (206, 207) and it is probably these centers that provide the basis of the interaction with the anion. Typical metal–anion distances (from crystallographic study) are approximately 3.6 Å, too long to be formal coordinate bonds, and indicative of longer range intermolecular forces. Photolysis has been used to drive anion encapsulation in this kind of receptor cage

$$[H_4V_{18}O_{42}X]^{4-}$$

X = Cl, Br, I

**89**

$$[HV_{22}O_{54}(ClO_4)]^{6-}$$

**90**

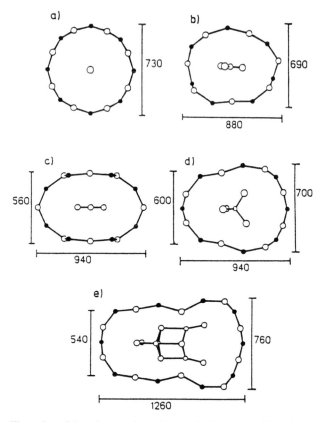

Figure 10. Illustration of the anion template effect on the formation of isopolyoxavanadate clusters. Shapes and dimensions (in picometers) of simplified V—O cluster shells are given. (a) $[H_4V_{18}O_{42}(X)]^{9-}$ X = Cl, Br, or I. (b) $[HV_{18}O_{44}(NO_3)]^{10-}$. (c) $[H_2V_{18}O_{44}(N_3)]^{5-}$. (d) $[HV_{22}O_{54}(ClO_4)]^{6-}$. (e) $[V_{34}O_{82}]^{10-}$. [Reprinted with permission of (205).]

(208,209). The chloride complexes of these receptors have been postulated to have particularly high anion coordination numbers, with the central anion being readily replaced. No binding strengths for this type of receptor have as yet been evaluated.

## 6.  Summary

Electrostatic interactions are of undoubted importance in the molecular recognition of anions (210). One of their key features is their nondirectional nature, which makes them useful general long-range forces but also makes it difficult for electrostatic forces alone to effect selective complexation unless they are built into synthetically sophisticated structural hosts. This factor has led to the development of aesthetically pleasing receptor molecules. The presence of hy-drophobic-binding cavities has also been used to enhance anion selectivity, al-though it is difficult to extend this selectivity beyond hydrophobic guest bind-ing.

Other intermolecular forces can also be used for the purpose of enhancing selectivity. One approach to enhancing electrostatic interactions in order to en-force a further degree of selectivity is to use hydrogen bonding. This method-ology is outlined in Section V.D.

## D.  Receptors Incorporating Hydrogen Bonding

As discussed in Section II.A and Section III, hydrogen bonding is very im-portant in the binding of anions by biological and organic receptors. Most of the organic receptors investigated incorporate some degree of hydrogen bonding in their mechanism of interaction with anions with both neutral and charged hydrogen-bond donors having been shown to be of importance. Hydrogen bond-ing provides a useful way of introducing selectivity to the anion-binding pro-cess, as simple organic transformations can be used to introduce a whole variety of hydrogen-bonding units into receptor species. Inorganic chemists have there-fore made extensive use of hydrogen-bonding groups in combination with metal centers to produce a wide range of exciting anion receptors.

### 1.  Cobaltocenium Based Receptors

The first ester functionalized cobaltocenium receptor (**74**) developed by us in 1989, was based solely on electrostatic interactions (as discussed earlier) (9, 186). It was therefore decided to append the cobaltocenium moiety with sec-ondary amide functionalized ''arms''. In this manner, the receptors became more resistant to hydrolysis (a recurrent problem with ester-based systems), and also incorporated neutral hydrogen-bond donors capable of coordinating anions. [An interaction between a secondary amide (acetamide) and the bromide anion was physically investigated (by IR spectroscopy) as long ago as 1961 (211)]. We made the first report of this novel type of receptor (**91** and **92**) in 1992 (212), which was, in fact, the first class of inorganic anion receptor that incor-porated hydrogen-bonding functionalities.

The binding of anionic guests to mono, 1,1'-bis, and tripodal substituted cobaltocenium receptors was initially investigated by H/NMR spectroscopy. On titration with tetrabutylammonium anion salts, large downfield shifts of receptor protons were observed. Particularly perturbed was the amide proton, indicative of strong hydrogen-bond formation. A considerable degree of hydrogen-bond formation was even observed in polar aprotic solvents such as acetonitrile and dimethyl sulfoxide (DMSO). In the case of hosts where the amide was tertiary (**91c**), however, no anion binding was observed, thus proving the essential na-

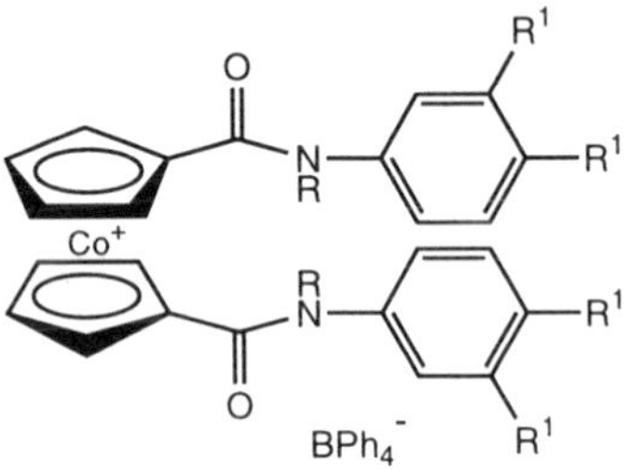

$R=R^1=H$          **91a**

$R=H\ R^1=OMe$     **91b**

$R=Me\ R^1=H$      **91c**

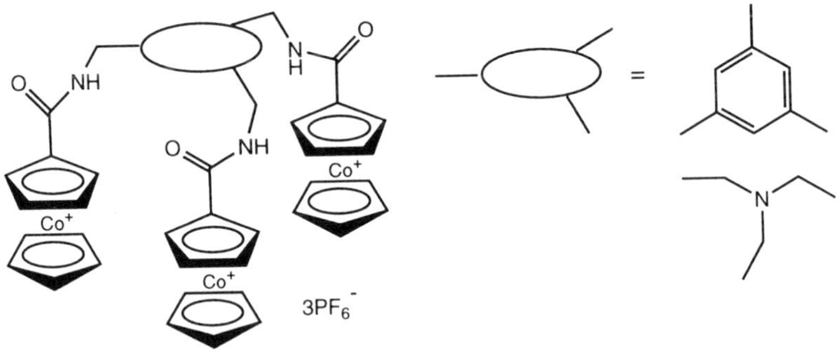

**92a**

**92b**

ture of the amide proton in the anion-binding process. Control experiments also showed the importance of the positive charge for anion-binding, with receptor **93** showing only very weak interaction with anions. This class of receptor therefore binds anions through a combination of hydrogen bonding with amide protons and electrostatic interaction with the positively charged cobaltocenium center.

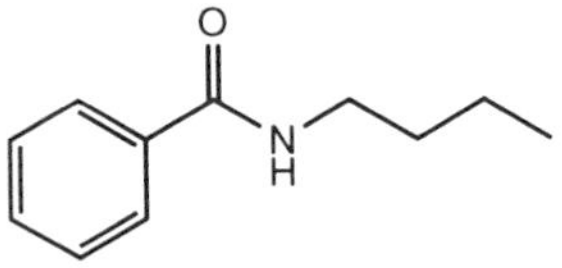

**93**

Cyclic voltammetric experiments indicated that not only was the anion bound by the host, but that the redox potential of the reversibly reducible cobaltocenium fragment was cathodically perturbed (Fig. 11) (the presence of a bound anion making the cobaltocenium unit more difficult to reduce). Tertiary amides showed no electrochemical effect on the addition of anions, ruling out the op-

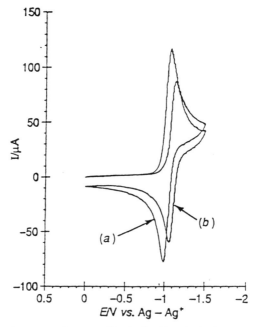

Figure 11.   Cyclic voltammetry of receptor **95a**: (*a*) free ligand, (*b*) plus 20 equivalents of chloride anion ($\Delta E = 60$ mV).

eration of any ion-pairing effects. These electrochemical results showed the potential of these receptors to act as redox sensors for anions.

Monosubstituted cobaltocenium receptors (**94**) were reported (213). The inductive effect of the aromatic ring substituent (X) was found to influence the degree of perturbation of the chemical shift of the amide proton, indicative of the more electron poor, acidic amides forming stronger hydrogen bonds with the bound anionic guest. The importance of hydrogen bonding was further emphasized in a recent paper (214). Receptors **95a–c** indicated that the strength of chloride ion binding is enhanced when additional amine–halide hydrogen-bonding interactions are sterically accessible, as is the case for **95a** and **b** but not

**94**

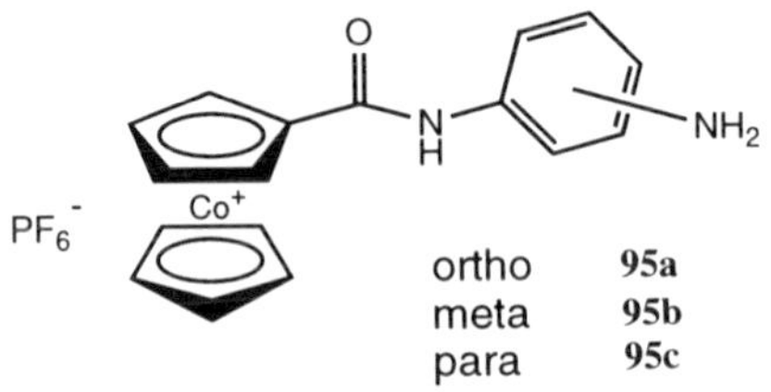

ortho    95a
meta    95b
para    95c

**95c**. The crystal structure of receptor **91a** with bound bromide anion revealed bromide hydrogen bonding to the amide proton and also to Cp and aryl protons (Fig. 12).

Aza-crown substituted cobaltocenium receptors (**96**) were also prepared and the electrostatic effect of the presence of alkali metal guests in the crown ether rings on the anion-binding process was investigated (215). Evidence for a co-operative anion-binding effect on the addition of sodium ions (due to electrostatic attraction) was presented. The ability of methylated pyridine substituents to electrostatically enhance the anion-binding process has also been reported (216).

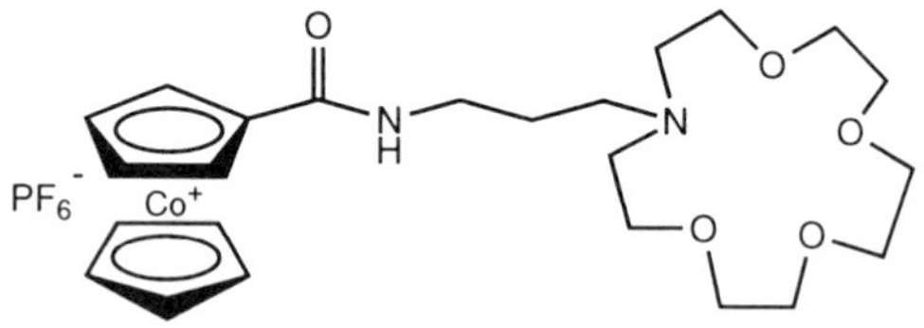

**96**

Following these preliminary results, Uno et al. reported the synthesis of a chiral, secondary amide functionalized cobaltocenium derivative (**97**). This receptor was proven (by $^1$H NMR spectroscopy in CDCl$_3$) to bind anions and showed a small enantioselectivity (estimated at 10%) for one chiral form of the optically active camphor-10-sulfonate anion.

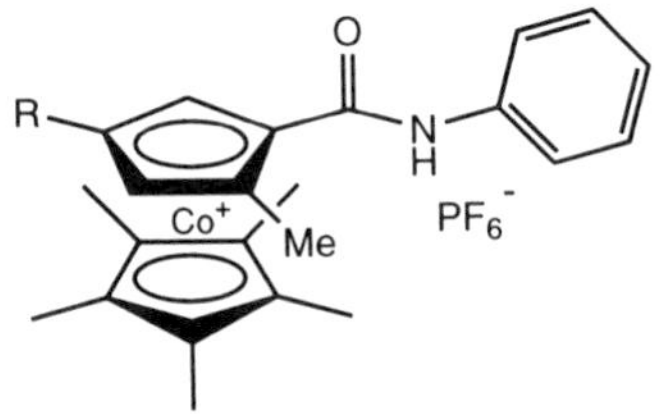

**97**

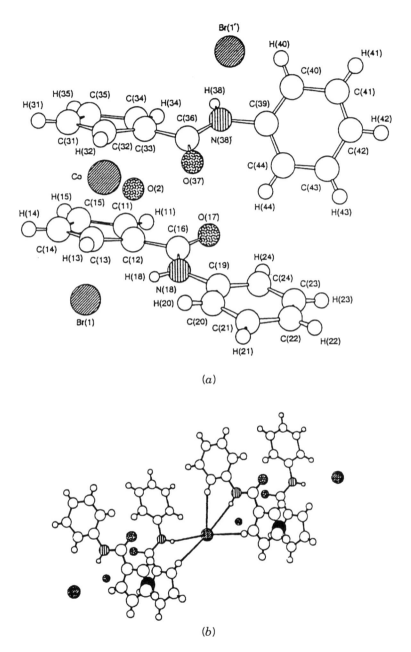

Figure 12. (*a*) Crystal structure showing receptor **91a** with bound bromide anion, including water of crystallization. (*b*) Structure of **91a**. Br⁻ showing the whole environment of the bromide anion and illustrating the importance of hydrogen bonding.

We produced stability constant data for receptors such as **91a**, which indicated particularly strong binding of the $H_2PO_4^-$ guest. This complemented the electrochemical experiments that showed this anion caused the largest cathodic perturbation of the redox wave. Attempts were therefore made to impart further selectivity and enhance complex stability for this class of anion receptor. This result was achieved by organic manipulation of the functionalized cobaltocenium "arms", with a series of ditopic bis(-cobaltocenium) systems (**98** and **99**) being prepared (219). The ${}^1$H NMR studies indicated that the alkyl-linked derivatives **99a–c** formed 1:1 complexes with halide anions. Stability constants showed that as the length of the alkyl chain increased, the general stability of the complex decreased, as did the degree of selectivity for chloride over bromide and iodide. Larger aryl or alkylamino spacers (**99d–e**) yielded complexes of 2:1 halide anion–receptor stoichiometry. All the bis(-cobaltocenium) systems showed electrochemical anion recognition with $H_2PO_4^-$ once again yielding the largest cathodic perturbations ($\Delta E = 250$ mV).

**98**

| -R- = | | |
|---|---|---|
| $-(CH_2)_2-$ | | **99a** |
| $-(CH_2)_3-$ | | **99b** |
| $-(CH_2)_4-$ | | **99c** |

**99d**

**99e**

A macrocyclic receptor (**100**) has also recently been prepared and its crystal structure was elucidated (220). In comparison with its acyclic analogue **101**, an anion macrocyclic effect was observed, the stability constants for chloride complex formation [in DMSO] being $K = 250\ M^{-1}$ (**100**) and $K = 20\ M^{-1}$ (**101**). Receptor **102** was shown to act as a switchable cobaltocenium based chloride-binding host (221). The free receptor binds chloride anions, but on the addition of potassium ions, the binding is switched off. This effect is probably due to the ability of the potassium ion to form a sandwich complex with the two crown ether substituents, sterically hindering the anion-binding site.

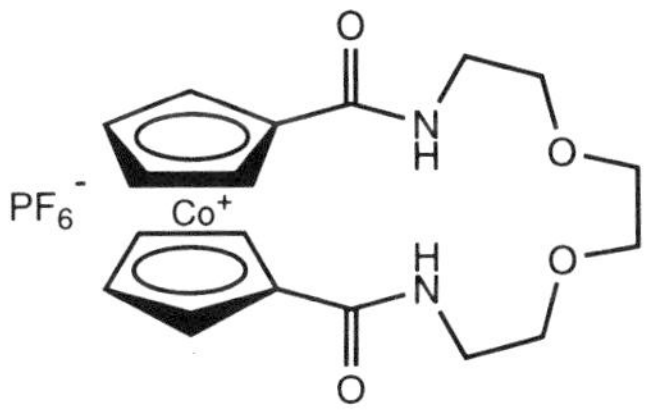

**100**

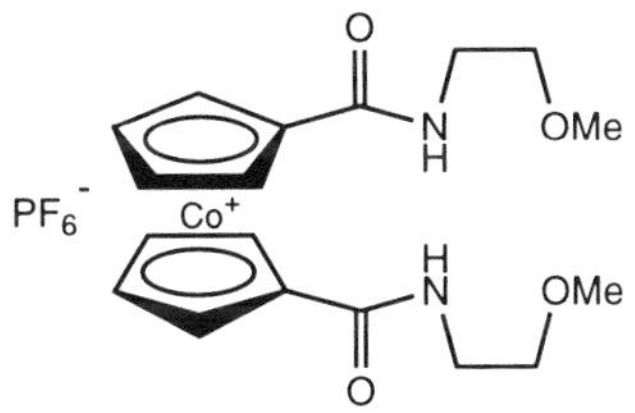

**101**

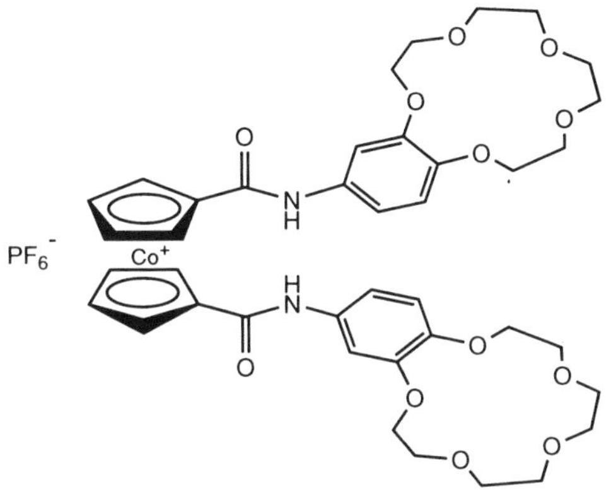

**102**

We have also synthesized a ditopic cobaltocenium host based on a calix[4]arene framework (**103**) and reported its crystal structure (Fig. 13) (219, 222). This receptor was shown to form extremely stable 1 : 1 anion complexes in polar DMSO solutions as well as with the adipate anion in acetone.

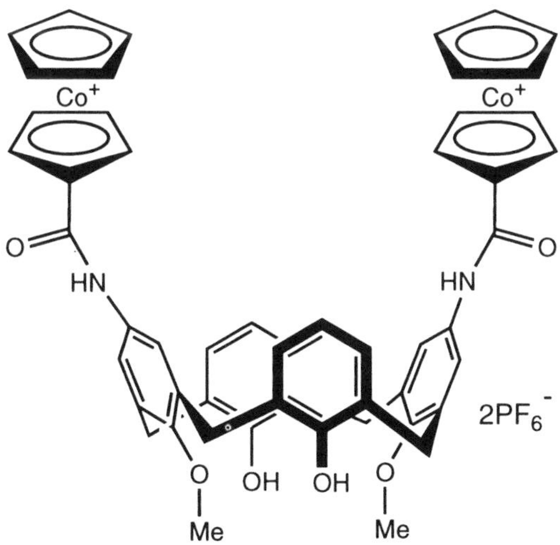

**103**

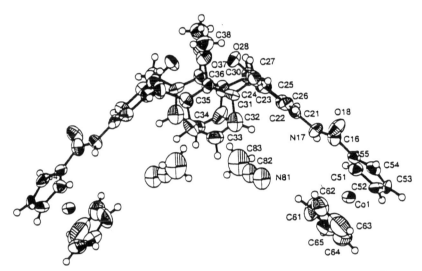

Figure 13. Crystal structure of cobaltocenium functionalized calix[4]arene receptor **103**. Aceto-nitrile is incorporated in the crystal structure. [Reprinted from reference 219]

Interestingly, this receptor displayed an uncommon selectivity preference for chloride ($K = 5035\ M^{-1}$) over dihydrogenphosphate ($K = 2800\ M^{-1}$). Altering the functionality on the lower rim of this receptor, however, dramatically altered the anion coordination properties. For example, receptor **104** exhibits selectivity for $H_2PO_4^-$ over $Cl^-$, the reverse of **103** (223).

Presumably, the bulky tosyl groups alter the topology of the upper rim anion-binding site. This result shows the capacity for facile-binding site manipulation in this class of receptor. Receptor **105** has been prepared and exhibits remarkable selectivity for $H_2PO_4^-$ ($K = 1200\ M^{-1}$) over $Cl^-$ ($K = 70\ M^{-1}$) (224). All these receptors have been shown to electrochemically recognize the presence of anions through cathodic shifts in their cobaltocenium redox wave.

A series of novel cobaltocenium–porphyrin receptors (**106**) have been prepared and shown to spectrally and electrochemically sense anions (225, 226). Free porphyrin does not, in its own right, recognize the presence of anions, indicating the essential presence of the cobaltocenium groups. Notable anion selectivities are displayed by the various atropisomers, highlighting the importance of the relative positions of the cobaltocenium amide moieties in the anion recognition process. For example, the cis-$\alpha,\alpha,\alpha,\alpha$-atropisomer (**106a**) exhibits the selectivity trend $Cl^- > Br^- \gg NO_3^-$. In contrast the $\alpha,\alpha$- (**106b**) and $\alpha,\alpha,\beta,\beta$- (**106c**) atropisomers display the rare selectivity sequence $NO_3^- > Br^- > Cl^-$, indicating a complementary host cavity exists for the environmen-

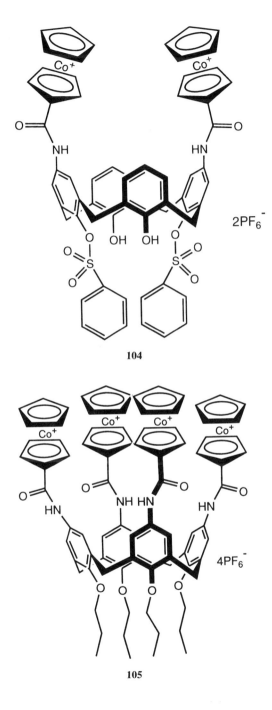

**104**

**105**

106a

106b

106c

tally sensitive nitrate anion. Metalloporphyrins have also been functionalized with hydrogen-bonding groups and these receptors will be discussed later.

Cobaltocenium based systems have therefore proved very versatile in the field of anion coordination chemistry. The role of the cobaltocenium unit is to enhance interaction with the bound anion and to function as a sensing unit. The structure of the receptor can then be controlled by simple organic manipulations of the Cp groups. In this manner, novel anion selectivity can readily be incorporated into these receptors.

In extension to the cobaltocenium anion receptor protocol, experiments in our laboratory led to the report of a series of receptors with anion-binding ability (107–109) (227). These hosts combined positively charged and neutral or-

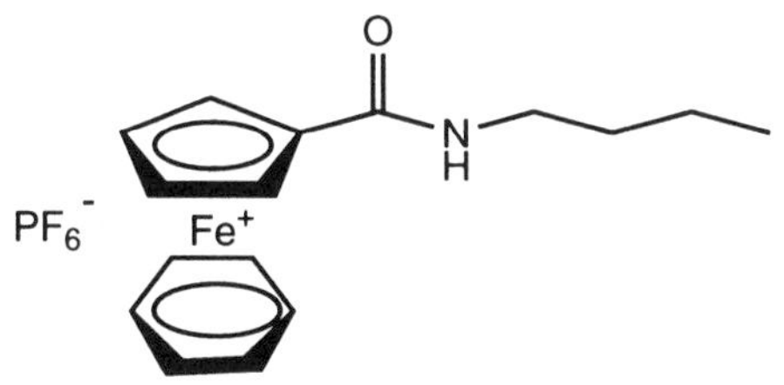

107

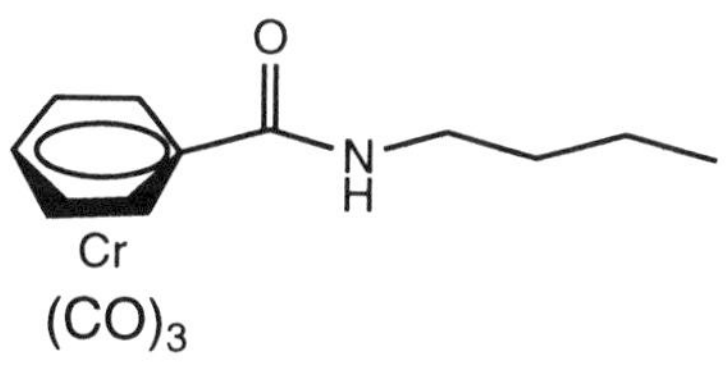

108

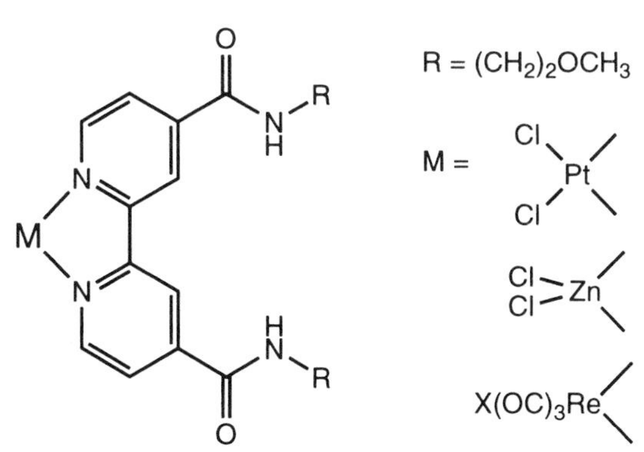

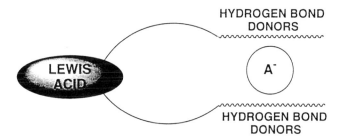

HYDROGEN BOND
DONORS

LEWIS
ACID

A⁻

HYDROGEN BOND
DONORS

Figure 14.   Schematic representation of our anion-binding protocol.

ganometallic and transition metal Lewis acidic units in combination with secondary amide-binding sites. Proton NMR anion titration experiments provided evidence for anion binding and established the generality of the principles of operation for this novel class of metal based hydrogen-bonding anion receptor (Fig. 14). This protocol for development of hydrogen-bond based inorganic receptors has been of considerable general utility, as the following sections indicate.

## 2.   Ferrocene Based Receptors

Ferrocene is extensively used as a redox unit capable of sensing the presence of bound cationic substrates (62). Recently, however, it has been successfully applied to the sensing of anionic guests. We have reported water-soluble polyaza ferrocene macrocycles (**110** and **111**) (228, 229). The anion-binding function of these receptors operates at pH 7 and below when they are partially protonated. Anions are bound through a combination of electrostatic and hydrogen-bonding interactions, analogous to the organic polyaza receptors discussed earlier. The binding process was readily followed using $^{31}P$ NMR techniques. The ferrocene unit is also capable of electrochemically reporting on the binding of biologically important anions such as $H_2PO_4^-$ and $ATP^{4-}$ by means

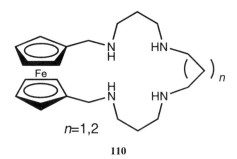

n=1,2

**110**

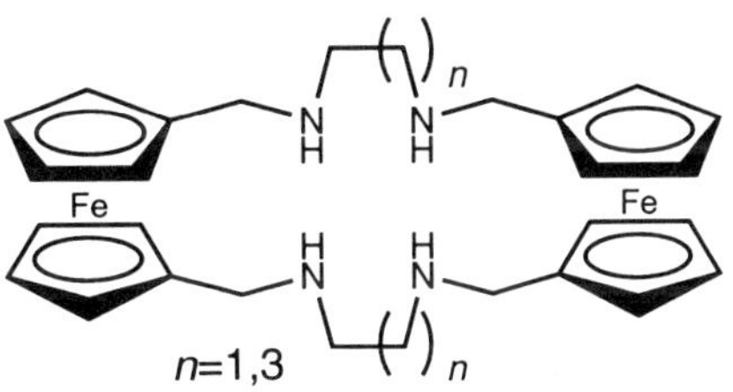

**111**

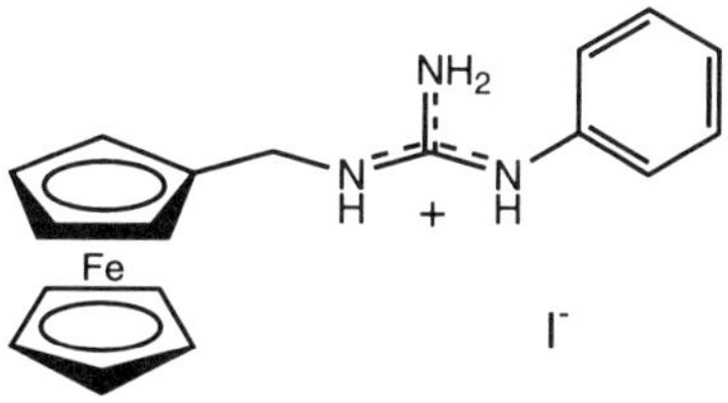

**112**

of large ($\approx 100$ mV) cathodic shifts. Ferrocene has also recently been combined with the guanidinium moiety to synthesize receptor **112** (230). This molecule has been shown to electrochemically recognize the biologically important pyrophosphate anion in 50:50 water/methanol, with cathodic redox shifts larger than 100 mV. The ability of these receptors to function as sensors in biologically important media is of great importance.

Ferrocene units appended with secondary amides have also been used for anion recognition (**113–115**) (231). Being neutral, unlike cobaltocenium based systems, these receptors have no inherent electrostatic attraction making the NMR stability constants much lower in magnitude than for the analogous cobaltocenium systems. Electrostatic interactions can, however, be switched on by oxidation of ferrocene to ferrocenium and consequently these molecules show interesting electrochemical effects and have a potential as amperometric anion sensors. Of interest to this development of sensor technology were the novel results of electrochemical competition experiments. These results demonstrated

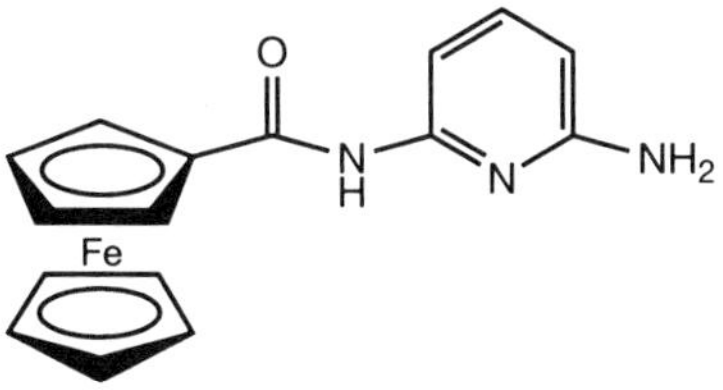

**113**

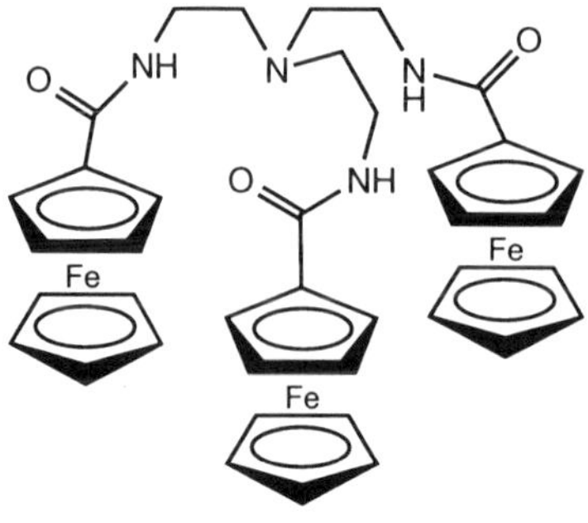

**114**

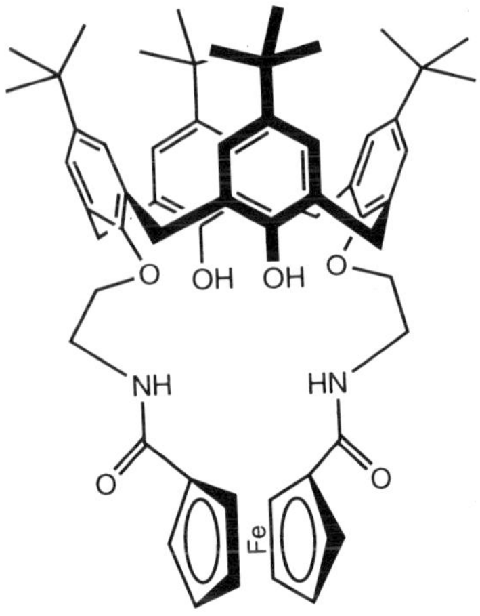

**115**

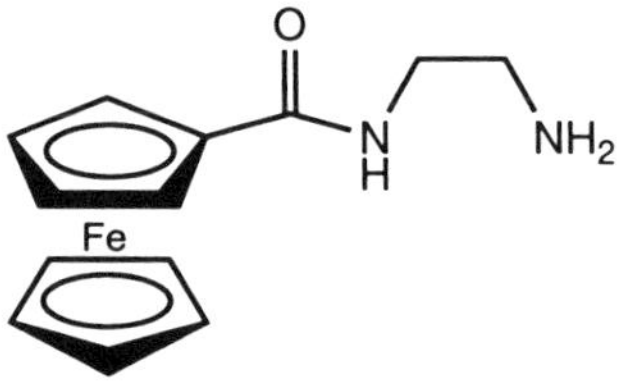

**116**

that **113–115** were capable of detecting $H_2PO_4^-$ anions in the presence of a 10-fold excess of $HSO_4^-$ and $Cl^-$ ions. Receptor **116**, however, has recently been shown to reverse this selectivity, binding $HSO_4^-$ selectively in the presence of $H_2PO_4^-$ (solvent: MeCN) (230). This novel anion selectivity is due to the presence of the basic amine functionality, which is protonated by the acidic hydrogen sulfate anion. This protonated receptor then shows a high-binding affinity for the dinegative sulfate anions produced, which invoke a marked electrochemical reductive stripping response.

We have synthesized receptor **117** by clipping the acyclic diphosphine **118** with the metal carbonyl group (232). This macrocyclic host has been shown to bind halide ions 10 times more strongly ($K = 70\ M^{-1}$, $CD_2Cl_2$) than its acyclic analogue **118** ($K = 7\ M^{-1}$). This effect is probably due to the higher degree of preorganization for binding possessed by the macrocyclic system. A crystal structure of the macrocyclic host has also been resolved.

The ditopic receptor **119** has been shown to bind both inorganic anions and alkali metal cations simultaneously (233). Selectivity was observed for $HSO_4^-$ over $Cl^-$ and the binding of alkali metal cations in the crown ethers was proven to substantially enhance the magnitude of anion binding (through electrostatic interaction).

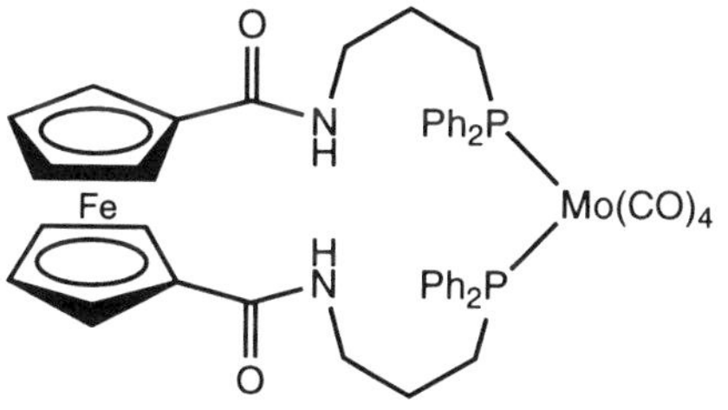

117

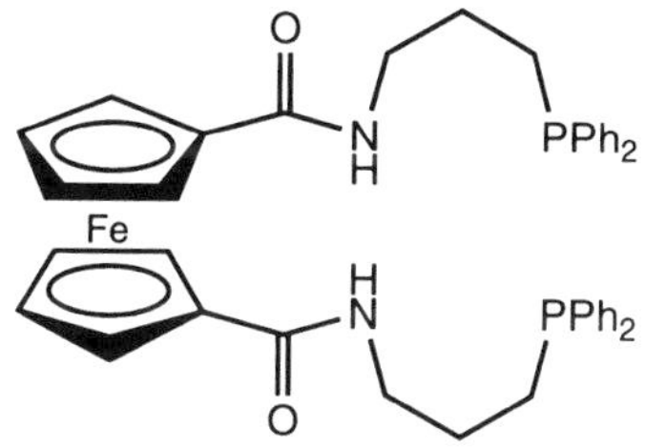

118

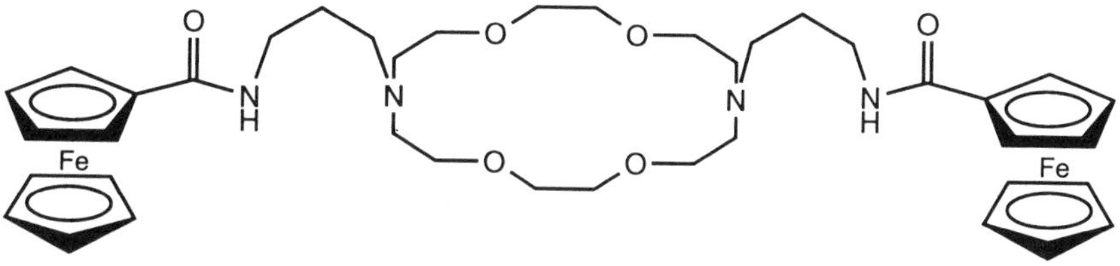

119

Delavaux-Nicot et al. (234) also reported a ferrocene macrocycle (**120**) capable of binding anions through hydrogen-bonding interactions. This receptor, the first ferrocenyl-phosphorus macrocycle, showed an EC (electron transfer followed by chemical reaction) electrochemical response and shifts in the anodic peak potential on the addition of anions [$\Delta E(\mathrm{H_2PO_4^-}) = 230$ mV]. Some evidence for dihydrogen phosphate selectivity (over hydrogen sulfate and chloride ions) in dichloromethane was presented. Molecular modeling studies indicated the most stable free receptor conformation had a 5-Å cavity defined by the four inwardly pointing N-H groups, exactly the right size for $\mathrm{H_2PO_4^-}$ encapsulation.

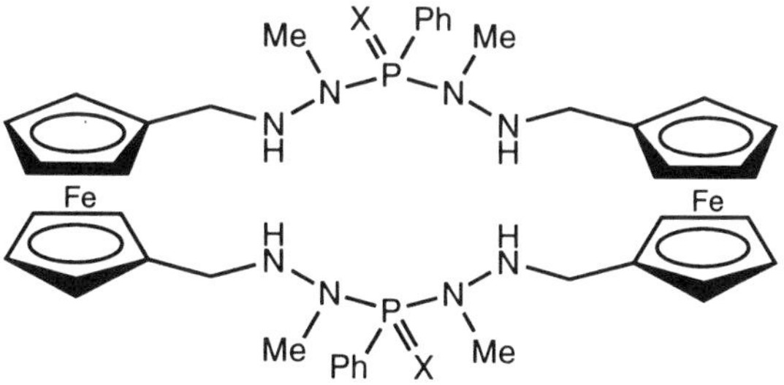

120

A recent communication reported an interesting ferrocene–anion solid state interaction between Compound **121** and polyoxometalate anions (Fig. 15) (235). An electrostatic interaction between the trimethylammonio group and the an-

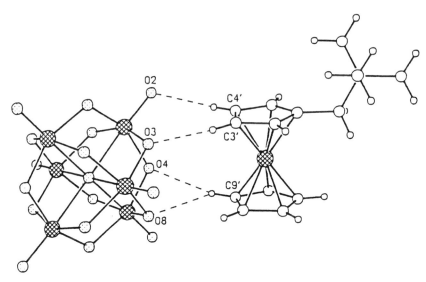

Figure 15. Crystal structure of Compound **121** illustrating the importance of hydrogen bonding between cyclopentadienyl protons and the negatively charged polyoxomolybdate anion. [Reprinted from *J. Organomet. Chem.*, *488*, C4 (1995) P. L. Veya and J. K. Kochi, Structural and spectral characterization of novel charge transfer salts of polyoxometalates and the cationic ferrocenyl donor, Copyright © 1995, with kind permission from Elsevier Science SA, Lausanne, Switzerland.]

ionic polyoxometalate was evidenced as was hydrogen bonding between the anion and the cyclopentadienyl protons [also observed in our cobaltocenium crystal structure (Fig. 12)]. This kind of crystallographic anion binding study is of increasing importance in solid phase engineering studies.

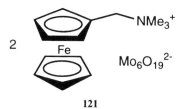

**121**

### 3. *Uranyl Salene Based Receptors*

Reinhoudt and co-workers (236, 237) first reported independently anion receptors (e.g. **122**) based on this kind of methodology in 1992. These receptors contained a Lewis acidic neutral uranyl salene group and hydrogen-bonding secondary amide functionalities. Receptors such as **123**, containing solely the

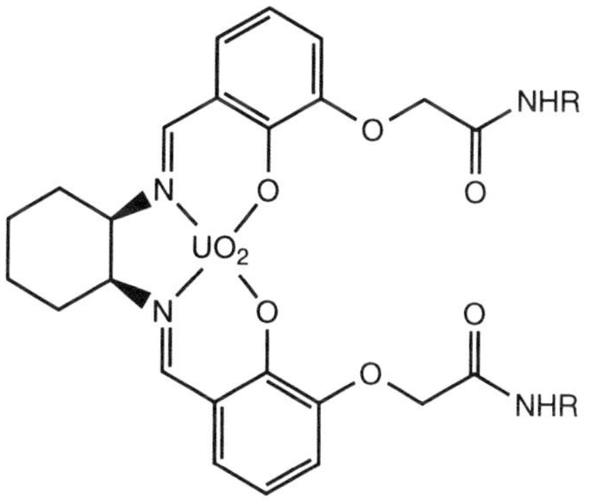

**122**

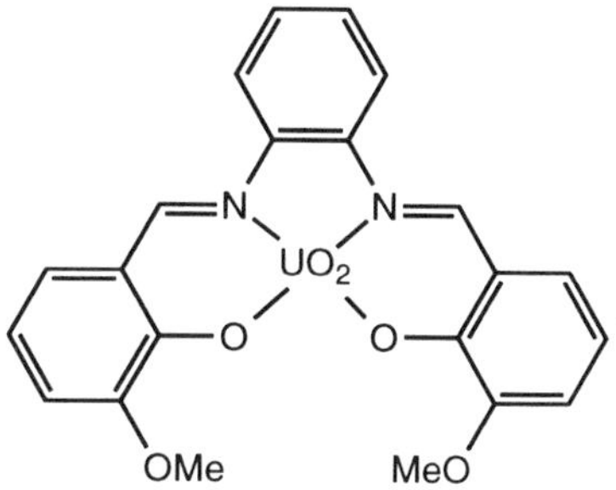

**123**

uranyl group, were observed to bind anions such as chloride with stability constants of 400 $M^{-1}$ in acetonitrile/DMSO (99:1). Crystallography showed the chloride ion to be bound to the Lewis acidic uranyl center ($r_{U-Cl} = 2.76$ Å). The incorporation of secondary amide functionalities markedly enhanced the binding of anions and, notably, dihydrogen phosphate was bound in polar organic solvents with stability constants as high as $10^5 M^{-1}$. This result was remarkable for a neutral receptor system. Crystallographic analysis of receptor **124** with bound dihydrogen phosphate (Fig. 16), showed the anionic guest was bound by a strong bond to the uranium atom ($r_{U-O} = 2.28$ Å) and secondary amide hydrogen-bond participation was also observed.

A further modification saw the introduction of crown ethers at the end of the side arms creating a ditopic receptor (**125**) (238). This modification imparted the ability to bind cations and anions simultaneously, introducing an additional electrostatic component to the anion-binding interaction. This receptor exhibited significant transport of hydrophilic KH$_2$PO$_4$ through a supported liquid membrane. Electrochemical studies indicated that H$_2$PO$_4^-$ perturbed the elec-

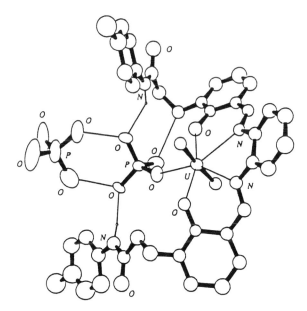

Figure 16.   Crystal structure of receptor **124** with two bound dihydrogen phosphate anions (tetra-butylammonium cations and solvent are omitted for clarity). [Reprinted with permission from D. W. Rudkevich, W. Verboom, Z. Brzozka, M. J. Palys, P. R. V. Stauthamer, G. J. von Hummel, S. M. Franken, S. Harkema, J. F. Engbersen, and D. N. Relnhoudt, *J. Am. Chem. Soc.,* *116*, 4341 (1994). Copyright © American Chemical Society.]

**124**

125

trochemistry of the proximate uranyl salene moiety, while the more distantly bound potassium ion left it unperturbed.

### 4. Crown Ether Based Receptors

Crown ethers have also been incorporated into other anion receptors. Lockhart and co-workers (239, 240) reported receptors such as **126**. This type of receptor, when protonated, has been shown to simultaneously bind cations in the crown ether rings and anions at the central linker portion (through hydrogen bonding and electrostatic attraction). Chloride binding was probed using $^{35}Cl$ NMR techniques. Potassium ions formed a 1:1 sandwich complex, while sodium ions formed a complex of 2:1 stoichiometry. No evidence was provided to demonstrate that anion and cation binding were complementary processes.

We reported a bis-crown ether functionalized calix[4]arene (**127**) (241). The unmetalated form of this receptor shows negligible interaction with anions, while the potassium bound "sandwich" complex binds anions effectively (chloride, $K = 3500\ M^{-1}$, $CD_3CN$). $^1H$ NMR investigations proved the involvement of amide hydrogen bonding in the anion-binding process. Simultaneous anion and

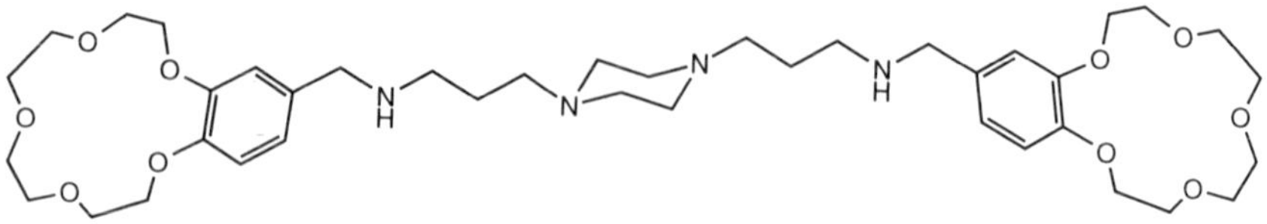

**126**

cation binding is, therefore, a complementary, electrostatically enhanced process for this receptor.

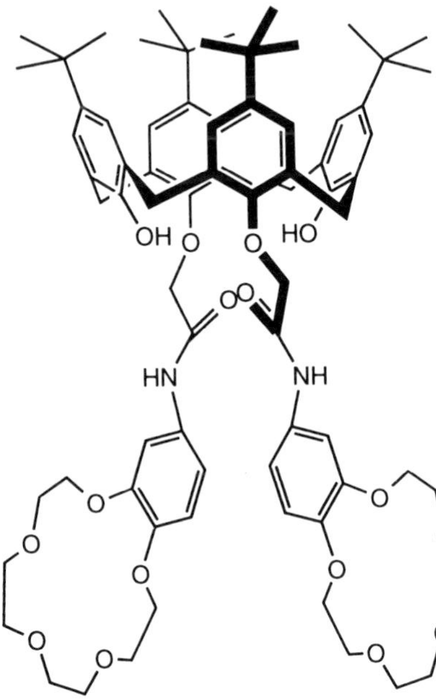

**127**

### 5.  Late Transition Metal Based Receptors

As mentioned in Section V.D.1 we prepared platinum and zinc coordination compounds functionalized with amides (e.g., **109a**) which were proven to bind anions through a combination of hydrogen bonding and electrostatics (226,242). Lanza et al. (243) also reported a platinum based anion-binding agent (**128**). The N—H protons of the coordinated neutral dithioxamide are proposed to act as the hydrogen-bonding anion-binding agent on the basis of low-temperature NMR studies. Unfortunately, the complex was nonisolable due to its instability at ambient temperature, methane being evolved and chloride becoming directly bound to the platinum center.

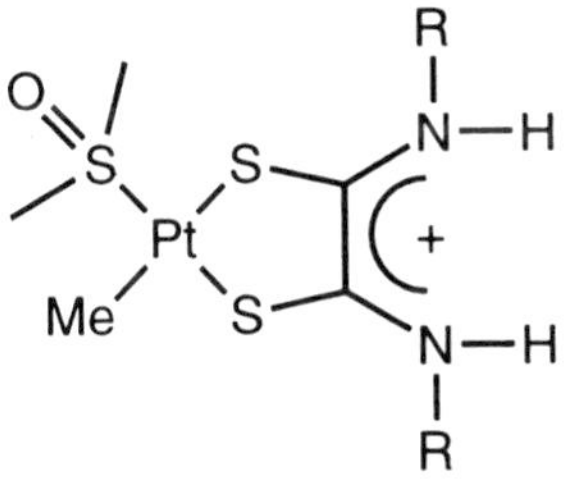

**128**

Burrows et al. (244) also recently reported a crystal engineered nickel complex, dependent on hydrogen bonding of an appended ligand to a cobound anion (**129**). They plan to vary both the cation and the dicarboxylate anion to investigate the effect on the hydrogen-bonded network. This kind of solid state engineered material could show interesting magnetic properties with potential practical applications.

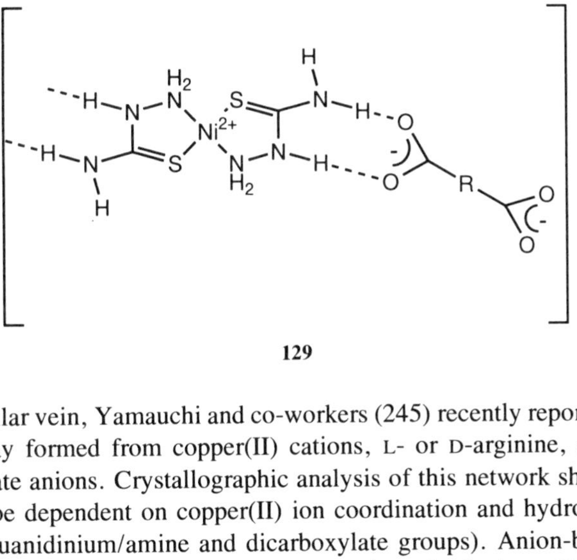

**129**

In a similar vein, Yamauchi and co-workers (245) recently reported a double-helical array formed from copper(II) cations, L- or D-arginine, and aromatic dicarboxylate anions. Crystallographic analysis of this network showed its formation to be dependent on copper(II) ion coordination and hydrogen bonding (between guanidinium/amine and dicarboxylate groups). Anion-binding interactions are consequently critical in the formation of this three-dimensional array. The helicity of the system was found to be dependent on the chirality of the arginine subunit used in its construction.

### 6. Metalloporphyrin Based Receptor

A metalloporphyrin has also been functionalized with hydrogen-bonding groups and has been reported as an anion-binding agent (**130**) (246). In this

**130**

case, the sources of the interaction with anions are coordination of the anion to the porphyrin bound zinc ion, two hydrogen bonds from the receptor to the amide group of the bound carboxy peptide, and generalized van der Waals forces. The importance of hydrogen bonding was clearly established by NMR and IR investigations. The chirality of the receptor enabled enantioselective extraction experiments to be performed (chloroform/water), with an enantio-selectivity ratio as good as 96:4 for N-benzyloxycarbonyl/valinate. Compound (**130**) is one of the few receptors to address the difficult problem of enantiose-lection combined with biological anion recognition. Corey–Pauling–Koltun (CPK) modeling indicated that steric hinderance was the key factor in achieving chiral discrimination.

### 7.  Ruthenium Based Receptors

There is much interest in transition metal polypyridyl complexes, largely due to their numerous applications in a variety of fields (247–250). In particular, ruthenium(II) tris(2,2′-bipyridyl) has been one of the most extensively studied complexes of the last decade due to its chemical stability, redox properties, excited-state reactivity, and luminescent emission (251, 252).

We recently incorporated the ruthenium(II) bipyridyl moiety into acyclic, macrocyclic, and lower rim calix[4]arene structural frameworks to produce a new class of anion receptor capable of optical and electrochemical sensing (226, 253, 254). Stability constant determinations in DMSO using $^1$H NMR titration techniques demonstrated that these acyclic receptors (**131** and **132**) form strong complexes with chloride and dihydrogen phosphate anions (stronger than with analogous monopositive cobaltocenium based receptors). The ruthenium ion is dipositive and hence the electrostatic interactions are particularly favorable. The 4,4′-substituted ruthenium bipyridyls were observed to bind anions more

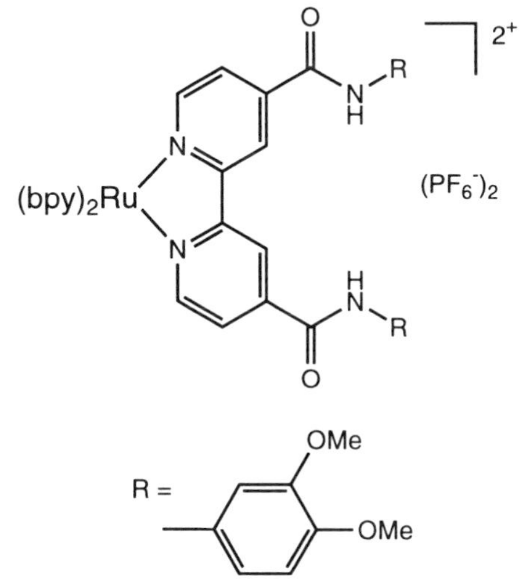

131

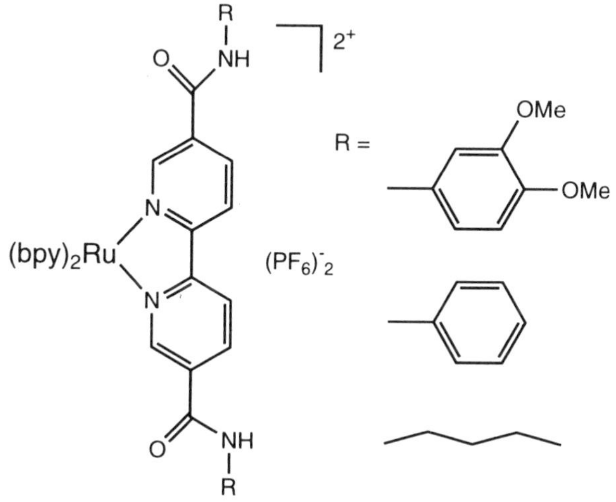

132a

132b

132c

strongly than the 5,5'-substituted analogues. This stronger binding is probably due to the more favorable alignment of the secondary amides for forming hydrogen-bonding interactions with the bound anions. The presence of proximate phenolic O—H groups has also been shown to enhance the strength of anion complex formation, presumably due to additional hydrogen-bond formation (255).

The macrocyclic receptors 133 and 134 form highly selective, strong complexes with $H_2PO_4^-$. Single-crystal X-ray structures of 131·Cl⁻ and

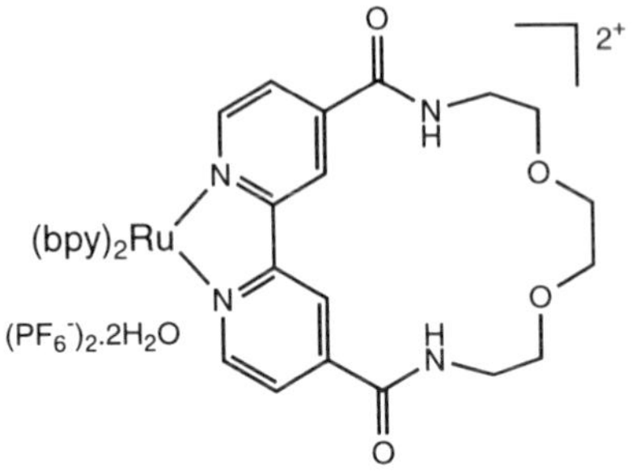

133a

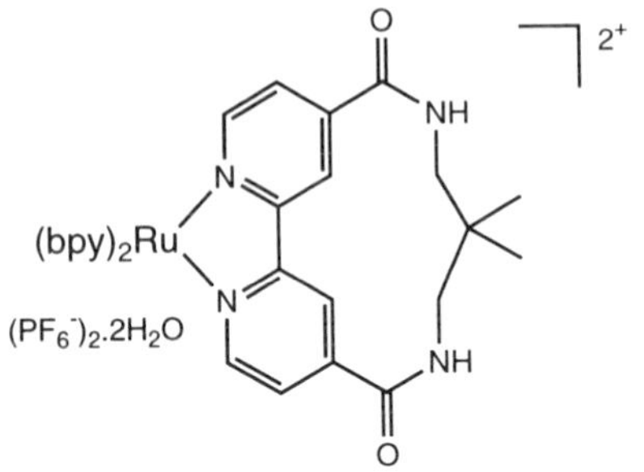

133b

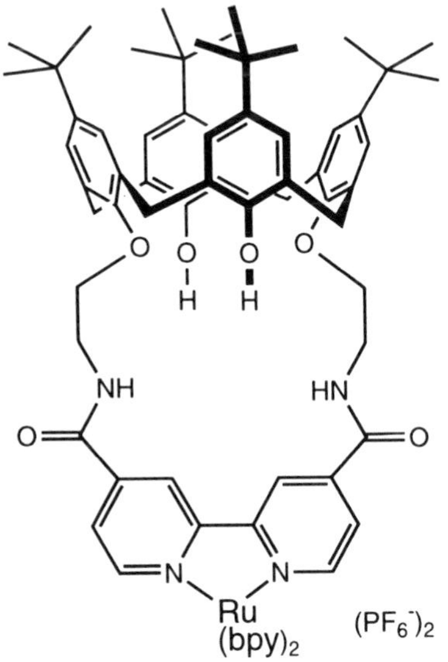

134

$134 \cdot H_2PO_4^-$ complexes were refined. These once again indicated the key importance of hydrogen bonding in this class of receptor. In the former complex (Fig. 17), six hydrogen bonds (two amide and four C—H groups) stabilize the chloride anion, and in the latter, three hydrogen bonds (two amide and one calix[4]arene hydroxyl) effect $H_2PO_4^-$ complexation.

The ruthenium(II) bipyridyl moiety is also capable of functionally sensing the presence of bound anion. Electrochemical anion recognition experiments showed substantial anion induced cathodic perturbation of the ligand centered amide substituted 2,2′-bipyridine (bpy) reduction redox couple. These perturbations were in agreement with stability constant values, with **134** sensing $H_2PO_4^-$ in the presence of 10-fold excess of $HSO_4^-$ and $Cl^-$. Fluorescence emis-

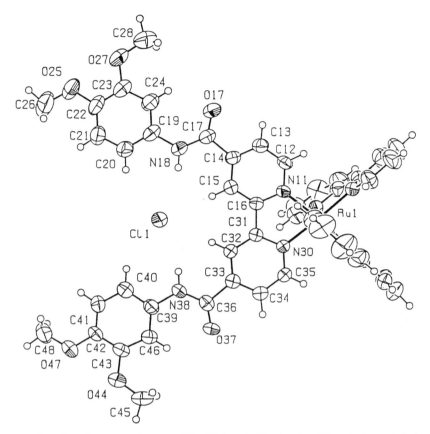

Figure 17.   Crystal structure of receptor **131** with bound chloride anion. The anionic guest sits in the plane of the receptor and forms hydrogen-bonding contacts with six preorganized protons.

sion measurements were also undertaken to probe the anion-binding process (254, 256). All receptors exhibited significant blue shifts in the metal–ligand charge transfer (MLCT) $\lambda_{max}$ emission band on addition of $Cl^-$ and $H_2PO_4^-$ with **134** displaying the largest perturbation (16 nm). These changes are *not* observed with unfunctionalized $[Ru(bpy)_3]^{2+}$. These shifts were accompanied by large increases in emission intensity. It was proposed that this could be due to the bound anion rigidifying the receptor and inhibiting vibrational and rotational relaxation modes.

We also prepared an acyclic mixed-ruthenium(II) bpy-ferrocene receptor **135** (257). The emission of the ruthenium center in the free receptor is quenched by the ferrocene units. However, on addition of dihydrogen phosphate anion the emission increases 20-fold, being switched on by the presence of the anion

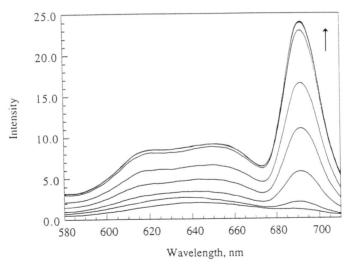

135

(Fig. 18). This effect is not observed with $Cl^-$ or $HSO_4^-$ ions. Therefore, we reported the first case of specific dihydrogen phosphate sensing by emission.

Upper rim calix[4]arenes functionalized with two and four ruthenium(II) bipyridyl amide groups (**136** and **137**) have very recently been prepared (258). These receptors also sense the presence of dihydrogen phosphate anions selectively.

Figure 18.   The effect of dihydrogen phosphate on the emission spectrum of receptor **135**. Emission is induced most strongly at 690 nm by anion addition.

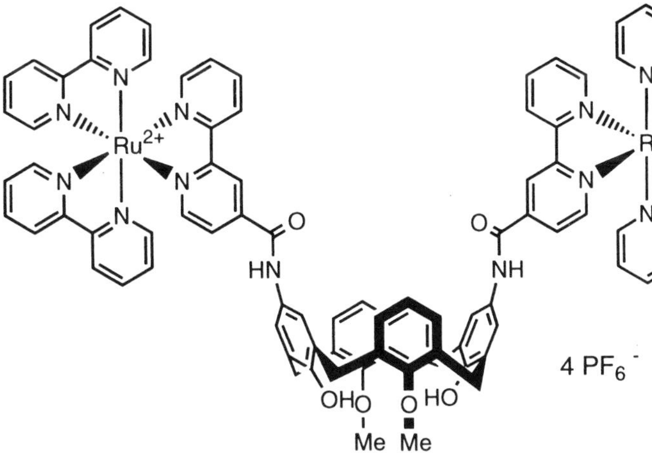

**136**

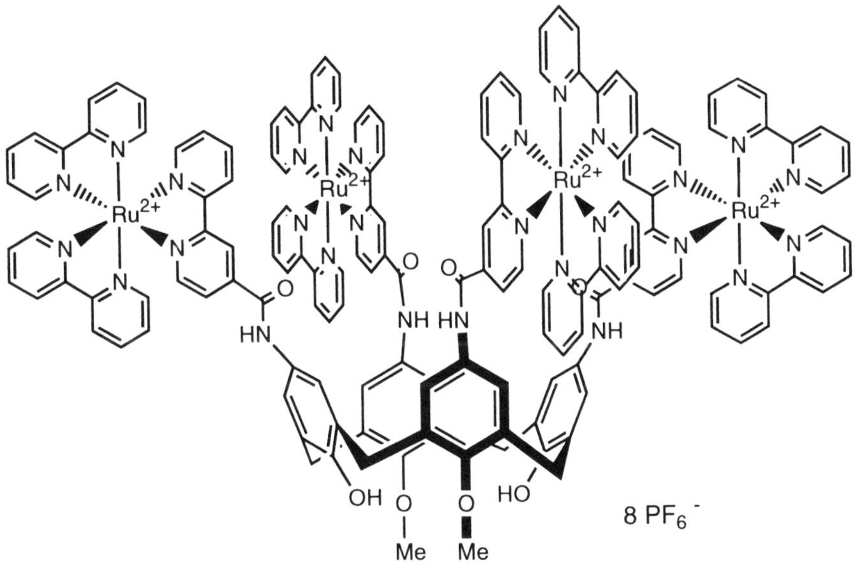

**137**

Nocera and co-workers (259) reported an amidinium functionalized ruthenium(II) bipyridyl system (**138**) binding 3,5-dinitrobenzoate (259). On binding the carboxylate anion, the chemical shift of the amidinium protons changes by >2.0 ppm (in DMSO). Electron-transfer kinetics in dichloromethane were investigated in order to elucidate the role of proton motion within the salt bridge

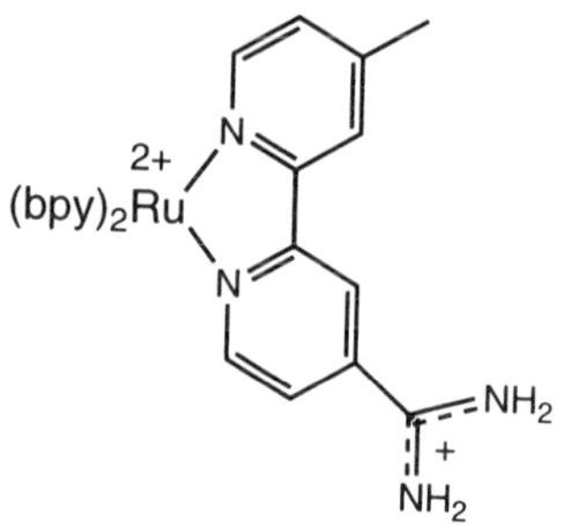

**138**

interface. This is, therefore, an example of coupling anion binding with an investigation of the functional role of a ruthenium bipyridyl system.

Hamilton et al. (260) also recently reported an anion-binding ruthenium terpyridyl system (**139**). This receptor showed strong binding to bis(tetrabutylammonium) pimelate in DMSO ($K > 10^4 M^{-1}$). Even with $5\% D_2O$ present, the stability constant was still $6000\ M^{-1}$. The role of the metal ion is twofold. First, it imparts a favorable chelate effect by recruiting two thiourea hydrogen-bonding sites. Second, the dipositive metal center once again provides an additional electrostatic stabilization of the complex. Hamilton et al.

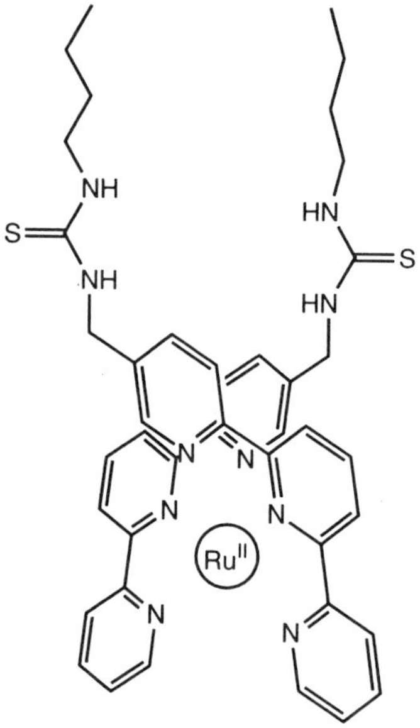

**139**

(260) hopes to build up a novel metal–template approach for the formation of artificial receptor libraries.

It is noteworthy that few of the receptors discussed so far exhibit specific binding and sensing of the chloride anion, yet this substrate is crucial for a large number of biological processes (261). The novel macrocyclic bis[ruthenium(II)bipyridyl] and ruthenium(II) bipyridyl-metallocene receptors (140–142) have been prepared. The $^1$H NMR titrations indicated that each re-

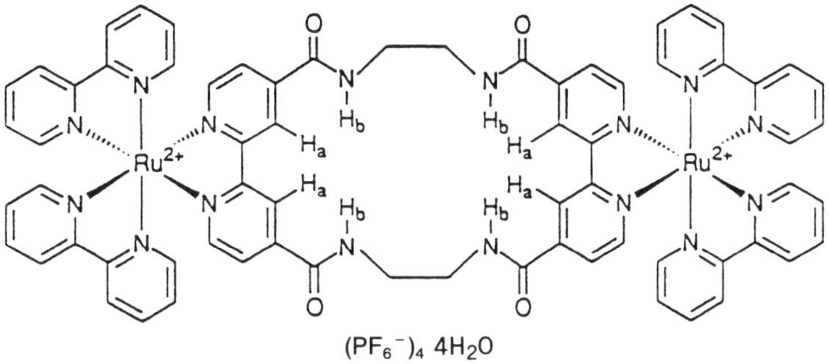

$(PF_6^-)_4$ $4H_2O$

**140**

$(PF_6^-)_2$ $2H_2O$

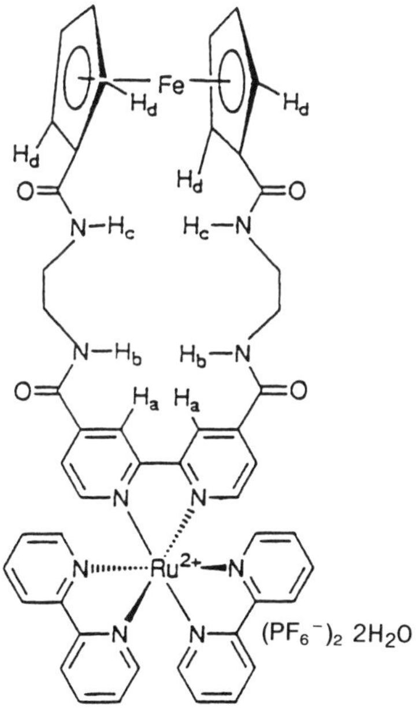

**141**

**142**

ceptor formed an extremely stable 1 : 1 stoichiometric complex with chloride in $d_6$-DMSO solutions (262). In fact, the magnitudes of these stability constants are among the largest known for any anion–abiotic receptor complex, two orders of magnitude larger than observed for the acyclic analogue. The $^1$H and $^{31}$P NMR experiments proved there was *no* dihydrogen phosphate anion binding in DMSO solution. This unique and remarkable selectivity may be attributed to the inherently rigid structures of the macrocycles (as the acyclic analogue binds $H_2PO_4^-$ more strongly than $Cl^-$). The MM2 molecular modeling calculations suggest that the minimized structure of **140** has all the amide and 3,3'-bipyridyl protons lying in a coplanar arrangement, creating a host cavity of similar dimensions to the chloride anion. The larger size, and tetrahedral shape of $H_2PO_4^-$ make this anion noncomplementary for the receptor's cavity. Fluorescence studies indicated a blue-shift response to chloride with significant intensity enhancement, but no response to dihydrogen phosphate. Receptors **140–142** are therefore prototype chloride selective sensory reagents.

## 8. Summary

These receptors, incorporating metal centers and hydrogen-bonding functionalities, show strong anion-binding effects even in polar organic solvents and sometimes in water. The metal centers increase interaction with anions and often play a structural role in holding the hydrogen-bonding ligating groups in a suitable orientation for binding. They also play a key functional role in sensing the presence of the anionic guest.

Anion binding is reversible and the precise interactions between receptor and anionic guest can be finely controlled by synthetic development of subtlety modified hydrogen-bond functional anion-binding cavities. This methodology for easy functionalization means these receptors have been modified to show unusual anion selectivities. They therefore have a great potential for incorporation into anion sensory, switchable, and functional devices.

## VI. CONCLUSIONS

As we have illustrated, metal based systems have introduced a huge variety of receptors to the field of anion binding, which function through a wide range of molecular interactions. This inorganic approach to anion binding has addressed several very important goals:

Lewis acidic hosts (Section V.A) illustrated important theoretical concepts such as the chelate effect and binding cooperativity, which have now been shown to exist for anion as well as cation binding. This work has also resulted in the crystallographic determination of eye-catching solid state receptor–anion complexes, while heteroelement NMR has allowed an accurate means of probing the solution phase structure of these complexes. Already, multinuclear tin systems are being built into functioning anion selective electrodes.

Enzyme behavior has been particularly effectively mimicked by those receptors discussed in Section V.B with the metal ions beginning to perform functional catalytic roles. The selectivity of binuclear hosts for a variety of anionic substrates has been investigated and, in particular, by combination of coordination interactions with intermolecular forces, novel anion selectivities have been observed.

The primarily electrostatic receptors described in Section V.C have shown both synthetic sophistication and the development of solvent shielded, hydrophobic, electron-deficient cavities. This research has yielded potentially useful pertechnetate extraction agents as well as receptors containing functioning metal centers, which possess potential for future technological application.

The incorporation of hydrogen bonding into the receptors as described in Section V.D allows anion selectivities to be carefully controlled, with novel selectivity orders being observed. The $^1$H NMR investigations allow accurate probing of solution-phase structures while crystallography has determined elegant solid phase structures with bound anions. A macrocyclic anion-binding effect has been reported. Anion binding is strong, often in biologically important polar media and the metal centers in these receptors are functional, capable of optical and electrochemical sensing properties. This combination of designed, controllable selectivity and functional ability gives this class of receptor great potential for future development.

In the near future, it must be hoped that anion coordination chemists can provide further devices, be they catalytic, sensing, or medicinal; and that biological anions will be able to be fully addressed, both efficiently and selectively in their natural environment. Hopefully, the field of anion recognition will reach a similar stage to cation recognition where, when a particular cation is to be bound, the coordination chemist reaches out for a well established and specific receptor.

Molecular recognition is a truly multidisciplinary field relying on the talents of workers from diverse scientific backgrounds. We have limited this chapter to those systems containing metal centers, but many of the anion receptors we have discussed are biologically inspired and incorporate inorganic chemistry, coupled with organic framework synthesis and the physical studies of their activity properties. With this kind of input from across the scientific spectrum, we can rest assured that the outlook for anion recognition is most exciting.

## REFERENCES

1. C. H. Park and H. E. Simmons, *J. Am. Chem. Soc.*, *90*, 2431 (1968).

2. R. A. Bell, G. G. Christoph, F. R. Fronczek, and R. E. Marsh, *Science*, *190*, 151 (1976).

3. D. F. Shriver and M. J. Biallas, *J. Am. Chem. Soc.*, *89*, 1078 (1967).

4. B. Dietrich, in *Inclusion Compounds*, J. L. Atwood, J. E. D. Davies, and D. D. MacNicol, Eds., Academic, New York, 1984, Vol. 2, p. 373.

5. J.-L. Pierre and P. Baret, *Bull. Soc. Chim. Fr.*, 367 (1983).

6. F. Vogtle, H. Seiger, and W. M. Muller, *Top. Curr. Chem.*, *98*, 107 (1981).

7. J.-M. Lehn, *Angew. Chem. Int. Ed. Eng.*, *27, 89 (1988).*

8. F. P. Schmidtchen, *Nachr. Chem. Tech. Lab.*, *36*, 8 (1988).

9. P. D. Beer, *Chem. Soc. Rev.*, *18*, 409 (1989).

10. H. E. Katz, in *Inclusion Compounds*, J. L. Atwood, J. E. D. Davies, and D. D. MacNicol, Eds., Oxford University Press, New York, 1991, Vol. 4, p. 391.

11. R. M. Izatt, *Chem. Rev.*, *91*, 1721 (1991).

12. P. D. Beer, *Adv. Inorg. Chem.*, *39*, 79 (1992).

13. B. Dietrich, *Pure and Appl. Chem.*, *65*, 1457 (1993).

14. D. E. Kaufmann and A. Otten, *Angew. Chem. Int. Ed. Engl.*, *33*, 1832 (1994).

15. J. Hodacova, *Chem. Listy*, *88*, 99 (1994).

16. R. M. Harrison, *Pollution: Causes Effects and Control*, Royal Society of Chemistry, London, 1983.

17. P. M. Quinton, *FASEB J.*, *4*, 2709 (1990).

18. C. A. Watson, Ed.; *Official and Standardized Methods of Analysis*; Vol. 3, The Royal Society of Chemistry, Cambridge, UK, 1994.

19. K. Renkawek and G. J. C. G. M. Bosman, *Neuroreport*, *6*, 929 (1995).

20. L. G. Lang, J. F. Riordan, and B. L. Vallee, *Biochemistry*, *13*, 4361 (1974).

21. J. W. Pflugrath and F. A. Quiocho, *Nature (London)*, *314*, 257 (1985).

22. H. Luecke and F. A. Quiocho, *Nature (London)*, *347*, 402 (1990).

23. P. Chakrabarti, *J. Mol. Biol.*, *234*, 463 (1993).

24. R. R. Copley and G. J. Barton, *J. Mol. Biol.*, *242*, 321 (1994).

25. J. J. R. F. da Silva and R. J. P. Williams, *The Biological Chemistry Of The Elements*, Clarendon, Oxford, UK, 1991.

26. K. Scheffzek, C. Klebe, K. Fritz-Wolf, W. Kabsch, and A. Wittinghofer, *Nature (London)*, *374*, 378 (1995).

27. E. N. Baker, B. F. Anderson, H. M. Baker, M. Haridas, G. E. Norris, S. V. Rumball, and C. A. Smith, *Pure Appl. Chem.*, *62*, 1067 (1990).

28. J. S. Richardson, K. A. Thomas, B. H. Rubin, and D. C. Richardson, *Proc. Natl. Acad. Sci. USA*, *72*, 1349 (1975).

29. E. Hough, L. K. Hansen, B. Birkness, K. Jynge, S. Hansen, A. Hordvik, C. Little, E. Dodson, and Z. Derewenda, *Nature (London)*, *338*, 357 (1989).

30. B. Dietrich, M. W. Hosseini, J.-M. Lehn, and R. B. Sessions, *J. Am. Chem. Soc.*, *103*, 1282 (1981).

31. B. Dietrich, M. W. Hosseini, J.-M. Lehn, and R. B. Sessions, *Helv. Chim. Acta*, *66*, 1262 (1983).

32. E. Kimura, M. Kodama, and T. Yatsunami, *J. Am. Chem. Soc.*, *104*, 3182 (1982).

33. E. Kimura, *Top. Curr. Chem.*, *128*, 113 (1985).

34. E. Kimura, A. Sakonaka, T. Yatsunami, and M. Kodama, *J. Am. Chem. Soc.*, *103*, 304 (1981).

35. E. Suet and H. Handel, *Tetrahedron Lett.*, *25*, 645 (1984).

36. M. W. Hosseini and J.-M. Lehn, *Helv. Chim. Acta*, *69*, 587 (1986).

37. M. W. Hosseini and J.-M. Lehn, *J. Am. Chem. Soc.*, *104*, 3525 (1982).

38. B. Dietrich, J. Guilhem, J.-M. Lehn, C. Pascard, and E. Sonveaux, *Helv. Chim. Acta*, *67*, 91 (1984).

39. R. J. Motekaitis, A. E. Martell, B. Dietrich, and J.-M. Lehn, *Inorg. Chem.*, *21*, 4253 (1982).

40. E. H. Serpersu, D. Shortle, and A. S. Mildvan, *Biochemistry*, *26*, 1289 (1987).

41. D. J. Weber, E. H. Serpersu, D. Shortle, and A. S. Mildvan, *Biochemistry*, *29*, 8632 (1990).

42. B. Dietrich, D. L. Fyles, T. M. Fyles, and J.-M. Lehn, *Helv. Chim. Acta*, *62*, 2763 (1979).

43. A. Echavarren, A. Galan, J. de Mendoza, A. Salmeron, and J.-M. Lehn, *Helv. Chim. Acta*, *71*, 685 (1988).

44. A. Echavarren, A. Galan, J.-M. Lehn, and J. de Mendoza, *J. Am. Chem. Soc.*, *111*, 4994 (1989).

45. R. S. Hutchins, P. Molina, M. Alajarin, A. Vidal, and L. G. Bachas, *Anal. Chem.*, *66*, 3188 (1994).

46. J. Sanchez-Quesada, C. Seel, P. Prados, J. de Mendoza, I. Dalcol, and E. Giralk, *J. Am. Chem. Soc.*, *118*, 277 (1996).

47. J. L. Sessler, M. J. Cyr, V. Lynch, E. McGhee, and J. A. Ibers, *J. Am. Chem. Soc.*, *112*, 2810 (1990).

48. J. L. Sessler, V. Kral, and H. Furuta, *J. Am. Chem. Soc.*, *114*, 8704 (1992).

49. F. P. Schmidtchen, *Chem. Ber.*, *113*, 864 (1980).

50. F. P. Schmidtchen, *Chem. Ber.*, *114*, 597 (1981).

51. F. P. Schmidtchen and G. Muller, *J. Chem. Soc. Chem. Commun.*, 1115, (1984).

52. K. Worm, F. P. Schmidtchen, A. Schier, A. Schafer, and M. Hesse, *Angew. Chem. Int. Ed. Engl.*, *33*, 327 (1994).

53. K. Worm and F. P. Schmidtchen, *Angew. Chem. Int. Ed. Engl.*, *34*, 65 (1995).

54. P. B. Savage, S. K. Holmgren, and S. H. Gellman, *J. Am. Chem. Soc.*, *116*, 4069.

55. E. Kimura, H. Anan, T. Koike, and M. Shiro, *J. Org. Chem.*, *54*, 3998 (1989).

56. R. A. Pascal, J. Spergel, and D. V. Engen, *Tetrahedron Lett.*, *27*, 4099 (1986).

57. J. S. Albert, and A. D. Hamilton, *Tetrahedron Lett.*, *34*, 7363 (1993).

58. S. C. Hirst, P. Tecilla, S. J. Geib, E. Fan, and A. D. Hamilton, *Isr. J. Chem.*, *32*, 105 (1992).

59. S. Nishizawa, P. Buhlmann, M. Iwao, and Y. Umezawa, *Tetrahedron Lett.*, *36*, 6483 (1995).

60. D. N. Reinhoudt, S. Valiyaveettil, J. F. J. Engbersen, and W. Verboom, *Angew. Chem. Int. Ed. Engl.*, *32*, 900 (1993).

61. B. C. Hamann, and N. R. Branda, *Tetrahedron Lett.*, *34*, 6837 (1993).

62. P. D. Beer, in *Chemical Sensors*, T. E. Edmonds, Ed., Blackie, London and Glasgow, 1988.

63. D. E. Fenton, *Pure Appl. Chem.*, *58*, 1437 (1986).

64. M. S. Front, and J. W. Ross, *U.S. Patent No.* 3406102 (1968).

65. H. V. Pham, E. Pretsch, K. Fluri, A. Bezegh, and W. Simon, *Helv. Chim. Acta*, *73*, 1894 (1990).

66. U. Wuthier, H. Pham, R. Zund, D. Welti, R. Funck, A. Bezegh, D. Ammann, E. Pretsch, and W. Simon, *Anal. Chem.*, *56*, 535 (1984).

67. Y. Azuma and M. Newcomb, *Organometallics*, *3*, 9 (1984).

68. M. Newcomb, A. M. Madonik, M. T. Blanda, and J. K. Judice, *Organometallics*, *6*, 145 (1987).

69. M. Newcomb, M. T. Blanda, Y. Azuma, and T. J. Delord, *J. Chem. Soc. Chem. Commun.*, 1159 (1984).

70. M. Newcomb, J. H. Horner, and M. T. Blanda, *J. Am. Chem. Soc.*, *109*, 7878 (1987).

71. M. Newcomb and M. T. Blanda, *Tetrahedron Lett.*, *29*, 4261 (1988).

72. M. Newcomb, J. H. Horner, M. T. Blanda, and P. J. Squattrito, *J. Am. Chem. Soc.*, *111*, 6294 (1989).

73. M. T. Blanda and M. Newcomb, *Tetrahedron Lett.*, *30*, 3501 (1989).

74. J. H. Horner and M. Newcomb, *Organometallics*, *10*, 1732 (1991).

75. K. Jurkschat, A. Ruhleman, and A. Tzschach, *J. Organomet. Chem.*, *381*, C53 (1990).

76. K. Jurkschat, H. G. Kuivila, S. Liu, and J. A. Zubieta, *Organometallics*, *8*, 2755 (1989).

77. A. Appel, C. Kober, C. Neumann, H. Noth, M. Schmidt, and W. Storch, *Chem. Ber.*, *129*, 175 (1996).

78. J. K. Tsagatakis, N. A. Chaniotakis, and K. Jurkschat, *Helv. Chim. Acta*, *77*, 2191 (1994).

79. H. E. Katz, *J. Am. Chem. Soc.*, *107*, 1420 (1985).

80. H. E. Katz, *J. Org. Chem.*, *50*, 5027 (1985).

81. H. E. Katz, *Organometallics*, *6*, 1134 (1987).

82. M. T. Reetz, C. M. Niemeyer, and K. Harms, *Angew. Chem. Int. Ed. Engl.*, *30*, 1472 (1991).

83. M. T. Reetz, C. M. Niemeyer, and K. Harms, *Angew. Chem. Int. Ed. Engl.*, *30*, 1474 (1991).

84. S. Jacobson and R. Pizer, *J. Am. Chem. Soc.*, *115*, 11216 (1993).

85. C. Dusemund, K. R. A. S. Sandanayake, and S. Shinkai, *J. Chem. Soc. Chem. Commun.*, 333 (1995).

86. H. E. Katz, *J. Am. Chem. Soc.*, *108*, 7640 (1986).

87. M. E. Jung and H. Xiu, *Tetrahedron Lett.*, *29*, 297 (1988).

88. K. Tamao, T. Hayashi, and Y. Ito, *J. Organomet. Chem.*, *506*, 85 (1996).

89. K. Ogawa, S. Aoyagi, and Y. Takeuchi, *J. Chem. Soc. Perkin Trans. 2*, 2389 (1993).

90. S. Aoyagi, K. Tanaka, I. Zicmane, and Y. Takeuchi, *J. Chem. Soc. Perkin Trans. 2*, 2217 (1992).

91. S. Aoyagi, K. Tanaka, and Y. Takeuchi, *J. Chem. Soc. Perkin Trans. 2*, 1549 (1994).

92. S. Aoyagi, K. Ogawa, K. Tanaka, and Y. Takeuchi, *J. Chem. Soc. Perkin Trans. 2*, 355 (1995).

93. M. F. Hawthorne, X. Yang, and Z. Zheng, *Pure Appl. Chem.*, *66*, 245 (1994).

94. A. L. Beauchamp, M. J. Olivier, J. D. Wuest, and B. Zacharie, *J. Am. Chem. Soc.*, *108*, 73 (1986).

95. J. D. Wuest and B. Zacharie, *Organometallics*, *4*, 410 (1985).

96. J. D. Wuest and B. Zacharie, *J. Am. Chem. Soc.*, *109*, 4714 (1987).

97. M. Rothmaier and W. Simon, *Anal. Chim. Acta*, *271*, 135 (1993).

98. V. B. Shur, A. Tikhonova, P. V. Petrovskii, and M. E. Vol'pin, *Organomet. Chem. USSR*, *2*, 759 (1989).

99. V. B. Shur, I. A. Tikhonova, A. I. Yanovsky, Y. T. Struchkov, P. V. Petrovskii, S. Y. Panov, G. G. Furin, and M. E. Vol'pin, *J. Organomet. Chem.*, *418*, C29 (1991).

100. V. B. Shur, I. A. Tikhonova, F. M. Dolushin, A. I. Yanovsky, Y. T. Struchkov, A. Y. Volkonsky, E. V. Solodova, S. Y. Panov, P. V. Petrovskii, and M. E. Vol'pin, *J. Organomet. Chem.*, *443*, C19 (1993).

101. X. Yang, C. B. Knobler, and M. F. Hawthorne, *Angew. Chem. Int. Ed. Engl.*, *30*, 1507 (1991).

102. Z. Zheng, X. Yang, C. B. Knobler, and M. F. Hawthorne, *J. Am. Chem. Soc.*, *115*, 5320 (1993).

103. X. Yang, C. B. Knobler, and M. F. Hawthorne, *J. Am. Chem. Soc.*, *114*, 380 (1992).

104. X. Yang, Z. Zheng, C. B. Knobler, and M. F. Hawthorne, *J. Am. Chem. Soc.*, *115*, 193 (1993).

105. A. A. Zinn, Z. Zheng, C. B. Knobler, and M. F. Hawthorne, *J. Am. Chem. Soc.*, *118*, 70 (1996).

106. Z. Zheng, C. B. Knobler, and M. F. Hawthorne, *J. Am. Chem. Soc.*, *117*, 5105 (1995).

107. Z. Zheng, M. Diaz, C. B. Knobler, and M. F. Hawthorne, *J. Am. Chem. Soc.*, *117*, 12338 (1995).

108. M. W. Gobel, *Angew Chem. Int. Ed. Engl.*, *33*, 1141 (1994).

109. N. H. Pilkington and R. Robson, *Aust. J. Chem.*, 2225 (1970).

110. R. W. Stotz and R. C. Stoufer, *J. Chem. Soc. Chem. Commun.* 1682 (1970).

111. K. Travis and D. H. Busch, *J. Chem. Soc. Chem. Commun.* 1041 (1970).

112. R. Robson, *Aust. J. Chem.*, *23*, 2217 (1970).

113. R. Robson, *Inorg. Nucl. Chem. Lett.*, *6*, 125 (1970).

114. W. D. McFadyen, R. Robson, H. Schaap, *Inorg. Chem.*, *11*, 1777 (1972).

115. W. D. McFadyen and R. Robson, *J. Coord. Chem.*, *5*, 49 (1976).

116. I. E. Dickson and R. Robson, *Inorg. Chem.*, *13*, 1301 (1974).

117. T. Ichinose, Y. Nishida, H. Okawa, S. Kida, *Bull. Chem. Soc. Jpn.*, *47*, 3045 (1974).

118. H. Okawa, T. Tokii, Y. Nonaka, Y. Muto, and S. Kida, *Bull. Chem. Soc. Jpn.*, *46*, 1462 (1973).

119. S. E. Groh, *Isr. J. Chem.*, *15*, 277 (1976/1977).

120. M. Wall, R. C. Hynes, and J. Chin, *Angew. Chem. Int. Ed. Engl.*, *32*, 1633 (1993).

121. J.-M. Lehn, *Pure Appl. Chem.*, *49*, 857 (1977).

122. J.-M. Lehn, *Pure Appl. Chem.*, *50*, 871 (1978).

123. J.-M. Lehn, *Pure Appl. Chem.*, *52*, 2441 (1980).

124. P. K. Coughlin, J. C. Dewar, S. J. Lippard, E. Watanabe, and J.-M. Lehn, *J. Am. Chem. Soc.*, *101*, 265 (1979).

125. K. G. Strothkamp and S. J. Lippard, *Biochemistry*, *20*, 7488 (1981).

126. P. K. Coughlin and S. J. Lippard, *J. Am. Chem. Soc.*, *106*, 2328 (1984).

127. P. K. Coughlin, A. E. Martin, J. C. Dewar, E. I. Watanabe, J. E. Rulkowski, J.-M. Lehn, and S. J. Lippard, *Inorg. Chem.*, *23*, 1004 (1984).

128. P. Comarmond, P. Plumere, J.-M. Lehn, Y. Agnus, R. Louis, R. Weiss, O. Kahn, and I. J. Morgenster-Badarau, *J. Am. Chem. Soc.*, *104*, 6330 (1982).

129. Y. Agnus, R. Louis, and R. Weiss, *J. Am. Chem. Soc.*, *101*, 3381 (1979).

130. P. E. Jurek, A. E. Martell, R. J. Motekaitis, and R. D. Hancock, *Inorg. Chem.*, *34*, 1823 (1995).

131. R. J. Motekaitis, W. B. Utley, and A. E. Martell, *Inorg. Chim. Acta*, *212* 15 (1993).

132. M. G. B. Drew, M. McCann, and S. M. Nelson, *J. Chem. Soc. Chem. Commun.*, 481 (1979).

133. R. J. Motekaitis, A. E. Martell, J. P. Lecomte, and J.-m. Lehn, *Inorg. Chem.*, *22*, 609 (1983).

134. A. E. Martell and R. J. Motekaitis, *J. Chem. Soc. Chem. Commun.* 915 (1988).

135. A. E. Martell and R. J. Motekaitis, *J. Am. Chem. Soc.*, *110*, 8059 (1988).

136. A. E. Martell and R. J. Motekaitis, *Inorg. Chem.*, *33*, 1032 (1994).

137. D. H. Vance and A. W. Czarnik, *J. Am. Chem. Soc.*, *115*, 12165 (1993).

138. R. J. Motekaitis, A. E. Martell, B. Dietrich, and J.-M. Lehn, *Inorg. Chem.*, *23*, 1588 (1984).

139. J. Jazwinski, J.-M. Lehn, D. Lilienbaum, R. Ziessel, J. Guilhelm, C. Pascard, *J. Chem. Soc. Chem. Commun.*, 1691 (1987).

140. D. McDowell and J. Nelson, *Tet. Lett.*, *29*, 385 (1988).

141. M. G. B. Drew, J. Hunter, D. J. Marrs, J. Nelson, and C. Harding, *J. Chem. Soc. Dalton Trans.*, 3235 (1992).

142. R. Menif, J. Reibenspies, and A. Martell, *Inorg. Chem.*, *30*, 3446 (1991).

143. L. Fabbrizzi, P. Pallavicini, L. Parodi, and A. Taglietti, *Inorg. Chim. Acta*, *238*, 5 (1995).

144. P. D. Beer, O. Kocian, R. J. Mortimer, and P. Spencer, *J. Chem. Soc. Chem. Commun.*, 602 (1992).

145. P. Spencer, Ph.D. Thesis, Polyaza Redox Active Acyclic and Macrocyclic Compounds Designed to Bind Cations and Anions, University of Oxford, Oxford, UK, (1994).

146. A. Zinke and E. Ziegler, *Chem. Ber.*, *77*, 264 (1944).

147. J. W. Cornforth, P. D'Arcy-Hart, G. A. Nicholls, R. J. W. Rees, and J. A. Stock, *Br. J. Pharmacol*, *10*, 73 (1955).

148. C. D. Gutsche and J. A. Levine, *J. Am. Chem. Soc.*, *104*, 2652 (1982).

149. R. Ungaro, A. Pochini, and G. D. Andretti, *J. Inclusion Phenom*, *2*, 199 (1984).

150. M. A. McKervey, E. M. Seward, G. Ferguson, and B. L. Ruhl, *J. Org. Chem.*, *51*, 3581 (1986).

151. C. D. Gutsche, *Pure Appl. Chem.*, *60*, 483 (1988).

152. P. D. Beer, M. G. B. Drew, P. B. Leeson, K. Lyssenko, and M. I. Ogden, *J. Chem. Soc. Chem. Commun.*, 929 (1995).

153. W. Xu, J. J. Vittal and R. J. Puddephatt, *J. Am. Chem. Soc.*, *115*, 6456 (1993).

154. W. Xu, J. J. Vittal and R. J. Puddephatt, *J. Am. Chem. Soc.*, *117*, 8362 (1995).

155. W. L. Gladfetter and H. B. Gray, *J. Am. Chem. Soc.*, *102*, 5909 (1980).

156. M. S. Lah, and V. L. Pecoraro, *Comments Inorg. Chem.*, *11*, 59 (1990).

157. S. M. Lah, M. L. Kirk, W. Hatfield, and V. L. Pecoraro, *J. Chem. Soc. Chem. Commun.*, 1606 (1989).

158. S. M. Lah and V. L. Pecoraro, *J. Am. Chem. Soc.*, *111*, 7528 (1989).

159. S. M. Lah and V. L. Pecoraro, *Inorg. Chem.*, *30*, 878 (1991).

160. B. R. Gibney, A. J. Stemmler, S. Pilotek, J. W. Kampf, and V. L. Pecoraro, *Inorg. Chem.*, *32*, 6008 (1993).

161. B. R. Gibney, D. P. Kessisoglou, J. W. Kampf, and V. L. Pecoraro, *Inorg. Chem.*, *33*, 4840 (1994).

162. A. J. Stemmler, J. W. Kampf, and V. L. Pecoraro, *Inorg. Chem.*, *34*, 2271 (1995).

163. R. Provencher and P. D. Harvey, *Inorg. Chem.*, *32*, 612 (1993).

164. T. Zhang, M. Drouin, and P. D. Harvey, *Chem. Soc. Chem. Commun.*, 877 (1996).

165. S. Kitagawa, M. Kondo, S. Kawata, S. Wada, M. Maekawa, and M. Munakata, *Inorg. Chem.*, *4*, 1455 (1995).

166. D. P. Smith, E. Baralt, B. Morales, M. M. Olmstead, M. F. Maestre, and R. H. Fish, *J. Am. Chem. Soc.*, *114*, 10647 (1992).

167. D. P. Smith, E. Kohen, M. F. Maestre, and R. H. Fish, *Inorg. Chem.*, *32*, 4119 (1993).

168. H. Chen, M. F. Maestre, and R. H. Fish, *J. Am. Chem. Soc.*, *117*, 3631 (1995).

169. H. Chen, M. M. Olmstead, M. F. Maestre, and R. H. Fish, *J. Am. Chem. Soc.*, *117*, 9097 (1995).

170. I. Tabushi, N. Shimizu, T. Sugimoto, M. Shiozuk, and K. Yamamura, *J. Am. Chem. Soc.*, *99*, 7100 (1977).

171. I. Tabushi, Y. Kuroda, and T. Mizutani, *Tetrahedron*, *40*, 545 (1984).

172. A. W. Schwabacher, J. Lee, and H. Lei, *J. Am. Chem. Soc.*, *114*, 7597 (1992).

173. J. H. Lee and A. W. Schwabacher, *J. Am. Chem. Soc.*, *116*, 8382 (1994).

174. M. Shionoya, T. Ikeda, E. Kimura, and M. Shiro, *J. Am. Chem. Soc.*, *116*, 3848 (1994).

175. G. D. Santis, L. Fabbrizzi, M. Licchelli, A. Poggi, and T. Taglietti, *Angew. Chem. Int. Ed. Engl.*, *35*, 202 (1996).

176. P. Schulthess, D. Ammann, W. Simon, C. Caderas, R. Stepanek, and B. Krauther, *Helv. Chim. Acta*, *67*, 1026 (1984).

177. D. Ammann, M. Huser, B. Krauther, B. Rusterholz, P. Schulthess, B. Lindemann, E. Halder, and W. Simon, *Helv. Chim. Acta*, *69*, 849 (1986).

178. N. A. Chaniotakis, A. M. Chasser, M. E. Meyerhoff, J. T. Groves, *Anal. Chem.*, *60*, 188 (1988).

179. G. R. Seely, D. Gust, T. A. Moore, and A. L. Moore, *J. Phys. Chem.*, *98*, 10659 (1994).

180. F. Bedioui, J. Devynck, and C. Biedcharreton, *Acc. Chem. Res.*, *28*, 30 (1995).

181. F. Hofmeister, *Exp. Pathol. Pharm.*, *24*, 247 (1888).

182. S. T. Yang and L. G. Bachas, *Talanta*, *41*, 963 (1994).

183. D. Gao, J.-Z. Li, R.-Q. Yu, and G.-D. Zheng, *Anal. Chem.*, *66*, 2245 (1994).

184. D. Gao, J. Gu, R.-Q. Yu, and G.-D. Zheng, *Analyst*, *120*, 499 (1995).

185. J. R. Allen, A. Florido, S. D. Young, S. Daunert, and L. G. Bachas, *Electroanalysis*, *7*, 710 (1995).

186. P. D. Beer and A. D. Keefe, *J. Organomet. Chem.*, *375*, C40 (1989).

187. M. Fujita, J. Yazaki, and K. Ogura, *J. Am. Chem. Soc.*, *112*, 5645 (1990).

188. M. Fujita, J. Yazaki, and K. Ogura, *Chem. Lett.*, 1031 (1991).

189. M. Fujita, J. Yazaki, and K. Ogura, *Tetrahedron Lett.*, *32*, 5589 (1991).

190. P. J. Stang and D. H. Cao, *J. Am. Chem. Soc.*, *116*, 4981 (1994).

191. P. J. Stang and J. A. Whiteford, *Organometallics*, *13*, 3776 (1994).

192. P. J. Stang, D. H. Cao, S. Saito, and A. M. Arif, *J. Am. Chem. Soc.*, *117*, 6273 (1995).

193. P. J. Stang and V. V. Zhdankin, *J. Am. Chem. Soc.*, *115*, 9808 (1993).

194. P. J. Stang and K. C. Chen, *J. Am. Chem. Soc.*, *117*, 1667 (1995).

195. H. Rauter, E. C. Hillgeris, A. Erxleben, and B. Lippert, *J. Am. Chem. Soc.*, *116*, 616 (1994).

196. R. V. Slone, D. I. Yoon, R. M. Calhoun, and J. T. Hupp, *J. Am. Chem. Soc.*, *117*, 11813 (1995).

197. M. Fujita, S. Nagao, and K. Ogura, *J. Am. Chem. Soc.*, *117*, 1649 (1995).

198. J. W. Steed, R. K. Juneja, R. S. Burkhalter, and J. L. Atwood, *J. Chem. Soc. Chem. Commun.*, 2205 (1994).

199. J. W. Steed, R. K. Juneja, and J. L. Atwood, *Angew. Chem. Int. Ed. Engl.*, *33*, 2456 (1994).

200. H. Zhang and J. L. Atwood, *J. Crystallogr. Spec. Res.*, *20*, 465 (1990).

201. K. T. Holman, M. M. Halihan, J. W. Steed, S. S. Jurisson, and J. L. Atwood, *J. Am. Chem. Soc.*, *117*, 7848 (1995).

202. J.-M. Lehn, J. Simon, and A. Moradpour, *Helv. Chim. Acta*, *61*, 2407 (1978).

203. V. W. Day, W. G. Klemperer, and O. M. Yaghi, *J. Am. Chem. Soc.*, *111*, 5959 (1989).

204. A. Muller, M. Penk, R. Rohlfing, E. Krickenmeyer, and J. Doring, *Angew. Chem. Int. Ed. Engl.*, *29*, 926 (1990).

205. A. Muller, E. Krickenmeyer, M. Penk, R. Rohlfing, A. Armatage, and H. Bogge, *Angew. Chem. Int. Ed. Engl.*, *30*, 1674 (1991).

206. A. Muller, E. Krickenmeyer, M. Penk, H.-J. Walberg, and H. Bogge, *Angew. Chem. Int. Ed. Engl.*, *26*, 1045 (1987).

207. A. Muller, E. Krickenmeyer, M. Penk, H.-J. Walberg, and H. Bogge, *Angew. Chem. Int. Ed. Engl.*, *27*, 1719 (1988).

208. T. Yamase and K. Ohkata, *J. Chem. Soc. Dalton Trans.*, 2599 (1994).

209. T. Yamase, K. Ohtaka, and M. Suzuki, *J. Chem. Soc. Dalton Trans.*, 283 (1996).

210. H.-J. Schneider, T. Blatter, A. Eliseev, V. Rudiger, and O. A. Raevsky, *Pure Appl. Chem.*, *65*, 2329 (1993).

211. J. Bulfalini and K. H. Stern, *J. Am. Chem. Soc.*, *83*, 4362 (1961).

212. P. D. Beer, D. Hesek, J. Hodacova, and S. E. Stokes, *J. Chem. Soc. Chem. Commun.*, 270 (1992).

213. P. D. Beer, C. Hazlewood, D. Hesek, J. Hodacova, and S. E. Stokes, *J. Chem. Soc. Dalton Trans.*, 1327 (1993).

214. P. D. Beer, M. G. B. Drew, A. R. Graydon, D. K. Smith, and S. E. Stokes, *J. Chem. Soc. Dalton Trans.*, 403 (1995).

215. P. D. Beer and A. R. Graydon, *J. Organomet. Chem.*, *466*, 241 (1994).

216. P. D. Beer and S. E. Stokes, *Polyhedron*, *14*, 873 (1995).

217. M. Uno, N. Komatsuzaki, K. Shirai, and S. Takahashi, *J. Organomet. Chem.*, *462*, 343 (1993).

218. N. Komatsuzaki, M. Uno, K. Shirai, Y. Takai, T. Tanaka, M. Sawada, S. Takahashi, *Bull. Chem. Soc. Jpn.* *69*, 17 (1996).

219. P. D. Beer, M. G. B. Drew, D. Hesek, J. Kingston, D. K. Smith, and S. E. Stokes, *Organometallics*, *14*, 3288 (1995).

220. P. D. Beer, M. G. B. Drew, J. Hodacova, and S. E. Stokes, *J. Chem. Soc. Dalton Trans.*, 3447 (1995).

221. P. D. Beer and S. E. Stokes, *Polyhedron*, *14*, 2631 (1995).

222. P. D. Beer, M. G. B. Drew, C. Hazlewood, D. Hesek, J. Hodacova, and S. E. Stokes, *J. Chem. Soc. Chem. Commun.*, 229 (1993).

223. P. D. Beer, M. G. B. Drew, D. Hesek, M. Shode, and F. Szemes, *J. Chem. Soc. Chem. Commun.*, 2161 (1996).

224. P. D. Beer, D. Hesek, and R. J. Mortimer, unpublished results.

225. P. D. Beer, M. G. B. Drew, D. Hesek, and R. Jagessar, *J. Chem. Soc. Chem. Commun.*, 1187 (1995).

226. P. D. Beer and R. Jagessar, unpublished results.

227. P. D. Beer, C. A. P. Dickson, N. C. Fletcher, A. J. Goulden, A. Grieve, J. Hodacova and T. Wear, *J. Chem. Soc. Chem. Commun.*, 828 (1993).

228. P. D. Beer, Z. Chen, M. G. B. Drew, J. Kingston, M. I. Ogden, and P. Spencer, *J. Chem. Soc. Chem. Commun.*, 1045 (1993).

229. P. D. Beer, Z. Chen, M. G. B. Drew, A. O. Johnson, D. K. Smith, and P. Spencer, *Inorg. Chim. Acta*, *246*, 143 (1996).

230. P. D. Beer and D. K. Smith, unpublished results. D. K. Smith, Ph.D. Thesis, "Redox-Active Receptors for Cationic and Anionic Guests," University of Oxford, UK, 1996.

231. P. D. Beer, Z. Chan, A. J. Goulden, A. R. Graydon, S. E. Stokes, and T. Wear, *J. Chem. Soc. Chem. Commun.*, 183 (1993).

232. J. E. Kingston, Ph.D. Thesis, "Redox-Active Host Molecules for Anion Recognition," University of Oxford, UK, 1996.

233. P. D. Beer, Z. Chen, M. I. Ogden, *J. Chem. Soc. Faraday Trans.*, *91*, 295 (1995).

234. B. Delavaux-Nicot, Y. Guari, B. Douziech, and R. Mathieu, *J. Chem. Soc. Chem. Commun.*, 585 (1995).

235. P. L. Veya and J. K. Kochi, *J. Organomet. Chem.*, *488*, C4 (1995).

236. D. M. Rudkevich, W. P. R. V. Stauthamer, W. Verboom, J. F. J. Engbersen, S. Harkema, and D. N. Reinhoudt, *J. Am. Chem. Soc.*, *114*, 9671 (1992).

237. D. M. Rudkevich, W. Verboom, Z. Brzozka, M. J. Palys, W. P. R. V. Stauthamer, G. J. van Hummel, S. M. Franken, S. Harkema, J. F. J. Engbersen, and D. N. Reinhoudt, *J. Am. Chem. Soc.*, *116*, 4341 (1994).

238. D. M. Rudkevich, Z. Brzozka, M. Palys, H. C. Visser, W. Verboom, and D. N. Reinhoudt, *Angew. Chem. Int. Ed. Engl.*, *33*, 467 (1994).

239. E. A. Arafa, K. I. Kinnear, and J. C. Lockhart, *J. Chem. Soc. Chem. Commun.*, 61 (1992).

240. K. I. Kinnear, D. P. Mousley, E. Arafa, and J. C. Lockhart, *J. Chem. Soc. Dalton Trans.*, 3637 (1994).

241. P. D. Beer, M. G. B. Drew, R. J. Knubley, and M. I. Ogden, *J. Chem. Soc. Dalton Trans.*, 3117 (1995).

242. N. C. Fletcher, Ph.D. Thesis, "A Novel Approach to Photo and Redox Active Anion Receptors," University of Oxford, U.K., 1994.

243. S. Lanza, L. M. Scolaro, and G. Rosace, *Inorg. Chim. Acta*, *227*, 63 (1994).

244. A. D. Burrows, M. P. Mingos, A. J. P. White, and D. J. Williams, *J. Chem. Soc. Chem. Commun.*, 97 (1996).

245. N. Ohata, H. Masuda, and O. Yamauchi, *Angew. Chem. Int. Ed. Engl.*, *35*, 531 (1996).

246. K. Konishi, K. Yahara, H. Toshishige, T. Aida, and S. Inoue, *J. Am. Chem. Soc.*, *116*, 1337 (1994).

247. C. Creutz and N. Sutin, *Proc. Natl. Acad. Sci. USA*, *72*, 2858 (1975).

248. C. Creutz, *Comments Inorg. Chem.*, *1*, 293 (1982).

249. L. A. Summers, *The Bipyridinium Herbicides*; Academic: New York, 1980.

250. A. Juris, V. Balzani, F. Barigalletti, S. Campagna, P. Belser, and A. von Zelewsky, *Coord. Chem. Rev.*, *84*, 85 (1988).

251. V. Balzani, F. Barigeletti, and L. D. Cola, *Top. Curr. Chem.*, *158*, 31 (1990).

252. R. J. Watts, *J. Chem. Educ.*, *60*, 834 (1983).

253. P. D. Beer, Z. Chen, A. J. Goulden, A. Grieve, D. Hesek, F. Szemes, and T. Wear, *J. Chem. Soc. Chem. Commun.*, 2021 (1994).

254. F. Szemes, D. Hesek, Z. Chen, S. W. Dent, M. G. B. Drew, A. J. Goulden, A. R. Graydon, A. Grieve, R. J. Mortimer, T. Wear, J. S. Weightman, and P. D. Beer, *Inorg. Chem.*, *35*, 5868 (1996).

255. P. D. Beer, S. W. Dent, and T. J. Wear, *J. Chem. Soc. Dalton Trans.*, 2341 (1996).

256. P. D. Beer, R. J. Mortimer, N. R. Stradiotto, F. Szemes, and J. S. Weightman, *Anal. Proc.*, *32*, 419 (1995).

257. P. D. Beer, A. R. Graydon, R. J. Mortimer, L. R. Sutton, and J. S. Weightman, *Polyhedron*, *15*, 2457 (1996).

258. P. D. Beer, D. Hesek, and R. J. Mortimer, unpublished results.

259. J. A. Roberts, J. P. Kirby, D. G. Nocera, *J. Am. Chem. Soc.*, *117*, 8051 (1995).

260. M. S. Goodman, V. Jubian, B. Linton, and A. D. Hamilton, *J. Am. Chem. Soc.*, *117*, 11610 (1995).

261. K. L. Kirk, *Biochemistry of the elemental halides*; Plenum, New York, 1991.

262. P. D. Beer and F. Szemes, *J. Chem. Soc. Chem. Commun.*, 2245 (1995).

# Copper(I), Lithium, and Magnesium Thiolate Complexes: An Overview with Due Mention of Selenolate and Tellurolate Analogues and Related Silver(I) and Gold(I) Species

**MAURITS D. JANSSEN, DAVID M. GROVE,** *and*
**GERARD VAN KOTEN**

*Department of Metal-Mediated Synthesis*
*Debye Institute*
*Utrecht University*
*The Netherlands*

CONTENTS

*Progress in Inorganic Chemistry, Vol. 46*, Edited by Kenneth D. Karlin.
ISBN 0-471-17992-2 © 1997 John Wiley & Sons, Inc.

# I. INTRODUCTION

It has long been known that thiolate ligands ($RS^-$), which are formally derived from thiols (RSH) by deprotonation, are well suited to form metal complexes (1). Specific reviews of this area have covered the structural chemistry of metal thiolates (1a), the coordination chemistry of metal thiolates from a bioinorganic perspective (1b), transition metal thiolates (1c), and, most recently, early transition metal thiolates (1d). Because of potential thione–thiol tautomerism, a review dealing with complexes of heterocylic thione donors (1e) is also relevant. This chapter concentrates on thiolate complexes of copper(I), lithium, and magnesium, but we also mention, for comparison or contrast, many related species of silver(I) and gold(I) and of some complexes that contain selenolate and tellurolate ligands. However, it should be emphasized that we have not attempted a comprehensive literature search outside the primary field of interest.

This chapter on thiolate ligands was motivated by the fact that metal thiolate species are finding increasing application in fields such as medicine, materials science, and organic synthesis. In particular, the recent use of thiolates in the copper-mediated reactions of organolithium and Grignard reagents (i.e., RLi and RMgX) makes an overview that links copper, lithium, and magnesium opportune (2). In this area, the unique capabilities of the sulfur atoms of a thiolate ligand for bonding as a terminal or a bridging center provides an interesting variety of structures which, when the underlying structural themes are understood, should provide a basis for the development of new thiolate ligands for application in enantioselective synthesis. In this chapter, we present our ideas regarding these underlying themes.

From a historic perspective, in 1936 Gilman and Straley (3) discovered that catalytic amounts of a copper salt lead to an increased reactivity of Grignard reagents. A little later, in 1941, Kharasch and Twaney (4) found that Grignard reagents react with $\alpha,\beta$-unsaturated ketones to afford products that result from 1,4-addition instead of the normally observed 1,2-addition (Eq. 1). Since then, various researchers have studied the application of cuprates [LiCuR$_2$] in numerous organic C—C bond formation reactions (5–9).

$$1,4\text{-Addition} \qquad 1,2\text{-Addition}$$

(1)

After the first preparation of the Gilman reagent, [LiCuMe$_2$], House et al. (10) used cuprates in organic synthesis, and found that only one-half of the organic groups are transferred. In order to reduce the loss of the often valuable organic groups, mixed cuprates [LiCuR$^r$R$^t$] were introduced in which R$^t$ is the valuable transferable group and R$^r$ is a cheap and expendable nontransferable group that remains in the copper reagent. Many modifications of the latter R$^r$ group have been reported over the years and examples of such groups include alkynyl (11), cyclopentadienyl (11b), halide (12), cyanide (13), silyl (14), hydride (15), amide (16, 17), phosphide (18), alcoholate (16, 19), and thiolate ligands (16).

Although Posner et al. (16) reported in 1973 that copper(I) thiophenolate, [Cu(SPh)], can be used as a stoichiometric catalyst in cuprate-mediated reactions in order to prevent the loss of the valuable group R$^t$ (i.e., PhS$^-$ is the nontransferable group R$^r$), the idea of using a well-chosen nontransferable arenethiolate group in cuprate-catalyzed organic synthesis found its first application as late as 1989 (2a). In later publications, Knotter et al. (2b, 20) reported the use of catalytic amounts of copper(I) arenethiolates with potentially intramolecularly coordinating amine functionalities in the enantioselective conjugate 1,4-addition of Grignard reagents to an $\alpha,\beta$-unsaturated enone (i.e., benzylidene acetone). Since then, copper(I) arenethiolate catalysts have also been applied in various heterocuprate mediated organic reactions and recent examples of the successful use of copper(I) arenethiolate catalysts with Grignard reagents in organic syntheses include: (a) the 1,4-addition to acyclic (2b, 20, 21) and cyclic (22) $\alpha,\beta$-unsaturated enones, (b) the 1,6-addition to $\alpha,\beta,\gamma,\delta$-unsaturated enyne esters (23), (c) the selective $\alpha$- or $\gamma$-nucleophilic substitution of allylic acetates (24), and (d) cross-coupling reactions with primary alkyl halides (25).

The successful application of copper(I) arenethiolate catalysts in organic chemistry thus led to a rapid growth of this field of research. However, whereas the influence of the arenethiolate backbone on the selectivity of organic conversions was studied in detail (2, 20–25), little is known about the influence of this backbone on the structure of the copper(I) arenethiolate itself, let alone on that of the kinetically active (key)intermediates.

It is well known that organocopper complexes [CuR] react with organolithium reagents [LiR'] to afford the corresponding cuprates [LiCuRR'] (Eq. 2) and that these reactions are most likely equilibria (2, 26).

$$CuR \ + \ LiR' \ \rightleftharpoons \ [LiCu(R)R'] \ \rightleftharpoons \ CuR' \ + \ LiR \qquad (2)$$

Consequently, it is reasonable to assume that copper(I) arenethiolates react with organolithium reagents to afford heterocuprates [LiCu(SAr)R] (Eq. 3) (2, 16).

$$Cu(SAr) \ + \ LiR \ \rightleftharpoons \ [LiCu(SAr)R] \ \rightleftharpoons \ CuR \ + \ Li(SAr) \qquad (3)$$

The presence of equilibria in the former reaction (Eq. 2) as well as the similarity between organocopper and copper(I) arenethiolate complexes leads one to the assumption that in the reaction of a copper arenethiolate with an organolithium reagent it must be also possible to have reverse formation of the cuprate species (i.e. reaction of organocopper with lithium arenethiolate as shown in Eq. 3). A similar line of reasoning is applicable to reactions involving organomagnesium and magnesium arenethiolate complexes.

Complexes of lithium and magnesium with monoanionic thiolate ligands are scarce, and the influence of intramolecular coordination on the structure and stability of these species has not been systematically studied. In this area, there is one report of a unique mixed-organo(arenethiolato) cluster containing copper and magnesium, that is $[Cu_4Mes_4][Mg(SAr)_2]_2$ (Fig. 1) (20), and its description as either a cuprate, $[Cu_4Mes_4(SAr)_2]]Mg(SAr)]_2$, or a coordination complex, $[Cu_4Mes_4][Mg(SAr)_2]_2$, is of interest not only from a structural point of view, but also for understanding the kinetically active species formed in a mixture of copper(I) arenethiolate catalyst, Grignard reagent, and enone substrate.

Within this overview of copper(I), lithium, and magnesium thiolate complexes, one of the aims is to highlight the influence that (intramolecular) nitrogen coordination exerts on the structure of such species, since intermolecular (auxiliary) coordination of heteroatoms often leads to poorly defined species. An example of this poor "species definition" is the reaction behavior of polymeric copper(I) thiophenolate, [Cu(SPh)], with triphenylphosphine: Several P-coordinated aggregates with variable stoichiometry were reported (Scheme 1) (27–29). This behavior makes it difficult to study one specific species in copper(I) thiolate catalyzed organic reactions since equilibria are likely to occur.

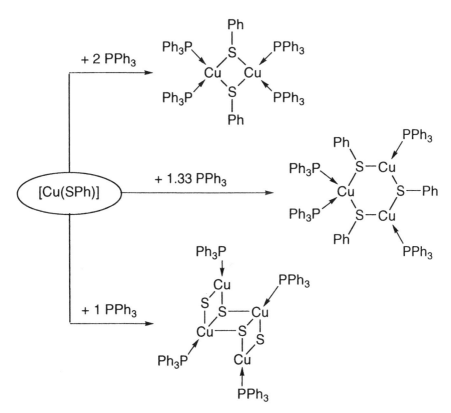

Figure 1.  Schematic representation and molecular structure of $[Cu_4Mes_4][Mg(SAr)_2]_2$ (hydrogen atoms are omitted for clarity).

Scheme 1.  Examples of $PPh_3$ coordinated copper(I) thiophenolate aggregates.

Finally, one anticipates that from this structural data, together with the accumulated data from spectroscopic and chemical studies on well-defined isolated complexes, that it should be possible to make meaningful postulates regarding the structures and reactivity of key intermediates in copper(I) arenethiolate catalyzed organic reactions.

## II. COPPER(I) THIOLATE COMPLEXES

### A. Copper, Silver, and Gold Thiolates in Perspective

The great interest in copper(I) arenethiolates stems, not only from their recent application in organic synthesis (16, 20–25) and from the unique and sometimes puzzling variation in their structural formats, but also from their relevance to the structure and function of the active sites in metalloproteins (30) such as the blue copper proteins, which involve $Cu-S$ binding. Moreover, copper arenethiolate complexes can sometimes behave as semiconductors (31), while thermolysis of such species can offer a low-temperature route to novel solid state chalcogenide materials (32). The complexation of thiolate ligands to silver(I) has been studied in detail (33–35), but only a few structurally characterized species are known and, where relevant, these have been included in this chapter. Gold(I) thiolates have been extensively studied in relation to chemotherapy for arthritis (36). These studies have spawned an enormous literature in this area (37). The mechanism of the action of gold-containing pharmaceuticals is not clear, but interaction of gold(I) centers with the thiol groups of proteins and enzymes seems critical. A limited number of simple gold(I) thiolate structures is known and reference is made to some of them where deemed necessary.

The organothiolate anion $RS^-$ is a ligand that is sometimes classified as a pseudohalide, comparable with ligands such as $Cl^-$, $Br^-$, and $I^-$, and its metal complexes are of fundamental chemical interest (1). One of the pseudohalide attributes of the $RS^-$ ligand is the formation of metal thiolate complexes that are similar to metal halides, but with the difference that thiolates are stronger Brønsted bases than halides. For example, there is a great similarity between the solid state structures of the dianionic complexes $[Cu_4(SPh)_6]^{2-}$ and $[Cu_4I_6]^{2-}$ (38). The thiolate ligand ($RS^-$) has the advantage over the related sulfido ligand ($S^{2-}$) that by variation of the group R the redox properties of the metal atom in its complexes can be tuned. In this way, complexes have been synthesized that are reported to mimic the blue copper protein and metallothionein proteins (39).

Thiolate ligands have a reputation for bridging metal atoms, a reputation engendered by the insolubility of copper(I) thiolate complexes $[Cu(SR)_n]_p$ in common organic solvents that is presumed to be the result of polymeric structures (1a). However, the great difficulty in obtaining structural information on (copper) thiolate complexes, that is caused by insolubility and poor crystal habit, has in a number of cases been overcome by either the use of auxiliary coordinating ligands or by appropriate variation of the group R of the thiolate ligand. The latter variations include the introduction of bulky (40) or intramolecularly coordinating (41) substituents as well as the introduction of stereogenic centers in cases where metal arenethiolates have been used as chiral catalysts (2, 20, 21, 42).

## B.  Synthesis of Copper(I) Thiolates

Several methods have been applied to prepare copper(I) thiolates in a pure form. As a result of the high thermal stability of such complexes, a copper(I) thiolate can generally be prepared in a number of ways, which is in contrast to the preparation of organocopper species where, due to to their low thermal stability, specific protocols are often necessary to successfully isolate the target compound (43).

### 1.  Synthesis via Deprotonation of Thiols

As a consequence of the inertness of copper(I) thiolates toward protic media, the synthesis of these complexes can often be carried out in solvents such as methanol or ethanol and a very simple synthetic procedure is the reaction of the thiol (RSH) with a suitable copper source (either $Cu^I$ or $Cu^{II}$) in the presence of a base (Eq. 4). A complication of this method is the tendency of copper to bind to more than one thiolate ligand and, therefore, instead of the neutral copper(I) thiolate [Cu(SR)] anionic species such as $[Cu(SR)_2]^-$ or $[Cu(SR)_3]^{2-}$ are frequently isolated (44, 45).

$$RSH + MX \longrightarrow MSR + HX$$

(4)

$$MX = Cu(O_2CMe)_2, Cu(NO_3)_2, AgNO_3, Cu_2O, CuO, CuCl$$
$$R = alkyl, aryl$$

Examples of the successful application of this method are the use of copper(II) acetate for the preparation of neutral complexes [Cu(SR)] (R = aryl, alkyl; no base needed) (46), and the use of copper(II) nitrate for the synthesis of anionic species $[Cu_4(SPh))_6]^{2-}$ and $[Cu_5(SPh)_7]^{2-}$ (47, 48), as well as of the cationic species $[Cu^I(SC_5H_4NH)_3]^+$ (49). The latter complex is interesting in that each thiolate ligand is neutral due to protonation of the pyridine nitrogen atom. Similarly, silver nitrate has been used to prepare several neutral alkylthiolato silver complexes (33). The use of copper(I) oxide was reported in the synthesis of a family of copper(I) arenethiolates containing intramolecularly coordinating amine substituents, one example being $[Cu_3(SC_6H_4(CH(R)NMe_2)-2)_3]$. Since $H_2O$ is formed in this synthetic route a base is not necessary, although one is actually present in the ortho substituent (20, 50). The use of copper(II) oxide was also reported, but in this case a mixed-valence species $[Cu_8^I Cu_6^{II}(SR)_{12}]Cl \cdot 3.5SO_4$ was isolated (51). Finally, in the preparation of $[Cu_8(SC_6H_2\text{-}i\text{-}Pr_3\text{-}2,4,6)_8]$, copper(I) chloride was used but, due to its low solubility, it was necessary to also use a base as a coordinating solvent (usually acetonitrile) and a high reaction temperature (reflux) (52). In many of these syntheses an inert atmosphere is required to prevent the oxidation of the thiol (31). The reaction of an arenethiol with CuCl has also been used to prepare the oligomeric complexes $[CuSC_6H_4S]_n$ and $[CuSC_6H_4SCu]_n$, which both display semiconducting properties (31). More recently, organocopper compounds $[CuR]_n$ and copper alkoxide complexes $[CuOR]_n$ have found employment in the preparation of copper(I) thiolates: the reaction of tert-amylmercaptide (Et-CMe_2SH) with mesitylcopper cleanly affords $[Cu(SCMe_2Et)]$ (53), while the reaction of arenethiols with copper tert-butoxide cleanly affords the corresponding copper arenethiolates $[Cu(SAr)]_n$ (54).

## 2. Electrochemical Synthesis

An extension of the preparative route via thiol deprotonation (Section II.B.1) is the electrochemical synthesis of copper(I) thiolates. In this latter method, acetonitrile solutions of either a thiol or a disulfide (and a coordinating ligand) are electrolyzed using a platinum cathode and a copper anode (Eq. 5) (55–57). This particular route can be easily extended to the synthesis of both silver(I) and gold(I) thiolates (55).

$$Pt_{(-)} / MeCN + RSH (+ L) / M_{(+)}$$

$$M = Cu, Ag, Au$$
$$L = phen, Ph_2PCH_2PPh_2, PPh_3$$

(5)

The yields in this type of synthesis are generally better than 90% (based on dissolved anode material) and the isolated material is of high purity (55). Anionic species $[Cu(SR)_n]^{1-n}$ with $n > 1$ are generally not isolated and both copper alkyl- and arenethiolates are accessible in this way. Moreover, this technique also allows access to copper selenolates $[Cu(SeR)]$ (57).

One example of a copper thiolate synthesis exists in which a disulfide reacts directly with metallic copper without the necessity of an externally applied potential: $CF_3SSCR_3$ and $Cu^0$ cleanly afford the copper(I) perfluoro-alkylthiolate complex $[Cu_{10}(SCF_3)_{10}(MeCN)_8]$ (58).

### 3. Synthesis via Transmetalation

Copper(I) thiolates can be prepared by transmetalation of Group 1 (IA) metal thiolate complexes with suitable copper halide salts whereby the halide ligand is selectively transferred to the more electropositive metal (Eq. 6).

$$M(SR) + CuX \longrightarrow Cu(SR) + MX$$

$$M = Li, Na, K; X = Cl, I, ClO_4, BF_4, 0.5 S_2C_4O_2$$

(6)

The copper salts most commonly used in this synthetic route are the sparingly soluble species CuCl (59), CuBr (60), and CuI (61). More soluble precursors, such as $[(PPh_3)CuCl]$ (62), $[Cu(MeCN)_4]ClO_4$ (63, 64), and $[Cu(MeCN)_4]BF_4$, have also been frequently used (65, 66). Less frequently used precursors include the copper bis(dithiosquarate) dianion $[Cu(S_2C_4O_2)_2]^{2-}$ (45) and copper(II) dichloride (67). As is the case in synthetic routes employing corresponding thiols (Sections II.B.1 and II.B.2), anionic species such as $[Cu(SR)_2]^-$, $[Cu(SR)_3]^{2-}$, and $[Cu_4(SR)_6]^{2-}$ are also formed, but in this case mixed-arenethiolatocopper copper halide aggregates, such as $[Cu_8(SR)_3Br_5]$ (60) and $[Cu_3(SR)_3]_4[CuI]_4$ (61), can also be isolated. Moreover, copper thiolates prepared this way sometimes retain the Group 1 (IA) metal salts MX, which are generated by this route, and, for example, in catalytic applications of these thiolate species the presence of such extraneous salts is highly undesirable (20, 50, 68).

### 4. Synthesis via Trimethylsilyl Arenethioethers

A synthetic procedure for copper(I) thiolates that avoids the formation of Group 1 (IA) metal salts is the reaction of trimethylsilyl arenethioethers $(Me_3SiSAr)$ with copper(I) chloride (Eq. 7) (68, 69). From this reaction the only side product is volatile chlorotrimethylsilane, which can be easily removed in vacuo.

A second advantage of this synthetic route is its wide applicability. Besides CuCl, other metal halide species such as MeZnCl (68), $ZnCl_2$ (70), $[PdCl_2(MeCN)_2]$ (71), and [AuCl(tht)] (72) can also be used, leading to the corresponding pure metal arenethiolate in virtually quantitative yield. However, when applied for the synthesis of copper(I) species, this method is limited to arenethiolate ligands containing potentially coordinating substituents (69). Fenske et al. (73) applied a similar approach to prepare copper(I) sulfide, selenide, and telluride cage clusters in the presence of auxiliary phosphine ligands and, alternatively, sulfide clusters $[M_6S_{17}]$ (M = Nb or Ta) have been prepared by reacting $M(OEt)_5$ with $(Me_3Si)_2S$ (74).

### 5. Miscellaneous

It was reported that copper dithiocarbanilato complexes [Cu(RSC(S)NPh)] [formed from CuCl or CuBr, RSC(S)N(H)Ph and NEt$_3$ in acetonitrile] can eliminate PhNCS both in solution and in the solid state to afford the corresponding copper(I) thiolates [Cu(SR)] (Eq. 8) (75).

$$
\left[ \begin{array}{c} \overset{R}{\underset{|}{S}} \\ Cu \overset{S}{\underset{N}{\diamond}} C{=}S \\ \underset{Ph}{\overset{|}{N}} \end{array} \right] \; \rightleftharpoons \; [Cu(SR)] + PhNCS \qquad (8)
$$

$$R = n\text{-Bu, Ph}$$

This method has not been extensively applied because the group R in the dithiocarbanilato copper(I) complexes has a large influence on the stability of the complexes [Cu(RSC(S)NPh)]. In the case where R is $n$-Bu, the elimination reaction is successful though rather slow (few hours), but with R being Me the dithiocarbanilato complex is stable and cannot be used to afford [Cu(SMe)]. In contrast, when R is Ph, the dithiocarbanilato copper complex is so unstable that [Cu(SPh)] is isolated quantitatively. Interestingly, in some cases the elimination reaction has been shown to be reversible since addition of an excess of PhNCS can lead to reformation of the dithiocarbanilato complex [Cu(RSC(S)NPh)] (75).

## C.  Structures of Copper(I) Arenethiolates with Intramolecular Coordination

In this chapter, the term copper(I) thiolate is taken in its broadest sense of being any species in which a Cu$^I$ ion is coordinated to one or more sulfur atoms of an SR unit. For comparative reasons, some examples of thione and thionate complexes (Scheme 2) are also included since in their deprotonated form thiones can be considered as anionic ligands ArS$^-$. The coordination behavior of thione ligands toward metals was reviewed some years ago (1d).

The various complexes covered here include $[Cu_n(SR)_n]$ species, which will

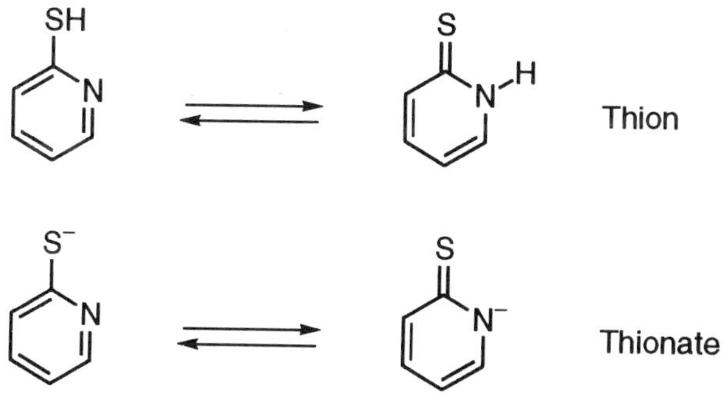

Scheme 2. Resonance equilibria of thione and thionate ligands.

be termed pure copper(I) thiolates, ionic species $[Cu_n(SR)_m]^{n-m}$ with $m > n$ (that often result when other ligands are absent) and species that also contain discrete auxiliary donor ligands such as phosphines or amines. However, the primary class of copper(I) thiolates to be discussed is that in which the thiolate ligand has a heteroatom containing function that can coordinate intramolecularly to the metal center. These intramolecularly coordinated copper(I) thiolates sometimes also contain auxiliary neutral donor ligands.

Copper(I) thiolates have a wide variety of structures and in almost all cases are found as aggregated species in which the sulfur atoms bridge two or more copper atoms. Table I lists structurally characterized neutral copper(I) thiolate

TABLE I

Neutral Copper(I) and Silver(I) Thiolate Complexes without Coordinating Ligands

| Entry | Compound[a] | References |
|-------|-------------|------------|
| 1 | $[Cu_3(SC(SiMe_3)_3)_3]$ | 76 |
| 2 | $[Ag_3(SC(SiMe_2Ph)_3)_3]$ | 33 |
| 3 | $[Cu_4(SC_6H_3(SiMe_3)_2\text{-}2,6)_4]$ | 77 |
| 4 | $[Cu_4(SC_6H_3\text{-}i\text{-}Pr_3\text{-}2,4,6)_4]$ | 52b |
| 5 | $[Cu_4(SSi(O\text{-}t\text{-}Bu)_3)_4]$ | 78 |
| 6 | $[Ag_4(SC(SiMe_3)_3)_4]$ | 33 |
| 7 | $[Ag_4(SCH(SiMe_3)_2)_4]$ | 33 |
| 8 | $[Ag_4(SC_6H_4SiMe_3\text{-}2)_4]$ | 40 |
| 9 | $[Au_4(SC(SiMe_3)_3)_4]$ | 79 |
| 10 | $[Cu_6(SeC_6H_2\text{-}i\text{-}Pr_3\text{-}2,4,6)_6]$ | 80 |
| 11 | $[Au_6(SC_6H_2\text{-}i\text{-}Pr_3\text{-}2,4,6)_6]$ | 81 |
| 12 | $[Cu_8(SC_6H_2\text{-}i\text{-}Pr_3\text{-}2,4,6)_8]$ | 52 |
| 13 | $[Ag_8(S\text{-}t\text{-}Bu)_8]$ | 82 |
| 14 | $[Ag_8(SC(Me)Et_2)_8]$ | 83 |
| 15 | $[Cu_{12}(SC_6H_4SiMe_3\text{-}2)_{12}]$ | 40, 84 |
| 16 | $[Ag_{12}(S\text{-}c\text{-}Hex)_{12}]$ | 85 |
| 17 | $[Ag(SC_6H_2\text{-}i\text{-}Pr_3\text{-}2,4,6)]_\infty$ | 86 |

[a]Isopropyl = $i$-Pr; *tert*-butyl = $t$-Bu; methyl = Me; ethyl = Et; cyclohexyl = $c$-hex; phenyl = Ph.

complexes of the type $[Cu_n(SR)_n]$, while the aggregates $[Cu_n(SR)_n]^{n-m}$ (usually anionic with $m > n$) in which the thiolate has no potentially coordinating functions are listed in Table II. An earlier review of complexes of this type appeared in 1986 (1a). Some examples of the structural formats found for these anionic cage aggregates are illustrated in Scheme 3. Tables I and II and Scheme 3 also include some closely related silver(I) and gold(I) species.

The structures of copper(I) thiolates in which the only other donor functions present are those of the auxiliary ligands (Table III) often resemble structures of those species in which the arenethiolate ligands have a heteroatom containing function that can coordinate intramolecularly to the metal center (Table IV). There are some structurally characterized mixed-organo(arenethiolato)copper(I) and arenethiolatocopper copper halide aggregates in which intramolecular coordination is present. These aggregates are collected in Table V, which allows some direct comparisons to be made between the two types of species. For ease

TABLE II

Ionic copper(I) and Silver(I) Thiolate Complexes without Coordinating Ligands

| Entry | Compound | References |
|---|---|---|
| 1 | $[Cu(SPh)_2]^-(N^nBu_4)^+$ | 44 |
| 2 | $[Cu(SC_6F_5)_2]^-(pyH)^+$ | 87 |
| 3 | $[Cu(SC_6HMe_4-2,3,5,6)_2]^-(N^nPr_4)^+$ | 65 |
| 4 | $[Ag(SC_6HMe_4-2,3,5,6)_2]^-(N_nPr_4)^+$ | 65 |
| 5 | $[Au(SC_6H_2-i-Pr_3-2,4,6)_2]^-(NH_4)^+$ | 81 |
| 6 | $[Cu(SPh)_3]^{2-}(NEt_4)_2^+$ | 62 |
| 7 | $[Cu(SPh)_3]^{2-}(PPh_4)_2^+$ | 45 |
| 8 | $[Cu(SC_6H_4N(H)C(O)-t-Bu)_3]^{2-}(NEt_4)_2^+$ | 88 |
| 9 | $[Cu(SC_5H_4NH)_3]^+(NO_3)^-$ | 49 |
| 10 | $[Cu_4(SPh)_6]^{2-}(NMe_4)_2^+$ | 89a |
| 11 | $[Cu_4(SPh)_6]^{2-}(NMe_4)_2^+ \cdot EtOH$ | 89b |
| 12 | $[Cu_4(SPh)_6]^{2-}(PPh_4)_2^+$ | 45, 48b |
| 13 | $[Cu_4(SMe)_6]^{2-}(NMe_4)_2^+$ | 89b, 90 |
| 14 | $[Cu_4(SEt)_6]^{2-}(PPh_4)_2^+$ | 91 |
| 15 | $[Cu_4(dmit)_3]^{2-}(N^nBu_4)_2^+$ | 64 |
| 16 | $[Cu_4(o-(SCH_2)_2C_6H_4)_3]^{2-}(PPh_4)_2^+$ | 66 |
| 17 | $[Cu_5(S-t-Bu)_6]^-(HNEt_3)^+$ | 92 |
| 18 | $[Cu_5(S-t-Bu)_6]^-(NEt_4)^+$ | 90, 92, 93 |
| 19 | $[Cu_5(SPh)_7]^{2-}(NMe_4)_2^+$ | 47, 48a, 90 |
| 20 | $[Ag_5(SPh)_7]^{2-}(NMe_4)_2^+$ | 47 |
| 21 | $[Ag_6(SPh)_8]^{2-}(NMe_4)_2^+$ | 94 |
| 22 | $[Cu_7(SEt)_8]^-(PPh_4)^+$ | 91 |
| 23 | $[Cu_8(DTS)_6]^{4-}(PPh_4)_4^+$ | 95 |
| 24 | $[Cu_8(DED)_6]^{4-}(NBu_4)_4^+$ | 95 |
| 25 | $[Ag_{12}(SPh)_{16}]^{4-}(NMe_4)_4^{+a}$ | 94 |

$^a$Comprises two linked $[Ag_6(SPh)_8]^{2-}$ cages.

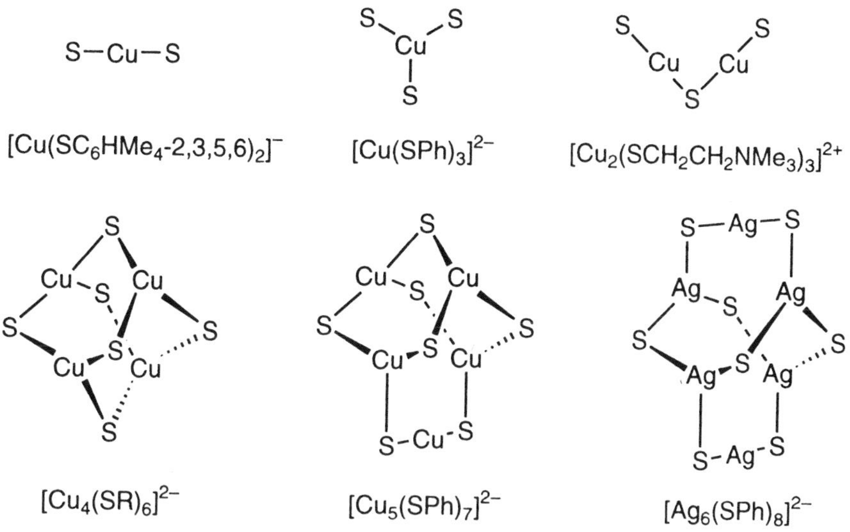

$[Cu(SC_6HMe_4\text{-}2,3,5,6)_2]^-$

$[Cu(SPh)_3]^{2-}$

$[Cu_2(SCH_2CH_2NMe_3)_3]^{2+}$

$[Cu_4(SR)_6]^{2-}$

$[Cu_5(SPh)_7]^{2-}$

$[Ag_6(SPh)_8]^{2-}$

Scheme 3.  Some structural formats of anionic copper(I) thiolates.

TABLE III

Copper(I), Silver(I), and Gold(I) Thiolates with Auxiliary Donor Ligands

| Entry | Compound | Reference |
|-------|----------|-----------|
| 1 | $[Cu(SPh)(SO_2)(PPh_2Me)_3]$ | 46 |
| 2 | $[Cu(SC_6F_5)(HB(3,5\text{-}i\text{-}Pr_2pz)_3)]$ | 96 |
| 3 | $[Cu(SCPh_3)(HB(3,5\text{-}i\text{-}Pr_2pz)_3)]$ | 96 |
| 4 | $[Cu(SC_6H_4NO_2\text{-}4)(HB(3,5\text{-}Me_2pz)_3)][KNO_3]$ | 59 |
| 5 | $[Au(SC_6H_2Et_3\text{-}2,4,6)(PPh_3)]$ | 97 |
| 6 | $[Au(SC_6H_2\text{-}i\text{-}Pr_3\text{-}2,4,6)(PPh_3)]$ | 97 |
| 7 | $[Cu_2(SPh)_2(PPh_3)_4]$ | 27 |
| 8 | $[Cu_2(SC_6H_4Me\text{-}2)_2(phen)_2]$ | 55 |
| 9 | $[Cu_2(SC_6H_4Me\text{-}2)_2(Me_2phen)_2]$ | 98 |
| 10 | $[Cu_2(SPh)_2(RC{\equiv}CR)_2]^a$ | 99 |
| 11 | $[Cu_2(SC_5H_4NH)_6]^{2+}$ | 100 |
| 12 | $[Cu_2(SePh)_2(PPh_3)_3]$ | 57 |
| 13 | $[Au_2(SPh)_2(PPh_3)_2]$ | 97 |
| 14 | $[Cu_3(SPh)_3(PPh_3)_4]$ | 28 |
| 15 | $[Cu_4(SPh)_4(PPh_3)_4]$ | 29 |
| 16 | $[Cu_4(S\text{-}t\text{-}Bu)_4(PPh_3)_2]$ | 101 |
| 17 | $[Cu_4(SC(Me)_2Et)_4(dppm)_2]$ | 56 |
| 18 | $[Cu_4(SC({=}S)C_6H_4Me\text{-}2)_4(dppm)_2]$ | 102 |
| 19 | $[Ag_6(SC_6H_4Cl\text{-}4)_6(PPh_3)_5]$ | 103 |
| 20 | $[Au_6(S_6C_6)(PPh_3)_6]^a$ | 104 |
| 21 | $[Ag_8(SC(Me)Et_2)_8(PPh_3)_2]$ | 105 |
| 22 | $[Cu_{10}(SCF_3)_{10}(MeCN)_8]$ | 58 |
| 23 | $[Ag_{14}(S\text{-}t\text{-}Bu)_{14}(PPh_3)_4]$ | 105 |
| 24 | $[Cu(SPh)(Me_2phen)]_\infty$ | 63 |

$^a$3,3,6,6-Tetramethyl-1-thiacyclohept-4-yne = $RC{\equiv}CR$ and $S_6C_6$ is benzenehexathiolate.

TABLE IV
Copper(I) and Silver(I) Thiolates with Intramolecular Coordination

| Entry | Compound | Reference |
|-------|----------|-----------|
| | *Neutral Complexes* | |
| 1 | $[Cu\{SC(=NPh)(OC_6H_3Me_2-2,6)\}(PPh_3)_2]$ | 106 |
| 2 | $[Cu_2\{SC(=NPh)(OC_6H_2-t-Bu_2-2,6-Me-4)\}_2(P(OMe)_3)_2]$ | 106 |
| 3 | $[Au_2(SCH_2CH_2PEt_2)_2]$ | 107 |
| 3 | $[Cu_3(SC_6H_4NMe_2-2)_3]$ | 69 |
| 4 | $[Cu_3\{SC_6H_4((R)-CH(Me)NMe_2-2)\}_3]$ | 50 |
| 5 | $[Cu_3\{SC(H)Me(\eta^5-C_5H_3PPh_2-2)Fe(\eta^5-C_5H_5)\}_3]$ | 54a |
| 6 | $[Cu_4(mimt)_4]$ | 108 |
| 7 | $[Cu_4\{SC_6H_4(CH_2N(Me)CH_2CH_2OMe-2)\}_4]$ | 68 |
| 8 | $[Cu_4(SC_3H_4NS)_4(py)]_\infty$ | 109 |
| 9 | $[Cu_4(SC_3H_4NS)_4(PPh_3)_2]$ | 109 |
| 10 | $[Ag_4(SC_3H_4NS)_4(PPh_3)_2]$ | 109 |
| 11 | $[Cu_2Ag_2(SC_3H_4NS)_4(PPh_3)_2]$ | 109 |
| 12 | $[Cu_4\{SC(=NPh)(OC_6H_3Me_2-2-6)\}_4]$ | 106 |
| 13 | $[Cu_6(SC_5H_4N)_6]$ | 110 |
| 14 | $[Cu_6(dmpymt)_6]$ | 111 |
| 15 | $[Cu_6(Me(OH)pymt)_6]^a$ | 112 |
| 16 | $[Cu_6(SC_5H_3N(SiMe_3)-6)_6]$ | 113 |
| 17 | $[Ag_6(SC_5H_3N(SiMe_3)-6)_6]$ | 113 |
| 18 | $[Cu_6(SC_9H_6N)_6]^a$ | 114 |
| 19 | $[Cu_6(SC(O)NEt_2)_6]$ | 115 |
| 20 | $[Cu_6(n-BuSC(S)NPh)_6]$ | 75 |
| 21 | $[Ag_6(SC(S)N-n-Pr_2)_6]$ | 116 |
| 22 | $[Cu_8(SC(Me)_2Et)_4(S_2CSC(Me)_2Et)_4]^b$ | 117 |
| 23 | $[Cu_9(S-1-C_{10}H_6NMe_2-8)_9]$ | 69 |
| 24 | $[Cu(SC_9H_7N)(ClO_4)]_\infty{}^c$ | 114 |
| | *Ionic Complexes* | |
| 25 | $[Cu_3(SCH_2CH_2S)_3]^{3-}$ $(Me_3NCH_2Ph)_2^+ Na^+ \cdot MeOH$ | 67 |
| 26 | $[(S_6)M(S_8)M(S_6)]^{4-}$ $(PPh_4)_4$ (M = Cu or Ag) | 118 |
| 27 | $[Cu_3(S_4)_3]^{3-}$ $(PPh_4)_2^+$ $(NH_4)^+$ | 118a |
| 28 | $[Cu_3(S_6)_3]^{3-}$ $(PPh_4)_2^+$ $(NH_4)^+$ | 119 |
| 29 | $[Ag_8(SC_5H_3N(SiMe_2Ph)-3)_6]^{2+}$ $[Ag(NO_3)_2]_2^-$ | 120 |
| | *Mixed-Valence Complexes* | |
| 30 | $[Cu_8^I Cu_6^{II}(SC(Me)_2CH_2NH_2)_{12}Cl]\cdot 3.5SO_4$ | 51 |

[a] 4-hydroxy-6-methylpyrimidine-2-thionate = Me(OH)pymt, quinoline-2-thionate = $SC_9H_6N$, quinoline-2-thione = $SC_9H_7N$.

[b] Formed by insertion of $CS_2$ into $[Cu(SC(Me)_2Et)]_n$; the Cu—S—C(S) unit can formally be interpreted as a four-membered chelate.

[c] Thione complex.

TABLE V
Mixed-Organo(arenethiolato)copper(I) and Arenethiolatocopper Copper Halide Species

| Entry | Compound | Reference |
|-------|----------|-----------|
| 1 | $[Cu_3(SC_6H_3(CH_2NMe_2)-2-Cl-5)_2(Mes)(PPh_3)]$ | 20 |
| 2 | $[Cu_4(SC_6H_3(CH_2NMe_2)-2-Cl-5)_2(Mes)_2]$ | 20 |
| 3 | $[Cu_6(SC_6H_4(CH_2NMe_2)-2)_4(C{\equiv}C\text{-}t\text{-}Bu)_2]$ | 21 |
| 4 | $[Cu_6\{SC_6H_4((S)\text{-}CH(Me)NMe_2\text{-}2)\}_4(C{\equiv}C\text{-}t\text{-}Bu)_2]$ | 121 |
| 5 | $[Cu_6(S\text{-}1\text{-}C_{10}H_6NMe_2\text{-}8)_4(C{\equiv}C\text{-}t\text{-}Bu)_2]$ | 69 |
| 6 | $[Cu_8(SC_6H_3(CH_2NMe_2)_2\text{-}2,6)_3Br_5]$ | 60 |
| 7 | $[Cu_3(S\text{-}oxazoline)_3]_4[CuI]_4$ | 61 |

of discussion, the complexes in this chapter are described in order of their aggregation state.

## 1. Monomeric Complexes

There are no known monomeric species, that is [Cu(SR)], when the thiolate ligand has no substituents present for intramolecular coordination. All monomeric species are of the type $[Cu(SAr)L_n]$ in which there is an auxiliary donor ligand L. Several species were reported where $n$ is 3 with ligands such as phosphines and amines [Fig. 2(a)]. The only reported example of a monomeric copper(I) thiolate in which, besides auxiliary coordination (of two $PPh_3$ ligands), there is also intramolecular coordination (of one N-donor group) is the complex $[Cu\{SC(=NPh)(OC_6H_3Me_2\text{-}2,6)\}(PPh_3)_2]$ (106). In all monomeric complexes $[Cu(SAr)L_3]$, the copper atom has a tetrahedral geometry. Some typical Cu—S bond lengths of these species are listed in Table VI.

Recently, we reported the use of the bis(alkynyl)titanocene complex $[(\eta^5\text{-}C_5H_4SiMe_3)_2Ti(C{\equiv}CSiMe_3)_2]$ as a molecular tweezer that functions as a bidentate ligand by using the two alkynyl functions to coordinate to a single cop-

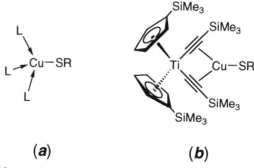

**(a)**                **(b)**

Figure 2.   Monomeric copper(I) thiolates with auxiliary donor ligand coordination.

TABLE VI
Geometrical Details of Monomeric Copper(I) Thiolate Complexes

| Entry | Compound | L | Cu—S ($\text{Å}$) | Reference |
|---|---|---|---|---|
| | | *Four-Coordinate Cu* | | |
| 1 | $[Cu(SPh)(SO_2)L_3]^a$ | $PPh_2Me$ | 2.404(2) | 46 |
| 2 | $[Cu(SC(=NPh)(OAr))L_2]^b$ | $PPh_3$ | 2.469(2) | 106 |
| 3 | $[Cu(SC_6H_4NO_2-4)L]$ | $HB(3,5-Me_2pz)_3$ | 2.19(1) | 59 |
| 4 | $[Cu(SC_6F_5)L]$ | $HB(3,5-i-Pr_2pz)_3$ | 2.176(4) | 96 |
| 5 | $[Cu(SCPh_3)L]$ | $HB(3,5-i-Pr_2pz)_3$ | 2.12(2) | 96 |
| | | *Three-Coordinate Cu* | | |
| 6 | $[Cu(SC_6H_4(CH_2NMe_2)-2)L_2]$ | $TiC\equiv CSiMe_3{}^c$ | 2.237(3) | 54b |

[a] The $SO_2$ is bonded to the $PhS^-$ ligand.
[b] $OAr = OC_6H_3Me_2-2,6$; a four-membered $Cu-S-C-N$ chelate ring is present.
[c] $L_2 = [(\eta^5-C_5H_4SiMe_3)_2Ti(C\equiv CSiMe_3)_2]$.

per(I) center. In this way, there can be formed monomeric species $[(\eta^5-C_5H_4SiMe_3)_2Ti(C\equiv CSiMe_3)_2CuX]$ in which X is an $\eta^1$-bonded inorganic (54b) or organic (122) ligand. This class of species includes the arenethiolate complexes, where X is $SC_6H_5$, $SC_6H_4NMe_2-2$, $SC_6H_4CH_2NMe_2-2$, and S-1-$C_{10}H_6NMe_2-8$ (54b). The copper atom in these species has a trigonal geometry (Fig. 2b) (54b, 122).

## 2.  Dimeric Complexes

Like monomeric copper(I) thiolates, all known dimeric thiolate species contain auxiliary coordinating donor ligands L. This class comprises complexes with two basic formulations, namely, $[Cu_2(SR)_2L_4]$ and $[Cu_2(SR)_2L_2]$, whose schematic structures are depicted in Figure 3. Reported examples of the former are $[Cu_2(SPh)_2(PPh_3)_4]$ (27), $[Cu_2(SC_6H_4Me-2)_2(phen)_2]$ (55), and $[Cu_2(SC_5H_4NH)_6(X)_2]$ (X = Cl or Br) (100), while for the latter there is only one example, that is $[Cu_2(SPh)_2(RC\equiv CR)_2]$ (RC$\equiv$CR = 3,3,6,6-tetramethyl-1-thiacyclohept-4-yne) (99). A special example of the type $[Cu_2(SR)_2L_2]$ is provided by the complex $[Cu_2\{SC(=NPh)(OC_6H_2-t-Bu_2-2,6-Me-4)\}_2\{P(OMe)_3\}_2]$ (106), which is a dimer formed through intermolecular coordination between two monomeric units [Fig. 3(c)]. Some structural details of these complexes are listed in Table VII.

## 3.  Trimeric Complexes

Most reported copper(I) thiolates in which there is intramolecular ligand coordination have aggregation states of three or higher, and to date three X-ray

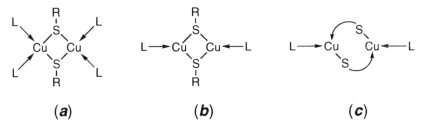

Figure 3. Schematic representation of copper(I) thiolate dimers: (a) $[Cu_2(SR)_2L_4]$, (b) $[Cu_2(SR)_2L_2]$, and (c) $[Cu_2(SR)_2L_2]$ with additional intramolecular coordination.

structures of trimeric examples exist (Fig. 4). These three structures are: $[Cu_3(SC_6H_4NMe_2-2)_3]$ which contains a five-membered chelate ring (69), chiral $[Cu_3(SC_6H_4((R)-CH(Me)NMe_2)-2)_3]$, which contains a six-membered chelate ring and which has some very interesting luminescence properties (50, 123), and a related species $[Cu_3(SC(H)Me(\eta^5-C_5H_3PPh_2-2)Fe(\eta^5-C_5H_5))_3]$, which is an alkylthiolato complex that not only has a six-membered chelate ring, but also a P-donor coordination (54a). The trinuclear structure of each of these complexes is characterized by the presence of a cyclohexane-like six-membered $Cu_3S_3$ heterocycle, which generally is in the chair conformation with the substituents on the thiolate sulfur atom occupying equatorial positions [Fig. 4(b)]. Geometrical details of some trimeric copper(I) thiolates are listed in Table VIII.

Apparently, it is the five- or six-membered chelate rings formed by intramolecular coordination that leads to species with the same aggregation state,

TABLE VII
Geometrical Details of Dimeric Copper(I) Thiolate Complexes

| Entry | Compound | Cu—S (Å) | Cu—S—Cu (deg) | Reference |
|---|---|---|---|---|
| | | *Four-Coordinate Cu* | | |
| 1 | $[Cu_2(SPh)_2(PPh_3)_4]$ | 2.344(4), 2.415(4) | 98.63(4), 102.75(4) | 27 |
| 2 | $[Cu_2(SC_6H_4Me-2)_2(phen)_2]$ | 2.304(4), 2.379(5) | 68.1(1), 67.8(1) | 55 |
| 3 | $[Cu_2(SC_5H_4NH)_6]Cl_2^a$ | 2.308(2), 2.498(3) | 74.3(1) | 100a |
| 4 | $[Cu_2(SC_5H_4NH)_6]Cl_2^a$ | 2.320(3), 2.538(4) | 74.6(1) | 100b |
| 5 | $[Cu_2(SC_5H_4NH)_6]Br_2^a$ | 2.297(3), 2.534(3) | 73.8(1) | 100b |
| 6 | $[Cu_2(SC(=NPh)(OAr))_2(L)_2]^b$ | 2.258(2), 2.268(2) | $^c$ | 106 |
| | | *Three-Coordinate Cu* | | |
| 7 | $[Cu_2(SPh)_2(RC≡CR)_2]^d$ | 2.283(1)–2.295(1) | 78.53(3), 78.95(4) | 99 |

$^a$Thione complex: the Cu—S distances of the terminal 2(1H)-pyridine-2-thione ligands are somewhat shorter: 2.274(3)–2.326(3) Å.
$^b$OAr = $OC_6H_2$-$t$-$Bu_2$-2,6-Me-4; L = $P(OMe)_3$.
$^c$Not present: see text.
$^d$3,3,6,6-Tetramethyl-1-thiacyclohept-4-yne = RC≡CR.

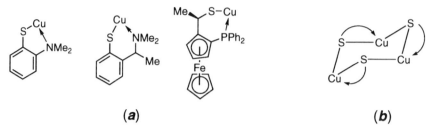

Figure 4. (a) Trimeric copper(I) thiolate complexes with intramolecular ortho-coordination. (b) The $Cu_3S_3$ heterocycle.

and the differences in either coordinating heteroatom (N vs. P) or the nature of the thiolate backbone (arenethiolate vs. alkylthiolate) do not seem to play an important role in determining the complex nuclearity. However, it must be noted that either a more rigid thiolate backbone or a smaller chelate ring size can affect both the aggregation state and the overall structure in a dramatic way (Sections II.C.5 and II.C.7).

The phosphine coordinated copper(I) thiophenolate $[Cu_3(SPh)_3(PPh_3)_4]$ (28) has an analogous trimeric structure to those species with intramolecular coordination, although in this case the fourth extra phosphine ligand results in a distorted tetrahedral coordination of one of the copper atoms and a distortion of the $Cu_3S_3$ six-membered heterocycle (Fig. 5).

When no potentially coordinating ligands are present either as auxiliaries or as thiolate substituents, a trimeric aggregation state is unusual. To our knowl-

TABLE VIII
Geometrical Details of Trinuclear Copper(I) Thiolate Complexes

| Entry | Compound | $Cu-S$ (Å) | $Cu-S-Cu$ (deg) | Reference |
|-------|----------|-----------|-----------------|-----------|
| 1 | $[Cu_3(SC_6H_4NMe_2-2)_3]$ | 2.198(1), 2.2219(9) | 75.18(4) | 69 |
| 2 | $[Cu_3(SC_6H_4(CH(Me)NMe_2)-2)_3]$ | 2.186(2), 2.231(2) | 79.63(7) | 50 |
| 3 | $[Cu_3(SC(H)Me(Fc)PPh_2-2)_3]^a$ | 2.178(6), 2.251(6) | 92.3(2), 98.3(2) | 54a |
| 4 | $[Cu_3(SPh)_3(PPh_3)_4]$ | 2.23–2.28 | 87 | 28 |
| | | $2.39–2.40^b$ | $105, 124^b$ | |
| 5 | $[Cu_3(SCH_2CH_2S)_3]^{3-}$ | 2.208(2)–2.313(2) | 74.7(1)–78.4(1) | 67 |
| | | $2.245(2)–2.270(2)^c$ | | |
| 6 | $[Cu_3(SAr)_2(Mes)(PPh_3)]^a$ | 2.234(1), 2.246(1) | 75.13(4), 76.87(4) | 20 |
| | | $2.267(1), 2.280(1)^d$ | | |

$^a$SC(H)Me(Fc)PPh_2-2 = SC(H)Me($\eta^5$-C_5H_3PPh_2-2)Fe($\eta^5$-C_5H_5), SAr = SC_6H_4(CH_2NMe_2-2)(Cl-5).
$^b$Involving the four-coordinate Cu atom.
$^c$Contacts outside the $Cu_3S_3$ six-membered ring, See also Fig. 5.
$^d$Involving the P-coordinated Cu atom.

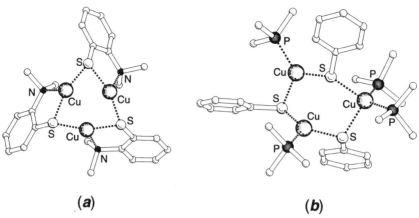

**(a)**                                              **(b)**

Figure 5. (a) Molecular structures of [Cu$_3$(SC$_6$H$_4$NMe$_2$-2)$_3$] and (b) [Cu$_3$(SPh)$_3$(PPh$_3$)$_4$] (hydrogen atoms are omitted for clarity).

edge, the only trimeric structures with no further coordinating ligands are those of [Cu$_3$\{SC(SiMe$_3$)$_3$\}$_3$] (76) and [Ag$_3$\{SC(SiMe$_2$Ph)$_3$\}$_3$] (33). The structure of the latter contains a slightly puckered almost planar Ag$_3$S$_3$ six-membered ring with an unusual coordination of two of the silver atoms; whereas silver(I) is usually almost linear, in this structure the S—Ag—S angles are 149.9(1)°, 153.6(1)°, and 163.1(1)°. It has been proposed that these distortions arise from the steric bulk of the SC(SiMe$_2$Ph)$_3$ ligand and from the constraints imposed by ring closure (33).

Interestingly, an anionic dithiolato copper(I) aggregate [Cu$_3$(SCH$_2$CH$_2$S)$_3$]$^{3-}$ was also reported (67) and it possesses a structure, depicted in Figure 6, that is similar to that of trimeric copper(I) thiolates (Table VIII, entries 1–3). The sodium counterion in this complex has a tetrahedral geometry that results from coordination of the three sulfur atoms not involved in the Cu$_3$S$_3$ six-membered ring and the oxygen atom of a methanol molecule. Structural elements similar to that depicted in Figure 6 are also known for the copper(I) sulfido complexes [Cu$_3$(S$_4$)$_3$]$^{3-}$ and [Cu$_3$(S$_6$)$_3$]$^{3-}$ (118, 119), although aggregates based on other motives are also known (124).

A trinuclear mixed-organo(arenethiolato)copper(I) aggregate [Cu$_3$\{SC$_6$H$_3$-(CH$_2$NMe$_2$)-2-Cl-5\}$_2$(Mes)(PPh$_3$)] was also structurally characterized (20) and interestingly, like the trimeric copper(I) arenethiolates, it has a structure with a six-membered core. However, the puckering of the six-membered Cu$_3$S$_2$C ring is of a boat type with the aryl ring of the arenethiolates bonded equatorially to the sulfur atoms in the ring and one of the three bridging atoms is the two-electron three-center (2e–3c) bonded mesityl C$_{ipso}$ atom (20).

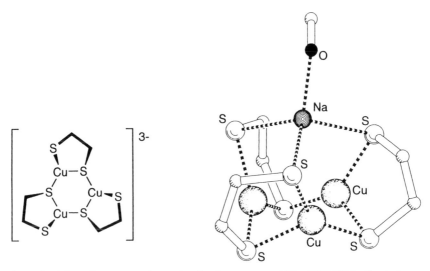

Figure 6.   Schematic representation and molecular structure of $[Cu_3(SCH_2CH_2S)_3]^{3-}$ (hydrogen atoms are omitted for clarity).

## 4.   Tetrameric Complexes

Tetrameric copper(I) thiolate aggregates are common, not only with intramolecularly coordinating ligands and auxiliary donor ligands, but also without donor ligands of any type. In the complexes $[Cu_4\{SC_6H_4(CH_2N(Me)CH_2CH_2OMe-2)\}_4]$ (68), $[Cu_4(S\text{-}t\text{-}Bu)_4(PPh_3)_2]$ (101), $[Cu_4(SCMe_2Et)_4(dppm)_2]$ (56), $[Cu_4(SC_6H_3(SiMe_3)_2\text{-}2,6)_4]$ (77), $[Cu_4(SC_6H_2\text{-}i\text{-}Pr_3\text{-}2,4,6)_4]$ (52b), $[Cu_4(SSi(O\text{-}t\text{-}Bu)_3)_4]$ (78), $[Ag_4(SC(SiMe_3)_3)_4]$ (33), $[Ag_4(SCH(SiMe_3)_2)_4]$ (33), and $[Ag_4(SC_6H_4SiMe_3\text{-}2)_4]$ (40) the tetranuclear aggregate contains an open $M_4S_4$ (M = Cu or Ag) eight-membered heterocycle (Type Ia, Fig. 7); this heterocycle can be based on Group 11 (IB) metal centers that are either solely two-coordinate, solely three-coordinate Group 11 (IB), or a mixture of two and three coordinate. In the mixed-organo-(arenethiolato)copper(I) aggregate $[Cu_4\{SC_6H_3(CH_2NMe_2)\text{-}2\text{-}Cl\text{-}5\}_2(Mes)_2]$ one also finds an open tetranuclear structure, although in this case there are two types of bridging atoms leading to a central $Cu_4S_2C_2$ eight-membered heterocycle (Type 1b, Fig. 7) (20).

The Group 11 (IB) metal atoms in the structures of $[Cu_4(1\text{-methylimidazoline-2-thionate})_4]$ (108), $[Cu_4(SC_3H_4NS)_4(py)]_\infty$ (109), $[M_2M_2'(SC_3H_4NS)_4(PPh_3)_2]$ (109), (M = M' = Cu; M = M' = Ag; M = Cu, M' = Ag) and $[Cu_4\{SC(=NPh)(OC_6H_3Me_2\text{-}2,6)\}_4]$ (106) are arranged in a tetrahedral form leading to a different structural format (Type II, Fig. 7).

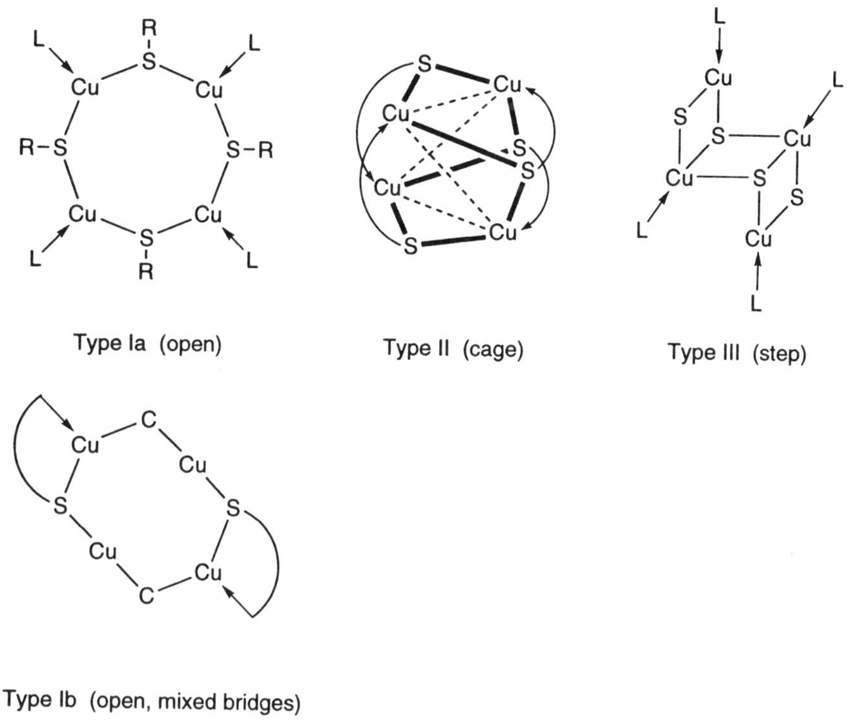

Type Ia  (open)            Type II  (cage)            Type III  (step)

Type Ib  (open, mixed bridges)

Figure 7.   Tetrameric copper(I) thiolate aggregates.

Type II structural format is a cage structure derived from the Type Ia structure by folding of the eight-membered $Cu_4S_4$ ring. The reasons for this folding can be steric in origin but, as is evident in the structure of [Cu$_4$(1-methylimidazoline-2-thionate)$_4$] (108), can also arise from intramolecular coordination. Moreover, in the structure of [Cu$_4$(SPh)$_4$(PPh$_3$)$_4$] (29) the reduction of the $Cu_8S_8$ core size, resulting in a step structure (Type III, Fig. 7), was proposed to arise from steric factors. Some geometrical details of tetranuclear copper(I) thiolates are listed in Table IX.

## 5.   Hexameric Complexes.

The next increase in aggregation state of Group 11 (IB) metal thiolates affords pentameric structures, but to the best of our knowledge a structurally characterized species of this type is unknown. In organocopper(I) chemistry, the only example is that of mesitylcopper [Cu$_5$(C$_6$H$_2$Me$_3$-2,4,6)$_5$] (125), which has a structure based on an open $Cu_5C_5$ 10-membered heterocycle.

TABLE IX
Geometrical Details of Tetrameric Copper(I) Thiolate Complexes

| Entry | Compound[a] | Cu—S (Å) | Cu—S—Cu (deg) | Reference |
|---|---|---|---|---|
| | | *Open Structures* | | |
| 1 | [Cu$_4$(SAr)$_4$] | 2.1991(8)–2.2463(9) | 74.12(2)–74.82(2) | 68 |
| 2 | [Cu$_4$(S-$t$-Bu)$_4$(PPh$_3$)$_2$] | 2.145(2)–2.167(2) 2.245(2)–2.263(2)[b] | 79.9(1)–89.9(1) | 101 |
| 3 | [Cu$_4$(SC(Me)$_2$Et)$_4$(dppm)$_2$] | 2.221(2)–2.268(2) | 85.9(1)–98.5(1) | 56 |
| 4 | [Cu$_4$(SC$_6$H$_3$(SiMe$_3$)$_2$-2,6)$_4$] | 2.134(8)–2.164(9) | 82.9(3)–92.4(3) | 77 |
| 5 | [Cu$_4$(SSi(O-$t$-Bu)$_3$)$_4$] | 2.161(2)–2.175(2) | 82.1(1) | 78 |
| | | *Cage Structures* | | |
| 6 | [Cu$_4$(Simid)$_4$] | 2.233(2)–2.307(2) | 72.1(1), 96.5(2) | 108 |
| 7 | [Cu$_4$(SC$_3$H$_4$NS)$_4$(PPh$_3$)$_2$] | 2.191(2), 2.265(2) 2.394(2), 2.424(2)[b] | 77.5, 81.6 | 109 |
| 8 | [Cu$_4${SC(=NPh)(OAr)}$_4$] | 2.252(1)–2.319(2) | 70.10(5)–73.01(6) | 106 |
| | | *Step Structure* | | |
| 9 | [Cu$_4$(SPh)$_4$(PPh$_3$)$_4$] | 2.277(2)–2.482(2) | 70.8(1)–130.1(1) | 29 |

[a]SAr = SC$_6$H$_4$(CH$_2$N(Me)CH$_2$CH$_2$OMe)-2 and OAr = OC$_6$H$_3$Me$_2$-2,6.
[b]Involving P-coordinated Cu.

Hexameric structures are generally only found with thiolate ligands in which potentially intramolecularly coordinating substituents are present. The structures of hexameric copper(I) thiolate species [Cu(SR)]$_6$ can best be described as resulting from the alternate stacking of two trimeric units [Cu(SR)]$_3$, similar to the trimeric aggregates described in Section II.C.3, with coordination of a heteroatom (nitrogen) from a thiolate ligand in one trimeric unit to a copper atom in the opposite unit (Fig. 8). In this way, the thiolate ligand is S bonded

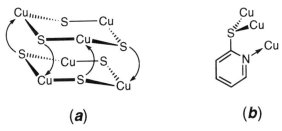

Figure 8. (*a*) Schematic structure of a hexameric copper(I) thiolate aggregates (ligands on the back of the structure omitted for clarity). (*b*) The bonding of the pyridine-2-thiolate to three copper atoms.

to two Cu atoms and $N$ bonded to a third Cu atom [Fig. 8(b)]. When the alternate stacking of the two trimeric units of the structure is regular, the structure can also be described as an octahedral arrangement of six Cu atoms in which the thiolate ligands each bridge two Cu atoms.

Hexameric-type structures are found not only for the thionate complexes $[Cu_6(SC_5H_4N)_6]$ (110), $[Cu_6(4,6\text{-}dimethylpyrimidine\text{-}2\text{-}thionate)_6]$ (111), $[Cu_6\text{-} (4\text{-}hydroxy\text{-}6\text{-}methyl\text{-}pyrimidine\text{-}2\text{-}thionate)_6]$ (112), $[Cu_6(SC_5H_3N(SiMe_3)\text{-}6)_6]$ (113), $[Ag_6(SC_5H_3N(SiMe_3)\text{-}6)_6]$ (113), and $[Cu_6(quinoline\text{-}2\text{-}thionate)_6]$ (114) but also for the $N,N$-diethylthiocarbamate complex $[Cu_6(SC(O)NEt_2)_6]$ (115) and the $S$-butyldithiocarbanilato species $[Cu_6(n\text{-}BuSC(S)=NPh)_6]$ (75). Interestingly, dipropyldithiocarbamate silver(I) also has a hexameric structure in which the dithiocarbamate ligands S-bridge two silver atoms and S'-coordinate to a third silver atom (116). The bonding nature of these $S,N$-chelating ligands resembles that of the $N,N$-chelating amido ligand in the amidocopper(I) species $[Cu_6(2\text{-}N(SiMe_3)C_5H_3N\text{-}6\text{-}Me)_4Cl_2]$ (126).

The smallest possible chelate ring in these hexameric species is a Cu—S—C—N unit, and such a four-membered ring is present in $[Cu\text{-} \{SC(=NPh)(OC_6H_3Me_2\text{-}2,6)\}(PPh_3)_2]$ (106), though absent in its parent species $[Cu_4\{SC(=NPh)(OC_6H_3Me_2\text{-}2,6)\}_4]$ (106). Finally, a rare example of a hexanuclear structure with auxiliary ligands is that of the silver thiolate triphenylphosphine complex $[Ag_6(SC_6H_4Cl\text{-}4)_6(PPh_3)_5]$ (103). Selected data of hexameric copper(I) thiolates are given in Table X.

## 6. Octameric Complexes

As far as we can ascertain there are no heptanuclear Cu(I) thiolate aggregates known. Octanuclear aggregates with intramolecularly coordinating thiolate ligands have not been reported, although there are some examples either without

TABLE X
Geometrical Details of Hexameric Copper(I) Thiolate Complexes

| Entry | Compound | Cu—S ($\text{Å}$) | Cu—S—Cu (deg) | Reference |
|---|---|---|---|---|
| 1 | $[Cu_6(SC_5H_4N)_6]$ | 2.220(2)–2.257(2) | 86.6[a] | 110 |
| 2 | $[Cu_6(dmpymt)_6]$ | 2.212(3)–2.279(3) | 82.8, 92.4[a] | 111 |
| 3 | $[Cu_6(Me(OH)pymt)_6]^{b}$ | 2.219(3)–2.272(4) | 85.8(1)–99.8(1) | 112 |
| 4 | $[Cu_6(SC_5H_3N(SiMe_3)\text{-}6)_6]$ | 2.236(2)–2.254(2) | 82.6(1)–88.5(1) | 113 |
| 5 | $[Cu_6(Squin)_6]$ | 2.243(3)–2.260(3) | 87.97(9)–88.67(9) | 114 |
| 6 | $[Cu_6(SC(O)NEt_2)_6]$ | 2.239(3) | 78(1) | 115b |
| 7 | $[Cu_6\{(SCS\text{-}n\text{-}Bu)=NPh\}_6]$ | 2.253(2) | 79.39(6) | 75 |

[a]Average from the Cambridge Crystallographic Structural Database.
[b]4-hydroxy-6-methylpyrimidine-2-thionate = Me(OH)pymt.

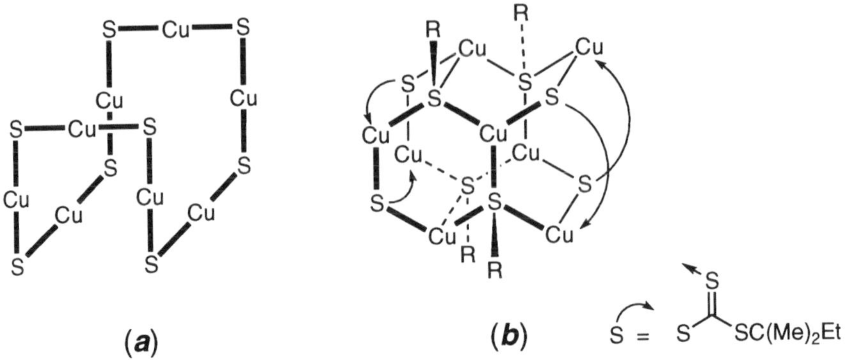

Figure 9.   Octanuclear structural types: (a) open structure of $[Cu_8(SC_6H_2{}^iPr_3\text{-}2,4,6)_8]$ (b) cage structure of $[Cu_8(SR)_4(S_2CSR)_4]$ (R = $C(Me)_2Et$).

coordinating ligands or with only auxiliary coordinating ligands. These octa-nuclear species can have an open $Cu_8S_8$ 16-membered core that is folded to form a U-shaped structure in which the sulfur atoms are approximately at the corners of a cube. Moreover, some complicated cage structures are known. Examples representing complexes containing the former structural type are $[Cu_8(SC_6H_2\text{-}i\text{-}Pr_3\text{-}2,4,6)_8]$ (52) and $[Ag_8(SCMe_2Et)_8]$ [Fig. 9(a)] (82); a complex having a complicated cage structure is $[Cu_8(SR)_4(S_2CSR)_4]$ (R = $CMe_2Et$) [Fig. 9(b)] (117).

A different theme is met in the octanuclear structure of $[Ag_8\text{-}(SCMeEt_2)_8(PPh_3)_2]$ (105), which is built from one tetranuclear $Ag_4S_4$ hetero-cycle, similar to that in $[Cu_4(SC_6H_3(SiMe_3)_2\text{-}2,6)_4]$ (77), that does not contain auxiliary coordinating ligands and one $Ag_4S_4$ heterocycle that is similar to that in $[Cu_4(S\text{-}t\text{-}Bu)_4(PPh_3)_2]$ (101) with two $PPh_3$ ligands coordinated to opposite Ag atoms. Close contacts between these tetranuclear units from silver in one unit to sulfur in the opposite unit and vice versa (range 2.86–2.97 Å) result in the formation of the overall octameric structure (105). Selected geometrical details of some octameric copper(I) arenethiolate aggregates have been collected in Table XI.

## 7.   Higher Aggregation States

When one goes to highly aggregated species one finds that intramolecular coordination plays a less dominant but still important role. For example, in the nonameric aggregate $[Cu_9(S\text{-}1\text{-}C_{10}H_6NMe_2\text{-}8)_9]$ (69), only six of the nine amine donor groups are involved in intramolecular coordination, in the decanuclear aggregate $[Cu_{10}(SCF_3)_{10}(MeCN)_8]$ (58), auxiliary coordination occurs from eight MeCN ligands, and in dodecanuclear aggregates $[Cu_{12}(SC_6H_4SiMe_3\text{-}2)_{12}]$

TABLE XI
Geometrical Details of Octameric Copper(I) Thiolate Complexes

| Entry | Compound | Cu—S (Å) | Cu—S—Cu (deg) | Reference |
|-------|----------|----------|---------------|-----------|
| 1 | $[Cu_8(SC_6H_2\text{-}i\text{-}Pr_3\text{-}2,4,6)_8]$ | $2.162^a$ | 77.2–92.5 | 52 |
| 2 | $[Cu_8(SR)_4(S_2CSR)_4]^b$ | 2.237(5)–2.301(5) | $119.7$–$120.4^c$ | 117 |
| | | | $86.7$–$89.7^c$ | |
| | | | $71.5$–$75.1^c$ | |
| | | $2.263(5)$–$2.292(5)^d$ | $74.2$–$77.8^c$ | |

$^a$Average.
$^b$R = C(Me)$_2$Et; formed by insertion of CS$_2$ into the Cu—S bond of $[Cu(SR)]_n$; the Cu—S—C(S) unit can formally be interpreted as a four-membered chelate ring.
$^c$Obtained from the Cambridge Crystallographic Structural Database.
$^d$The Cu—S (alkyl trithiocarbonate).

(40, 84) and $[Ag_{12}(S\text{-}c\text{-}Hex)_{12}]$ (85), donor coordination is totally absent. The core structures and the subsequent cage expansion of the nonamer, the decamer, and the dodecamer, are depicted in Figure 10.

In these highly aggregated species, one can easily recognize one or more six-membered Cu$_3$S$_3$ subunits, and such rings seem very stable structural units in most of the structurally characterized copper(I) thiolates.

The highest Group 11 (IB) metal thiolate aggregate that does not have a polymeric structure is tetradecanuclear $[Ag_{14}(S\text{-}t\text{-}Bu)_{14}(PPh_3)_4]$ (105). The structure of this aggregate has a central open 28-membered Ag$_{14}$S$_{14}$ heterocycle with only four PPh$_3$ ligands.

The structure of polymeric $[Cu(SMe)]_\infty$ was determined by means of a Rietveld refinement of powder diffraction data and is depicted in Figure 11 (127). The determined structure can be described as two strings of fused Cu$_3$S$_3$ six-membered rings that are interconnected by bonding from a copper atom in one string to a sulfur atom in the opposite string. In this way, a structure is formed that contains Cu$_3$S$_3$ rings, which occur in both the chair and boat conformations. The copper atoms have a close to ideal trigonal planar coordination, whereas the sulfur atoms have a distorted tetrahedral geometry. In this structure of $[Cu(SMe)]_\infty$, there are two acute Cu—S—Cu angles of 83.4(6)° and one wide Cu—S—Cu angle of 128.1(9)°, though a possible explanation for these different angles was not proposed in the original report (127).

## D. Mixed-Organo(arenethiolato)copper and Mixed-Arenethiolatocopper Copper Halide Aggregates

Besides the pure copper thiolate aggregates $[Cu(SR)]_n$, there are several reported examples of mixed-organo(arenethiolato)copper(I) aggregates in which

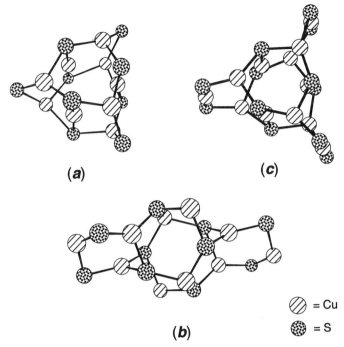

Figure 10.    Cage structures of higher aggregates: (*a*) nonamer, (*b*) decamer, and (*c*) dodecamer.

intramolecularly coordinating arenethiolate ligands are present. Such complexes are a topic of major interest because of their possible occurrence in catalyzed organic reactions. Mixed-organo(arenethiolato) copper(I) aggregates can generally be prepared by reacting a pure copper(I) arenethiolate, $[Cu(SAr)]_n$, with either an organolithium or an organocopper reagent (e.g., $LiC \equiv C\text{-}t\text{-}Bu$, $CuC \equiv C\text{-}t\text{-}Bu$, or CuMes).

It is important to note that all known examples of mixed-organo-(arenethiolato)copper(I) aggregates possess additional coordination from a donor substituent. A typical example is trinuclear $[Cu_3\{SC_6H_3(CH_2NMe_2)\text{-}2\text{-}Cl\text{-}$

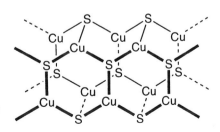

Figure 11.    Part of the structure of polymeric $[Cu(SMe)]_\infty$ (Me groups are omitted for clarity).

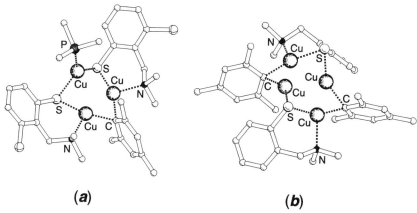

Figure 12.   Mixed organo(arenethiolato)copper(I) aggregates $[Cu_3(SAr)_2(R)(PPh_3)]$ (*a*) and $[Cu_4(SAr)_2(R)_2]$ (*b*) (Ph groups of PPh$_3$ and hydrogen atoms are omitted for clarity).

$5\}_2(Mes)(PPh_3)]$ (20), which is an aggregate of the type $[Cu_3(SAr)_2(R)(L)]$ [Fig. 12(*a*)], comprising two arenethiolate entities: one organic (mesityl) group and one auxiliary coordinating donor ligand L that completes the coordination of the third copper atom. The structure of this aggregate contains a central $Cu_3S_2C$ ring that has a boat conformation; a feature that is seldom encountered in structures of pure copper(I) arenethiolates.

Other examples include aggregates of the type $[Cu_4(SAr)_2(R)_2]$ [R = Mes; SAr = $SC_6H_3(CH_2NMe_2)$-2-Cl-5 (20), S-1-$C_{10}H_6NMe_2$-8 (69)] in which the formal ratio CuSAr/CuR is 1:1 [Fig. 12(*b*)]. A general structural feature in these aggregates is a central $Cu_4S_2C_2$ ring with the sulfur (and carbon) atoms on mutually opposite sides of the ring. Interestingly, these aggregates react with triphenylphosphine to afford the corresponding trinuclear aggregates $[Cu_3(SAr)_2(R)(L)]$ which coexist in solution with free pentameric mesitylcopper(I) (Eq. 9).

(9)

A third type of mixed aggregate is that of the form $[Cu_6(SAr)_4(R)_2]$, which is obtained when the R group is an alkynyl ligand (e.g., $C{\equiv}C\text{-}t\text{-}Bu$) rather than an aryl group. These hexanuclear aggregates can formally be described as dimers of the first type of aggregate $[Cu_3(SAr)_2(R)]$, brought about by alkynyl coordination from one such trinuclear unit to a copper atom in the second unit. An astonishing feature of these hexanuclear aggregates is their insensitivity toward changes in the arenethiolate backbone: the presence of the arenethiolate ligands $SC_6H_4(CH_2NMe_2)\text{-}2$ (122), $SC_6H_4((S)\text{-}CH(Me)NMe_2)\text{-}2$ (122), or the more rigid 1,8-disubstituted naphthalenethiolate ligand $S\text{-}1\text{-}C_{10}H_6NMe_2\text{-}8$ (69), hardly affects the structure of these species.

Furthermore, there are reports of two examples of mixed-arenethiolatocopper copper halide aggregates, namely, $[Cu_8\{SC_6H_3(CH_2NMe_2)_2\text{-}2,6\}_3Br_5]$ (60), in which there are three CuSAr units and five CuBr units, and a species of overall composition $[Cu_3(S\text{-oxazoline})_3]_4[CuI]_4$ (61), which comprises four trimeric copper arenethiolate entities connected to a central $Cu_4I$ unit. It should be noted that in both structures the copper halide units do not occur as separate parts of the structure, but form part of an intimate assembly that produces one well-defined mixed aggregate.

From the structure of these mixed aggregates it is quite clear that, as regards its structural and bonding properties, an arenethiolate ligand is quite similar to a halide ligand. In fact, in the structure of $[Cu_8\{SC_6H_3(CH_2NMe_2)_2\text{-}2,6\}_3Br_5]$ (60) (see Fig. 13) one can distinguish not only different bridging modes of an

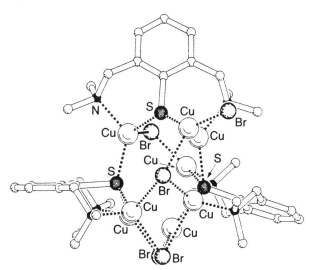

Figure 13.   Mixed-arenethiolatocopper(I) copper halide $[Cu_8(SC_6H_3(CH_2NMe_2)_2\text{-}2,6)_3Br_5]$ (hydrogen atoms are omitted for clarity).

arenethiolate ligand (i.e., $\mu^3$-S and $\mu^4$-S), but also different ways of halide bonding (i.e., $\eta^1$-Br, $\mu^2$-Br, and $\mu^3$-Br).

The latter mixed aggregate is so thermodynamically stable that even in the presence of excess of lithium arenethiolate, which is used during the synthesis, one does not obtain a pure copper(I) arenethiolate $[Cu(SAr)]_n$ (60).

## E.  Mixed-Metal Copper Thiolate Complexes

Interestingly, several examples were reported of mixed-metal copper thiolate complexes in which one or more copper thiolate entities are present within an aggregate that also contains another metal (Fig. 14, Table XII) (128–135).

For example, the structures of the complexes $[(S_2MoS_2)Cu(SPh)]^{2+}$ (128) and $[(PhS)Cu(S_2MoS_2)Cu(SPh)]^{2+}$ (128) can be described as comprising a tetrathiomolybdate coordinating to one or two copper thiolate entities, respectively [Fig. 14(a), entries 1 and 2 in Table XII]. This structural description is furthermore validated by the absence of direct Cu—Mo bonding.

In other cases, the thiolate ligand is an intimate part of the mixed-metal assembly, and examples of these structural frameworks are complexes in which two thiolate ligands bridge the metal atoms, namely, $[Cp_2Ti(SEt)_2Cu(PR_3)]^+$

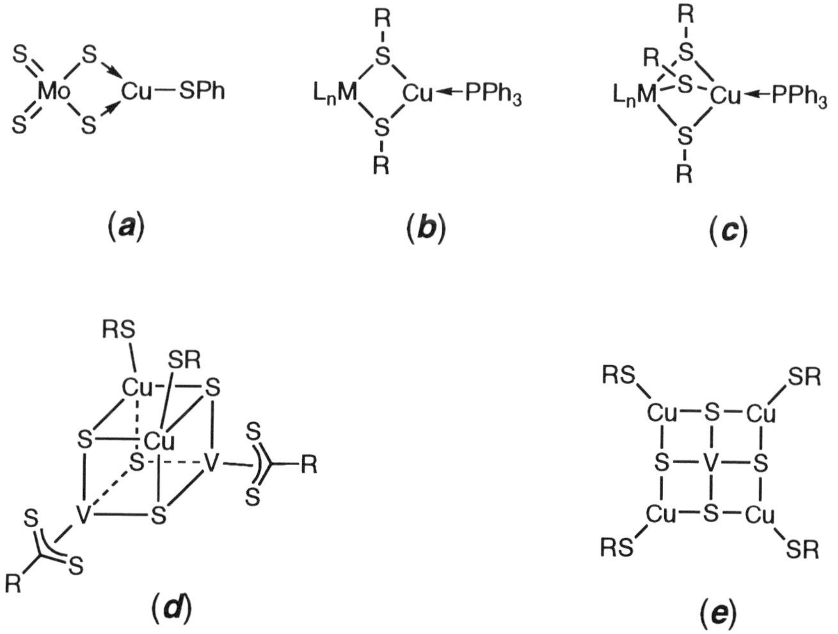

Figure 14.    Mixed-metal copper thiolate complexes.

TABLE XII
Structurally Characterized Mixed-Metal Copper Thiolates

| Entry | Compound | Reference |
|:---:|:---:|:---:|
| 1 | $[(S_2MoS_2)Cu(SPh)]^{2+}$ $(N-n-Pr_4)_2^{+a}$ | 128 |
| 2 | $[(PhS)Cu(S_2MoS_2)Cu(SPh)]^{2+}$ $(N-n-Pr_4)_2^{+b}$ | 128 |
| 3 | $[Cp_2Ti(SEt)_2Cu(PR_3)]^+$ $(PF_6)^{-c}$ | 129 |
| 4 | $[(RS)_2In(SR)_2Cu(PPh_3)_2]^d$ | 130 |
| 5 | $[(Ph_3P)Cu(ArS)_3Mo(SAr)_3Cu(PPh_3)]^e$ | 131 |
| 6 | $[(Ph_3P)Cu(ArS)_3W(SAr)_3Cu(PPh_3)]^e$ | 132 |
| 7 | $[V_2Cu_2S_4(SPh)_2(S_2CR)_2]^{2-}$ $(NEt_4)_2^{+f}$ | 133 |
| 8 | $[V_2Ag_2S_4(SPh)_2(S_2CR)_2]^{2-}$ $(NEt_4)_2^{+f}$ | 133 |
| 9 | $[VCu_4S_4(SPh)_4]^{3-}$ $(NEt_4)_3^+$ | 134 |
| 10 | $[VCu_4S_4(SPh)_3(S_2CR)]^{3-}$ $(NEt_4)_3^{+f}$ | 134 |
| 11 | $[VCu_4S_4(SPh)_2(S_2CR)_2]^{3-}$ $(NEt_4)_3^{+f}$ | 134 |
| 12 | $[Cu_6In_3(SEt)_{16}]^-$ $(PPh_4)^+$ | 135 |

[a] The $S_2MoS_2$ unit acts once as a bidentate ligand.
[b] The $S_2MoS_2$ unit acts twice as a bidentate ligand.
[c] R = Ph or c-Hex.
[d] R = Et or sec-Bu.
[e] The $(SAr)_6$ unit (M = Mo or W) acts as two tridentate ligands, where SAr = $SC_6H_4Me-4$.
[f] R = $NC_4H_8O$.

(R = Ph or c-Hex) (129) [Fig. 14(b)] and related $[(RS)_2In(SR)_2Cu(PPh_3)_2]$ having a tetrahedrally coordinated copper atom (130).

A variation of this structural type is one in which three thiolates bridge the two metals [Fig. 14(c)], examples being $[(Ph_3P)Cu(ArS)_3Mo(SAr)_3Cu(PPh_3)]$ (131) and $[(Ph_3P)Cu(ArS)_3W(SAr)_3Cu(PPh_3)]$ (132), which are of interest as model complexes for nitrogenase proteins. Theoretical calculations on these latter species show that the major bonding component between the $M(SR)_6$ and the $Cu(PPh_3)$ groups is electrostatic in character, though there is evidence for a weak direct M—Cu bonding interaction. The latter interaction has been substantiated by electrochemical investigations (131).

More complicated aggregate structures are found in the vanadium–copper and vanadium–silver species $[V_2M_2S_4(SPh)_2(S_2CR)_2]^{2-}$ (M = Cu or Ag) (133), in which the metal sulfido core is in the shape of a cube [Fig. 14(d)]; bulk magnetic susceptibility measurements have shown that there is probably a direct V—V bond within these species. These and related complexes are of interest since a vanadium-containing nitrogenase has been discovered (136).

As we have already seen, tetrathiometalates can act as bidentate ligands (entries 1 and 2 in Table XII) and tetrathiovanadate has even been found to coordinate to four Cu(SPh) groups thus leading to aggregates such as $[VCu_4S_4(SPh)_4]^{3-}$ [Fig. 14(e)] (134). The thiophenolate groups in this aggre-

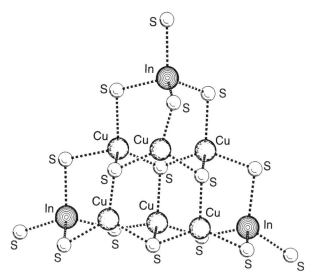

Figure 15. The indium copper thiolate aggregate $[Cu_6In_3(SEt)_{16}]^-$ (Et groups are omitted for clarity).

gate can also be replaced by a dithiocarbamate grouping and a series of complexes $[VCu_4S_4(SPh)_{4-n}(S_2CR)_n]^{3-}$ ($n$ = 0, 1, or 2) has been subject to X-ray crystallographic techniques (134). From X-ray photoelectron spectroscopy it was concluded that the copper atoms in these species are monovalent $Cu^I$ (134).

The nonanuclear complex $[Cu_6In_3(SEt)_{16}]^-$ (Fig. 15) (135) cannot be easily placed into one of the frameworks above, and is unique in that it combines several structural features from pure copper thiolate species with those from indium salts. In the adamantoidal framework structure of $[Cu_6In_3(SEt)_{16}]^-$ (135), it is possible to recognize several structural features that are present in the single adamantoidal aggregates $[Cu_4(SR)_6]^{2-}$ (consisting of four fused six-membered $Cu_3S_3$ rings with alternating copper and sulfur atoms as in pure trimeric copper thiolates; Section II.C.3), where R is either Me (89b, 90), Et (91), or Ph (52, 48b, 89) and in the related tris-dithiolate aggregates (64, 66) shown in Table II. In this respect, $[Cu_6In_3(SEt)_{16}]^-$ (135) can be seen as a polymeric adamantoidal framework structure that is terminated by three $In(SEt)_n$ units.

Complexes of copper thiolates and indium were mainly studied as single source precursors for $CuInS_2$, which can be used in thin-film solar cells, and the thermal decomposition of $[(RS)_2In(SR)_2Cu(PPh_3)_2]$ (130) indeed leads to the selective formation of $CuInS_2$. However, this material has a much lower photovoltaic performance than the corresponding selenide $CuInSe_2$. Based on the structure of nonanuclear $[Cu_6In_3(SEt)_{16}]^-$ (135), which is reminiscent of the chalcopyrite structure of $CuInS_2$ (also constructed of fused adamantane units), a proposal has been made for the lower photovoltaic performance of $CuInS_2$.

In this proposal, the $[Cu_4(SEt)_6]$ clusters in $[Cu_6In_3(SEt)_{16}]^-$ that may also occur in $CuInS_2$, generate "dark-conducting" paths that short the photovoltaic
device. The copper atoms in $CuInSe_2$ have larger $Cu\cdot\cdot\cdot Cu$ distances than in
$CuInS_2$ and, therefore, there is a lower tendency for diffusion of the copper
atoms, thus leading to a better photovoltaic performance.

## F. The Coordination Geometry of Copper Atoms

The crystallographically found geometries of the copper atoms in copper
arenethiolate structures are, at first sight, different to what one might expect for
two- and three-coordinate copper atoms. However, when the copper atoms participate in 2e–3c bonds, the direction of the copper orbital involved with the
bridging ligand is not directly clear from the molecular structure (Fig. 16). The
2e–3c description of the $Cu-X-Cu$ system is an alternative for a bonding
scheme involving a direct $Cu-Cu$ interaction (137) and can provide a good
explanation for the $Cu-S-Cu$ angles often found in the structures of copper(I)
thiolate complexes. In these bridged systems, $Cu-X-Cu$, the bonding orbital
on Cu that binds to the bridging ligand is generally positioned between the
neighboring copper atom and the bridging ligand X rather than along the $Cu-X$
or $Cu-Cu$ vectors.

Knotter et al. (20) developed a trigonometric method to determine the position of this orbital from X-ray crystallographic data, using a model system (Fig.
16) that has a metal atom $Cu_c$ that is surrounded by a neighboring metal atom,
$Cu_n$, a 2e–2c bonded auxiliary or intramolecularly coordinating ligand L, and
a ligand X, which is bridging $Cu_c$ and $Cu_n$ in a 2e–3c bond. Furthermore, one
assumes: (1) that the vector of the bonding orbital of the bridging ligand is

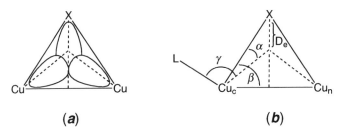

**(a)**                                        **(b)**

Figure 16.   Two-electron three-center bonds. (a) Bonding model in which the solid lines are the
connections between the atoms found crystallographically, while the dotted lines are the direction
of the orbitals, which are indicative for the real geometry around the metal atoms. (b) A model for
the calculation of the geometry around the metal center $Cu_c$. The angle $\gamma$ between the two orbitals
on $Cu_c$, which are responsible for bonding to ligand L and the bridging ligand X, is the summation
of the angles $\alpha$ and X-$Cu_c$-L.

perpendicular to the $Cu_c \cdot \cdot \cdot Cu_n$ vector and is thus in the plane defined by the atoms $Cu_c$, X, and $Cu_n$, and (2) that the metal orbitals are directed toward the point of the highest electron density of the bridging ligand orbital.

The position of this orbital on the copper involved in the bridging multicenter interaction is defined as making an angle $\alpha$ with the $Cu_c—X$ bond; this angle is a fraction of the crystallographically determined $Cu_n—Cu_c—X$ angle $\beta$. If the point of the highest electron density is at a distance $D_e$ away from the bridging atom X, then the angle $\alpha$ is given by Eq. 10.

$$\alpha = \beta - \arctan \left[ \frac{d(Cu_c\text{-}X) \sin \beta - D_e}{d(Cu_c\text{-}X) \cos \beta} \right] \qquad (10)$$

This equation can be solved by estimating $D_e$ as one-half the bond distance either of a $C=C$ bond if an $sp^2$ carbon is bridging ($D_e = 0.669$ Å), or of $S_2$ if an $sp^2$ sulfur atom is bridging ($D_e = 0.994$ Å). The calculated values of $\alpha$ when added to the angle $L—Cu_c—X$ provide an estimate for the interorbital angle $\gamma$ which reflects the coordination geometry of copper.

The viability of this method has been illustrated by the results obtained from mixed-organo(arenethiolato)copper(I) aggregates [$Cu_3\{SC_6H_3(CH_2NMe_2)$-2-Cl-5$\}_2$(Mes)(PPh$_3$)], [$Cu_4\{SC_6H_3(CH_2NMe_2)$-2-Cl-5$\}_2$(Mes)$_2$], [$Cu_6\{SC_6$-$H_4(CH_2NMe_2)$-2$\}_4(C≡C$-$t$-Bu)$_2$] and [$Cu_6\{SC_6H_4((S)$-CH(Me)NMe$_2)$-2)$\}_4$-$(C≡C$-$t$-Bu)$_2$], which show that the copper atoms in these species have an almost perfect trigonal or linear arrangement of their bonding orbitals, that is, $sp^2$ or $sp$ hybridization. In those instances, where large deviations from $120°$ (or $180°$) occur, the explanation can be found in steric strain caused by one or two of the ligands (20).

## III. LITHIUM AND MAGNESIUM THIOLATE COMPLEXES

### A. Lithium and Magnesium Thiolates in Perspective

Reports that thermolysis of metal chalcogenolate complexes offers a low-temperature route to the synthesis of novel solid state materials (32) is also stimulating research in this area. As is the case with copper(I) thiolates, lithium thiolates are often aggregated species (Section III.C) and from the few magnesium thiolate structures known (Section III.D) one sees a preference for monomeric and dimeric formulations when bulky substituents are present.

Whereas studies of the structures of transition metal thiolate complexes are quite numerous, analogous studies with complexes of the more electropositive elements (e.g., alkali, alkaline earths, and the $f$-block elements) are less common. Of these latter elements the metals lithium and magnesium are unique from a structural and bonding perspective, as well as providing the most useful reagents in synthetic chemistry (138, 139).

## B.  Synthesis of Lithium and Magnesium Thiolates

### 1.  Synthesis via Deprotonation of Thiols

The preparation of lithium and magnesium thiolates commonly comprises the reaction of a suitable lithium or magnesium base with the corresponding thiol at low temperature (Eq. 11).

$$RSH + BuLi \longrightarrow LiSR + BuH$$
$$2\,RSH + Bu_2Mg \longrightarrow Mg(SR)_2 + 2\,BuH$$

(11)

The commonly used bases in this reaction are commercially available solutions of $n$-BuLi (in hexane) (140–146), or $n$-Bu$_2$Mg (in heptane) (144); freshly prepared lithium amide LiNH$_2$ has also been used for the synthesis of lithium methylthiolate (147). Similarly, several bulky arenethiols (e.g., HSC$_6$H$_2$Ph$_3$-2,4,6) have been successfully deprotonated with either a 1:1 mixture of $n$-Bu$_2$Mg and $sec$-Bu$_2$Mg in heptane or Mg{N(SiMe$_3$)$_2$}$_2$ in toluene (148). A major shortcoming of this procedure is that thiols that are not commercially available have to be prepared via metal thiolate species—clearly a roundabout approach.

This type of deprotonation procedure can also be applied to prepare lithium or magnesium areneselenolates (143, 148) and sterically hindered silyltellurolate derivatives of the Group I (IA) metals Li, Na, and K (149). An alternative procedure based on the reduction of a diselenide (RSeSeR) with [LiHBEt$_3$] in THF can also afford the desired lithium areneselenolate (32).

### 2.  Synthesis via Sulfur Insertion into Metal–Carbon Bonds

An alternative synthetic procedure to lithium arenethiolates is the insertion reaction of elemental sulfur (S$_8$) into the lithium–carbon bond of a suitable organolithium reagent in THF at low temperature (Eq. 12).

(12)

The insertion of sulfur into a reactive metal–carbon bond has been known in organic chemistry for some time, but was mainly used to prepare arenethiols (150). However, this method has only recently found its use in the synthesis of pure lithium arenethiolate complexes, which contain potentially intramolecularly coordinating heteroatom donor functions (50, 151). It has since then been reported that this method is also applicable for the preparation of lithium areneselenolates and arenetellurolates (152). This route has also been used for the synthesis of bis(arenethiolato)magnesium complexes with intramolecular coordination (153).

## C.   Structures of Lithium Thiolates

### 1.   Monomeric, Dimeric, and Trimeric Complexes

Structurally characterized monomeric species with a general formulation [Li(SAr)L$_3$] (Fig. 17) are included in Table XIII. These species have auxiliary donor ligands that are usually thf (141, 142) or py (140).

However, situations where more weakly coordinating ligands such as Et$_2$O (144) are present or where less than stoichiometric amounts (3 equiv) of thf are used (146), already lead to the formation of dimeric aggregates of the type [Li(SAr)L]$_2$ (Fig. 18). Similarly, polynuclear lithium areneselenolates [Li(SeAr)]$_n$ react with bpy to afford dimeric aggregates [Li(SeAr)(bpy)]$_2$ (32). A trimeric structure, for which details were not reported, has very recently been found for the thiolato mono thf complex [Li(thf)(SC$_6$H$_2$-$t$-Bu$_3$-2,4,6)]$_3$ (154); the structure of the analogous selenolato complex is also trimeric (155).

Interestingly, we have characterized the first representative of dinuclear mixed-arenethiolatolithium     lithium     halide     aggregates,     namely, [Li$_2$\{SC$_6$H$_3$(CH$_2$NMe$_2$)$_2$-2,6\}I(thf)$_2$] (151). In this complex, the two lithium atoms are essentially symmetrically bridged by the sulfur atom of the arenethiolate ligand and the iodide anion. The tetrahedral coordination geometry of the lithium atoms is completed by intramolecular coordination of the dimethylamino groups and auxiliary coordination of two molecules of thf (Fig. 19).

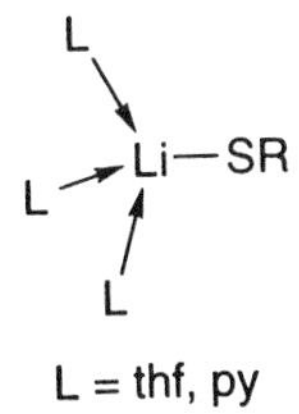

L

L—Li—SR

L

L = thf, py

Figure 17.   Monomeric lithium arenethiolates [Li(SAr)L$_3$].

TABLE XIII
Structurally Characterized Lithium Thiolate, Selenolate, and Tellurolate Species

| Entry | Compound | Reference |
|---|---|---|
| 1 | $[Li(SMe)]_n{}^a$ | 147 |
| | *Complexes with Auxiliary Donor Ligands* | |
| 2 | $[Li(SC_6H_4Me-2)(py)_3]$ | 140 |
| 3 | $[Li(SC_6H_2Ph_3-2,4,6)(thf)_3]$ | 141 |
| 4 | $[Li(SC_6H_2-t-Bu_3-2,4,6)(thf)_3]$ | 142 |
| 5 | $[Li(SeC_6H_2-t-Bu_3-2,4,6)(thf)_3]$ | 143 |
| 6 | $[Li_2(SC_6H_3Mes_2-2,6)_2(Et_2O)_2]$ | 144 |
| 7 | $[Li_2(SePh)_2(bipy)_2]$ | 32 |
| 8 | $[Li_2(SeC_5H_4N)_2(bipy)_2]$ | 32 |
| 9 | $[Li_2(SC(O)Ph)_2(tmeda)_2]$ | 145 |
| 10 | $[Li_2(SC(SiMe_3)_3)_2(thf)_{3.5}]$ | 146 |
| 11 | $[Li_2(SCH(SiMe_3)_2)_2(thf)_4]$ | 146 |
| 12 | $[Li_2(SC(NHPh)C(NR_2)CO_2Et)_2(Et_2O)_2]{}^b$ | 156 |
| 13 | $[Li_3(SeC_6H_2-t-Bu_3-2,4,6)_3(thf)_3]$ | 155 |
| 14 | $[Li(\mu^2-SPh)(py)_2]_\infty$ | 140 |
| 15 | $[Li(\mu^3-SCH_2Ph)(py)]_\infty$ | 140 |
| | *Complexes with Intramolecular Coordination* | |
| 16 | $[Li\{Te(\eta^5-C_5H_3(CH_2NMe_2)-2)Fe(\eta^5-C_5H_5)\}(dme)]$ | 152 |
| 17 | $[Li_2(SC_6H_4(CH_2N(Me)CH_2CH_2OMe)-2)_2]$ | 151 |
| 18 | $[Li_2(SC_6H_3(CH_2NMe_2)_2-2,6)I(thf)_2]$ | 151 |
| 19 | $[Li_6(SC_6H_4((R)-CH(Me)NMe_2)-2)_6]$ | 151 |
| 20 | $[Li_6(SC_6H_3(CH_2NMe_2)_2-2,6)_6]$ | 151 |

$^a$Powder diffraction.
$^b$R$_2$ = Me$_2$SiCH$_2$CH$_2$SiMe$_2$.

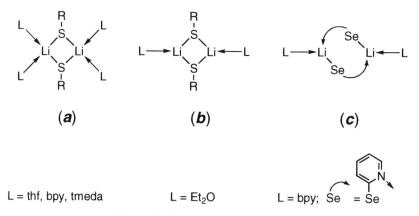

L = thf, bpy, tmeda        L = Et$_2$O        L = bpy;  Se  = Se

Figure 18.   Dimeric lithium arenethiolates.

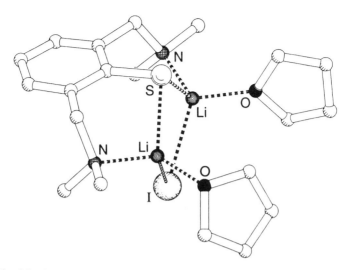

Figure 19.   Mixed-arenethiolatolithium lithium halide species $[Li_2\{SC_6H_3(CH_2NMe_2)_2\text{-}2,6\}I(thf)_2]$ (hydrogen atoms are omitted for clarity).

It should be noted that, as is the case with the mixed-arenethiolatocopper copper halide aggregates, the metal halide unit forms part of an intimate assembly that produces one well-defined mixed aggregate.

## 2.   Hexameric Complexes

There are no reported structures of lithium thiolate complexes with a tetrameric aggregation state. This situation contrasts sharply with the copper(I) thiolates where such aggregation states are well exemplified; pentameric lithium thiolate species are also unknown. However, hexameric structures are possible and can be formed when there is intramolecular coordination by the nitrogen donor functions in the arenethiolate ligand (151). This hexameric formulation has two structural appearances, namely, a flat 12-membered $Li_6S_6$ ring in $[Li\{SC_6H_3(CH_2NMe_2)_2\text{-}2,6\}]_6$, which looks like a regular hexagon due to the T-shaped geometry of sulfur [Figs. 20(a) and 21] and, alternatively, a $Li_6S_6$ cage structure in $[Li\{SC_6H_4((R)\text{-}CH(Me)NMe_2)\text{-}2\}]_6$ in which two $Li_3S_3$ six-membered rings are connected by $\mu^3$-$S$ bonded arenethiolate ligands [Fig. 20(b)]. In both structural types, intramolecular coordination completes the tetrahedral coordination of the lithium atoms.

The latter hexanuclear structural type [Fig. 20(b)] is different from the copper(I) arenethiolate hexamers (Section II.C.5), since in these copper structures only $\mu^2$-$S$ bonded arenethiolate ligands are present, that is, the hexanuclear

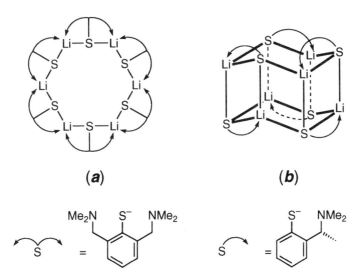

Figure 20.   Hexameric lithium arenethiolates. (*a*) Planar $Li_6S_6$ 12-membered ring structure. (*b*) $Li_6S_6$ cage structure.

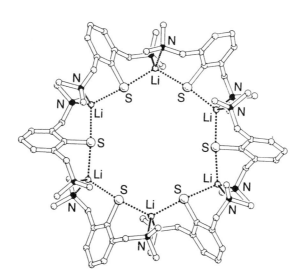

Figure 21.   Molecular structure of planar hexanuclear $[Li_6\{SC_6H_3(CH_2NMe_2)_2\text{-}2,6\}_6]$ (hydrogen atoms are omitted for clarity).

structure is brought about through intramolecular coordination from one $Cu_3S_3$ layer to the other.

The bonding mode of the sulfur atoms in lithium thiolate complexes is often not easily identifiable from the coordination geometry of these atoms (156). Ab initio calculations on the bonding in complexes [MSH] (M = alkali and alkaline earth metal) indicate that this bonding is highly ionic (157).

### 3. Polymeric Complexes

As well as well-defined aggregates, there is ample evidence for polymeric lithium thiolate structures, and species that have been studied by X-ray diffraction techniques include $[Li(SMe)]_n$, $[Li(SPh)(py)_2]_\infty$, and $[Li(SCH_2Ph)(py)]_\infty$ (see Table XIII). The latter two polynuclear structures are similar, but in the former structure, with two py ligands, the thiolate sulfur atoms bridge the lithium atoms in a $\mu^2$-fashion, whereas in the latter structure, with only one py, the sulfur atoms are involved in $\mu^3$-S bridges (Fig. 22). It is very striking that the structures resulting from coordination of two py ligands (i.e., polymeric ones) are very different from those obtained with two molecules of diethyl ether (i.e., dimeric ones). This result cannot be explained only by the Lewis-base strength of these ligands.

### 4. Mixed-Metal Lithium Thiolate Complexes

Interestingly, four examples exist of mixed-metal thiolate complexes in which one or more lithium thiolate moieties are present within an aggregate that also contains a lanthanide or actinide metal thiolate unit (Table XIV). These species

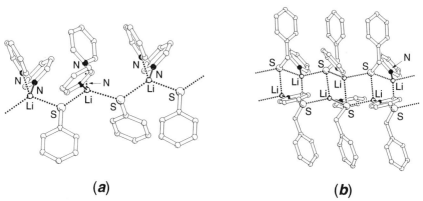

(a)     (b)

Figure 22.    Molecular structures of $[Li(\mu^2\text{-SPh})(py)_2]_\infty$ (a) and $[Li(\mu^3\text{-SCH}_2Ph)(py)]_\infty$ (b) (hydrogen atoms are omitted for clarity).

TABLE XIV
Structurally Characterized Mixed-Metal Lithium Thiolates

| Entry | Compound | Reference |
|-------|----------|-----------|
| 1 | [Li(S-$t$-Bu)$_2$Lu($\eta^5$-C$_5$Me$_5$)$_2$(thf)$_2$] | 158 |
| 2 | [Li$_3$(S-$t$-Bu)$_6$Yb(tmeda)$_3$] | 159 |
| 3 | [Li$_3$(S-$t$-Bu)$_6$Sm(tmeda)$_3$] | 159 |
| 4 | [Li$_4$(SCH$_2$CH$_2$S)$_4$U(dme)$_4$] | 160 |

can be seen as lutethium, ytterbium, samarium, and uranium -ate complexes that are stabilized by lithium cations.

One should note that in these neutral monomeric complexes [Li(S-$t$-Bu)$_2$Lu($\eta^5$-C$_5$Me$_5$)$_2$(thf)$_2$] (158) (Fig. 23), [Li$_3$(S-$t$-Bu)$_6$M(tmeda)$_3$] (M = Yb or Sm) (159), and [Li$_4$(SCH$_2$CH$_2$S)$_4$U(dme)$_4$] (160) the Li atoms are still bonded to the thiolate ligands with corresponding Li−S distances of 2.44, 2.38, and 2.42 Å (average), respectively. In this area of $f$-block chemistry, the analogy between thiolates and halides does not hold well. For example, the incorporation of metal salts (e.g., NaCl) in lanthanum phenoxide complexes was reported and has a large influence on the overall structure of these species (161). In contrast, lithium arenethiolates [Li(SAr)], containing intramolecularly coordinating nitrogen-donor atoms, have been reported to react with YbCl$_3$ to form complexes [Yb(SAr)$_2$Cl] and [Yb(SAr)$_3$], where incorporation of lithium salts does not occur (162).

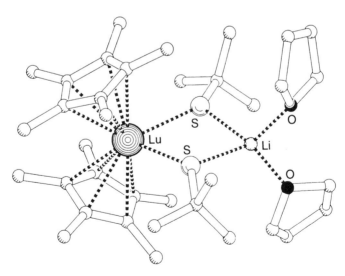

Figure 23.   Molecular structure of [Li(S-$t$-Bu)$_2$Lu($\eta^5$-C$_5$Me$_5$)$_2$(thf)$_2$] (hydrogen atoms are omitted for clarity).

## D.  Structures of Magnesium Thiolates

Although magnesium arenethiolates, $Mg(SAr)_2$, have long been known to be intermediates in the preparation of arenethiols by insertion of $S_8$ into the $Mg-C$ bond of diorganomagnesium reagents (150), it is only very recently that a few structures of these and related materials have appeared in the literature (Fig. 24). The first of these reports in 1992 was of the unprecedented aggregate $[Mg_2\{SC_6H_4((R)\text{-}CH(Me)NMe_2)\text{-}2\}_4Cu_4(Mes)_4]$ formed by reaction of the copper(I) arenethiolate and mesitylcopper [see Fig. 24(d)] (20). Table XV lists the few structurally characterized magnesium thiolate complexes that are known.

In the presence of Lewis bases (e.g., thf or $Et_2O$), magnesium arenethiolate complexes are encountered as monomeric species $[Mg(SR)_2L_2]$ [Fig. 24(a)] in which the coordination geometry of magnesium is tetrahedral. However, there is only one example of a magnesium arenethiolate in which there are no additional neutral donor ligands, and this complex, $[Mg_2(SC_6H_2Ph_3\text{-}2,4,6)_4]$, has a dimeric structure [Fig. 24(b)], which was described as having three-coordinate

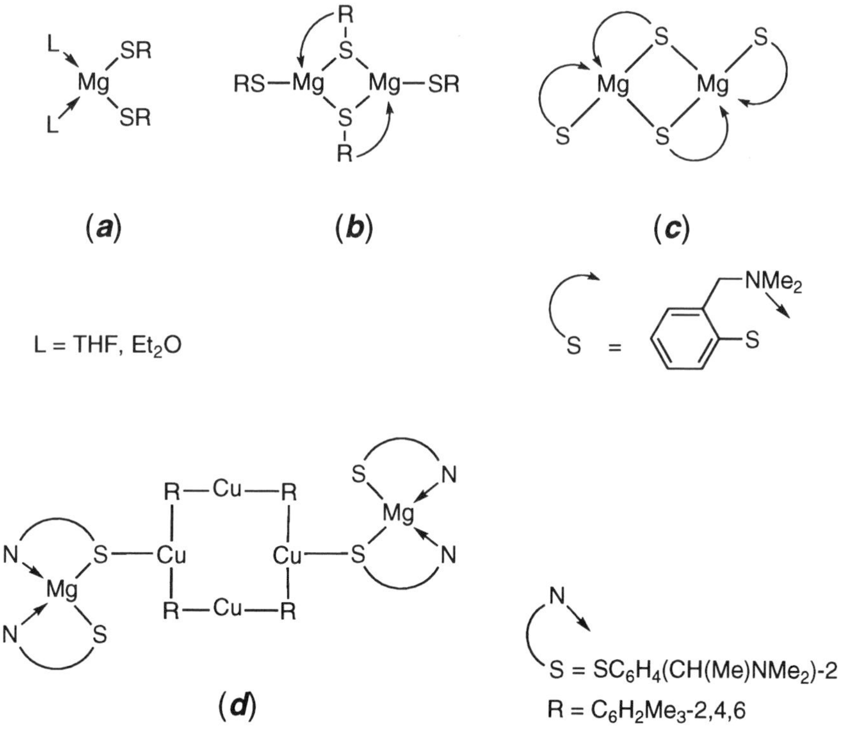

L = THF, $Et_2O$

$S = SC_6H_4(CH(Me)NMe_2)\text{-}2$
$R = C_6H_2Me_3\text{-}2,4,6$

Figure 24.   General structures of magnesium arenethiolates.

TABLE XV
Structurally Characterized Magnesium Thiolates and Selenolates

| Entry | Compound | Reference |
|-------|----------|-----------|
| 1 | [Mg(SC$_6$H$_2$-$t$-Bu$_3$-2,4,6)$_2$(Et$_2$O)$_2$] | 147 |
| 2 | [Mg(SeC$_6$H$_2$-$t$-Bu$_3$-2,4,6)$_2$(thf)$_2$] | 147 |
| 3 | [Mg$_2$(SC$_6$H$_2$Ph$_3$-2,4,6)$_4$] | 147 |
| 4 | [Mg$_2$(SC$_6$H$_4$(CH$_2$NMe$_2$)-2)$_4$] | 153 |
| 5 | [Mg$_2${SC$_6$H$_4$(($R$)-CH(Me)NMe$_2$-2)}$_4$Cu$_4$(Mes)$_4$] | 20 |

magnesium centers (148). However, since there is an additional intramolecular interaction between magnesium and one of the ortho phenyl substituents, we feel that this species is better described as having a distorted pseudotetrahedral (four-coordinate) geometry.

Interestingly, the magnesium bis(arenethiolate) complex based on the [2-(dimethylamino)methyl]phenylthiolate ligand that contains N-donor ortho-substituents, is not a monomer, but is instead a dimer [Mg$_2$(SC$_6$H$_4$(CH$_2$NMe$_2$)-2)$_4$] [Figs. 24(c) and 25] in which the magnesium atoms are five coordinate in the solid state (153). However, from variable temperature $^1$H NMR experiments as well as from cryoscopic molecular weight determinations it was concluded that the latter species is a monomer in solution, that is, the magnesium atoms are now four coordinate (153).

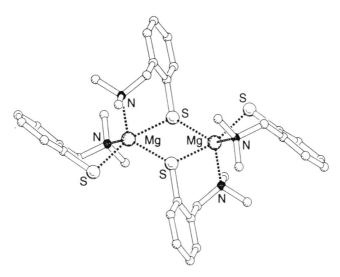

Figure 25.   Molecular structure of [Mg$_2$(SC$_6$H$_4$(CH$_2$NMe$_2$)-2)$_4$] (hydrogen atoms are omitted for clarity).

## IV.  CONCLUSIONS

This chapter has clearly shown that the structures of copper(I), lithium, and magnesium thiolate complexes and, more especially, the aggregation state of these species are highly dependent on the nature of the thiolate ligand used. In particular, the use of thiolate ligands with potential for intramolecular coordination seems to be especially advantageous since this leads to a corresponding complex with a well-defined solid state and solution structure. Furthermore, the organic backbone, connecting the sulfur atom to the coordinating donor atom, plays an important role in the aggregate (self-)assembling process. This process is exemplified by the structures of trimeric $[Cu\{SC_6H_4((R)\text{-}CH(Me)NMe_2)\text{-}2\}]_3$ and nonameric $[Cu(S\text{-}1\text{-}C_{10}H_6NMe_2\text{-}8)]_9$, where the more rigid naphthalene backbone leads to the formation of a higher aggregate. In the latter aggregate, the potential for coordination of the dimethylamino groups seems less important as only six of the nine available nitrogen atoms coordinate to copper. Higher aggregates are thus less influenced by intramolecularly or auxiliary coordinating ligands: in the largest reported aggregate, that is, $[Cu(SC_6H_4SiMe_3\text{-}2)]_{12}$, coordinating ligands are even completely absent.

As far as the mixed-organo(arenethiolato)copper(I) aggregates are concerned, it seems that there are two basic building blocks, namely, a trinuclear $[Cu_3(SAr)_2(R)]$ and a tetranuclear $[Cu_4(SAr)_2(R)_2]$ entity, which are hardly influenced by the nature of the thiolate ligand. These species are more dependent on the nature of the organic grouping R (alkyl vs. aryl) and are thus best classified as organocopper copper (pseudo)halide aggregates (26b).

In the area of mixed-arenethiolatocopper copper halide aggregates, there is still much exploration to be done since so far only two species are known. However, these species have structures that combine several features of both copper arenethiolates and copper halides and may therefore become interesting model compounds in the near future.

The most evident characteristic of thiolate ligands, not only in the structure of the copper(I) thiolates, but also in those of lithium and magnesium thiolates, is that they are, through the variable hybridization of sulfur, ideally suited for bridging both equivalent and nonequivalent metal centers. This ability of sulfur to adapt to the further ligand environment of a metal center means that resulting aggregate structures and the (self-)assembling process are influenced primarily by steric factors (e.g., ligand rigidity) and the presence of other potentially coordinating groupings. The latter can be either thiolate substituents, where chelate ring size also plays a role, or auxiliary donor ligands.

These characteristics of thiolate ligands means they are ideally suited for applications where two or more metal centers are to be brought together in a well-defined way. Such is the case in a number of metal complex catalyzed organic reactions where organometallic reagents are employed for $C-C$ coupling reactions (2, 5–25). The successful use of arenethiolates for copper(I) catalyzed reactions, mentioned in the introduction, provides a specific example that illustrates the validity of this conclusion. Arenethiolates would seem to be particularly useful in that the aryl nucleus can be easily substituted to provide not only specific donor-atom containing groups with correctly chosen properties, but also chiral information. An example of this is $SC_6H_4((R)\text{-}CH(Me)NMe_2)\text{-}2$, which has been used, among others, as a nontransferable group in enantioselective conjugate 1,4-addition reactions of Grignard reagents to benzylidene acetone and in regio- and enantioselective allylic substitution reactions (2, 20–25).

Based on the structural characteristics of thiolate ligands, as summarized in this chapter, we were able to make a reasonable postulate for the structure of the key intermediate in the latter catalytic system as one where the thiolate ligand bridges a heterodinuclear (Cu or Mg) unit (163). Molecular modeling shows that this postulated system is realistic and that the spatial orientation of the $CH(Me)NMe_2$ ortho substituent is important in determining the aggregate stereochemistry that is important for relaying the information of the chiral center, so that *re-* and *si-*coordination of the enone substrate is well discriminated (163). Support for this type of structural postulate (Fig. 26) was also obtained from EXAFS measurements on cuprates derived from a copper(I) arenethiolate [Cu(SAr)] and an organometallic reagent (LiR or RMgI) (164). These measurements indicate that in the key intermediate the nitrogen coordination is to magnesium rather than to copper and that a magnesium atom is in close proximity to copper (164).

It is apparent that thiolate chemistry of Cu, Li, and Mg is a field of significant current interest and that structural studies in this area are very relevant to the development of new and better catalytic systems. We hope that this chapter, in bringing the most important structural elements to light, will be of use in stimulating further research in this area.

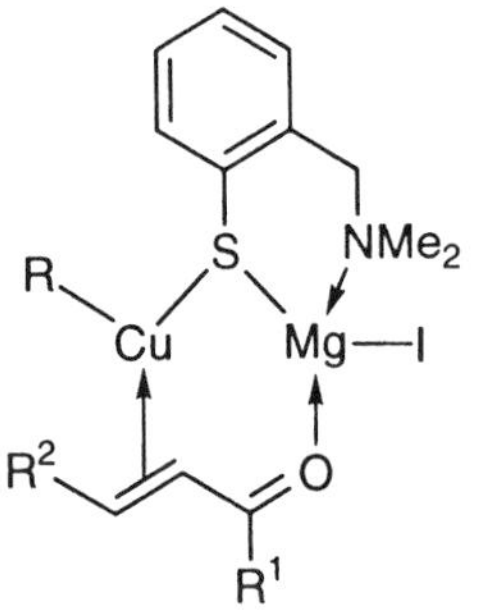

Figure 26.   Proposed key-intermediate in copper(I) arenethiolate catalyzed reactions.

## NOTE ADDED IN PROOF

After preparation of this manuscript, the following structure was brought to our attention: The complex $[Cu_4(SCH_2CH_2N(H)CH_2\text{-}2\text{-}C_5H_4N)_4]$ contains a flattened $Cu_4$ tetrahedron with bridging sulfur atoms and with the secondary amine nitrogen atoms coordinating to copper (Fig. 7). Interestingly, the 2-pyridyl groups participate in a hydrogen-bonding network with the metal-bound secondary amine groups (165).

## ABBREVIATIONS

| | |
|---|---|
| bpy | 2,2′-Bipyridine |
| DED | 1,1-Dicarboethoxy-2,2-ethylenedithiolate |
| dme | 1,2-Dimethoxyethane |
| dmit | 4,5-Dimercapto-1,3-dithiole-2-thionato-2 |
| dmpymt | 4,6-Dimethylpyrimidine-2-thionate |
| dppm | Bis(diphenylphosphino)methane, $Ph_2PCH_2PPh_2$ |
| DTS | Dithiosquarate |
| EXAFS | Extended X-ray absorption fine structure |
| $^1$H NMR | Proton nuclear magnetic resonance |
| Me(OH)pymt | 4-Hydroxy-6-methylpyrimidine-2-thionate |
| Me$_2$phen | 2,9-Dimethyl-1,10-phenantroline |
| Mes | Mesityl = 2,4,6-trimethylphenyl = $C_6H_2Me_3$-2,4,6 |
| mimt | 1-Methylimidazoline-2-thionate |
| phen | 1,10-Phenanthroline |
| py | Pyridine |
| Simid | 1-Methylimidazoline-2-thionate |
| Squin | Quinoline-2-thionate |
| thf | Tetrahydrofuran (ligand) |
| THF | Tetrahydrofuran (solvent) |
| tht | Tetrahydrothiophene |
| tmeda | N,N,N′,N′-Tetramethylethylenediamine |

## REFERENCES

1. (a) I. G. Dance, *Polyhedron, 5,* 1037 (1986). (b) P. J. Blower and J. R. Dilworth, *Coord. Chem. Rev., 76,* 121 (1987). (c) E. Block and J. Zubieta, *Adv. Sulfur Chem., 1,* 133 (1994). (d) D. W. Stephan and T. T. Nadasdi, *Coord. Chem. Rev., 147,* 147 (1996). (e) E. S. Raper, *Coord. Chem. Rev., 61,* 115 (1985).

2. (a) D. M. Knotter, A. L. Spek, and G. van Koten, *J. Chem. Soc., Chem. Commun.*, 1738 (1989). (b) F. Lambert, D. M. Knotter, M. D. Janssen, M. van Klaveren, J. Boersma, and G. van Koten, *Tetrahedron: Asymm.*, *2*, 1097 (1991).

3. H. Gilman, and J. M. Straley, *Recl. Trav. Chim. Pays-Bas*, *55*, 821 (1936).

4. M. S. Kharasch and P. O. Twaney, *J. Am. Chem. Soc.*, *63*, 2308 (1941).

5. (a) G. H. Posner, *Org. React.*, *19*, 1 (1972). (b) G. H. Posner, *An Introduction to Synthesis Using Organocopper Reagents*, Wiley, New York, 1980. (c) Y. Yamamoto, *Angew. Chem.*, *98*, 945 (1986). (d) B. H. Lipschutz, *Synthesis*, 325 (1987). (e) J. A. Kozlowski, in *Comprehensive Organic Synthesis*, B. M. Trost and I. Fleming, Eds., Pergamon, Oxford, UK, 1991, Vol. 4, p. 169.

6. A. Alexakis, S. Mutti, and J. F. Normant, *J. Am. Chem. Soc.*, *113*, 6332 (1991).

7. N. Krause and S. Arndt, *Chem. Ber.*, *126*, 261 (1993).

8. (a) C. C. Tseng, S. D. Paisley, and H. L. Goering, *J. Org. Chem.*, *51*, 2884 (1986). (b) H. L. Goering and T. L. Underiner, *J. Org. Chem.*, *56*, 2563 (1991). (c) J.-E. Bäckvall, M. Séllen, and B. Grant, *J. Am. Chem. Soc.*, *112*, 6615 (1990).

9. (a) M. Tamura and J. K. Kochi, *J. Organomet. Chem.*, *42*, 205 (1972). (b) C. R. Johnson and G. A. Dutra, *J. Am. Chem. Soc.*, *95*, 7777 (1973).

10. H. O. House, W. L. Respess, and G. M. Whitesides, *J. Org. Chem.*, *31*, 3138 (1966).

11. (a) W. H. Mandeville and G. M. Whitesides, *J. Org. Chem.*, *39*, 400 (1974). (b) E. J. Corey and D. J. Beames, *J. Am. Chem. Soc.*, *94*, 7210 (1972).

12. (a) N. T. Luong-Thi and H. Riviere, *Tetrahedron Lett.*, *11*, 1583 (1970). (b) N. T. Luong-Thi, and H. Riviere, *Tetrahedron Lett.*, *12*, 587 (1971).

13. (a) B. H. Lipschutz, *Synlett*, 119 (1990). (b) B. H. Lipschutz, R. S. Wilhelm, and D. M. Floyd, *J. Am. Chem. Soc.*, *103*, 7672 (1981). (c) S. H. Bertz and G. Dabbagh, *J. Chem. Soc. Chem. Commun.*, 1030 (1982).

14. I. Fleming and T. W. Newton, *J. Chem. Soc., Perkin Trans. 1*, 1805 (1984).

15. E. C. Ashby and A. B. Goel, *Inorg. Chem.*, *16*, 3043 (1977).

16. G. H. Posner, C. E. Whitten, and J. J. Sterling, *J. Am. Chem. Soc.*, *95*, 7788 (1973).

17. (a) Y. Yamamoto, N. Asao, and T. Uyehara, *J. Am. Chem. Soc.*, *114*, 5427 (1992). (b) S. H. Bertz, G. Dabbagh, and G. Sundararajan, *J. Org. Chem.*, *51*, 4953 (1986). (c) B. E. Rossiter and M. Eguchi, *Tetrahedron Lett.*, *31*, 965 (1990). (d) R. K. Dieter and M. Tokles. *J. Am. Chem. Soc.*, *109*, 2040 (1987).

18. (a) S. F. Martin, J. R. Fishpaugh, J. M. Power, D. M. Giolando, R. A. Jones, C. M. Nunn, and A. H. Cowley, *J. Am. Chem. Soc.*, *110*, 7226 (1988). (b) S. H. Bertz, G. Dabbagh, and G. M. Villacorta, *J. Am. Chem. Soc.*, *104*, 5824 (1982). (c) S. H. Bertz and G. Dabbagh, *J. Org. Chem.*, *49*, 1119 (1984).

19. E. J. Corey, R. Naef, and F. J. Hannon, *J. Am. Chem. Soc.*, *108*, 7114 (1986).

20. D. M. Knotter, D. M. Grove, W. J. J. Smeets, A. L. Spek, and G. van Koten, *J. Am. Chem. Soc.*, *114*, 3400 (1992).

21. (a) M. van Klaveren, F. Lambert, D. J. F. M. Eijkelkamp, D. M. Grove, and G. van Koten, *Tetrahedron Lett.*, *35*, 6135 (1994).

22. Q. Zhou and A. Pfaltz, *Tetrahedron Lett.*, *34*, 7725 (1993).

23. A. Haubrich, M. van Klaveren, G. van Koten, G. Handke, and N. Krause, *J. Org. Chem.*, *58*, 5849 (1993).

24. (a) M. van Klaveren, E. S. M. Persson, D. M. Grove, J.-E. Bäckvall, and G. van Koten, *Tetrahedron Lett.*, *35*, 5931 (1994). (b) M. van Klaveren, E. S. M. Persson, A. del Villar, D. M. Grove, J.-E. Bäckvall, and G. van Koten, *Tetrahedron Lett.*, *36*, 3059 (1995). (c) M. van Klaveren, E. S. M. Persson, D. M. Grove, J.-E. Bäckvall, and G. van Koten, *Chem. Eur. J.*, *1*, 351 (1995).

25. M. van Klaveren, M. C. Goossens, D. M. Grove, and G. van Koten, to be published.

26. (a) G. van Koten and J. G. Noltes, in *Comprehensive Organometallic Chemistry*, G. Wilkinson, F. G. A. Stone, and E. W. Abel, Eds., Pergamon, Oxford, U.K, 1984, Vol. 1, Chapter 14, pp. 709–763. (b) G. van Koten, S. L. James, and J. T. B. H. Jastrzebski, in *Comprehensive Organometallic Chemistry II*, E. W. Abel, F. G. A. Stone, and G. Wilkinson, Eds., Pergamon, Oxford, U.K., 1995, Vol. 3, Chapter 2, pp. 57–133. (c) P. P. Power, *Progress in Inorganic Chemistry*, Wiley-Interscience, New York, 1991, Vol. 39, p. 75.

27. I. G. Dance, P. J. Guerney, A. D. Rae, and M. L. Scudder, *Inorg. Chem.*, *22*, 2883 (1983).

28. I. G. Dance, L. J. Fitzpatrick, and M. L. Scudder, *J. Chem. Soc. Chem. Commun.*, 546 (1983).

29. I. G. Dance, M. L. Scudder, and L. J. Fitzpatrick, *Inorg. Chem.*, *24*, 2547 (1985).

30. W. Kaim and B. Schwederski, *Bioanorganische Chemie*, Teubner, Stuttgart, Germany, 1991.

31. G. N. Schrauzer and H. Prakash, *Inorg. Chem.*, *14*, 1200 (1975).

32. D. V. Khasnis, M. Buretea, T. J. Emge, and J. G. Brennan, *J. Chem. Soc. Dalton Trans.*, 45 (1995), and references cited therein.

33. K. Tang, M. Aslam, E. Block, T. Nicholson, and J. Zubieta, *Inorg. Chem.*, *26*, 1488 (1987).

34. P. González-Duarte and J. Vives, *J. Chem. Soc. Dalton Trans.*, 2477 (1990).

35. P. González-Duarte and J. Vives, *J. Chem. Soc. Dalton Trans.*, 2483 (1990).

36. (a) A. A. Aslab and P. J. Sadler, *J. Chem. Soc. Dalton Trans.*, 1657 (1981). (b) A. A. Aslab and P. J. Sadler, *J. Chem. Soc. Dalton Trans.*, 135 (1982). (c) C. F. Shaw, *J. Inorg. Biochem.*, *14*, 267 (1981).

37. S. J. Lippard, Ed., *Platinum, Gold and Other Metal Chemotherapeutic Agents: Chemistry and Biochemistry*, ACS Symposium Series, No. 209, American Chemical Society, Washington, DC, 1983.

38. G. A. Bowmaker, G. R. Clark, and D. K. P. Yuen, *J. Chem. Soc. Dalton Trans.*, 2329 (1976).

39. D. Swenson, N. C. Baenziger, and D. Coucouvanis, *J. Am. Chem. Soc.*, *100*, 1932 (1978).

40. E. Block, M. Gernon, H. Kang, G. Ofori-Okai, and J. Zubieta, *Inorg. Chem.*, *28*, 1263 (1989).

41. D. M. Knotter, G. van Koten, H. L. van Maanen, D. M. Grove, and A. L. Spek, *Angew. Chem., Int. Ed. Engl., 101*, 341 (1989).

42. G. van Koten, *Pure Appl. Chem., 61*, 1681 (1989).

43. (a) G. van Koten, A. J. Leusink, and J. G. Noltes, *J. Organomet. Chem., 84*, 117 (1975). (b) G. van Koten and J. G. Noltes, *J. Organomet. Chem., 84*, 419 (1975). (c) G. van Koten, *J. Organomet. Chem., 400*, 283 (1990). (d) M. D. Janssen, M. A. Corsten, A. L. Spek, D. M. Grove, and G. van Koten, *Organometallics, 15*, 2810 (1996).

44. G. A. Bowmaker and B. C. Dobson, *J. Chem. Soc. Dalton Trans.*, 267 (1981).

45. D. Coucouvanis, C. N. Murphy, and S. K. Kanodia, *Inorg. Chem., 19*, 2993 (1980).

46. P. G. Eller and G. J. Kubas, *J. Am. Chem. Soc., 99*, 4346 (1977).

47. I. G. Dance, *Aust. J. Chem., 31*, 2195 (1978).

48. (a) I. G. Dance, *J. Chem. Soc. Chem. Commun.*, 103 (1976). (b) M. Baumgartner, W. Bensch, P. Hug, and E. Dubler, *Inorg. Chim. Acta, 136*, 139 (1987).

49. S. C. Kokkou, S. Fortier, P. J. Rentzeperis, and P. Karagiannidis, *Acta Crystallogr. Sect. C, 39*, 178 (1983).

50. D. M. Knotter, H. L. van Maanen, D. M. Grove, A. L. Spek, and G. van Koten, *Inorg. Chem., 30*, 3309 (1991).

51. H. J. Schugar, C.-C. Ou, J. A. Thich, J. A. Potenza, T. R. Felthouse, M. S. Haddad, D. N. Hendrickson, W. Furey, and R. A. Lalancette, *Inorg. Chem., 19*, 543 (1980).

52. (a) Q. Yang, K. Tang, H. Liao, Y. Han, Z. Chen, and Y. Tang, *J. Chem. Soc. Chem. Commun.*, 1076 (1987). (b) I. Schröter-Schmid and J. Strähle, *Z. Naturforsch. B, 45*, 1537 (1990).

53. T. Tsuda, T. Yazawa, K. Watanabe, T. Fujii, and T. Saegusa, *J. Org. Chem., 46*, 192 (1981).

54. (a) A. Togni, G. Rihs, and R. E. Blumer, *Organometallics, 11*, 613 (1992). (b) M. D. Janssen, M. Herres, L. Zsolnai, A. L. Spek, D. M. Grove, H. Lang, and G. van Koten, *Inorg. Chem., 35*, 2476 (1996).

55. R. K. Chadha, R. Kumar, and D. G. Tuck, *Can. J. Chem., 65*, 1336 (1987).

56. M. A. Khan, R. Kumar, and D. G. Tuck, *Polyhedron, 7*, 49 (1988).

57. J. Kampf, R. Kumar, and J. P. Oliver, *Inorg. Chem., 31*, 3626 (1992).

58. A. L. Rheingold, S. Munavalli, D. I. Rossman, and C. P. Ferguson, *Inorg. Chem., 33*, 1723 (1994).

59. J. S. Thompson, T. J. Marks, and J. A. Ibers, *J. Am. Chem. Soc., 101*, 4180 (1979).

60. M. D. Janssen, A. L. Spek, D. M. Grove, and G. van Koten, *Inorg. Chem., 35*, 4078 (1996).

61. A. Pfaltz, private communication.

62. C. D. Garner, J. R. Nicholson, and W. Clegg, *Inorg. Chem., 23*, 2148 (1984).

63. O. P. Anderson, K. K. Brito, and S. K. Laird, *Acta Crystallogr. Sect. C, 46*, 1600 (1990).

64. G. Matsubayashi and A. Yokozawa, *J. Chem. Soc. Chem. Commun.*, 68 (1991).

65. S. A. Koch, R. Fikar, M. Millar, and T. O'Sullivan, *Inorg. Chem.*, *23*, 121 (1984).

66. J. R. Nicholson, I. L. Abrahams, W. Clegg, and C. D. Garner, *Inorg. Chem.*, *24*, 1092 (1985).

67. Ch. Pulla Rao, J. R. Dorfman, and R. H. Holm, *Inorg. Chem.*, *25*, 428 (1986).

68. D. M. Knotter, M. D. Janssen, D. M. Grove, W. J. J. Smeets, E. Horn, A. L. Spek, and G. van Koten, *Inorg. Chem.*, *30*, 4361 (1991).

69. M. D. Janssen, J. G. Donkervoort, S. B. van Berlekom, A. L. Spek, D. M. Grove, and G. van Koten, *Inorg. Chem.*, *35*, 4752 (1996).

70. (a) E. Rijnberg, J. T. B. H. Jastrzebski, M. D. Janssen, J. Boersma, and G. van Koten, *Tetrahedron Lett.*, *35*, 6521 (1994). (b) E. Rijnberg, A. W. Kleij, J. T. B. H. Jastrzebski, M. D. Janssen, J. Boersma, A. L. Spek, and G. van Koten, *Organometallics*, 1997, in press.

71. M. D. Janssen, D. M. Grove, G. van Koten, and A. L. Spek, *Recl. Trav. Chim. Pays-Bas*, *115*, 286 (1996).

72. M. D. Janssen, unpublished results.

73. (a) D. Fenske, H. Krautscheid, and S. Balter, *Angew. Chem., Int. Ed. Engl.*, *29*, 796 (1990). (b) D. Fenske and J.-C. Steck, *Angew. Chem., Int. Ed. Engl.*, *32*, 238 (1993). (c) H. Krautscheid, D. Fenske, G. Baum, and M. Semmelman, *Angew. Chem., Int. Ed. Engl.*, *32*, 1303 (1993). (d) D. Fenske, J. Ohmer, J. Hachgenei, and K. Merzweiler, *Angew. Chem.*, *100*, 1300 (1988).

74. J. Sola, Y. Do, J. M. Berg, and R. H. Holm, *J. Am. Chem. Soc.*, *105*, 7784 (1983).

75. J. Willemse, W. P. Bosman, J. H. Noordik, and J. A. Cras, *Recl. Trav. Chim. Pays-Bas*, *102*, 477 (1983).

76. K. Tang, Q. Yang, J. Yang, and Y. Tang, *Beijing Dax. Xue., Zir. Kex.*, *24*, 398 (1988); *Chem. Abs.*, *110*, 184758x (1989).

77. E. Block, H. Kang, G. Ofori-Okai, and J. Zubieta, *Inorg. Chim. Acta*, *167*, 147 (1990).

78. B. Becker, W. Wojnowski, K. Peters, E.-M. Peters, and H. G. von Schnering, *Polyhedron*, *9*, 1659 (1990).

79. P. J. Bonasia, D. E. Gindelberger, and J. Arnold, *Inorg. Chem.*, *32*, 5126 (1993).

80. D. Ohlmann, H. Pritzkow, H. Grützmacher, M. Anthamatten, and R. Glaser, *J. Chem. Soc. Chem. Commun.*, 1011 (1995).

81. (a) I. Schröter and J. Strähle, *Chem. Ber.*, *124*, 2161 (1991). (b) P. A. Bates and J. M. Waters, *Acta Crystallogr. Sect. C*, *41*, 862 (1985).

82. (a) S. Åkerstrom, *Acta Chem. Scand. Ser. A*, *18*, 1308 (1964). (b) S. Åkerstrom, *Arkiv. Kemi.*, *24*, 505 (1965).

83. I. G. Dance, L. J. Fitzpatrick, A. D. Rae, and M. L. Scudder, *Inorg. Chem.*, *22*, 3785 (1983).

84. E. Block, M. Gernon, H. Kang, S. Liu, and J. Zubieta, *J. Chem. Soc. Chem. Commun.*, 1031 (1988).

85. (a) S.-H. Hong, A. Olin, and R. Hesse, *Acta Chem. Scand. Ser. A, 29*, 583 (1975). For a reinterpretation of this structure see: (b) I. G. Dance, *Inorg. Chim. Acta, 25*, L17 (1977).

86. K. Tang, J. Yang, Q. Yang, and Y. Tang, *J. Chem. Soc. Dalton Trans.*, 2297 (1989).

87. (a) M. Muller, R. J. H. Clark, and R. S. Nyholm, *Transition Met. Chem., 3*, 369 (1978). (b) W. Beck, K. H. Stetter, S. Tadros, and K. E. Schwarzhans, *Chem. Ber., 100*, 3944 (1967).

88. T.-A. Okamura, N. Ueyama, A. Nakamura, E. W. Ainscough, A. M. Brodie, and J. M. Waters, *J. Chem. Soc. Chem. Commun.*, 1658 (1993).

89. (a) I. G. Dance and J. C. Calabrese, *Inorg. Chim. Acta, 19*, L41 (1976). (b) I. G. Dance, G. A. Bowmaker, G. R. Clark, and J. K. Seadon, *Polyhedron, 2*, 1031 (1983).

90. G. A. Bowmaker and L.-C. Tan, *Aust. J. Chem., 32*, 1443 (1979).

91. M. Baumgartner, H. Schmalle, and E. Dubler, *Polyhedron, 9*, 1155 (1990).

92. G. A. Bowmaker, G. R. Clark, J. K. Seadon, and I. G. Dance, *Polyhedron, 3*, 535 (1984).

93. I. G. Dance, *J. Chem. Soc. Chem. Commun.*, 68 (1976).

94. I. G. Dance, *Inorg. Chem., 20*, 1487 (1981).

95. F. J. Hollander and D. Coucouvanis, *J. Am. Chem. Soc., 99*, 6268 (1977).

96. N. Kitajima, K. Fujisawa, M. Tanaka, and Y. Moro-oka, *J. Am. Chem. Soc., 114*, 9232 (1992).

97. M. Nakamoto, W. Hiller, and H. Schmidbaur, *Chem. Ber., 126*, 605 (1993).

98. A. F. Stange, E. Waldhor, M. Moscherosch, and W. Kaim, *Z. Naturforsch. B, 50*, 115 (1995).

99. F. Olbrich, J. Kopf, E. Weiss, A. Krebs, and S. Müller, *Acta Crystallogr. Sect. C, 46*, 1650 (1990).

100. (a) E. C. Constable and P. R. Raithby, *J. Chem. Soc. Dalton Trans.*, 2281 (1987). (b) G. A. Stergioudis, S. C. Kokkou, P. J. Rentzeperis, and P. Karagiannidis, *Acta Crystallogr. Sect. C, 43*, 1685 (1987).

101. I. G. Dance, L. J. Fitzpatrick, D. C. Craig, and M. L. Scudder, *Inorg. Chem., 28*, 1853 (1989).

102. A. M. M. Lanfredi, F. Ugozzoli, A. Camus, and N. Marsich, *Inorg. Chim. Acta, 99*, 111 (1985).

103. I. G. Dance, L. J. Fitzpatrick, and M. L. Scudder, *Inorg. Chem., 23*, 2276 (1984).

104. H. K. Yip, A. Schier, J. Riede, and H. Schmidbaur, *J. Chem. Soc. Dalton Trans.*, 2333 (1994).

105. I. G. Dance, L. J. Fitzpatrick, M. L. Scudder, and D. C. Craig, *J. Chem. Soc. Chem. Commun.*, 17 (1984).

106. S. P. Abraham, N. Narasimhamurthy, M. Nethaji, and A. G. Samuelson, *Inorg. Chem., 32*, 1739 (1993).

107. W. S. Crane and H. Beall, *Inorg. Chim. Acta, 31*, L469 (1978).

108. E. S. Raper, J. R. Creighton, and W. Clegg, *Inorg. Chim. Acta, 183*, 179 (1991).

109. J. P. Fackler, C. A. López, R. J. Staples, S. Wang, R. E. P. Winpenny, and R. P. Lattimer, *J. Chem. Soc. Chem. Commun.*, 146 (1992).

110. S. Kitagawa, M. Munakata, H. Shimono, S. Matsuyama, and H. Masuda, *J. Chem. Soc. Dalton Trans.*, 2105 (1990).

111. R. Castro, M. L. Durán, J. A. García-Vázquez, J. Romero, A. Sousa, E. E. Castellano, and J. Zukerman-Schpector, *J. Chem. Soc. Dalton Trans.*, 2559 (1992).

112. S. Kitagawa, Y. Nozaka, M. Munakata, and S. Kawata, *Inorg. Chim. Acta, 197*, 169 (1992).

113. E. Block, M. Gernon, H. Kang, and J. Zubieta, *Angew. Chem., 100*, 1389 (1988).

114. S. Kitagawa, S. Kawata, Y. Nozaka, and M. Munakata, *J. Chem. Soc. Dalton Trans.*, 1399 (1993).

115. (a) H. Dietrich, *Acta Crystallogr. Sect A, 34*, S126 (1978). (b) H. Dietrich, W. Storck, and G. Manecke, *Makromol. Chem., 182*, 2371 (1981).

116. R. Hesse and L. Nilson, *Acta Chem. Scand., 23*, 825 (1969).

117. R. Chadha, R. Kumar, and D. G. Tuck, *J. Chem. Soc. Chem. Commun.*, 188 (1986).

118. (a) A. Muller, F.-W. Baumann, H. Bogge, M. Romer, E. Krickemeyer, and K. Schmitz, *Angew. Chem. Int. Ed. Engl., 23*, 632 (1984). (b) A. Muller, M. Romer, H. Bogge, E. Krickemeyer, F.-W. Baumann, and K. Schmitz, *Inorg. Chim. Acta, 89*, L7 (1984). (c) G. Gattow and T. Dingeldein, *Z. Anorg. Allg. Chem., 590*, 127 (1990).

119. A. Muller and U. Schimanski, *Inorg. Chim. Acta, 77*, L187 (1983).

120. E. Block, D. Macherone, S. N. Shaikh, and J. Zubieta, *Polyhedron, 9*, 1429 (1990).

121. D. M. Knotter, A. L. Spek, D. M. Grove, and G. van Koten, *Organometallics, 11*, 4083 (1992).

122. (a) M. D. Janssen, M. Herres, A. L. Spek, D. M. Grove, H. Lang, and G. van Koten, *J. Chem. Soc. Chem. Commun.*, 925 (1995). (b) M. D. Janssen, M. Herres, L. Zsolnai, D. M. Grove, A. L. Spek, H. Lang, and G. van Koten, *Organometallics, 14*, 1098 (1995). (c) M. D. Janssen, W. J. J. Smeets, A. L. Spek, D. M. Grove, H. Lang, and G. van Koten, *J. Organomet. Chem., 505*, 123 (1995). (d) M. D. Janssen, K. Köhler, M. Herres, A. Dedieu, W. J. J. Smeets, A. L. Spek, D. M. Grove, H. Lang, and G. van Koten, *J. Am. Chem. Soc., 118*, 4817 (1996).

123. D. M. Knotter, G. Blasse, J. P. M. van Vliet, and G. van Koten, *Inorg. Chem., 31*, 2196 (1992).

124. G. Henkel, P. Betz, and B. Krebs, *J. Chem. Soc. Chem. Commun.*, 314 (1984).

125. E. M. Meyer, S. Gambarotta, C. Floriani, A. Chiesi-Villa, and C. Guastini, *Organometallics, 8*, 1067 (1989).

126. L. M. Engelhardt, G. E. Jacobsen, W. C. Patalinghug, B. W. Skelton, C. L. Raston, and A. H. White, *J. Chem. Soc. Dalton Trans.*, 2859 (1991).

127. M. Baumgartner, H. Schmalle, and C. Baerlocher, *J. Solid State Chem.*, *107*, 63 (1993).

128. S. R. Acott, C. D. Garner, J. R. Nicholson, and W. Clegg, *J. Chem. Soc. Dalton Trans.*, 713 (1983).

129. T. A. Wark and D. W. Stephan, *Inorg. Chem.*, *26*, 363 (1987).

130. W. Hirpo, S. Dhingra, A. C. Sutorik, and M. G. Kanatzidis, *J. Am. Chem. Soc.*, *115*, 1597 (1993).

131. P. M. Boorman, H.-B. Kraatz, M. Parvez, and T. Ziegler, *J. Chem. Soc. Dalton Trans.*, 433 (1993).

132. J. M. Ball, P. M. Boorman, J. F. Fait, and T. Ziegler, *J. Chem. Soc. Chem. Commun.*, 722 (1989).

133. Y. Yang, Q. Liu, L. Huang, B. Kang, and J. Lu, *J. Chem. Soc. Chem. Commun.*, 1512 (1992).

134. Y. Yang, Q. Liu, L. Huang, D. Wu, B. Kang, and J. Lu, *Inorg. Chem.*, *32*, 5431 (1993).

135. W. Hirpo, S. Dhingra, and M. G. Kanatzidis, *J. Chem. Soc. Chem. Commun.*, 557 (1992).

136. B. J. Hales, E. E. Case, J. E. Morningstar, M. F. Dzeda, and L. A. Mauterer, *Biochemistry*, *25*, 7251 (1986).

137. P. K. Mehrotra and R. Hoffmann, *Inorg. Chem.*, *17*, 2187 (1978).

138. (a) B. J. Wakefield, *The Chemistry of Organolithium Compounds*, Pergamon, Oxford, UK, 1974. (b) B. J. Wakefield, *Organolithium Methods*, Academic, Orlando, FL, 1988.

139. (a) P. R. Markies, O. S. Akkerman, F. Bickelhaupt, W. J. J. Smeets, and A. L. Spek, *Adv. Organomet. Chem.*, *32*, 147 (1991). (b) P. R. Markies, R. M. Altink, A. Villena, O. S. Akkerman, and F. Bickelhaupt, *J. Organomet. Chem.*, *402*, 289 (1991).

140. A. J. Banister, W. Clegg, and W. R. Gill, *J. Chem. Soc. Chem. Commun.*, 850 (1987).

141. K. Ruhlandt-Senge and P. P. Power, *Bull. Soc. Chim. Fr.*, *129*, 594 (1992).

142. G. A. Sigel and P. P. Power, *Inorg. Chem.*, *26*, 2819 (1987).

143. K. Ruhlandt-Senge and P. P. Power, *Inorg. Chem.*, *30*, 3683 (1991).

144. J. J. Ellison and P. P. Power, *Inorg. Chem.*, *33*, 4231 (1994).

145. D. R. Armstrong, A. J. Banister, W. Clegg, and W. R. Gill, *J. Chem. Soc. Chem. Commun.*, 1672 (1986).

146. M. Aslam, R. A. Bartlett, E. Block, M. M. Olmstead, P. P. Power, and G. E. Sigel, *J. Chem. Soc. Chem. Commun.*, 1674 (1985).

147. E. Weiss and U. Joergens, *Chem. Ber.*, *105*, 481 (1972).

148. K. Ruhlandt-Senge, *Inorg. Chem.*, *34*, 3499 (1995).

149. P. J. Bonasia, D. E. Gindelberger, B. O. Dabbousi, and J. Arnold, *J. Am. Chem. Soc.*, *114*, 5209 (1992).

150. Houben-Weyl, *Methoden der Organischen Chemie*, Georg Thieme Verlag, Stuttgart, Germany, 1973, 4th ed., Vol. 13/2a.

151. M. D. Janssen, E. Rijnberg, C. A. de Wolf, M. P. Hogerheide, D. Kruis, H. Kooijman, A. L. Spek, D. M. Grove, and G. van Koten, *Inorg. Chem.*, *35*, 6735 (1996).

152. H. Gornitzka, S. Besser, R. Herbst-Irmer, U. Kilimann, and F. T. Edelmann, *Angew. Chem., Int. Ed. Engl.*, *31*, 1260 (1992).

153. M. D. Janssen, R. van der Rijst, A. L. Spek, D. M. Grove, and G. van Koten, *Inorg. Chem.*, *35*, 3436 (1996).

154. K. Ruhlandt-Senge and U. Englich, *J. Chem. Soc. Chem. Commun.*, 147 (1996), and references cited therein.

155. K. Ruhlandt-Senge and P. P. Power, *Inorg. Chem.*, *32*, 4505 (1993).

156. H. L. van Maanen, J. T. B. H. Jastrzebski, H. Kooijman, A. L. Spek, and G. van Koten, *Tetrahedron*, *50*, 11509 (1994).

157. J. A. Pappas, *J. Am. Chem. Soc.*, *100*, 6023 (1978).

158. (a) H. Schumann, I. Albrecht, and E. Hahn, *Angew. Chem., Int. Ed. Engl.*, *24*, 985 (1985). (b) H. Schumann, I. Albrecht, M. Gallagher, E. Hahn, C. Muchmore, and J. Pickardt, *J. Organomet. Chem.*, *349*, 103 (1988).

159. K. Tatsumi, T. Amemiya, H. Kawaguchi, and K. Tani, *J. Chem. Soc. Chem. Commun.*, 773 (1993).

160. K. Tatsumi, I. Matsubara, Y. Inoue, A. Nakamura, R. E. Cramer, G. J. Tagoshi, J. A. Golen, and J. W. Gilje, *Inorg. Chem.*, *29*, 4928 (1990).

161. (a) M. P. Hogerheide, J. T. B. H. Jastrzebski, J. Boersma, W. J. J. Smeets, A. L. Spek, and G. van Koten, *Inorg. Chem.*, *33*, 4431 (1994). (b) M. P. Hogerheide, S. N. Ringelberg, D. M. Grove, J. T. B. H. Jastrzebski, J. Boersma, W. J. J. Smeets, A. L. Spek, and G. van Koten, *Inorg. Chem.*, *35*, 1195 (1996).

162. H. Gornitzka, F. T. Edelmann, and K. Jacob, *J. Organomet. Chem.*, *436*, 325 (1992).

163. M. van Klaveren, unpublished results.

164. (a) M. D. Jansen, M. van Klaveren, G. van Koten, B. L. Mojet, and D. C. Koningsberger, *SRS Annual Reports of the Daresbury Laboratory*, *1*, 124 (1995); *2*, 414 (1995). (b) M. D. Janssen, M. van Klaveren, B. L. Mojet, D. C. Koningsberger, and G. van Koten, to be published.

165. A. F. Stange, K.-W. Klinkhammer, and W. Kaim, private communication.

# The Role of the Pyrazolate Ligand in Building Polynuclear Transition Metal Systems

**GIROLAMO LA MONICA** *and* **G. ATTILIO ARDIZZOIA**

*Dipartimento di Chimica Inorganica*
*Metallorganica e Analitica and Centro C.N.R.*
*Università di Milano*
*Milano, Italy*

CONTENTS

*Progress in Inorganic Chemistry, Vol. 46*, Edited by Kenneth D. Karlin.
ISBN 0-471-17992-2 © 1997 John Wiley & Sons, Inc.

# I. INTRODUCTION

The synthesis of multimetallic transition metal complexes where the metals are held at specific distances from each other is an important objective because of their potential role in multimetal-centered catalysis in both biological and industrial reactions (1). Moreover, such systems, through cooperative electronic and/or steric effects between metal centers, might give rise to distinct reactivity patterns for both their stoichiometric and catalytic reactions, which are not available to their monometallic analogues (2). Of the ligands that are able to maintain the metal centers in close proximity, the pyrazolate ion ($pz^*$) appears to be a particularly suitable candidate. Pyrazoles ($Hpz^*$) are weak bases (3, 4) and behave as 2-monohapto ligands.

$\eta^1$-Coordination (**A**) for neutral pyrazoles has been extensively described and several X-ray crystal structural studies have been reported (4–9). Factors affecting the maximum number of pyrazole molecules that may coordinate to a metal center have been thoroughly explored (6). The 1-unsubstituted pyrazoles are also weak acids (3, 4). Indeed, they react with alkali metals affording $M^+(pz^*)^-$ (10), with Grignard reagents (PhMgX) yielding $(pz^*)MgX$ (11), or with alkali metal borohydrides to give the corresponding poly(pyrazolyl-1-yl)borates anions (scorpionate ligands) (12).

The nucleophilicity of N2 and the acid character of $Hpz^*$ may be modified by introducing appropriate substituents in the 3-, 4- or 5-positions of the het-

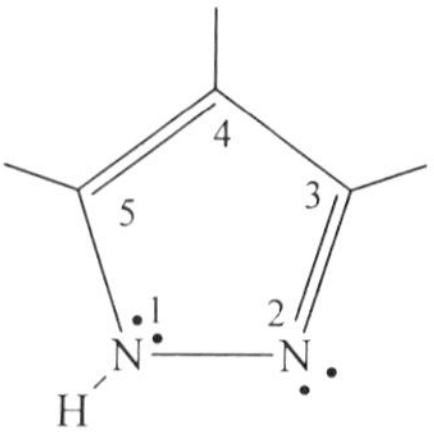

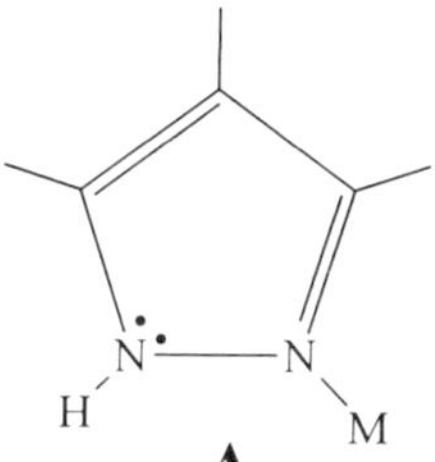

**A**

erocyclic ring (3, 4). By far, the pyrazolate anion (pz*)⁻ **(B)** is the best studied related species. This species can be easily obtained from Hpz* by deprotonation.

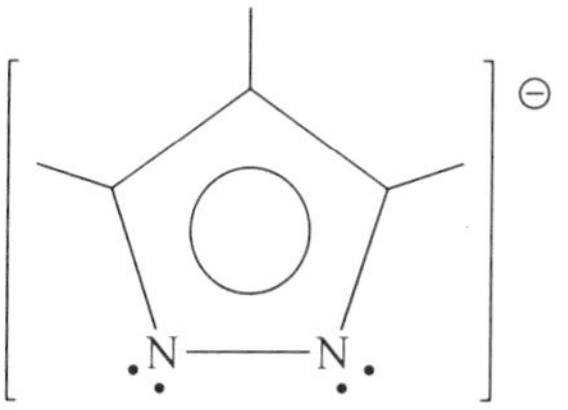

**B**

The coordination chemistry of Hpz* and pz* has been thoroughly examined by Trofimenko (5, 6) in his notable reviews dealing with the coordination chemistry of pyrazole-derived ligands. The pyrazolate ligand pz* is known to exhibit three coordination modes on binding metal centers (Fig. 1).

It can act as an anionic monodentate (pyrazolate-$N$), exo-bidentate (pyrazolate-$N, N'$), or endo-bidentate ligand. In most cases, the pz* ligand coordinates in an exo-bidentate fashion, thus linking two metal centers that may be identical or different. For these complexes, singly bridged $M(\mu$-pz*)M or $M(\mu$-pz*)M', doubly bridged $M(\mu$-pz*)$_2$M or $M(\mu$-pz*)$_2$M' and the much less common triply bridged $M(\mu$-pz*)$_3$M or $M(\mu$-pz*)$_3$M' and quadruply bridged $M(\mu$-pz*)$_4$M species were reported.

Heterobridged complexes having an usually puckered five-membered ring, $M(\mu$-pz*)$(\mu$-X)M or $M(\mu$-pz*)$(\mu$-X)M' (X = single atom donor ligand), are also known. Much work has been done with metallocycles containing the $M(\mu$-pz*)$_2$M ring. Various exchange reactions can be carried out on the terminal ligands leaving the central $M(\mu$-pz*)$_2$M core intact.

Singly bridged complexes are less numerous than doubly bridged ones. Representative examples are the homoleptic species $[M(pz^*)]_n$ [$M$ = Cu(I), Ag(I), and Au(I)]. Notably, the number of these binary complexes increased in the last years as well as their X-ray crystal structures (13).

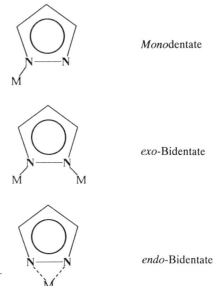

*Mono*dentate

*exo*-Bidentate

*endo*-Bidentate

Figure 1.  Coordination modes of the pyrazolate anion.

The usefulness of pyrazolate mononuclear complexes containing pz* as a monodentate ligand in the synthesis of dinuclear and trinuclear heterometallic species was demonstrated (14). Apart from a few examples of monodentate behavior, the pz* groups are, generally, such good exo-bidentate donors that when a suitable acceptor is not immediately available, coordination polymers or oligomers formed. Only a few examples of complexes containing the endo-bidentate pyrazolate ligand were also reported.

Research on the coordination chemistry of the pyrazolate ligand has progressed very rapidly as the potentiality of the pz* groups became evident following two comprehensive reviews on this topic (5, 6). Recent studies recognized the good catalytic activity, under mild conditions, of some polynuclear heterobridged pyrazolate complexes, thus encouraging the exploration of the unusual and specific features of the pyrazolate ligand.

This chapter covers the years 1985–1995; it is intended to update those of Trofimenko (5, 6), which should be consulted for most earlier references.

## II.  HOMOLEPTIC PYRAZOLATE COMPLEXES

### A.  Singly Bridged M($\mu$-pz*)M Complexes

Binary compounds of the general formula [M(pz*)]$_n$ are known only for Group 11 (IB) metals [M = Cu(I), Ag(I), or Au(I); pz* acts as a singly bridging ligand]. These few structurally characterized systems were found to possess a variety of stoichiometries, molecular arrangements, and oligomeric or polymeric structures (13). In a few cases, oligomers of different nuclearities were selectively prepared and characterized (13, 15–17).

In 1889, Büchner (18) reported the formation of the insoluble silver(I) pyrazolate $[Ag(pz)]_n$, **1a**, which he simply denoted as a "silver salt". The oligometric or polymeric nature of this species was not clarified until 1994, although it was always referred to as being a polynuclear pyrazolate compound (6). A similar structure was claimed for $[Cu(pz)]_n$, **2**, on the basis of IR evidences (6, 19).

Recently, efforts were made in order to investigate the possibility of solving simple structures in the field of polymeric metallorganic materials, which usually appear as fine powdered samples. For this class of compounds single crystals cannot be grown because of their negligible solubility in common organic solvents (13). Indeed, this is the case for $[Cu(pz)]_n$ and $[Ag(pz)]_n$.

In 1994, the ab initio X-ray crystal stucture determination from powder diffraction data of $[Cu(pz)]_n$ and $[Ag(pz)]_n$ were reported (13). During these studies it was ascertained that copper(I) and silver(I) pyrazolates, depending on the synthetic method used, each appear in two distinct crystalline phases. The $\alpha$-$[Cu(pz)]_n$, **2a**, and $\beta$-$[Cu(pz)]_n$, **2b**, consist of infinite chains of linearly coordinated copper atoms, bridged by exo-bidentate pyrazolate anions (Fig. 2).* The two phases differ mainly in the interchain Cu· · ·Cu contacts, which are 3.34 Å in $\alpha$-$[Cu(pz)]_n$ and 2.97 Å in $\beta$-$[Cu(pz)]_n$.

The $\alpha$ phase was easily obtained by reacting $[Cu(MeCN)_4](BF_4)$ with Hpz in acetone in the presence of triethylamine (13, 18). The complex $\beta$-$[Cu(pz)]_n$ was prepared via the high-yield, high-temperature reaction of $Cu_2O$ with molten pyrazole (13). Two crystalline phases were also found for silver(I) pyrazolate (13). The first, $[Ag(pz)]_n$, **1a**, was shown to be isomorphous and isostructural with $\alpha$-$[Cu(pz)]_n$, (Ag· · ·Ag interchain distances are 3.23 Å). This complex was quantitatively obtained by using the synthetic procedure originally reported by Büchner (18). The second silver(I) pyrazolate phase, $[Ag(pz)]_3$, **1b**, appeared as a truly molecular compound whose topology closely resembles those of other Group 11 (IB) metal trimeric pyrazolates. Complex **1b** was obtained by removing $PPh_3$ from the dimeric silver(I) pyrazolate complex $[(PPh_3)Ag(pz)]_2$, **38** (21). A detailed study of the IR spectra of **1a**, **1b**, **2a**, and **2b** was performed (13), which is diagnostic for the identification of each phase (Fig. 3).

*The following figure is a key for atom symbols used in the figures in this chapter.

 Metals 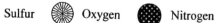 Sulfur  Oxygen  Nitrogen  Others

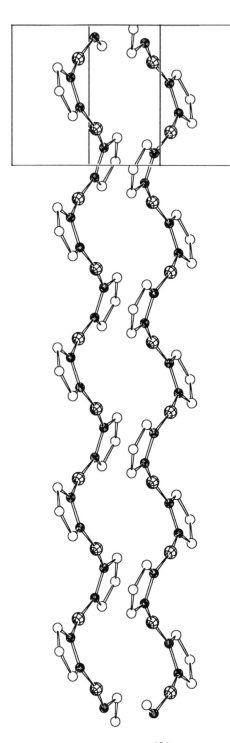

Figure 2. Structure drawing of two adjacent $\alpha$-[Cu(pz)]$_n$ chains. At the resolution of the draws $\beta$-[Cu(pz)]$_n$ chains look similar. [Based on data from (13)]. (All the structure drawings reproduced in this review are obtained by the program SCHAKAL (20), from the data reported in the literature. Throughout the paper, metal, S, O, and N atoms are sketched by cross-hatched, dashed, starred, and dotted circles, respectively. See the footnote on page 155.)

156

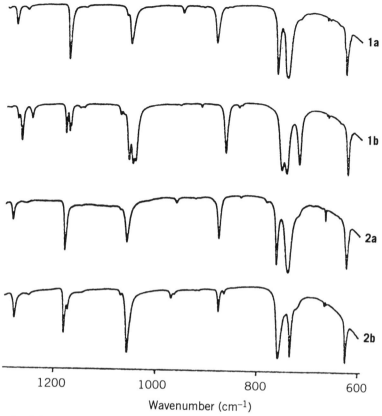

Figure 3. Infrared spectra (Nujol mulls) of [Ag(pz)]$_n$, **1a**, [Ag(pz)]$_3$, **1b**, $\alpha$-[Cu(pz)]$_n$, **2a** and $\beta$-[Cu(pz)]$_n$, **2b**. [Adapted from (13).]

Cyclic trimeric pyrazolate complexes of Group 11 (IB) metals are now somewhat common. In the last few years, several species having such nuclearity were reported. In 1988, Fackler and co-workers (15) reported the synthesis and X-ray crystal structures of [Ag(dppz)]$_3$, **3,** and [Au(dppz)]$_3$, **4.** These structures were obtained by reacting Na(dppz) with AgNO$_3$ and Au(tht)Cl, respectively. In **3,** the three Ag positions do not form an equilateral triangle owing to the significantly different Ag $\cdots$ Ag distances [3.305(2), 3.362(2), and 3.486(2) Å)]. Moreover, the nine-membered Ag$_3$N$_6$ ring is not planar. On the contrary, the gold trimer **4** appears as a rigorously planar nine-membered inorganic ring. The intermolecular Au $\cdots$ Au distances in **4** are in excess of 2.25 times that of the intramolecular distances.

A low-yield synthesis of [Cu(dppz)]$_3$, **5,** from CuCl and Na(dppz) in the presence of AgNO$_3$ was reported, and its X-ray crystal structure was determined

(16). The structure of **5** consists of a nine-membered $(Cu-N-N)_3$ metallocyclic ring similar to the ones of the isostructural silver(I) and gold(I) compounds (15). Significant deviations from planarity, as found in the silver(I) trimer, are present.

A related copper(I) complex containing exo-bidentate 3,5-dimethylpyrazolate has been known since 1971 (22) and thought to be a polymeric species, that is $[Cu(dmpz)]_n$. The X-ray crystal structure, published in 1990 (23), evidenced the trimeric nature of this copper(I) complex, that is $[Cu(dmpz)]_3$, **6.**

It is now widely accepted that the nuclearity of Group 11(IB) univalent metal binary pyrazolate complexes is strongly affected by the nature of the substituents on the heterocyclic ligand. In some cases, depending on the preparative method, complexes having different nuclearities have been selectively obtained. The existence of the two quite different phases for silver(I) pyrazolate complexes, that is, $[Ag(pz)]_3$, **1a** and $[Ag(pz)]_3$, **1b,** is in this respect indicative. Besides the trimeric complex $[Au(dppz)]_3$, **4,** the hexamer $[Au(dppz)]_6$, **7,** was obtained by reaction of $Au(PPh_3)Cl$ with $Na(dppz)$ in the presence of silver benzoate and was crystallographically characterized (15). Complex **7** contains an 18-membered inorganic ring in the shape of a two-bladed propeller. The geometry of the six gold centers is best described as an edge-sharing bitetrahedron (Fig. 4).

The preparative method reported for the synthesis of $\alpha$-$[Cu(pz)]_n$, **2a,** was extended to pyrazoles having different substituents in the 3,5-positions (24). In the case of 3,5-diphenylpyrazole, the tetranuclear complex $[Cu(dppz)]_4$, **8,** was obtained (17). In complex **8,** the four copper atoms, which lie at an average nonbonding distance of 3.12 Å, are strictly coplanar, while the four bridging dppz units, each one spanning the ideal Cu $\cdot\cdot\cdot$ Cu edges, stem out from the two sides in an alternate fashion, giving the whole complex an idealized $D_2$, rather than a fourfold, symmetry. The Cu—N bond distances compare well with those found in $[Cu(dmpz)]_3$, **6,** (23) but are much shorter than in the trimeric species $[Cu(dppz)]_3$ (**5**), which contains the same pyrazolate group (16). Trimeric copper(I) complexes $[Cu\{(p\text{-}ClC_6H_4)_2pz\}]_3$, **9,** and $[Cu\{(pCH_3\text{-}C_6H_4)_2pz\}]_3$, **10,** were obtained by reacting 3,5-di-($p$-chlorophenyl)pyrazole and 3,5-di-($p$-tolyl)pyrazole with $[Cu(MeCN)_4](BF_4)$ in the presence of triethylamine (21). While for **9,** an X-ray crystal structure determination definitively confirmed its trinuclear formulation, for **10** the nucleaity was argued on the basis of electron-impact mass spectrometry.

The reaction of molten Htmpz with copper metal shot in air yielded a mixture of the trinuclear copper(I) complex $[Cu(tmpz)]_3$, **11,** and the trinuclear copper(I/II) complex $[Cu(3\text{-}CO_2\text{-}4,5\text{-}Me_2pz)(Htmpz)]_2Cu$ in which the $3\text{-}CO_2\text{-}4,5\text{-}Me_2pz$ ligand, derived from the oxidation of one of the three methyl groups of the tmpz, is present (25). Alternatively, **11** was prepared by reacting cuprous iodide with Htmpz in the presence of triethylamine (26). The structure of **11** is

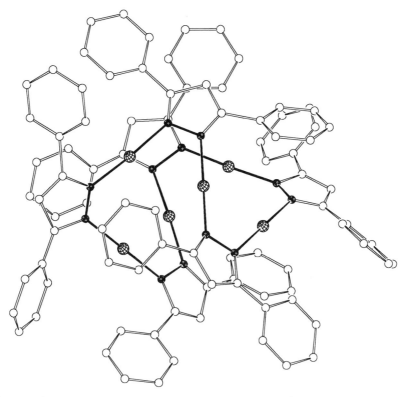

Figure 4.   Structure drawing of the hexanuclear gold(I) complex [Au(dppz)]$_6$, **7.** [Based on data from (15).]

very similar to that reported for [Cu(dmpz)]$_3$ (23). The latter synthetic method was extended to other pyrazoles of the general formula 4-XdmpzH (X = Cl, Br, or I) from which copper(I) pyrazolates formulated as trimers were prepared (26). This formulation is supported by the fact that they exhibit electron-impact mass spectra in which the most intense peak is that due to the molecular ion expected for the trimer. The isotopic patterns are in excellent agreement with those predicted. The X-ray crystal structure of the copper(I/II) mixed-oxidation state trinuclear species [Cu(3-CO$_2$-4,5-Me$_2$pz)(Htmpz)]$_2$Cu was also reported (25). The 3-CO$_2$-4,5-Me$_2$pz ligands chelate the central copper(II) atom via the carboxylate oxygen atom and the adjacent ring nitrogen atom. A similar reaction involving 4-BrdmpzH gave the dinuclear copper(II) complex [Cu(3-CO$_2$-4-Br-mpz)(4-BrdmpzH)$_2$]$_2$ (25).

The reaction of [Cu(Me$_3$CN)$_4$](BF$_4$) with 3,5-dicarbomethoxypyrazole in the presence of triethylamine was reported (27). A copper(I) pyrazolate, [Cu-

(dcmpz)]$_n$, **12**, was obtained. Complex **12** was formulated as an oligomeric complex. Its IR spectrum exhibits only one $\nu(C=O)$ at 1723 cm$^{-1}$, suggesting the equivalence of the carbonyl groups of the two COOMe substituents in the 3,5-positions of the pyrazole ring. This absorption lies quite at the same frequency as for the dimeric species [Cu(dcmpz)(RNC)]$_2$, **34** (R = cyclohexyl), whose X-ray crystal structure ruled out interactions between the C = O groups and the copper centers (27).

The reaction of (Me)$_2$SAuCl with a pyrazole having bulky substituents in the 3,5-position or with an asymmetrically substituted pyrazole in the presence of a strong base, was described (28). In the case of Hdbpz, a mixture containing two different pyrazolate complexes that were formulated on the basis of NMR data and by electron impact mass spectrometry as trimeric and tetrameric species were obtained (28). By reacting Hdmepz, the formation of a mixture of two oligomers, probably a trimer and a tetramer, was again claimed. The interpretation of the NMR spectra was particularly complicated owing to the concomitant presence of isomers.

The reaction of [Cu(MeCN)$_4$](BF$_4$) or AgNO$_3$ with 3,5-dimethyl-4-nitro-pyrazole in the presence of NEt$_3$ was recently studied (21). Two different copper(I) pyrazolate complexes were obtained, depending on the Cu/4-NO$_2$dmpzH/NEt$_3$ molar ratio. Reactions employing a 1:1:1 molar ratio gave the white [Cu(4-NO$_2$dmpz)]$_3$, **13**, characterized by a powder X-ray structure determination (21). In the presence of excess pyrazole and NEt$_3$, the orange product [Cu$_4$(4-NO$_2$dmpz)$_6$](Et$_3$NH)$_2$, **14**, was isolated. An X-ray crystal structure analysis confirmed its formulation (21). The related trinuclear silver(I) compound, [Ag(4-NO$_2$dmpz)]$_3$ was obtained by using AgNO$_3$. The nuclearity of such species was based on the identity of its IR spectrum with that of **13** (21).

Many structural features observed for Group 11 (IB) pyrazolate complexes can be found in the realm of purely organic pyrazoles, where hydrogen bonding between different molecules is capable of generating, in the solid state, a variety of Hpz* clusters, ranging from dimers up to polymeric chains. This analogy goes even further, as it can be observed that a fairly straight parallelism between pyrazoles (if N—H· · ·N links are considered) and Cu(I)-pyrazolates exists: pyrazole and [Cu(pz)]$_n$ ($\alpha$ and $\beta$ phases) are polymeric (13, 29, 30), 3,5-dimethylpyrazole, 4-nitro-3,5-dimethylpyrazole, [Cu(dmpz)]$_3$, and [Cu(4-NO$_2$-dmpz)]$_3$ are trimers (21, 23, 31, 32); while 3,5-diphenylpyrazole and [Cu(dppz)]$_4$ are tetramers (17, 33, 34) of idealized $D_{2d}$ symmetry. This result might suggest that the static N—Cu—N (with a N· · ·N distance of $\sim$ 3.7 Å) and the dynamic (tautomeric) N—H· · ·N (with a N· · ·N distance of $\sim$ 2.9 Å) bonds, in spite of their different nature and strength, are linear hinges about which the pyrazolate rings are essentially free to rotate. The different steric demands of different substituents in the 3,5-positions of the heterocycle drive the preferred dihedral angle to specific values, thus determining the nuclearity of the oligomers. The analogy of the linear coordination at the $H^+$ and $Cu^+$ acidic centers impose local packing of the single molecules but does not determine the supramolecular arrangement in the crystal, so that crystal systems and space groups of the H and Cu derivatives do not match.

## B.  Doubly Bridged M(μ-pz*)₂M Complexes

Binary pyrazolate complexes containing divalent metals bridged by two pz* groups are known for many transition metals. The increasing interest for this class of complexes stems from the expectation that they may provide useful insights in the field of magnetostructural correlations as well as in multimetal centered catalysis.

Reactions of copper metal shot with molten pyrazole or substituted pyrazoles in the presence of dioxygen were successfully employed by Ehlert et al. (35–37) in the synthesis of binary copper(II) pyrazolates of the general formula $[Cu(pz^*)_2]_n$ (Hpz* = Hpz, 4-ClpzH, 4-BrpzH, or 4-MepzH). The complex $[Cu(pz)_2]_n$, **15,** was obtained in 79% yield as a bright green powder when the reaction mixture was heated at 110°C in the presence of air (35). The polymeric nature of **15** was ascertained by an X-ray crystal structure analysis. The Cu atoms have a $D_2$ distorted tetrahedral coordination geometry with Cu—N = 1.957(2) Å and N—Cu—N = 94.3(1)–139.5(1)°. In an earlier paper by Vos and Groeneveld (38), reactions of copper(II) salts with Hpz in aqueous base were reported to yield a black material of composition $[Cu(pz)_2]_n$. The latter reactions were checked by Ehlert et al. (35) who attributed the composition $[Cu_2(pz)_3(OH)]_n$ to the black material, rather than $[Cu(pz)_2]_n$.

Magnetic susceptibility studies (4.2–299 K) on **15** revealed very strong antiferromagnetic exchange between copper(II) centers in the extended linear chains (35). A comparison of the low-temperature single-crystal X-ray diffraction study of $[Cu(pz)_2]_n$ with the room temperature study (35) showed that in copper(II) pyrazolate extended-chain polymers interchain distances may be affected by changes in temperature (37). This fact appeared to have little effect on interchain bond lengths and angles, but may significantly alter the dihedral angle between fused Cu—(N—N)₂—Cu and $N_2C_3$ (pyrazole) rings and the dihedral angle between consecutive Cu—(N—N)₂—Cu planes.

In addition to the copper/molten ligand method, the syntheses of [Cu(4-Xpz)₂]ₙ (X = Cl, **16,** and Br, **17**) on treatment of Cu(OH)₂ or Cu₂O with excess 4-XpzH in refluxing xylene were described (36). Two forms of the 4-ClpzH derivative were obtained, one green and one brown. It was apparent from the X-ray powder diffraction pattern that the two forms are structurally distinct. While details of the structure of the brown form still remain unknown, comparison of its solubility and thermal properties with those of the structurally characterized pyrazolate complexes, suggest a polymeric nature for this material also (36).

The crystal structures of $[Cu(4\text{-}Clpz)_2]_n$, **16**, (green form) and $[Cu(4\text{-}CH_3pz)_2]_n$, **18**, confirmed their polymeric nature. In addition, for Complexes **16** and **18**, magnetic susceptibility studies over the temperature range 2–300 K revealed a relatively strong antiferromagnetic coupling (36).

Recently, quite different products were claimed to be formed when $Cu(OH)_2$ reacted with molten 4-XpzH. A brown powder having the empirical formula $[Cu_2(OH)(4\text{-}Xpz)_3(4\text{-}XpzH)_2]$ (X = Cl or Br) was isolated with excess molten pyrazole, while the dinuclear species $[Cu_2(OH)(4\text{-}Xpz)_3]$ (X = H, Me, Cl, and Br) formed when a stoichiometric amount of 4-XpzH was used (39).

On the basis of their IR spectra, complexes $[Cu_2(OH)(4\text{-}Xpz)_3]$ were formulated as containing the copper(II) ions bridged alternatively by a pair of 4-Xpz groups and a combination of 4-Xpz and OH groups to form linear-chain structures. A series of new copper(II) pyrazolate complexes, $[Cu(4\text{-}Xdmpz)_2]_n$ (X = H, **19**; Cl, **20**; Br **21**; Me, **22**) were synthesized and characterized (26). As a group these compounds differ from those previously studied (35, 36) in that they have bulky methyl substituents in the more sterically sensitive 3,5-positions. Compound **19** could be obtained from the metal/molten ligand reaction employed earlier in the preparation of $[Cu(pz)_2]_n$, (**15**), (35); for the other complexes this reaction led to impure products. A successful route to the required copper(II) compounds starting from the corresponding copper(I) derivatives, $[Cu(4\text{-}Xdmpz)]_3$, was developed (26). This method, which gave high yields in all cases, consists in the reaction at high temperature of the trinuclear copper(I) complexes $[Cu(4\text{-}Xdmpz)]_3$ with the pyrazole 4-XdmpzH under a dioxygen atmosphere.

Complexes **19–22** were reported to most likely have structures involving extended chains of copper ions linked by double pyrazolate bridges in analogy with the structures of species $[Cu(4\text{-}Xpz)_2]_n$, where X = H, Me, and Cl (35, 36). However, it is possible that the presence of methyl substituents in the 3,5-positions of the heterocyclic ring, causing steric hindrance, prevents the formation of an infinite polymeric chain.

Pyridine abstraction from $[Cu(dcmpz)_2(py)_2]$, **229**, via azeotropic distillation with benzene, allowed the isolation of the linear trinuclear complex $[Cu_3(dcmpz)_6]$, **23**, whose X-ray crystal structure was determined (40) (Eq. 1).

$$3[Cu(dcmpz)_2(py)_2] \rightleftharpoons [Cu_3(dcmpz)_6] + 6py \qquad (1)$$

In complex **23**, each copper atom is roughly square planar, with apical contacts involving some of the carbonylic oxygen atoms of the dcmpz ligands for the *external* copper(II) centers. The synthesis, characterization, and magnetic properties of a series of Cobalt(II) complexes containing the pyrazolate anion and several of its C-substituted derivatives have been reported (41). These compounds possess the general formula $[Co(pz^*)_2]_n$ (where $Hpz^* = Hpz$, Hmpz, Hdmpz, Htmpz, 4-CldmpzH, and 4-BrdmpzH).

The complex $[Co(pz)_2]_n$ (**24**), was prepared by reacting $CoCl_2$ with Hpz in the presence of aqueous NaOH under anaerobic conditions. The remaining five binary Cobalt(II) (substituted)pyrazolates were synthesized with the metal/molten pyrazole method: Cobalt metal was combined with the appropriate substituted pyrazole and the mixture was heated in a dioxygen atmosphere. In all cases, the cobalt(II) pyrazolates were obtained as purple powders (41). The materials are microcrystalline and, although none could be obtained in a form suitable for single-crystal X-ray diffraction studies, indirect evidences supported polymeric doubly pyrazolate-bridged chain structure with pseudotetrahedral cobalt centers. Magnetic studies at variable temperature revealed anisotropy in the susceptibilities and the presence of significant antiferromagnetic exchange (41).

A limited number of $[M(pz^*)_2]_n$ species (M = Zn, Pt, or Pd) were obtained by removing the neutral coordinated $Hpz^*$ ligands from complexes of the general formula $[M(pz^*)_2(Hpz^*)_x]_2$ by thermolytic methods.

Heating $[Zn(dmpz)_2(Hdmpz)]_2$, **31**, at 240°C (see Fig. 5) caused the loss of Hdmpz and the formation of the species $[Zn(dmpz)_2]_n$, **25** (42). Complex **25** is assumed to be polymeric, with an infinite chain structure, by analogy with the known structure of other binary copper(II) pyrazolates (35).

The trimer $[Pt(pz)_2]_3$, **26**, was obtained mixed with the polymeric $[Pt(pz)_2]_n$, **27**, on decomposition of the dimeric complex $[Pt(pz)_2(Hpz)_2]_2$, **244** (43, 44). For **26**, the X-ray crystal structure confirmed that the Pt atoms are each bridged by two pyrazolate ligands in a ring configuration (43). The polymeric nature of complex **27** was only suggested.

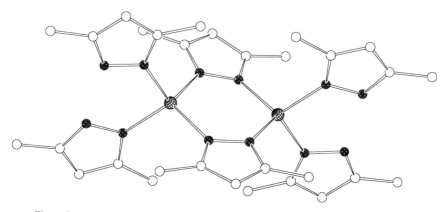

Figure 5.   Structure drawing of $[Zn_2(dmpz)_2(Hdmpz)]_2$, **31**. [Based on data from (42).]

TABLE I
List of Structurally Characterized Binary Metal Pyrazolates

| Cu | Reference | Ag | Reference | Au | Reference | Pt | Reference |
|---|---|---|---|---|---|---|---|
| $[Cu(pz)]_n$ | 13 | $[Ag(pz)]_n$ | 13 | $[Au(dppz)]_3$ | 15 | $[Pt(pz)_2]_3$ | 43 |
| $[Cu(dmpz)]_3$ | 23 | $[Ag(pz)]_3$ | 13 | $[Au(dppz)]_6$ | 15 | | |
| $[Cu(dppz)]_3$ | 16 | $[Ag(dppz)]_3$ | 15 | $[Au(dfmpz)]_3$ | 47 | | |
| $[Cu(dppz)]_4$ | 17 | $[Ag(dmpz)]_3$ | 21 | | | | |
| $[Cu\{(p\text{-}ClC_6H_5)_2pz\}]_3$ | 21 | | | | | | |
| $[Cu(tmpz)]_3$ | 26 | | | | | | |
| $[Cu(4\text{-}NO_2dmpz)]_3$ | 21 | | | | | | |
| $[Cu_4(4\text{-}NO_2dmpz)_6]^=$ | 21 | | | | | | |
| $[Cu(pz)_2]_n$ | 35 | | | | | | |
| $[Cu(dcmpz)_2]_3$ | 40 | | | | | | |
| $[Cu(4\text{-}Mepz)_2]_n$ | 36 | | | | | | |
| $[Cu(4\text{-}Clpz)_2]_n$ | 36 | | | | | | |
| | | $[Pd_2Ag_4(dmpz)_8]$ | 45 | | | | |

A related thermal reaction, carried out on the palladium(II) complexes $[Pd(pz)_2(Hpz)_2]_2$, **245,** and $[Pd(dmpz)_2(Hdmpz)_2]_2$, **246,** gave $[Pd(pz)_2]_n$, **28,** and $[Pd(dmpz)_2]_3$, **29,** respectively (45). An alternative synthesis of $[Pd(pz)_2]_n$ was already reported (46). It was obtained by reacting $PdCl_2$ with sodium pyrazolate, and was formulated as a polymeric species. For the palladium(II) Complex **29,** a trimeric structure has been suggested in analogy with the platinum(II) derivative **26** (43, 45).

The unique bimetallic homoleptic pyrazolate complex $[Pd_2Ag_4(dmpz)_8]$, **30,** was recently isolated and its X-ray crystal structure was determined (45). The hexanuclear complex **30** was obtained by reacting **246** with $AgNO_3$. Its molecular structure revealed close analogy with the precursor; indeed, Complex **30** can be thought of as derived by the formal substitution of the isolobal $H^+$ and $Ag^+$ fragments. The related Pd(II)/Cu(I) bimetallic complex $[Pd_2Cu_4(dmpz)_8]$ was analogously obtained (45), but its formulation was based only on IR evidences. Table I reports the structurally characterized binary metal pyrazolates.

## III. COMPLEXES CONTAINING THE $M(\mu\text{-}pz^*)_2M$ CORE AND ADDITIONAL NEUTRAL OR CHARGED LIGANDS

In these complexes, identical or different metal centers are, generally, connected by two pyrazolate ligands coordinated in an exo-bidentate fashion. In most cases, the $M(\mu\text{-}pz^*)_2M$ core exhibits a boat conformation. Doubly bridged pyrazolate complexes are usually dimeric; however, in the last years some homo- (MMM) and heterotrinuclear (MM'M) species containing the metallocycles $M(\mu\text{-}pz^*)_2M(\mu\text{-}pz^*)_2M$ and $M(\mu\text{-}pz^*)_2M'(\mu\text{-}pz^*)_2M$ were reported.

## A. Group 12:Zn

Cleaned zinc metal shot was reported to react with excess Hdmpz in the presence of dioxygen at 90°C affording the dimeric complex [Zn(dmpz)$_2$(Hdmpz)]$_2$, **31**, in virtually quantitative yields (42). The molecular structure of **31** consists of two zinc atoms bridged by two dmpz groups with each zinc center being end-capped by a monodentate pyrazolate ion and a neutral Hdmpz molecule (Fig. 5). Strong hydrogen bonding occurs between the capped ligands.

The unexpected bis(dmpz) bridged tetrachloride di-zinc(II) anionic complex [Zn$_2$(dmpz)$_2$Cl$_4$]$^=$, was obtained by metal ion induced breakdown of a pyrazole-containing ligand, debd, when reacted with ZnCl$_2$ in methanol. The X-ray crystal structure of [Zn(debd)Cl]$_2$[Zn$_2$(dmpz)$_2$Cl$_4$] was also determined (48).

## B. Group 11: Cu, Ag, or Au

Dinuclear complexes were obtained by reacting some binary copper(I) and silver(I) homoleptic pyrazolate complexes with neutral ligands. The trimeric [Cu(dmpz)]$_3$ (23) readily reacted with phen or RNC (R = cyclohexyl) to give the doubly bridged species [(phen)Cu($\mu$-dmpz)$_2$Cu(phen)], **32**, (49) or [(RNC)Cu($\mu$-dmpz)$_2$Cu(RNC)], **33** (50). The dimeric nature of **32** was argued from its spectroscopic and chemical properties, while **33** was characterized by an X-ray crystal structure analysis (50).

The Cu($\mu$-dmpz)$_2$Cu core in Complex **33** was found in an uncommon planar conformation. The related compound [(RNC)Cu(pz)]$_2$ was obtained from [Cu(pz)]$_n$ (**2a** or **2b** isomers) (13) on treatment with RNC (50). Dinuclear copper(I)-cyclohexylisocyanide complexes were also obtained by reacting [Cu(dcmpz)]$_n$ (27) or [Cu(dppz)]$_4$ (17) with RNC. The Complex [(RNC)-Cu(dcmpz)]$_2$, **34**, is topologically equivalent to the dmpz analogue **33**; however, the pattern of bond distances and angles of the Cu—(N—N)$_2$—Cu ring, as well as its conformation, is markedly different (27). The X-ray crystal structure of [(RNC)Cu(dppz)]$_2$, **35**, was reported and the results were compared with those of the parent Complexes **33** and **34** (17). Complex **35** possesses a Cu—[N—N]$_2$—Cu ring in a chairlike conformation, which has only a few precedents (51–53) (Fig. 6).

The reactions of **33** with some heterocumulenes were described (50). The products result from *formal* nucleophilic addition of the dmpz anion to the central carbon of the heterocumulene to give a new bidentate ligand.

Carbon disulfide reacted with **33** in the presence of free RNC affording [Cu{dmpz-CS$_2$}(RNC)$_2$], **36**. The $\nu$(C=S) in **36** appeared at 1330 cm$^{-1}$, suggesting an NS coordination of the dithiocarboxylate anion. This mode of coordination was suggested in 1968 by Trofimenko (54) in the case of metal(II)

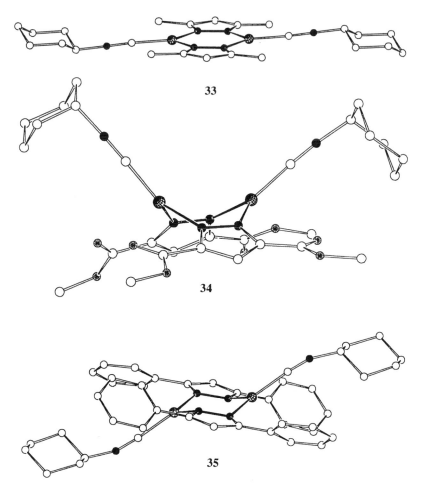

Figure 6. Structure drawings of [Cu(dmpz)(RNC)]$_2$, **33,** [Cu(dcmpz)(RNC)]$_2$, **34,** and [Cu(dppz)(RNC)]$_2$, **35** (R = cyclohexyl). Note the different conformations of the Cu-[N-N]$_2$-Cu rings. [Based on data from (17, 27, 50).]

pyrazolecarbodithioates. The X-ray crystal structure determination of Complex **36** substantiated such a hypothesis (55).

An unexpected reaction was reported to occur on treatment of [Cu(dcmpz)]$_n$, **12,** with CO in pyridine (27). The dinuclear monocarbonyl complex [Cu$_2$(dcmpz)$_2$(py)$_2$(CO), **37,** was obtained and its X-ray crystal structure was determined. An analogy was drawn between the CO-hemocyanin adduct and the dinuclear carbonyl Complex **37.** The observed stoichiometry of CO bonding

$$RNC-Cu \quad Cu-CNR \xrightarrow{\;Y=C=Z\;} 2\;RNC-Cu\underset{Z}{\overset{Y}{\diagup}}C=Y$$

R = cyclohexyl

Y = O, Z = NR
Y = S, Z = NPh
Y = O, Z = S

to hemocyanin has been interpreted in terms of different actual coordination of the copper atoms (56). The formation of the carbonyl complex **37** was attributed to steric effects due to the pyridine ligands, which prevent the bonding of a second molecule of carbon monoxide.

Two $Ag^{(I)}/PPh_3$ pyrazolate complexes, $[(PPh_3)Ag(pz)]_2$, **38**, and $[(PPh_3)_3$-$Ag_2(pz)_2]$, **39**, were isolated by reacting $[Ag(pz)]_n$, **1a**, with $PPh_3$ (21). The molar ratio $PPh_3/Ag$ plays a determining role in order to isolate the two species in a pure form. The X-ray crystal structure was determined for both complexes (21).

The dinuclear species **38** contains the two silver(I) centers joined by two pz groups, giving rise to a planar $Ag-[N-N]_2-Ag$ ring conformation. Each silver(I) center coordinates a $PPh_3$ ligand. In Complex **39**, the two metal centers exhibit different coordination numbers (3 and 4, respectively). The geometries of the two silver atoms are planar trigonal and distorted tetrahedral, giving rise to an $Ag-[N-N]_2-Ag$ ring in a distorted boatlike conformation (21).

It was reported that $[Au(dppz)]_3$, **4**, on treatment with aqua regia formed a gold(III) dimeric species formulated as $[Au(4\text{-}Cldppz)]_2Cl_4$ on the basis of analytical data, molecular weight measurements, and spectroscopic properties (57). This reaction was reinvestigated and a crystallographically characterized product was isolated (58). The latter was shown to be a mixed-valence $Au_2^I/Au^{III}$ complex in which the pyrazolate rings have been chlorinated, that is, $[Au_3(Cldppz)_3Cl_2]$, **40**. The molecule consists of three gold atoms bridged by three pyrazolate ligands, forming a $(Au-N-N)_3$ nine-membered planar metallocycle. Two of the metal atoms are two-coordinate linear $Au^I$ centers and the third one is a square planar trans $Au^{III}$ bearing the two chlorine ligands (58). A related mixed-valence $Au_2^I/Au^{III}$ pyrazolate complex, $[Au(dppz)]_3Cl_2$, **41**, surprisingly formed by reacting the monomeric gold(III) species $AuCl_3(py)$ with $Na(dppz)$ (59). This result was interpreted in terms of an unusual stability for the $d^{10}d^{10}d^8$ configuration of the pyrazolate metallocycle. The X-ray crystal structure (59) and an XPS study (58) of the heterovalent gold-pyrazolate trimer **41** was also reported.

## C. Group 10: Ni, Pd, or Pt

The considerable nucleophilicity of the OH bridges in the anionic complexes $[\{R_2M(\mu\text{-}OH)\}_2](NBu_4)_2$ (M = Ni, Pd, or Pt) is indicated by their reactivity toward weak acids such as pyrazoles, with formation of homodinuclear complexes of the formula $[\{R_2M(\mu\text{-}pz^*)\}_2](NBu_4)_2$ (Scheme 1) (60–66).

The di-$\mu$-pyrazolate complexes reported in Scheme 1 are obtained in methanol using a 1 : 2 molar ratio of reactants. Their formulation was ascertained on the basis of analytical, spectroscopic, and conductance data. Moreover, the same complexes are obtained when the mixed-bridged $\mu$-hydroxo-$\mu$-pyrazolates species (see Section IV.A and Scheme 1) are reacted with Hpz*. This fact suggested the possible intermediate formation of complexes containing both $\mu$-pz* and $\mu$-OH bridges, from which the doubly bridged pyrazolate products are formed (60, 66).

Neutral pyrazolate-bridged dinuclear palladium(II) complexes of the general formula $[Pd_2X_2(\mu\text{-}pz^*)_2(PR_3)]$ (Hpz* = Hpz or Hdmpz; X = Cl, Br, I; $PR_3$ = $PBu_3$, $PMe_2Ph$, or $PMePh_2$) were prepared (67). These complexes were char-

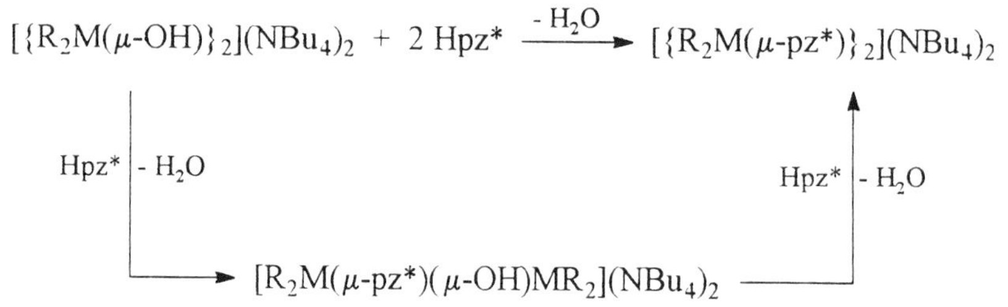

$$[\{R_2M(\mu\text{-}OH)\}_2](NBu_4)_2 \;+\; 2\;Hpz^* \xrightarrow{\;-\;H_2O\;} [\{R_2M(\mu\text{-}pz^*)\}_2](NBu_4)_2$$

$Hpz^*$ | $-\;H_2O$               $Hpz^*$ | $-\;H_2O$

$$[R_2M(\mu\text{-}pz^*)(\mu\text{-}OH)MR_2](NBu_4)_2$$

| M | R | Hpz* |
|----|--------|-------------------|
| Ni | $C_6F_5$ | Hpz, Hmpz |
| Pd | $C_6F_5$ | Hpz, Hmpz, Hdmpz |
| Pd | $C_6Cl_5$ | Hpz, Hmpz |
| Pd | $C_6F_3H_2$ | Hpz, Hmpz, Hdmpz |
| Pt | $C_6F_5$ | Hpz, Hmpz, Hdmpz |

Scheme 1. Exploiting the nucleophilicity of the OH bridges in the anionic complexes $[\{R_2M(\mu\text{-}OH)\}_2](NBu_4)_2$ (M = Ni, Pd, or Pt) is possible to prepare a large number of doubly bridged pyrazolate complexes (see text for references).

acterized by elemental analysis, $^1$H and $^{31}$P NMR data and, in the case of $[Pd_2Cl_2(\mu\text{-dmpz})_2(PMe_2Ph)_2]$, **42,** by single-crystal X-ray diffraction methods. Attempts to synthesize platinum(II) analogues by using the same method of preparation as for the palladium(II) species, that is, by reaction of $[Pt_2Cl_2(\mu\text{-}Cl)_2(PR_3)_2]$ with Hpz* in the presence of methanolic sodium hydroxide, gave either $(\mu\text{-Cl})(\mu\text{-pz*})$ complexes or chloro-bridged derivatives containing terminal pyrazolate ligands (68, 69). However, dinuclear pyrazolate-bridged platinum(II) complexes of the type $[Pt_2Cl_2(\mu\text{-pz*})_2(PR_3)_2]$ (Hpz* = Hpz or Hdmpz; $PR_3$ = $PEt_3$, $PMe_2Ph$, or $PMePh_2$) were isolated in high yield from the reaction of $[Pt_2Cl_2(\mu\text{-}O_2CCH_3)_2(PR_3)_2]$ with Hpz* (70). A single-crystal structure of the representative complex $[Pt_2Cl_2(\mu\text{-dmpz})_2(PMePh_2)_2]$ was also reported (70).

A trinuclear nickel(II) complex, $[Ni_3(dmpz)_4(acac)_2]$, **43,** has been obtained by reacting $Ni(acac)_2$ with an equimolar amount of $Na(dmpz)$ (71). Its X-ray crystal structure emphasized that the geometry of each nickel atom is square planar and these planes provide the zigzag mode. The terminal and center nickel atoms are bridged by nitrogen atoms of two dmpz groups. Furthermore, the terminal metal centers are coordinated to two oxygen atoms of acac.

## D.   Group 9: Co, Rh, or Ir

The dimeric $[Co(dmpz)_2(Hdmpz)]_2$, **44,** and the trimetallic $[Co(dmpz)_2\text{-}Cl(Hdmpz)]_2Co$, **45,** complexes were synthesized. These complexes were studied magnetically and by single-crystal X-ray diffraction (72). Complex **44** was shown to be isomorphous and isostructural with the zinc analogue $[Zn(dmpz)_2(Hdmpz)]_2$, **31** (42). The IR data for **44** reflect the presence of strong hydrogen bonding in the molecule with the appearance of two broad $\nu$(NH) bands at about 2380 and 1870 cm$^{-1}$, and a $\gamma$(NH) band appearing at 865 cm$^{-1}$. Complex **45** crystallizes as a trimetallic molecule with the Co$^{II}$ ions in an approximately linear arrangement. The terminal cobalt atoms are linked to the central cobalt by double dmpz bridges. The central cobalt is coordinated by four pyrazolate nitrogen atoms in a pseudotetrahedral fashion. The terminal Co$^{II}$ ions are also coordinated by one Hdmpz and one chloride ion in a distorted tetrahedral manner.

In 1981, Usón et al. (73) reported a general route to the synthesis of pyrazolato bridged dirhodium(I) complexes of the formula $[(PR_3)(CO)Rh(\mu\text{-pz*})]_2$. It consists in the treatment of $[(cod)Rh(\mu\text{-Cl})]_2$ with an alkali pyrazolate followed by reaction with CO and tertiary phosphines (Eq. 2).

$$[(cod)Rh(\mu\text{-Cl})]_2 \xrightarrow{\text{pz}^-} [(cod)Rh(\mu\text{-pz})]_2 \xrightarrow{\text{CO}}$$

$$[(CO)_2Rh(\mu\text{-pz})]_2 \xrightarrow{\text{PR}_3} [(PR_3)(CO)Rh(\mu\text{-pz})]_2 \qquad (2)$$

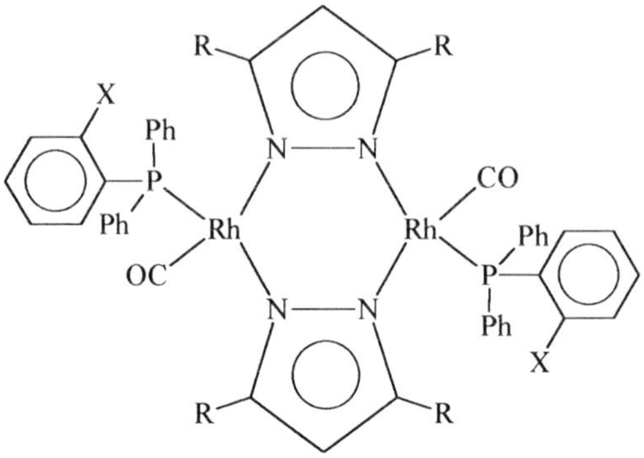

R = H,   X = COOH, **46**

R = Me, X = COOH, **47**

R = H,   X = CHO, **48**

R = Me, X = CHO, **49**

The aforementioned synthetic strategy was found also to be applicable to the synthesis of Complexes **46–49** (74, 75). The X-ray crystal structure of **49** was reported (75).

Moreover, it was reported that the complex [Rh($\mu$-pz)(CO)(PCBr)]$_2$, **50** [PCBr = P($o$-BrC$_6$F$_4$)Ph$_2$], prepared, accordingly Eq. 2, by treating [Rh($\mu$-pz)(CO)$_2$]$_2$ with a stoichiometric amount of PCBr, undergoes orthometalation by a 2c–2e oxidative–addition reaction to give low yields of a new dinuclear compound containing a rhodium–rhodium bond, [Rh$_2$($\mu$-pz)$_2$($\mu$-PC)Br(CO)-(PCBr)] (**51**). This complex was structurally characterized by X-ray crystallography (76).

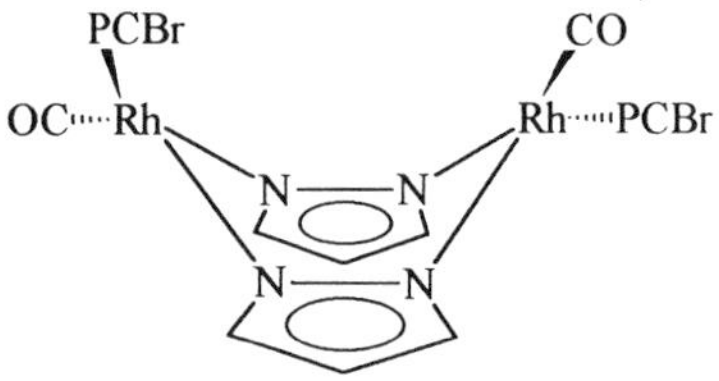

**50** 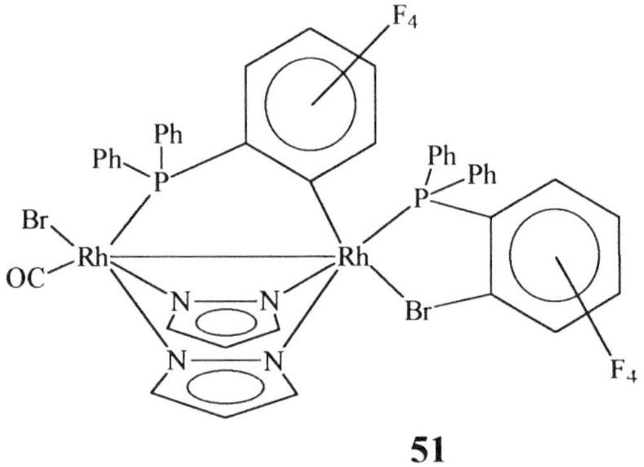                                            **51**

These results have been extended to other complexes of the **50** type; it was ascertained that when substituted pyrazolate ligands were used, the orthometalation reaction becomes more selective (77). The X-ray crystal structure of [Rh$_2$($\mu$-dmpz)$_2$($\mu$-PC)Br(CO)(PCBr)], **52,** was also reported (77).

Dinuclear pyrazolate bridged iridium(I) complexes were prepared by employing a synthetic route analogue to that used for the rhodium(I) derivatives, namely, from $[Ir(cod)(\mu\text{-Cl})]_2$ by Cl displacement by Hpz* in basic medium (78) (Eq. 3):

$$[Ir(cod)(\mu\text{-Cl})]_2 \xrightarrow[\text{base}]{\text{Hpz*}} [Ir(cod)(\mu\text{-pz*})]_2 \tag{3}$$

$$(\text{base} = NEt_3, K(t\text{-BuO})$$

(Hpz* = 4-MepzH, Hmpz, Hdmpz, Hdppz, 3-Ph-5-MepzH,

3-CF$_3$-5-MepzH, Hdfmpz, Htmpz, 4-ClpzH, 4-IpzH, 4-NO$_2$pzH,

4-CldmpzH, 3-CF$_3$-5-PhpzH)

These products were characterized by $^1$H and $^{13}$C NMR spectroscopy. Among complexes incorporating unsymmetrically substituted bridging pyrazolates, $[Ir(cod)(\mu\text{-mpz})]_2$ and $[Ir(cod)(\mu\text{-3-CF}_3\text{-5-Mepz})]_2$ exist as diastereoisomeric mixtures (**C** and **D**) with an approximate $1:1$ ratio, while for $[Ir(cod)(\mu\text{-3-Ph-5-Mepz})]_2$ the diastereoisomer **D** predominates.

$$\textbf{C} \qquad\qquad\qquad \textbf{D}$$

The crystal and molecular structures of the 3-Ph-5-Mepz and tmpz derivatives, as well as the synthesis of the "mixed-bridge" analogue $[Ir(cod)(\mu\text{-pz})(\mu\text{-dfmpz})Ir(cod)]$ and the mixed-metal complex $[Rh(cod)(\mu\text{-pz})_2Ir(cod)]$, were reported (78). The reaction of $[\{Rh(C_5Me_5)\}_2(\mu\text{-OH})_3](ClO_4)$ with an excess of Hpz* and KOH, gave neutral complexes of the formula $[\{Rh(C_5Me_5)(pz*)\}_2(\mu\text{-pz*})_2]$ (Hpz* = Hpz, **53**; 4-BrpzH, **54**) (52). The X-ray structural analysis of Complex **53** confirmed the presence of either exo-bidentate or monodentate pyrazolate groups in the solid state. The addition of HClO$_4$ to Complex **54** resulted in the protonation of the monodentate pyrazolate ligand with formation of $[\{Rh(C_5Me_5)(4\text{-BrpzH})\}_2(\mu\text{-4-Brpz})_2](ClO_4)_2$. Complex **54** can be regenerated on treatment with KOH.

The existence of the well-known acac complex $[Rh(C_5Me_5)(acac)Cl]$ potentially provides a preparative route to dinuclear rhodium(III)–rhodium(I) species with two pz* groups acting as bridging ligands. Thus, its reaction with the cationic complexes [Rh(Hpz)$_2$(L)](ClO$_4$) (L = tfbb or cod) in the presence of stoichiometric amounts of KOH, yielded the corresponding neutral complexes [(C$_5$Me$_5$)ClRh($\mu$-pz)$_2$Rh(L)] (79). The latter complexes can be alternatively prepared by reacting the neutral [(C$_5$Me$_5$)Rh(acac)(pz)] with [RhCl(Hpz)(L)]. The reaction of [(C$_5$Me$_5$)Rh(acac)(pz)] with the cationic rhodium(I) complex [Rh(tfbb)(Hpz)$_2$](ClO$_4$) in the presence of KOH afforded [(C$_5$Me$_5$)(pz)Rh($\mu$-pz)$_2$Rh(tfbb)], which was reported to contain terminal and bridged pyrazolate groups on the basis of their $^1$H NMR spectra. Carbonylation of [(C$_5$Me$_5$)(Cl)Rh($\mu$-pz)$_2$Rh(L)] caused the displacement of the coordinated diolefin and the formation of the carbonyl complex [(C$_5$Me$_5$)(Cl)Rh($\mu$-pz)$_2$Rh(CO)$_2$], (55) (Scheme 2) (79).

The reactivity of Complex 55, of the related iodo derivative [(C$_5$Me$_5$)(I)Rh($\mu$-pz)$_2$Rh(CO)$_2$], 56, and of the heterobimetallic compound [(C$_5$Me$_5$)(Cl)Ir($\mu$-pz)$_2$Rh(CO)$_2$], 57, with monodentate tertiary phosphines, diphosphines, and diarsines was also investigated (Scheme 3) (80–82).

The structure of [(C$_5$Me$_5$)ClRh($\mu$-pz)$_2$($\mu$-CO)Rh(dppp)](BPh$_4$) was established by X-ray crystallography (81).

Only in the last decade has it become apparent that oxidative addition may result in attachment of substrate fragments at other than a single metal site, although such events have frequently been hypothesized in relation to cooperativity effects in biological systems, as well as in the context of heterogeneous catalytic reactions, especially reorganization of organic molecules at metal surfaces. Around 1980 it was found that doubly bridged pyrazolate iridium(I) dimers undergo facile reactions with halogens or alkyl halides yielding complexes

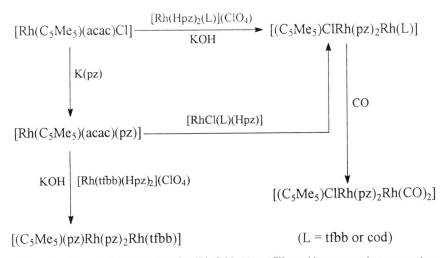

Scheme 2. The acetylacetonate complex [Rh(C$_5$Me$_5$)(acac)Cl] provides a convenient preparative route to doubly bridged Rh(I) pyrazolate derivatives (see text for references).

$[(C_5Me_5)M(\mu\text{-pz})_2(\mu\text{-CO})Rh(CO)(PR_3)]^+$

↑ (a)

$[(C_5Me_5)(Cl)M(\mu\text{-pz})_2Rh(CO)_2] \xrightarrow{\text{(b)}} [(C_5Me_5)M(\mu\text{-pz})_2(\mu\text{-CO})Rh(L\text{-}L)]^+$

↓ (c) ↑ (d)

$[(C_5Me_5)M(\mu\text{-pz})_2(\mu\text{-CO})Rh(CO)(L\text{-}L)]^+ \xrightarrow{\text{(d)}}$

Scheme 3. Reactivity of the species $[(C_5Me_5)(Cl)M(\mu\text{-pz})_2Rh(CO)_2]$. (M = Rh, or Ir) (a) $PR_3$ (R = Ph or Cy), $NaBPh_4$, MeOH; (b) L—L (dppm, dppp, dppb, or dpae), $AgBF_4$ (c) L—L, $NaBPh_4$, MeOH; (d) acetone or $CHCl_3$ or $CH_2Cl_2$ (see text for references).

formulated as the products of a two-center oxidative addition (83, 84). In late 1984, Atwood et al. (85) reported the spectral data and the full characterization by X-ray crystallography of the dinuclear complex $[Ir(PPh_3)(CO)(\mu\text{-pz})]_2$, **58**. Reactions of **58** with $I_2$, $Br_2$, or $Cl_2$ gave products formulated as $[Ir(PPh_3)(CO)(X)(\mu\text{-pz})]_2$ (X = Cl, **59**; Br, **60**; I, **61**), having each metal center in a formal II oxidation state (85). In the case of reaction with $Cl_2$, Cl substitution at the 4-position in the pyrazolyl ring was observed. The X-ray crystal structure analysis for **59** revealed that the Ir—Ir separation had indeed shortened to within bonding distance. Reaction of **58** with MeI was reported to yield a similar 1:1 adduct (85).

A series of two fragment, two-center additions of $CH_2I_2$ to the doubly bridged pyrazolate iridium(I) dimers $[IrL_2(\mu\text{-pz})]_2$ [$L_2 = (CO)_2$, **62**; cod, **63**; CO, $PPh_3$, **64**] was reported to form 1:1 adducts, $[Ir_2(L_2)_2(\mu\text{-pz})_2(I)(CH_2I)]$, which were formulated mainly on the basis of spectroscopic data (86). Interestingly, complexes $[Ir_2(CO)_4(\mu\text{-pz})_2(I)(CH_2I)]$, **65**, and $[Ir_2(CO)_2(PPh_3)_2(\mu\text{-pz})_2(I)(CH_2I)]$, **66**, readily undergo thermal "oxidative isomerization" to the di-iridium(III) methylene species $[Ir_2(CO)_4(\mu\text{-pz})_2(\mu\text{-CH}_2)(I)_2]$, **67**, and $[Ir_2(CO)_2(PPh_3)_2(\mu\text{-pz})_2(\mu\text{-CH}_2)(I)_2]$, **68** (86, 87).

An X-ray crystal structure determination for Compound **68** confirmed that

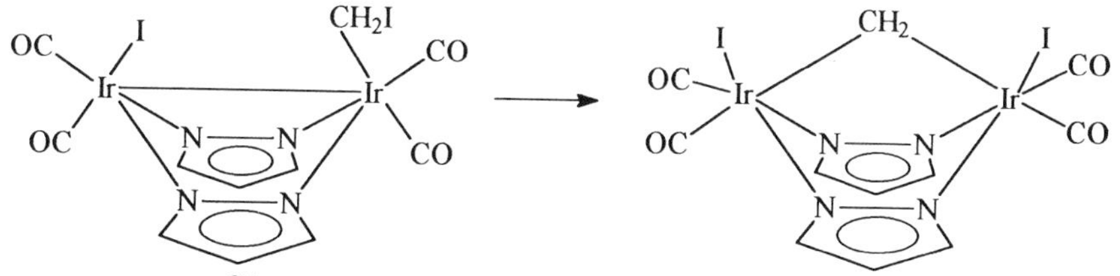

**65** → **67**

the two Ir centers lie outside the bonding range at 3.43 Å, with a bridging $CH_2$ group (87). The isomerization of Complexes **65** and **66** to the di-iridium(III) methylene species **67** and **68** lays emphasis to the remarkable role of the pyrazolate ligands. In fact, it was reported that the oxidative addition of $CH_2I_2$ to the neutral dinuclear iridium(I) complex $[Ir_2(CO)_4(\mu\text{-}C_5H_4NS)_2]$ ($C_5H_4NS$ = 2-pyridinethiolate) afforded $[Ir_2(CO)_4(\mu\text{-}C_5H_4NS)_2(I)(CH_2I)]$ as the final product, with no formation of a methylene–bridged complex $[\{Ir(\mu\text{-}C_5H_4NS)_2(I)(CO)_2\}(\mu\text{-}CH_2)]$, analogous to **67,** occurring (88).

Kinetic data for two-fragment, two-center addition of MeI to the iridium(I) dimers **62** and **63** were reported (89). Addition of MeI occurs irreversibly to **62** in two kinetically distinguishable steps; by contrast MeI and **63** enter into equilibrium.

The treatment with gaseous HCl of the cationic nitrosyl-pyrazolate complex $[(cod)(NO)Ir(\mu\text{-}pz)_2Ir(cod)](BF_4)$, **69** (obtained from $[Ir(cod)(\mu\text{-}pz)]_2$ and $NOBF_4$) led to the isolation of $[(cod)(Cl)Ir(\mu\text{-}pz)_2(\mu\text{-}NO)Ir(Cl)(cod)](BF_4)$, **70** (90). Complex **70** has been characterized by X-ray crystal structure determination (90). The reaction affording complex **70** has been described as an unprecedented two-center oxidative addition as a result of which a bent, terminal NO group takes up a bridging position.

Oxidative–addition reaction by iodine carried out in methanol or ethanol on complexes $[(C_5Me_5)(Cl)Rh(\mu\text{-}pz)_2Rh(CO)_2]$, **55,** or $[(C_5Me_5)(Cl)Ir(\mu\text{-}pz)_2Rh(CO)_2]$, **71,** in the presence of sodium salts such as $NaBPh_4$ or $NaI \cdot 2H_2O$, gave the alkoxycarbonyl complex $[(C_5Me_5)M(\mu\text{-}pz)(\mu\text{-}I)_2Rh(I)(CO_2R)(CO)]$ (M = Rh, R = Me or Et; M = Ir, R = Me). The molecular structure of the Ir/Rh derivative was also determined (82).

The alkoxocarbonyl complexes described above are thought to be the products of the nucleophilic attack of MeOH or EtOH on the dinuclear carbonyl Complexes **55** and **71,** which were previously oxidized by iodine, as the latter complex are indefinitely stable in alcohols (82).

The easy three-fragment, four-electron oxidative addition of chloroform and gem-dichloroalkanes to $[Rh(\mu\text{-}pz)(t\text{-}BuNC)_2]_2$, **72,** were prepared in good yield by replacement of cod in $[Rh(\mu\text{-}pz)(cod)]_2$ (91, 92), yielding functionalized methylene bridged complexes (93). Thus, molecules such as $CHCl_2R$ (R = H, Ph, COOMe, or Cl) react with Complex **72** to yield mainly $[\{Rh(\mu\text{-}pz)Cl(t\text{-}BuNC)_2\}_2(\mu\text{-}CHR)]$ and $[Rh(\mu\text{-}pz)Cl(t\text{-}BuNC)_2]_2$ as side products. A definitive proof of the structure of the methylene-bridged complexes comes from their $^1H$ and $^{13}C$ NMR spectra (93). No intermediates were observed in these reactions.

The reaction of **72** with 1,1,1-trichloroethane was also tested (92). A spontaneous elimination of HCl takes place and $[\{Rh(\mu\text{-}pz)Cl(t\text{-}BuNC)_2\}(\mu\text{-}C{=}CH_2)]$, **73,** containing a bridging vinylidene ligand, was obtained and structurally characterized (93). Interestingly, the subsequent reaction of **73** with $HBF_4 \cdot Et_2O$ and $O_2$ afforded the cationic dinuclear acyl derivative $[\{Rh(\mu\text{-}pz)(t\text{-}BuNC)_2\}_2(\mu\text{-}Cl)Cl(\eta^1\text{-}COCH_3)](BF_4)$, **74** (Scheme 4). The latter reaction in-

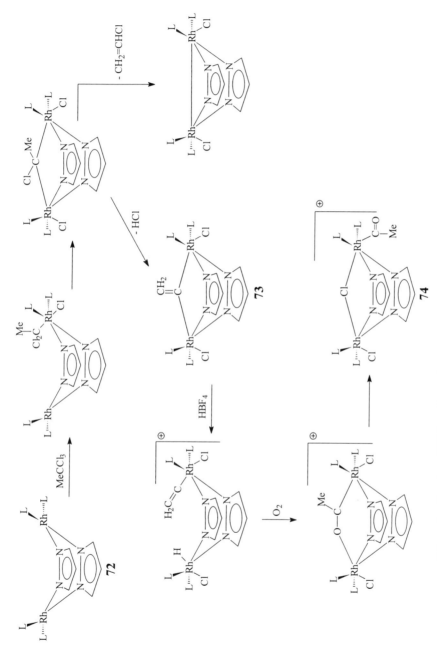

Scheme 4. Reaction of [Rh(μ-pz)(L)₂]₂, **72** (L = *t*-BuNC) with MeCCl₃ and proposed mechanism for the formation of the acyl derivative **74** (see text). [Adapted from (94).]

175

volves the oxidation of a vinylidene ligand by molecular oxygen, although the starting complex does not react with $O_2$ and vinylidene ligands are not easy to oxidize. A plausible mechanism for these reactions was proposed (94).

In 1986, the preparation and characterization of the bis(pyrazolate)pyrazole complex $[C_5Me_5)Ir(pz)_2(Hpz)]$, **75,** was described (95).

**75**

From a coordination point of view, Complex **75** and the related ruthenium complex $[(\eta^6\text{-}p\text{-cymene})Ru(pz)_2(Hpz)]$, **76** (7), are comparable to protonated polypyrazolylborates $RB(pz)_2(pzH)$ (R = H or pz) (12). The formal similarity between the tris(pyrazolyl)borate anion and the deprotonated form of the iridium complex **75** suggested the preparation of heterodinuclear complexes by using **75** as a building block. Thus, heterodinuclear ($\mu$-pz)$_2$ complexes of the formula $[(C_5Me_5)(pz)Ir(\mu\text{-pz})_2M(PPh_3)]$ (M = Cu, **77;** Ag, **78;** Au, **79**) were obtained (96).

Complexes **77–79** were characterized by elemental analysis, spectroscopic measurements and, for Complex **78,** by X-ray crystal structure determination. Complexes **77–79** exhibit fluxional properties: At room temperature the M(PPh$_3$) fragment exchanges fast among the three free nitrogen atoms of the pyrazolate ligands. The crystal structure of Complex **78** shows that, in the solid state, there is a free nitrogen atom whose lone-electron pair can be potentially used to further coordination. In effect, Complex **78** reacts with equimolar amounts of AgBF$_4$ and PPh$_3$, affording the trinuclear complex $[(C_5Me_5)Ir(\mu\text{-}pz)_3\{Ag(PPh_3)\}_2](BF_4)$, **80.** Spectral data of **80** have been interpreted as the existence of a fluxional process that renders the three pyrazolate and the two Ag(PPh$_3$) groups equivalent (96).

Iridium–rhodium heterometallic complexes of the formula $[(C_5Me_5)(pz)Ir(\mu\text{-}$

pz)$_2$RhL$_2$] [L$_2$ = cod, **81**; (CO)$_2$, **82**)] were also isolated (96). The free nitrogen atom of Complexes **81** and **82** is again capabale of coordination to the cationic moiety [Ag(PPh$_3$)]$^+$, which is isolobal to the proton. Consequently, heterotrinuclear complexes, Ir/Ag/Rh, were prepared (96). The dynamic behavior of the heterobimetallic compounds **78**, **80**, and [(C$_5$Me$_5$)(PPh$_3$)Ir($\mu$-pz)$_2$-Ag(PPh$_3$)](BF$_4$), **83**, were studied by multinuclear NMR spectroscopy (97). The results demonstrated that argentotropism accounts for the observed spectra of Compounds **78** and **80**; in the case of Compound **83**, the dynamic behavior corresponds to a flipping of the central heteroring Ir—[N—N]$_2$—Ag (97).

An electrochemical study of the behavior of the dinuclear complex [Rh$_2$(CO)$_2$(PPh$_3$)$_2$($\mu$-dmpz)$_2$], **84,** was described (98). This species undergoes two consecutive oxidation processes at a platinum electrode. The first oxidation gives the corresponding cationic species, [Rh$_2$(CO)$_2$(PPh$_3$)$_2$($\mu$-dmpz)$_2$]$^+$, as inferred from electrochemical and spectroscopic examination of the oxidized product.

The rates of photoinduced electron transfer (ET) reactions in a series of iridium (spacer)pyridinium complexes, [Ir($\mu$-dmpz)(CO)(Ph$_2$PO-CH$_2$-CH$_2$-py$^+$)]$_2$ and [Ir($\mu$-dmpz)(CO)(Ph$_2$PO-C$_6$H$_4$(CH$_2$)$_n$-py$^+$)]$_2$ ($n$ = 0 − 3), have been studied in acetonitrile solution at room temperature (99). The nuclear reorganization energies and electronic couplings in these systems have been evaluated.

The dimeric iridium(I) compound [Ir($\mu$-dmpz)(CO)(Ph$_2$PO-C$_6$H$_5$Me)]$_2$ has been synthesized by the reaction of [Ir($\mu$-dmpz)(CO)$_2$]$_2$ with the 4-tolyl-diphenylphosphinite ligand, Ph$_2$PO-C$_6$H$_4$Me (99) and its X-ray crystal structure has been determined (100). The photophysical and electrochemical properties of this iridium dimer complex have been investigated (100).

### E.   Group 8: Fe, Ru, or Os

By using [RuH(Cl)(CO)(Hpz)(PPh$_3$)$_2$] (101) as the starting reagent, a series of new mono- and dinuclear complexes containing two pyrazole groups can be prepared (102). Thus, treatment of [RuH(Cl)(CO)(Hpz)(PPh$_3$)$_2$] with stoichiometric amounts of KOH and Hpz afforded the neutral complex [RuH(pz)(CO)(Hpz)(PPh$_3$)$_2$], **85**. The latter complex reacted with [M($\mu$-OMe)(tfbb)]$_2$ yielding the dinuclear compounds [(PPh$_3$)$_2$(CO)HRu($\mu$-pz)$_2$M(tfbb)] (M = Ir, **86**; Rh, **87**). The diolefin tfbb, in Compounds **86** and **87**, can be easily displaced by carbon monoxide yielding [(PPh$_3$)$_2$(CO)HRu($\mu$-pz)$_2$M(CO)$_2$] (102) (Scheme 5).

The dinuclear complexes [Ru$_2$($\mu$-pz*)$_2$(CO)$_6$] (Hpz* = Hpz, **88**; Hdmpz, **89**; 4-MepzH, **90**, Hmpz, **91a**, and **91b**) were prepared by reaction of RuCl$_3 \cdot n$H$_2$O, the cheapest starting material for ruthenium compounds, with CO followed by reduction with zinc in the presence of Hpz* and CO (103, 104) or, alternatively, by reacting Ru$_3$(CO)$_{12}$ with the relevant Hpz* (105).

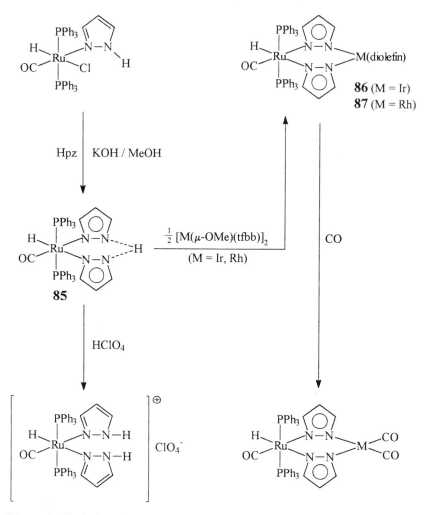

Scheme 5. Synthesis and some reactions of the pyrazole/pyrazolate Ru(II) derivative [RuH(px)(CO)(Hpz)(PPh₃)₂], **85**. [Adapted from (102).]

These complexes were fully characterized and the X-ray crystal structure of **89** was also reported. The reactivity of these dinuclear complexes toward a variety of nitrogen- and phosphorus-donor ligands was described (104–106) as well as oxidative–addition reaction with iodine (104).

With monodentate phosphines, monosubstituted or disubstituted products such as [Ru₂(μ-pz*)₂(CO)₅(PR₃)] or [Ru₂(μ-pz*)₂(CO)₄(PR₃)₂] were obtained depending on the nature of the pyrazolate ligand. Complex **97** (Table II) was

**91a**                 **91b**                 **97**

alternatively obtained by the reaction of Na(pz) with [Ru$_2$(CO)$_4$-(O$_2$CMe)(MeCN)$_2$] followed by addition of PPh$_3$ and its X-ray crystal structure was determined (107). A nitrogen-containing ligand such as MeCN or py gave [Ru$_2$($\mu$-dmpz)$_2$(CO)$_5$(L)] when it was reacted with **89** (104). The dmpz derivative (**89**) afforded the monosubstituted species even with dppm (Table II) (104). The reaction of **88** or **89** with carboxylic acids affords the carboxylato-bridged dimers [Ru$_2$(CO)$_4$($\mu$-O$_2$CR)$_2$(Hpz*)$_2$] (105). In this process, the heterocyclic bridging ligand is transformed into the monodentate Hpz* group. The unity of the system is maintained by the carboxylato bridges formed. The ligands dppe and cyclop react with **88** to give the diphosphine bridged monosubstitution products **98** and **99**.

On the contrary, with **89,** the coordination of dppe and cyclop proceed under ortho-metalation of the phenyl group to generate the hydrido clusters **100** and **101** (106). The X-ray crystal structure of **100** evidenced an almost symmetrical position between the Ru atoms for the bridging hydride ligand (106).

Complexes **88** and **89** reacted with I$_2$ to afford [Ru$_2$($\mu$-I)($\mu$-pz*)$_2$(CO)$_6$]I$_3$ or [Ru$_2$I$_2$($\mu$-pz*)$_2$(CO)$_6$] depending on the Ru/I ratio used. The crystal structure of

TABLE II
Dinuclear Complexes of Formula [Ru$_2$(CO)$_4$(pz*)$_2$(L)(L')]

| Complex | Hpz* | L | L' | Notes |
|---|---|---|---|---|
| **88** | Hpz | CO | CO | |
| **89** | Hdmpz | CO | CO | X-ray |
| **90** | 4-MepzH | CO | CO | |
| **91** | Hmpz | CO | CO | Two isomers |
| **92** | Hdmpz | CO | MeCN | |
| **93** | Hdmpz | CO | py | |
| **94** | Hdmpz | CO | PPh$_3$ | |
| **95** | Hdmpz | CO | P(C$_6$H$_{11}$)$_3$ | |
| **96** | Hdmpz | CO | dppm | |
| **97** | Hpz | PPh$_3$ | PPh$_3$ | X-ray |

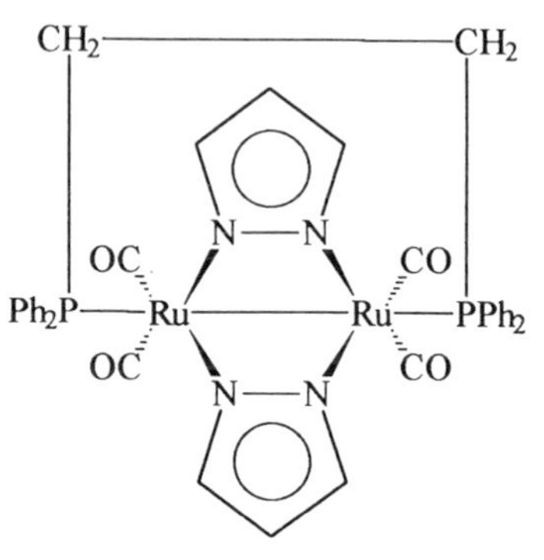

**98**

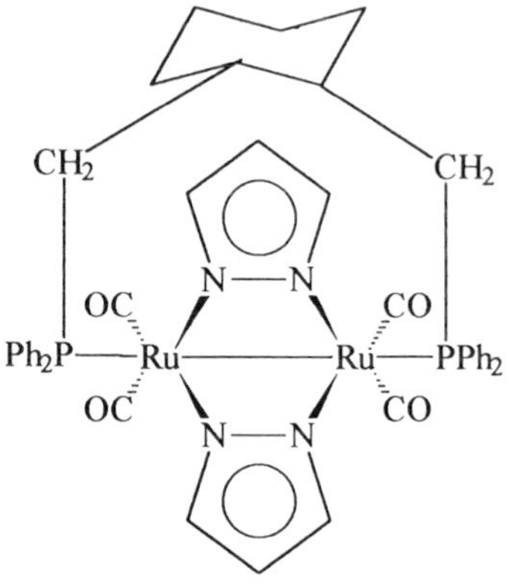

**99**

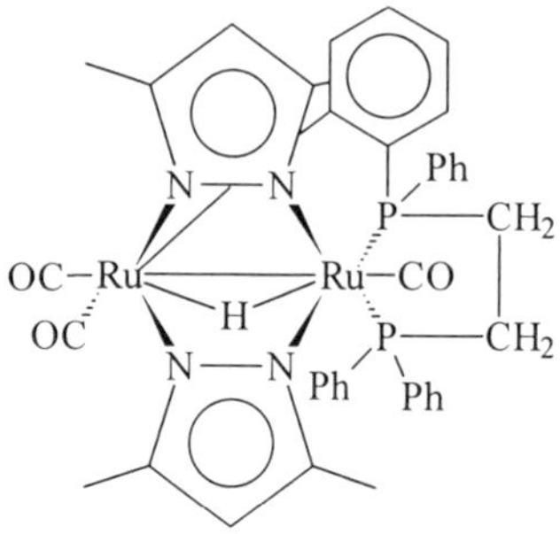

**100**

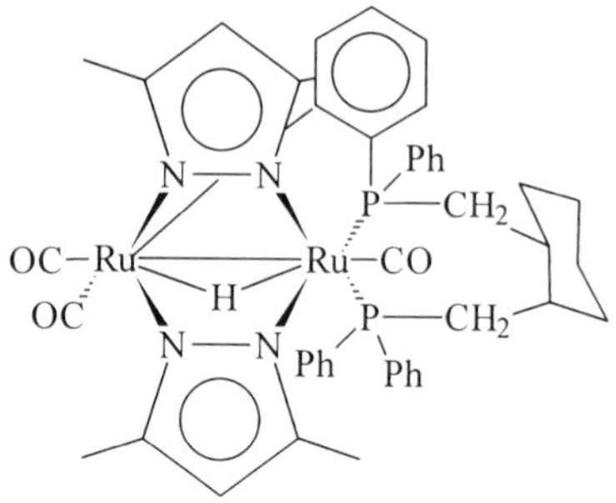

**101**

**89** and [Ru$_2$($\mu$-I)($\mu$-dmpz)$_2$(CO)$_6$]I$_3$ were determined by X-ray diffraction methods. The Ru—Ru distance in **89** is consistent with a metal–metal bond (104).

The reaction of the mononuclear ruthenium p-cymene pyrazolate complex [($\eta^6$-p-cymene)Ru(pz)$_2$(Hpz)], **76**(7) with the halide triphenylphosphine complexes [MCl(PPh$_3$)]$_x$ and KOH led to the corresponding heterodinuclear compounds [($\eta^6$-p-cymene)(pz)Ru($\mu$-pz)$_2$M(PPh$_3$)] (M = Cu, **102**; Ag, **103**; Au, **104**) (108). The $^1$H and $^{31}$P NMR data of **103** are comparable to those of [(C$_5$Me$_5$)(pz)Ir($\mu$-pz)$_2$Ag(PPh$_3$)], **78,** for which an X-ray crystal structure analysis was performed (96). It is therefore reasonable that both compounds adopt similar structures.

Complex **103** reacted with AgBF$_4$ and PPh$_3$ to yield the heterotrinuclear RuAg$_2$ complex [($\eta^6$-p-cymene)Ru($\mu$-pz)$_3$\{Ag(PPh$_3$)\}$_2$](BF$_4$) (108). The NMR data of this complex can be interpreted by assuming that a fluxional process consisting of the rotation of the two Ag(PPh$_3$) fragments around the three free nitrogen atoms of the ($\eta^6$-$p$-cymene)Ru(pz)$_3$ moiety was occurring in a concerted way.

The ruthenium complex **76** reacted with equimolar amounts of [M(acac)L$_2$] (M = Rh, L$_2$ = cod or (CO)$_2$; M = Ir, L = CO) to give the corresponding heterodinuclear complexes [($\eta^6$-$p$-cymene)(pz)Ru($\mu$-pz)$_2$ML$_2$]. The latter compounds gave cationic derivatives of formula [($\eta^6$-$p$-cymene)(Hpz)Ru($\mu$-pz)$_2$ML$_2$](BF$_4$) when treated with an equimolar amount of HBF$_4$. An X-ray crystal structure determination was performed for the cationic complex having M = Ir and L = CO (108).

## IV.  HETEROBRIDGED PYRAZOLATE COMPLEXES

### A.   Complexes Containing the ($\mu$-pz*)$_a$($\mu$-X)$_b$($\mu$-Y)$_c$ Fragment

A variety of heterobridged di- and trinuclear complexes containing, along with pyrazolate ligands, simple anions such as Cl, OH, OR, N$_3$, or SCN, in which single atoms are the bridging groups, were prepared (109–113). Among these complexes, the formation and the chemical properties of mixed-bridged complexes containing the ($\mu$-pz*)($\mu$-SR) fragment was thoroughly investigated (109–114, 117, 118). The increasing interest for these mixed-bridged systems arises also from the recognized good catalytic activity, under mild conditions, of the homobridged complex [Rh$_2$($\mu$-S-$t$-Bu)$_2$(CO)$_2${P(OMe)$_3$}$_2$] in the selective hydroformilation of alkenes (115).

A series of dinuclear rhodium(I) and iridium(I) complexes containing the M($\mu$-pz*)($\mu$-S-$t$-Bu)M core was obtained according to Eqs. 4 and 5 (109, 117, 118) (Table III):

$$[MCl(Hpz*)(cod)] \ + \ [M(acac)(cod)] \xrightarrow{\ -Hacac\ }$$

$$[M_2(\mu\text{-pz*})(\mu\text{-Cl})(cod)_2] \xrightarrow[-KCl]{KS\text{-}t\text{-Bu}} [M_2(\mu\text{-pz*})(\mu\text{-S-}t\text{-Bu})(cod)_2] \qquad (4)$$

$$[M_2(\mu\text{-pz*})(\mu\text{-S-}t\text{-Bu})(cod)_2] \xrightarrow{CO} [M_2(\mu\text{-pz*})(\mu\text{-S-}t\text{-Bu})(CO)_4] \xrightarrow{L}$$

$$[M_2(\mu\text{-pz*})(\mu\text{-S-}t\text{-Bu})(CO)_2L_2] \qquad (5)$$

In complexes containing the ($\mu$-S-$t$-Bu) group, the individual metal centers are joined by one exo-bidentate pyrazolate ligand and one tert-buthylthiol group, in such a way that the cyclic bridged M($\mu$-pz*)($\mu$-S-$t$-Bu)M core deviates from planarity, adopting a flexible bent conformation that results in a wide range of intermetallic separations (114).

The crystal and molecular structure of **111** was determined from a single

TABLE III
Homodinuclear Complexes Containing the $M(\mu\text{-pz*})(\mu\text{-S-}t\text{-Bu})M$ Core (M = Rh or Ir)

| Complex | Formula | Notes | Reference |
|---|---|---|---|
| 105 | [Rh$_2$($\mu$-pz)($\mu$-S-$t$-Bu)(cod)$_2$] | | 117 |
| 106 | [Ir$_2$($\mu$-pz)($\mu$-S-$t$-Bu)(cod)$_2$] | | 118 |
| 107 | [Ir$_2$($\mu$-dmpz)($\mu$-S-$t$-Bu)(cod)$_2$] | | 114 |
| 108 | [Rh$_2$($\mu$-pz)($\mu$-S-$t$-Bu)(CO)$_4$] | | 117 |
| 109 | [Ir$_2$($\mu$-pz)($\mu$-S-$t$-Bu)(CO)$_4$] | | 118 |
| 110 | [Ir$_2$($\mu$-dmpz)($\mu$-S-$t$-Bu)(CO)$_4$] | | 114 |
| 111 | [Rh$_2$($\mu$-pz)($\mu$-S-$t$-Bu)(CO)$_2${P(OMe)$_3$}$_2$] | X-ray | 117 |
| 112 | [Rh$_2$($\mu$-pz)($\mu$-S-$t$-Bu)(CO)$_2${P(OPh)$_3$}$_2$] | | 117 |
| 113 | [Rh$_2$($\mu$-pz)($\mu$-S-$t$-Bu)(CO)$_2$(PPh$_3$)$_2$] | | 117 |
| 114 | [Ir$_2$($\mu$-pz)($\mu$-S-$t$-Bu)(CO)$_2${P(OMe)$_3$}$_2$] | | 118 |
| 115 | [Ir$_2$($\mu$-dmpz)($\mu$-S-$t$-Bu)(CO)$_2${P(OMe)$_3$}$_2$] | | 114 |
| 116 | [Ir$_2$($\mu$-dmpz)($\mu$-S-$t$-Bu)(CO)$_2$(PPh$_3$)$_2$] | | 114 |
| r117 | [Rh$_2$($\mu$-pz)($\mu$-S-$t$-Bu)($\mu$-dppm)(CO)$_2$] | | 110 |
| 118 | [Ir$_2$($\mu$-pz)($\mu$-S-$t$-Bu)($\mu$-dppm)(CO)$_2$] | | 110 |
| 119 | [Ir$_2$($\mu$-dmpz)($\mu$-S-$t$-Bu)($\mu$-dppm)(CO)$_2$] | | 110 |
| 120 | [Ir$_2$($\mu$-pz)($\mu$-S-$t$-Bu)I$_2$(CO)$_2${P(OMe$_3$)}$_2$] | | 118 |
| 121 | [Ir$_2$($\mu$-pz)($\mu$-S-$t$-Bu)I$_4$(CO)$_2${P(OMe$_3$)}$_2$] | | 118 |
| 122 | [Ir$_2$($\mu$-pz)($\mu$-S-$t$-Bu)($\mu$-MeO$_2$C—C=C—CO$_2$Me)(CO)$_2${P(OMe$_3$)}$_2$] | X-ray | 118 |
| 123 | [Ir$_2$($\mu$-pz)($\mu$-S-$t$-Bu)($\mu$-MeO$_2$C—C=C—CO$_2$Me)I$_2$(CO)$_2${P(OMe$_3$)}$_2$] | X-ray | 118 |
| 124 | [Rh$_2$($\mu$-pz)($\mu$-S-$t$-Bu)(Me)I(CO)$_2${P(OMe$_3$)}$_2$] | | 114 |
| 125 | [Ir$_2$($\mu$-pz)($\mu$-S-$t$-Bu)(Me)I(CO)$_2${P(OMe$_3$)}$_2$] | | 114 |
| 126 | [Ir$_2$($\mu$-pz)($\mu$-S-$t$-Bu)($\mu$-CH$_2$)I$_2$(CO)$_2${P(OMe$_3$)}$_2$] | X-ray | 114 |
| 127 | [Ir$_2$($\mu$-dmpz)($\mu$-S-$t$-Bu)($\mu$-CH$_2$I)I(CO)$_2${P(OMe$_3$)}$_2$] | | 114 |
| 128 | [Ir$_2$($\mu$-pz)($\mu$-S-$t$-Bu)Cl$_2$(CO)$_2${P(OMe$_3$)}$_2$] | | 114 |
| 129 | [Ir$_2$($\mu$-pz)($\mu$-S-$t$-Bu)(SnCl$_2$)$_2$(CO)$_2${P(OMe$_3$)}$_2$] | | 114 |
| 130 | [Ir$_2$($\mu$-pz)($\mu$-S-$t$-Bu)(SnCl$_3$)$_2$(CO)$_2${P(OMe$_3$)}$_2$] | Two isomers X-ray for one | 114 |
| 131 | [Ir$_2$($\mu$-pz)($\mu$-S-$t$-Bu)($\mu$-SnCl$_2$)(CO)$_2${P(OMe$_3$)}$_2$] | | 114 |
| 132 | [Rh$_2$($\mu$-pz)($\mu$-S-$t$-Bu)($\mu$-dppm)I$_2$(CO)$_2$] | | 116 |
| 133 | [Ir$_2$($\mu$-pz)($\mu$-S-$t$-Bu)($\mu$-dppm)I$_2$(CO)$_2$] | | 116 |
| 134 | [Ir$_2$($\mu$-dmpz)($\mu$-S-$t$-Bu)($\mu$-dppm)I$_2$(CO)$_2$] | | 116 |
| 135 | [Ir$_2$($\mu$-pz)($\mu$-S-$t$-Bu)($\mu$-dppm)($\mu$-HgCl)Cl(CO)$_2$] | | 116 |
| 136 | [Ir$_2$($\mu$-dmpz)($\mu$-S-$t$-Bu)($\mu$-dppm)($\mu$-HgCl)Cl(CO)$_2$] | | 116 |

crystal by X-ray diffraction. This complex is dinuclear, with a pseudomirror plane passing through the S atom and bisecting the pyrazolate ring, with the carbonyl ligands (related to this mirror plane) in a cis conformation, and with the sulfur atom trans to the trimethylphosphite ligands (117).

It is of interest that the heterobridged complexes 108 and 111 were formed by mixing the homobridged complexes [Rh$_2$($\mu$-pz)$_2$(CO)$_2$L$_2$] and [Rh$_2$($\mu$-S-$t$-Bu)$_2$(CO)$_2$L$_2$] [L = P(OMe)$_3$ or CO]. No dismutation of the heterobridged complexes was observed (117). The reaction of dppm with Complexes 108–110 quantitatively yields [Rh$_2$($\mu$-pz)($\mu$-S-$t$-Bu)($\mu$-dppm)(CO)$_2$], 117, [Ir$_2$($\mu$-pz)($\mu$-S-

$t$-Bu)($\mu$-dppm)(CO)$_2$], **118**, and [Ir$_2$($\mu$-dmpz)($\mu$-S-$t$-Bu)($\mu$-dppm)(CO)$_2$], **119**, respectively (110). The structure of these triply heterobridged complexes was assigned on the basis of $^{31}$P NMR spectral data. Extensive studies were carried out on the chemical properties of some of the complexes reported above. In particular, attention has been paid to oxidative–addition reactions with alogens (I$_2$ or Cl$_2$), methyl iodide, di-iodomethane, and HgX$_2$ (X = Cl or I).

Compound **114** undergoes stepwise addition of I$_2$ to yield the complexes [Ir($\mu$-pz)($\mu$-S-$t$-Bu)(I)$_n$(CO)$_2${P(OMe)$_3$}$_2$] ($n$ = 2, **120** or 4, **121**) (Scheme 6). The increase of the formal oxidation number of the iridium center from I to II and, then to III, is paralleled by the increase of the carbonyl stretching frequency from 1985 to 2040 and 2082 cm$^{-1}$, respectively (118).

Reaction of complex [Ir$_2$($\mu$-pz)($\mu$-S-$t$-Bu)(CO)$_2${P(OMe)$_3$}$_2$], **114**, with the electron-withdrawing alkyne dimethyl-acetylenedicarboxylate (MeO$_2$C—C≡ C—CO$_2$Me) gave the triply heterobridged compound [Ir$_2$($\mu$-pz)($\mu$-S-$t$-Bu)($\mu$-MeO$_2$C—C=C—CO$_2$Me)(CO)$_2${P(OMe)$_3$}$_2$], **122** (118). Its X-ray crystal structure evidenced the cis-dimetalated olefinic bonding mode for the unsaturated ligand (118). Complex **122** undergoes further oxidative addition with I$_2$ affording **123**.

The oxidative addition of MeI to complexes **111** and **114** and of CH$_2$I$_2$ to **114** or **115** was also studied. From these reactions, Complexes **124–127** were obtained (111). The spectroscopic data provided the basis of a consistent interpretation of the mechanism of the reaction involving **114** and CH$_2$I$_2$.

The first step of the reaction is the formation of the intermediate species **E** (Scheme 7) containing a single iridium–iridium bond. The migration of the methylene group leads to the formation of the final Ir$^{III}$Ir$^{III}$ symmetric complex. The intermediate unsymmetrical species **E** was isolated in the reaction of CH$_2$I$_2$ with **115**.

The reaction of **114** with HgX$_2$ (X = Cl or I) gave the oxidation compounds [Ir$_2$($\mu$-pz)($\mu$-S-$t$-Bu)(X)$_2$(CO)$_2${P(OMe)$_3$}$_2$] (X = Cl, **128**; I, **120**). The addition of SnCl$_2$ to the latter complexes involves the insertion of tin(II) into the Ir—X bond and the formation of the trihalogeno/tin derivatives **129** and **130**.

A 1 : 1 adduct (**131**) was obtained on direct treatment of **114** with SnCl$_2$. The formation of these complexes was deduced by IR, $^{31}$P, and $^{119}$Sn NMR spectroscopy. In solution, [Ir$_2$($\mu$-pz)($\mu$-S-$t$-Bu)(SnCl$_3$)$_2$(CO)$_2${P(OMe)$_3$}$_2$] appears to be a mixture of an asymmetric **130a** and a symmetric **130b** isomer.

The crystal structure of **130a**, which was obtained by fractional crystallization, was reported. The Ir—Ir distance (2.72 Å) indicates the presence of a metal–metal bond (114). Complexes **117–119** react readily with an equimolar amount of I$_2$ to yield **132–134**, respectively, which on the basis of analytical and spectral data were formulated as metal(II)–metal(II) complexes (Table III) (110).

Further confirmation of a symmetrical structure for **132–134** came from their

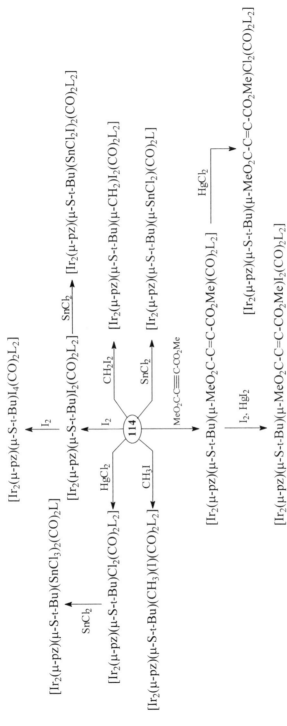

Scheme 6. Oxidative–addition reactions of the Ir(I) dinuclear pyrazolate complex $[Ir_2(\mu\text{-pz})(\mu\text{-S-t-Bu})(CO)_2L_2]$, **114** [L = P(OMe)$_3$] (see text for references).

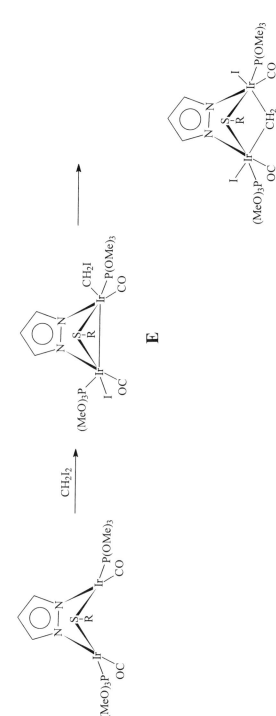

**E**

Scheme 7.  Proposed mechanism for the oxidative addition of $CH_2I_2$ to $[Ir_2(\mu\text{-}pz)(\mu\text{-}SR)(CO)_2\{P(OMe)_3\}_2]$, **114** (R = *t*-Bu). [Adapted from (111).]

185

**130a**                                              **130b**

[31]P NMR spectra. The terminal iodide ligands in this class of complexes can be removed by reaction with $AgNO_3$ (Scheme 8).

Moreover, Complexes **118** and **119** react with mercury(II) chloride to give the $Ir^{III}$–$Ir^I$ derivatives **135** and **136,** resulting from oxidative addition of $HgCl_2$ to one metal center. A related complex, $[Rh_2(\mu\text{-pz})_2(\mu\text{-HgCl})Cl(CO)_2(PPh_3)_2]$, which exhibits a mercury atom asymmetrically bridging the two rhodium centers, was previously characterized by X-ray diffraction (116).

The synthesis of ruthenium–palladium and ruthenium–rhodium complexes of the formula $[(p\text{-cymene})(X)Ru(\mu\text{-pz*})(\mu\text{-X})M(L_2)]$ (Table IV) were reported (119, 120).

This procedure used involves the reaction of a ruthenium(II) mononuclear pyrazole complex with a second metal acac, as indicated in Eq. 6

$$[(p\text{-cymene})RuX_2(Hpz^*)] + [M(acac)(L_2)] \longrightarrow$$

$$[(p\text{-cymene})(X)Ru(\mu\text{-pz*})(\mu\text{-X})M(L_2)]$$

$$(M = Pd \text{ or } Rh) \tag{6}$$

One of the complexes, $[(p\text{-cymene})(Cl)Ru(\mu\text{-pz})(\mu\text{-Cl})Pd(C_8H_{11})]$, **137,** was characterized by single-crystal X-ray diffraction (120).

The X-ray structure of the Ru/Rh derivative $[(p\text{-cymene})(Cl)Ru(\mu\text{-pz})(\mu\text{-Cl})Rh(tfbb)]$, **138** (119), exhibits the triply bridged arrangement $(\mu\text{-pz})(\mu\text{-Cl})_2$,

TABLE IV
Heterodinuclear Complexes of Formula
$[(p\text{-Cymene})(X)Ru(\mu\text{-pz*})(\mu\text{-X})M(L_2)]^a$

| M | X | $L_2$ | Hpz* |
|---|---|---|---|
| Pd | Cl | $C_3H_5$ | Hpz |
|  |  | $C_4H_7$ |  |
|  |  | $C_8H_{11}$ |  |
| Rh | Cl | tfbb | Hpz |
|  | I | nbd | Hmpz |

[a]References (119) and (120).

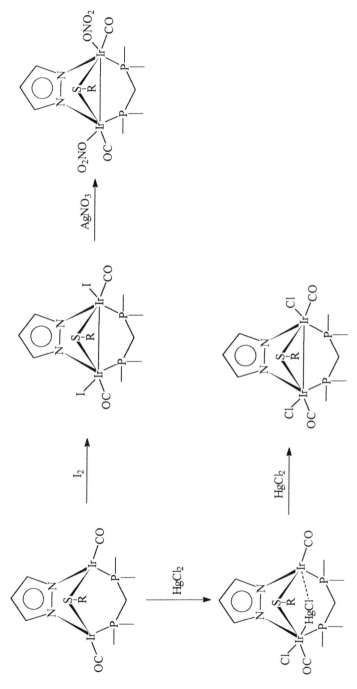

Scheme 8. Oxidative–addition reactions of iodine and HgCl$_2$ on [Ir$_2(\mu$-pz)($\mu$-SR)($\mu$-dppm)(CO)$_2$], **118** (R = $t$-Bu). [Adapted from (110).]

in contrast with the doubly bridged arrangement, $(\mu\text{-pz})(\mu\text{-Cl})$, observed for **137** and for the isoelectronic $[(C_5Me_5)(Cl)Rh(\mu\text{-pz})(\mu\text{-Cl})Rh(tfbb)]$, **139** (79). A related cationic complex, $[(p\text{-cymene})(Hpz)Ru(\mu\text{-pz})(\mu\text{-Cl})Rh(tfbb)](ClO_4)$, **140,** containing the bridging pyrazolate group as well as the neutral pyrazole ligand, was prepared by treating the cationic complex $[(p\text{-cymene})RuCl(Hpz)_2](ClO_4)_2$ with $[Rh(acac)(tfbb)]$. The reaction of **140** with Na(acac) afforded the neutral complex $[(p\text{-cymene})RuCl(\mu\text{-pz})_2Rh(tfbb)]$ presumably containing the pyrazolate groups as bridging ligands (120).

The synthetic strategy reported above (Eq. 6) has been previously employed for the preparation of mixed-valence $Rh^{III}Rh^{I}$ complexes containing the $(\mu\text{-pz}^*)(\mu\text{-Cl})$ moiety. Treatment of $[(C_5Me_5)RhCl_2(Hpz)]$ with $[Rh(acac)(tfbb)]$ afforded **139**, which was characterized by an X-ray crystal structure determination (79). This heterobridged complex contains two rhodium atoms in two different formal oxidation states ($Rh^{III}$ and $Rh^{I}$), with pseudooctahedral and square planar coordinations, respectively.

Reaction of the ruthenium compounds $[RuHCl(CO)(PR_3)_n]$ ($R = Ph$, $n = 3$; $R = i\text{-Pr}$, $n = 2$) with Hpz afforded the complexes $[RuHCl(CO)(HPz)(PR_3)_2]$ (101). Treatment of these products with transition metal complexes containing a hydrogen abstractor ($MeO^-$ or $acac^-$), such as $[M(\mu\text{-OMe})(diolefin)]_2$ ($M = Rh$ or Ir; diolefin = cod or tfbb) or $[Pd(acac)(\eta^3\text{-}C_3H_5)]$, gave a series of heterodinuclear Ru—Ir, Ru—Rh, and Ru—Pd complexes of the formula $[H(CO)(PR_3)_2Ru(\mu\text{-pz})(\mu\text{-Cl})M(L_2)]$ (Table V).

The structure of $[H(CO)(PPh_3)_2Ru(\mu\text{-pz})(\mu\text{-Cl})Ir(tfbb)]$, **141**, was also determined (101).

The reaction of $[Pt_2Cl_2(\mu\text{-Cl})_2(PMe_2Ph)_2]$ with 1 mol of Hpz in the presence of NaOH afforded the corresponding mono(chloro)-mono(pyrazolate) complex $[Pt_2Cl_2(\mu\text{-pz})(\mu\text{-Cl})(PMe_2Ph)_2]$, **142**, from which the $(\mu\text{-pz})(\mu\text{-OMe})$ derivative was obtained on treatment with an excess of sodium methoxide (68).

Moreover, dinuclear platinum(II) complexes containing both $\mu\text{-pz}$ and other Group 16 (VI A) donors as bridging ligands were obtained according to Eq. 7.

TABLE V
Heterodinuclear Complexes of the
General Formula
$[H(CO)(PR_3)_2Ru(\mu\text{-pz})(\mu\text{-Cl})M(L_2)]^a$

| M | R | $L_2$ |
|---|---|---|
| Rh, Ir | Ph | cod |
|  |  | tfbb |
| Rh | $i\text{-Pr}$ | tfbb |
| Pd | Ph | $C_3H_5$ |

$^a$See (101).

$$[Pt_2Cl_2(\mu\text{-ER})(\mu\text{-Cl})(PR'_3)_2] \xrightarrow{\text{Hpz/NaOH}} [Pt_2Cl_2(\mu\text{-ER})(\mu\text{-pz})(PR'_3)_2] \quad (7)$$

$$(E = S, Se, \text{ or } Te; R = \text{alkyl or aryl})$$

The $^{31}$P NMR data seem to indicate that complexes $[Pt_2Cl_2(\mu\text{-ER})(\mu\text{-pz})(PR'_3)_2]$ exist in the following cis form (68).

The preparation and properties of trinuclear heterobridged complexes M—Pd—M (M = Rh or Ir) containing the $(\mu\text{-pz})(\mu\text{-X})$ fragment (X = S-t-Bu, Cl) were reported (Table VI) (121–123).
The synthetic route utilized for the preparation of Complexes **146** and **152** employs $[PdCl_2(Hpz)_2]$ and $[M(acac)(cod)]$ as starting materials (Eq. 8).

$$2[M(acac)(cod)] + [PdCl_2(Hpz)_2] \xrightarrow{-2\,Hacac} [Pd\{M(\mu\text{-pz})(\mu\text{-Cl})(cod)\}_2] \longrightarrow$$

$$\xrightarrow{2K\text{-S-}t\text{-Bu}} [Pd\{M(\mu\text{-pz})(\mu\text{-S-}t\text{-Bu})(cod)\}_2] \quad (8)$$

TABLE VI
Trinuclear Species of Formula $[L_2M(\mu\text{-X})(\mu\text{-pz})Pd(\mu\text{-X})(\mu\text{-pz})ML_2]$

| Complex | M | X | $L_2$ | Notes | Reference |
|---|---|---|---|---|---|
| **143** | Rh | Cl | cod | | 123 |
| **144** | Rh | Cl | tfbb | | 123 |
| **145** | Rh | Cl | $(CO)_2$ | | 123 |
| **146** | Rh | S-t-Bu | cod | | 124 |
| **147** | Rh | S-t-Bu | $(CO)_2$ | X-ray | 124 |
| **148** | Rh | S-t-Bu | CO, PPh$_3$ | | 124 |
| **149** | Rh | S-t-Bu | CO, P(OMe)$_3$ | | 124 |
| **150** | Ir | Cl | cod | | 123 |
| **151** | Ir | Cl | $(CO)_2$ | | 123 |
| **152** | Ir | S-t-Bu | cod | | 121 |
| **153** | Ir | S-t-Bu | $(CO)_2$ | | 121 |
| **154** | Ir | S-t-Bu | CO, P(OMe)$_3$ | | 121 |

In principle, the heterometallic species **143, 144, 150,** and **151** may give three geometrical isomers:

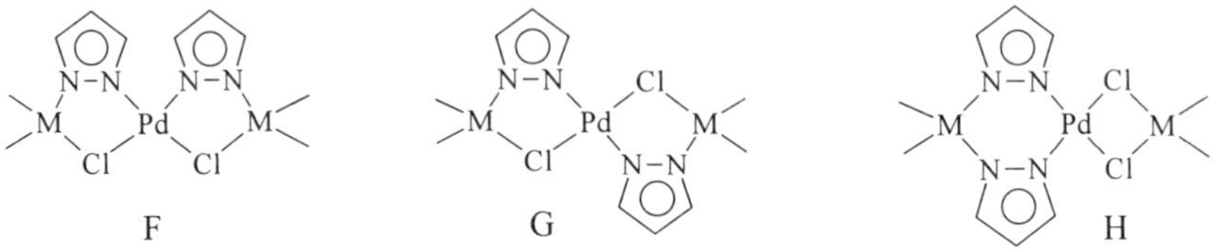

On the basis of $^1$H NMR spectra and taking into account the trans geometry of the starting [PdCl$_2$(Hpz)$_2$] complex, the trans arrangement (G) has been considered the most probable structure for this family of trinuclear compounds (123). A trans disposition of the bridging ligands has been observed in the rhodium complex **147.** The same trans conformation was also proposed for Compound **152** (121).

The oxidative–addition reactions carried out on some of the trinuclear complexes reported in Table VI (see Scheme 9) were particularly studied (121, 122, 124).

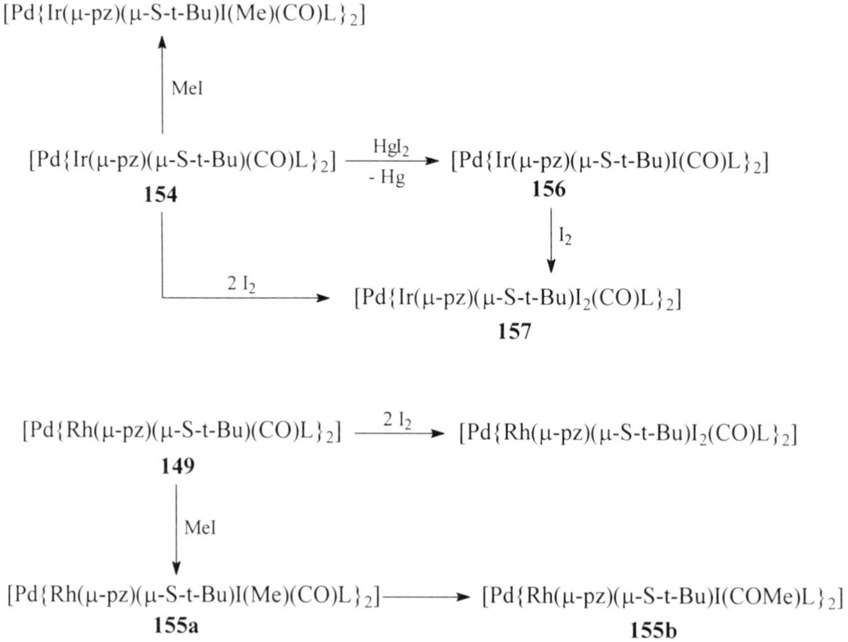

Scheme 9. Oxidative–addition reactions carried-out on the trinuclear species [Pd{Ir($\mu$-pz)($\mu$-S-t-Bu)(CO)L}$_2$, **154,** and [Pd{Rh($\mu$-pz)($\mu$-S-$t$-Bu)(CO)L}$_2$], **149** (see text for references).

The reaction involving Complex **149** and methyl iodide can be interpreted on the basis of an oxidative addition of a molecule of MeI to each rhodium center (122). The formation of the acetyl derivative **155b** was considered as arising from cis migration of a methyl group in isomer **155a,** as indicated in Scheme 9. It is interesting to point out the different behavior of the related iridium derivative **154,** which did not afford the corresponding acetyl compound. In this case, the formation of a metal–metal bond was explained in terms of a heterolytic addition of the iodide and of the methyl groups (121). Addition of mercury(II) iodide to **154** resulted in the formation of **156** and metallic mercury (Scheme 9). This reaction implies the addition of one iodine atom to each iridium center with concomitant formation of two Ir—Pd bonds (121). Compounds **154** and **156** present a very unusual bonding situation where the iridium(II) centers are candidates to further oxidative–addition reactions. Effectively, Compound **156** reacts with 1 mol of $I_2$ to yield **157,** which is also obtained by direct addition of 2 mol of iodine to the iridium(I) derivative **154.** Another family of homotrinuclear complexes was derived from two palladium precursors. The reaction of $[Pd(acac)(\eta^3\text{-allyl})]$ (allyl = $C_3H_5$ or $C_4H_7$) with $[PdCl_2(Hpz)_2]$ yielded $[(\eta^3\text{-allyl})Pd(\mu\text{-pz})(\mu\text{-Cl})Pd(\mu\text{-pz})(\mu\text{-Cl})Pd(\eta^3\text{-allyl})]$ (123).

The reactivity of dinuclear anionic hydroxo complexes of the type $[\{R_2M(\mu\text{-}OH)\}_2]^{2-}$ (M = Ni, Pd, or Pt) toward weak acids as pyrazoles was studied extensively. The chemical behavior is consistent with the high-field proton resonances of the OH bridges (65, 66). The ($\mu$-hydroxo)($\mu$-pyrazolate) or bis($\mu$-pyrazolate) complexes (the latter are described in Section III.C) were obtained, depending on the reactants molar ratio (M/Hpz* = 1 or 2, respectively) (60–65) (Table VII).

TABLE VII
Anionic Heterobridged Complexes of Formula $[R_2M(\mu\text{-pz*})(\mu\text{-OH})MR_2]$

| M | R | Hpz* | Notes | References |
|---|---|---|---|---|
| Ni | $C_6F_5$ | Hpz | X-ray | 60, 61 |
| | | Hmpz | | 60, 61 |
| | | Hdmpz | | 60, 61 |
| Pd | $C_6F_5$ | Hpz | | 64 |
| | | Hmpz | | 64 |
| | | Hdmpz | | 64 |
| | $C_6F_3H_2$ | Hdmpz | | 63 |
| | $C_6Cl_5$ | Hdmpz | | 62 |
| Pt | $C_6F_5$ | Hpz | | 66 |
| | | Hmpz | | 66 |
| | | Hdmpz | X-ray | 66 |

The reaction of $[\{R_2Pd(\mu\text{-OH})\}_2]^{2-}$ with Hpz* (1:1 molar ratio) did not afford pure $(\mu\text{-pz*})(\mu\text{-OH})$ complexes because of the persistent presence of impurities of the corresponding bis$(\mu\text{-pz*})$ complexes. Nevertheless, their formation is confirmed by IR and NMR data. The dmpz derivatives were prepared, in pure form, by a different strategy (62–64) (Eq. 9; Table VII).

$$[R_2Pd(\mu\text{-Cl})_2PdR_2](NBu_4)_2 + 2(NBu_4)OH + Hdmpz \longrightarrow$$

$$[R_2Pd(\mu\text{-dmpz})(\mu\text{-OH})PdR_2] + 2(NBu_4)Cl + H_2O \qquad (9)$$

Related dinuclear palladium(II) complexes of formula $[R(PPh_3)Pd(\mu\text{-pz})(\mu\text{-OH})Pd(PPh_3)R]$ (R = $C_6F_5$ or $C_6Cl_5$) were analogously prepared by reacting $[RPd(PPh_3)(\mu\text{-Cl})]_2$ with $(NBu_4)OH$ and Hpz (125). The NMR data show unambiguously that the $(\mu\text{-pz})(\mu\text{-OH})$-palladium(II) complexes exist in solution as the cis isomers, with two equivalent phosphines trans to the bridging OH group.

It has been reported that the 1:1 reaction of $[\{Pt(C_6F_5)_2(\mu\text{-OH})\}_2]^{2-}$ with Hpz* (Hpz* = Hpz, Hdmpz, or Hmpz) represents a convenient route for the preparation of the complexes $[\{Pt(C_6F_5)_2\}_2(\mu\text{-pz*})(\mu\text{-OH})]^{2-}$ (66). The latter compounds gave the corresponding mono(methoxo)–mono(pyrazolate) derivatives on treatment with methanol. Spectroscopic (IR, $^1$H, and $^{19}$F NMR) data were used for structural assignments, and an X-ray structure determination was carried out for $[\{Pt(C_6F_5)_2\}_2(\mu\text{-dmpz})(\mu\text{-OH})](NBu_4)_2$, **158,** which established the dinuclear nature of the anion (66).

Since the Ru$^{II}$(arene) unit is isoelectronic with the M$^{III}$(C$_5$Me$_5$) (M = Rh or Ir) moieties, studies have been performed in order to obtain ruthenium analogous complexes containing $(\mu\text{-OH})$ or $(\mu\text{-OR})$ groups as well as pyrazolate ligands but perhaps with new stoichiometries, which were not observed in rhodium or iridium chemistry (126).

The tris$(\mu\text{-hydroxo})$-diruthenium complexes $[\{p\text{-cymene})Ru\}_2(\mu\text{-OH})_3](A)$ (A = $BPh_4$, $BF_4$, or $PF_6$) react with equimolar amounts of Hpz* (Hpz* = Hpz, Hdmpz, or Hmpz) to give the heterobridged complexes $[\{(p\text{-cymene})Ru\}_2(\mu\text{-pz*})(\mu\text{-OH})_2](A)$ and/or $[\{(p\text{-cymene})Ru\}_2(\mu\text{-pz*})_2(\mu\text{-OH})](A)$, depending on the nature of Hpz*, the counteranion, and the reaction conditions (126). The tris$(\mu\text{-OMe})$-diruthenium complexes $[\{(p\text{-cymene})Ru\}_2(\mu\text{-OMe})_3](A)$ react in methanol with Hpz* to give the mono$(\mu\text{-pyrazolate})$ compounds $[\{(p\text{-cymene})Ru\}_2(\mu\text{-pz*})(\mu\text{-OMe})_2](A)$. The $(\mu\text{-pz*})_2(\mu\text{-OMe})$ analogues (A = $BF_4$ or $PF_6$) were obtained by treating the appropriate tris$(\mu\text{-OMe})$ complex with an excess of Hpz*. The structure of $[\{(p\text{-cymene})Ru\}_2(\mu\text{-pz})(\mu\text{-OH})_2](BF_4)$, **159,** was also determined (126). The synthesis and reactions of dinuclear pentamethyl-cyclopentadienyl-rhodium(III) complexes were studied extensively. Although several hydroxobridged compounds were reported, all attempts to isolate related alkoxobridged derivatives were unsuccessful. The latter complexes have been invoked as probable intermediates in the preparation of various hydroxo-bridged compounds (127). On the contrary, a series of dinuclear rhodium pyr-azolate complexes containing the $C_5Me_5$ group and methoxo- or hydroxobridg-ing ligands was reported (127). These compounds are of interest because there are still relatively few stable and accessible alkoxo- or hydroxocomplexes of the platinum metals.

Dinuclear complexes of formula $[\{Rh(C_5Me_5)\}_2(\mu\text{-}pz^*)(OMe)_2](ClO_4)$ were obtained by treating $[\{Rh(C_5Me_5)\}_2(OH)_3](ClO_4)$ with the appropriate Hpz* in methanol (Table VIII).

It is likely that the intermediate $[\{Rh(C_5Me_5)\}_2(OMe)_3]^+$, formed by reaction of the $(\mu\text{-}OH)_3$ derivative with methanol, is protonated by a Hpz* ligand to form a "methanol intermediate," which immediately reacts with the pz* groups to give Complexes **160–165**.

The presence of the bridging pyrazolate group seems to be an essential require-ment in order to stabilize the methoxobridges.

Dinuclear complexes of the formula $[\{Rh(C_5Me_5)\}_2(\mu\text{-}pz^*)_2(\mu\text{-}OH)](ClO_4)$ (Hpz* = Hpz, **166**, Hmpz, **167**) were obtained by reacting $[\{Rh(C_5Me_5)\}_2(\mu\text{-}OH)_3](ClO_4)$ with Hpz* (1 : 2) (127). The X-ray structural characterization of

TABLE VIII
Dinuclear Complexes of Formula
$[Rh(C_5Me_5)(\mu\text{-}pz^*)(OMe)_2](A)^a$

| Complex | Hpz* | A |
|---------|------|---|
| **160** | Hpz | $ClO_4$ |
| **161** | Hmpz | $ClO_4$ |
| **162** | Hdmpz | $ClO_4$ |
| **163** | BrmpH | $ClO_4$ |
| **164** | 4-BrpzH | $ClO_4$ |
| **165** | Hmpz | $BF_4$ |

$^a$Reference (127).

Complex **166** supported the triple-bridged formulation involving two exo-bidentate pyrazolate ligands and one bridging hydroxo group. The data available for Rh derivatives seem to indicate that complexes containing the $(\mu\text{-pz*})(\mu\text{-OMe})_2$ fragment are formed for all systems, but $(\mu\text{-pz*})_2(\mu\text{-OH})$ complexes are preferentially obtained when the pyrazoles are unsubstituted or have only a single 3-methyl substituent.

The synthetic strategy used for the preparation of the aforementioned heterobridged rhodium complexes containing pz* and MeO or OH bridging ligands was extended for related $C_5Me_5$–iridium complexes (128). Thus, complexes of formula $[\{Ir(C_5Me_5)\}_2(\mu\text{-pz*})_2(\mu\text{-OH})](BF_4)$ (Hpz* = Hpz or Hmpz) and $[\{Ir(C_5Me_5)\}_2(\mu\text{-pz*})(\mu\text{-OMe})_2](A)$ (A = $ClO_4$ or $BF_4$; Hpz* = Hpz, Hdmpz, or Hmpz) have been prepared.

The complexes $[\{Ir(C_5Me_5)\}_2(\mu\text{-pz})_2(\mu\text{-OH})](BF_4)$ and $[\{Ir(C_5Me_5)\}_2(\mu\text{-pz})(\mu\text{-OMe})_2](BF_4)$ and their rhodium analogues react with HCl to yield the dinuclear chloride complexes $[\{M(C_5Me_5)\}_2(\mu\text{-pz})_2(\mu\text{-Cl})](BF_4)$ and $[\{M(C_5Me_5)\}_2(\mu\text{-pz})(\mu\text{-Cl})_2](BF_4)$ (M = Rh or Ir). These complexes were also prepared from the corresponding neutral species $[M(C_5Me_5)(Cl)_2(Hpz)]$ (128, 129). Moreover, addition of $NEt_3$ to $[Rh(C_5Me_5)(Cl)_2(Hpz)]$ afforded $[Rh(C_5Me_5)(Cl)(\mu\text{-pz})]_2$, which was crystallographically characterized.

Treatment of $[(\{Rh(C_5Me_5)\}_2(\mu\text{-pz})(\mu\text{-Cl})_2](BF_4)$ with excess Zn dust and Cu sand gave a dirhodium(II) complex formulated as $[Rh(C_5Me_5)(\mu\text{-pz})]_2$, although this product has not yet been isolated as a pure solid. On the contrary, the $C_5H_5$ derivative, $[Rh(C_5H_5)(\mu\text{-pz})]_2$, was obtained as crystals that were suitable for X-ray diffraction (129). Reaction of the latter rhodium compound with $NOBF_4$ results in reoxidation to a structurally characterized dirhodium(III) nitrosyl complex, $[Rh_2(C_5H_5)_2(\mu\text{-pz})_2(\mu\text{-NO})](BF_4)$ (129).

The pyrazolate ligand plays an important role in the addition of $HPPh_2$ to complexes containing the $(C_5Me_5)Rh^{III}$ fragment, allowing new links to the pentamethylcyclopentadienyl–rhodium(III) phosphido chemistry, which is very scarcely represented.

Dinuclear complexes containing pyrazolate and diphenylphosphido bridging ligands were prepared by reacting $[\{Rh(C_5Me_5)\}_2(\mu\text{-pz*})(\mu\text{-OMe})_2](BF_4)$ with $HPPh_2$ (Table IX) (130, 131).

The X-ray crystal structures of $[\{Rh(C_5Me_5)\}_2(\mu\text{-pz})_2(\mu\text{-PPh}_2)](BF_4)$, **172**, and $[\{Rh(C_5Me_5)\}_2(\mu\text{-pz})(\mu\text{-Cl})(\mu\text{-PPh}_2)](BF_4)$, **176**, obtained by reacting $[\{Rh(C_5Me_5)\}_2(\mu\text{-pz})(\mu\text{-OH})(\mu\text{-PPh}_2)](BF_4)$ with hydrochloric acid, were also reported (130, 131).

The rhodium anionic bis(phosphonate) complex $[Rh(C_5Me_5)(I)\{PO(OMe)_2\}_3]^-$, can act as a tridentate ligand with the halogen atom as one of the donor centers (132). This behavior can be enhanced by introducing a pz* group instead of the halogen atom. Bimetallic derivatives of the formula $[Rh(C_5Me_5)\{PO(OMe)_2\}_2(\mu\text{-pz*})M]$ (M = Na or Tl) were prepared by addition

TABLE IX
Complexes Containing the $(\mu\text{-pz*})(\mu\text{-PPh}_2)$ Fragment[a]

| Complex | Formula | Notes |
|---|---|---|
| 168 | $[\{Rh(C_5Me_5)\}_2(\mu\text{-dmpz})(\mu\text{-OH})(\mu\text{-PPh}_2)](BF_4)$ | |
| 169 | $[\{Rh(C_5Me_5)\}_2(\mu\text{-pz})(\mu\text{-OH})(\mu\text{-PPh}_2)](BF_4)$ | |
| 170 | $[\{Rh(C_5Me_5)\}_2(\mu\text{-mpz})(\mu\text{-OH})(\mu\text{-PPh}_2)](BF_4)$ | |
| 171 | $[\{Rh(C_5Me_5)\}_2(\mu\text{-pz})(\mu\text{-PPh}_2)_2](BF_4)$ | |
| 172 | $[\{Rh(C_5Me_5)\}_2(\mu\text{-pz})_2(\mu\text{-PPh}_2)](BF_4)$ | X-ray |
| 173 | $[\{Rh(C_5Me_5)\}_2(\mu\text{-mpz})(\mu\text{-PPh}_2)_2](BF_4)$ | |
| 174 | $[\{Rh(C_5Me_5)\}_2(\mu\text{-bmpz})(\mu\text{-OH})(\mu\text{-PPh}_2)](BF_4)$ | |
| 175 | $[\{Rh(C_5Me_5)\}_2(\mu\text{-bmpz})(\mu\text{-PPh}_2)_2](BF_4)$ | |
| 176 | $[\{Rh(C_5Me_5)\}_2(\mu\text{-pz})(\mu\text{-Cl})(\mu\text{-PPh}_2)_2](BF_4)$ | X-ray |
| 202 | $[\{Rh(C_5Me_5)\}_2(\mu\text{-H})(\mu\text{-pz})(\mu\text{-PPh}_2)](BF_4)$ | |

[a]References (130, 131).

of NaH to the cationic pyrazole complexes $[Rh(C_5Me_5)\{PO(OMe)_2\}-\{P(OH)(OMe)_2\}(Hpz*)](ClO_4)$ or Tl(acac) to the neutral compounds $[Rh(C_5Me_5)\{PO(OMe)_2\}_2(Hpz*)]$ (Hpz* = Hpz, Hmpz, or Hdmpz).

The sodium or tallium compounds were used as starting materials for the formation of a series of cationic or neutral homo- and heterodinuclear complexes containing the $(\mu\text{-pz*})(\mu\text{-}\{PO(OMe)_2\})_2$ fragment (Eq. 10a–c) (132):

$$[Rh(C_5Me_5)\{PO(OMe)_2\}_2(\mu\text{-pz*})Na] \xrightarrow{(a)}$$

$$[Rh(C_5Me_5)\{\mu\text{-PO(OMe)}_2\}_2(\mu\text{-pz*})M(C_5Me_5)](ClO_4) \quad (10a)$$

$$[Rh(C_5Me_5)\{PO(OMe)_2\}_2(\mu\text{-pz*})Na] \xrightarrow{(b)}$$

$$[Rh(C_5Me_5)\{\mu\text{-PO(OMe)}_2\}_2(\mu\text{-pz*})Ru(arene)](ClO_4) \quad (10b)$$

$$[Rh(C_5Me_5)\{PO(OMe)_2\}_2(\mu\text{-pz*})Tl] \xrightarrow{(c)}$$

$$[Rh(C_5Me_5)\{\mu\text{-PO(OMe)}_2\}_2(\mu\text{-pz*})M(CO)_3] \quad (10c)$$

(a) $[Rh(C_5Me_5)Cl_2]_2$ or $[Ir(C_5Me_5)Cl_2]_2$, (b) $[Ru(arene)Cl_2]_2$, (c) $[Re(CO)_5Br]$ or $[Mn(CO)_5Br]$.

Furthermore, similar results were obtained by reacting the neutral pyrazole complex $[Rh(C_5Me_5)\{PO(OMe)_2\}_2(Hpz)]$ with the tetrameric platinum(IV) complex $[Pt(Me)_3I]_4$ in the presence of Tl(OEt), with the neutral complex $[Rh(C_5Me_5)\{\mu\text{-PO(OMe)}_2\}_2(\mu\text{-pz})Pt(Me)_3]$ being formed.

The crystal structure of the cationic heterobimetallic complex $[Rh(C_5Me_5)\{\mu\text{-PO(OMe)}_2\}_2(\mu\text{-pz})Ru(Ph)](ClO_4)$ confirmed that the metal center of the organometallic fragment exhibits a strong tendency to obey the 18-electron rule

(132). In this Rh/Ru complex, the phosphonate anions are coordinated to the rhodium through the phosphorous and to the ruthenium through the oxygen atom. The geometries around the metals are the normal pseudooctahedral three legged piano stool commonly found for six-coordinated $M(C_5Me_5)$ or M(arene) derivatives.

The preparation and properties of dinuclear complexes containing the pyrazolate and azide groups as bridging ligands were reported (113). Representative formulas are $[M_2(\mu\text{-pz})(\mu\text{-N}_3)(CO)_4]$, $[M_2(\mu\text{-pz})(\mu\text{-N}_3)(cod)_2]$ (M = Rh or Ir), $[(CO)_2Rh(\mu\text{-pz})(\mu\text{-N}_3)Rh(cod)]$, $[(CO)_2Rh(\mu\text{-pz})(\mu\text{-N}_3)Ir(CO)_2]$, and $[(\eta^3\text{-}C_3H_5)Pd(\mu\text{-pz})(\mu\text{-N}_3)Rh(CO)_2]$.

The IR spectra (solid state) of the carbonyl compounds were rather complex. This observation, along with the dark color and metallic luster of some of the complexes, suggested extended metal–metal interactions, which were confirmed by the X-ray crystal structure of the Pd/Rh derivative (113).

Two octamolybdenum compounds, containing both bridging pyrazolate and terminal pyrazole ligands, $[Mo_8(pz)_6(O)_{18}(Hpz)_6]$, **177**, and $[Mo_8(pz)_6(O)_{21}(Hpz)_6]$, **178**, were prepared by the reaction of molten pyrazole with molybdenum oxides and their structures were determined by single-crystal X-ray diffraction (133). The overall geometries of **177** and **178** are similar. Both complexes consist of an octametallic framework that may be described as two basally connected flattered trigonal pyramids.

Complex **177** contains both $Mo^{VI}$ and $Mo^V$ ions, with the latter linked in pairs by metal–metal single bonds. Compound **178** involves a single-valent molybdenum(VI) octamolybdenum species and is photosensitive.

The complex $[Zn(dmpz)_2(Hdmpz)]_2$, **31**, was reacted with $[Co(MeCN)_6](BF_4)_2$ in the presence of $NEt_3$ with the aim of obtaining the mixed-cobalt/zinc analogue of the cobalt(II) polymer $[Co(dmpz)_2]_n$ (41), that is, $[CoZn_2(dmpz)_6]_n$ (134). The unexpected formation of the tetranuclear complex $[Co_4(\mu\text{-dmpz})_6(\mu_4\text{-O})]$, **179**, was observed. Its X-ray crystal structure consists of a central O atom around which the four $Co^{II}$ ions are pseudotetraedrally arrayed. The six edges of the tetrahedron are bridged by dmpz ligands (134) (Fig. 8).

The trinuclear copper(II) derivative $[Cu_3(OH)(pz)_3(py)_2Cl_2] \cdot py$, **180**, was obtained by oxidation of the dimeric copper(I) pyrazole complex $[Cu(Hpz)_2Cl]_2$ with molecular oxygen in the presence of pyridine (135). A related compound, $[Cu_3(OH)(pz)_3(HPz)_2(NO_3)_2]$, was previously reported and its X-ray crystal structure was determined (136).

The structure of **180** consists of a triangular arrangement of $Cu^{II}$ atoms connected by $\mu$-pz groups. Three terminal ligands (one Cl and two py) one on each copper atom, and two bridging ligands ($\mu$-Cl and $\mu_3$-OH) complete the coordination spheres of the three metal centers.

Unexpected octanuclear mixed-bridged copper(II)–pyrazolate complexes, $[Cu_8(\mu\text{-OH})_8(\mu\text{-pz*})_8]$ (Hpz* = Hpz, **181**; Hdmpz, **182**), were isolated on treat-

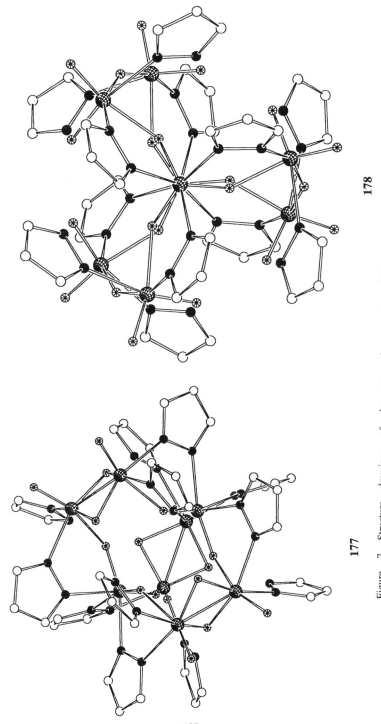

**178**

**177**

Figure 7. Structure drawings of the octanuclear species $[Mo_8(pz)_6(O)_{18}(Hpz)_6]$, **177** and $[Mo_8(pz)_6(O)_{21}(Hpz)_6]$, **178**. [Based on data from (133).]

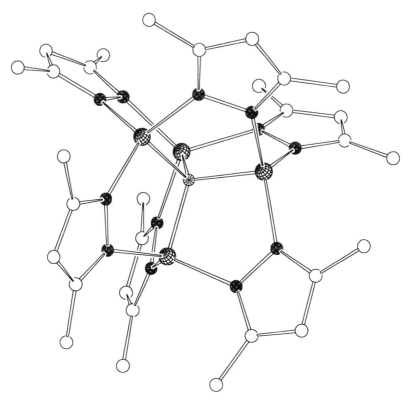

Figure 8.  Structure drawing of the tetranuclear cobalt(II) species $[Co_4(\mu\text{-dmpz})_6(\mu_4\text{-O})]$, **179**. [Based on data from (134).]

ment of the binary copper(I) complexes $[Cu(pz)]_n$, **2**, or $[Cu(dmpz)]_3$, **6**, with dioxygen in wet solvents (137). The X-ray crystal structure of **182** was determined. The eight metal atoms, which lie approximately at the vertices of a regular octagon, are connected by $\mu$-OH and $\mu$-dmpz ligands in an alternate "up and down" sequence (Fig. 9) (137).

Complex **182** reacted with $PPh_3$ at 60°C under an inert atmosphere. This reaction yielded 4 mole of $O{=}PPh_3$ per mole of complex. The metal was recovered in quantitative yield as $[Cu(dmpz)]_3$, according to Eq. 11.

$$[Cu_8(OH)_8(dmpz)_8] + 4PPh_3 \longrightarrow \tfrac{8}{3}[Cu(dmpz)]_3 + 4O{=}PPh_3 + 4H_2O$$

$$(11)$$

This uncommon behavior was explained by assuming that, in solution, the following equilibrium takes place (Eq. 12).

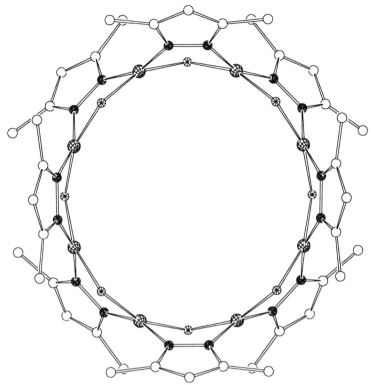

Figure 9. Structure drawing of the octanuclear copper(II) complex [Cu$_8$(dmpz)$_8$(OH)$_8$], **182.** [Based on data from (137).]

$$[Cu_8(OH)_8(dmpz)_8] \rightleftharpoons [Cu_8(dmpz)_8(O)_x(OH)_{8-2x}(H_2O)_x]$$

$$(x = 1\text{--}4) \tag{12}$$

The tautomeric species thus formed were thought to be the plausible active species in the phosphine oxidation. Complex **182** was shown to exhibit catalytic properties in the selective oxidation of various organic substrates (137).

In 1970, the term "dinucleating ligands" was introduced for a series of polydentate ligands capable of simultaneously securing two metasl ions (138). In the past few years, increasing attention has been paid to the synthesis of such ligands and related complexes, resulting in the publication of several papers on this subject (139). Dinuclear copper(II) complexes were especially studied because of the recognized dependence of magnetic exchange within the dinuclear system upon stereochemical factors and the nature of the bridging species. In bis($\mu$-hydroxo) and bis($\mu$-alkoxo) bridged dinuclear copper(II) complexes, an increase in the strength of antiferromagnetic exchange coupling with increasing Cu—O—Cu bridging angle in the range 90–105° is well documented (140). More recently, interest in complexes containing larger bridging angles (120–135°) and unsymmetric doubly bridged structures, has stimulated a closer look at the effect of geometry and the nature of the bridging ligands on the strength of the exchange interactions (140). Therefore, the synthesis, structure, and magnetic properties of several dinuclear copper(II) complexes containing bridging pyrazolate ions and neutral or anionic binucleating ligands (Fig. 10) were reported (140–146). It was thus possible to compare data for such complexes to those obtained for single-atom bridged analogues (i.e., OH$^-$, Cl$^-$, or OR$^-$). It was concluded that the order of the decreasing negative coupling constant $J$, when all is kept constant, is OR$^-$ $\approx$ pz* > MeCOO$^-$ $\approx$ OH$^-$ > Cl$^-$.

A closer treatment of this topic is out of the realm of this chapter. Representative examples of dinuclear copper(II) complexes employed in such studies are reported in Table X.

Palladium(II) and nickel(II) complexes containing the bridging pyrazolate ion as well as the thiolate based dinucleating ligand H$_3$L$^{12}$, [M$_2$(L$^{12}$)(pz*)] (Hpz* = Hpz or Hdmpz), were obtained (148). The complexes were easily prepared by treating the [M$_2$(L$^{12}$)(CH$_2$COO)] species with Hpz*.

A tetranuclear nickel(II) complex with the dinucleating ligand H$_3$L$^{13}$ [2,6-bis(salicylideneaminomethyl)-4-methyl-phenol], [Ni$_4$(L$^{13}$)$_2$(pz)$_2$(MeOH)] synthesized and characterized by electronic spectra, magnetic susceptibilities, and X-ray crystal structure analysis (149).

$$H_3L^{12}$$

## B.   Complexes Containing the ($\mu$-pz*)$_a$($\mu$-H)$_b$ Fragment

It is well known that hydride derivatives of transition metals can be prepared by hydrogen-transfer reactions from primary and secondary alcohols. Pentamethylcyclopentadienyl ($\mu$-pyrazolate)-($\mu$-hydrido) complexes of iridium and rhodium were obtained by using propan-2-ol as the hydride source (128, 150).

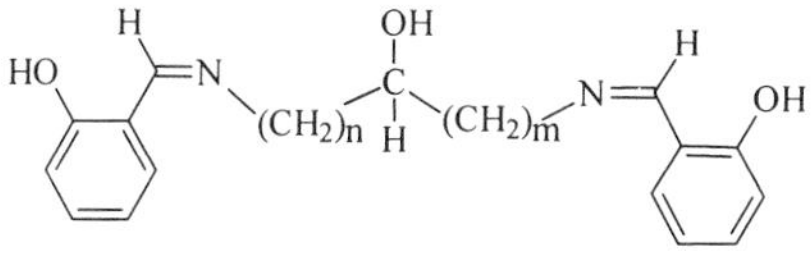

$n = 1$, $m = 1$, $\mathbf{H_3L^1}$
$n = 2$, $m = 2$, $\mathbf{H_3L^2}$
$n = 3$, $m = 1$, $\mathbf{H_3L^3}$

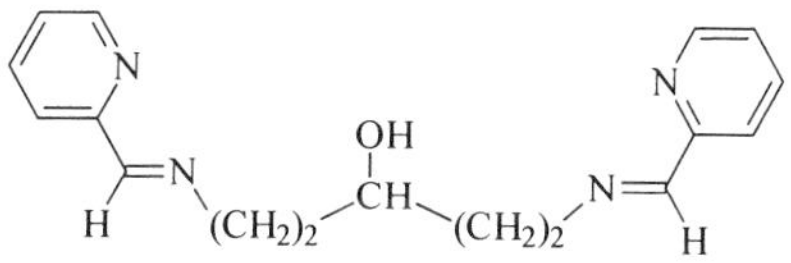

$\mathbf{HL^4}$

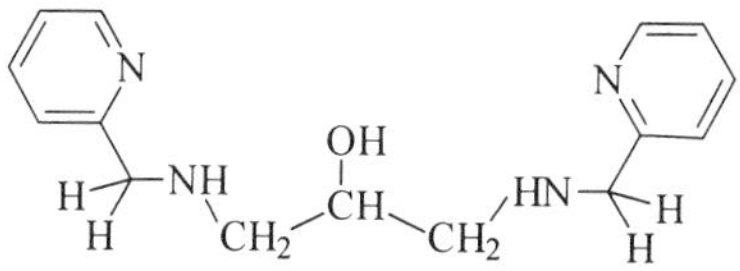

$\mathbf{HL^5}$

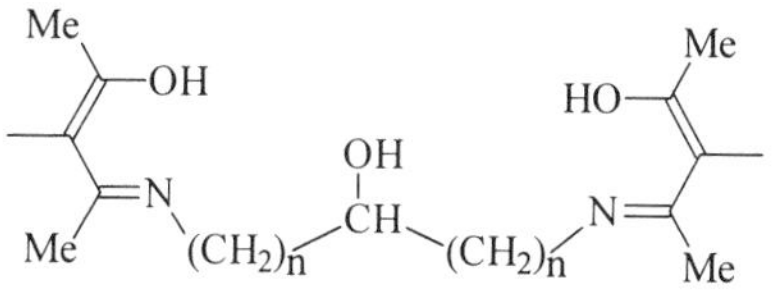

$n = 1$, $\mathbf{H_3L^6}$        $n = 2$, $\mathbf{H_3L^7}$

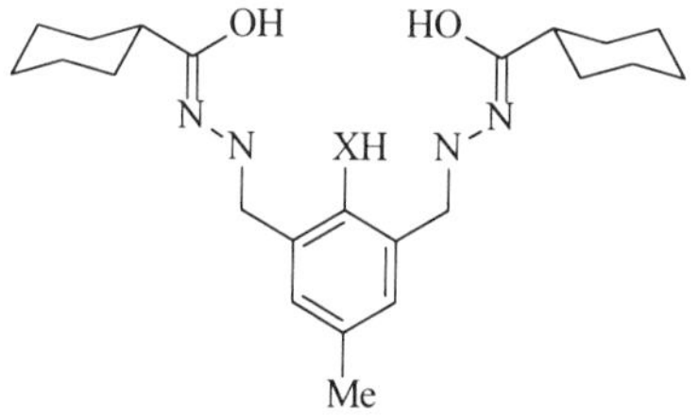

$X = O$, $\mathbf{H_3L^8}$    $X = S$, $\mathbf{H_3L^9}$

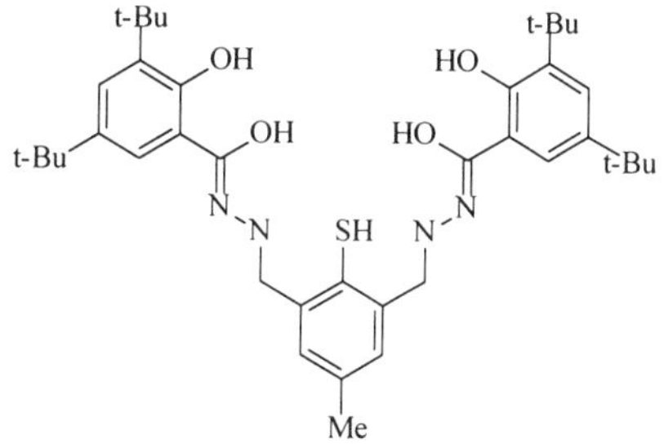

$\mathbf{H_3L^{10}}$

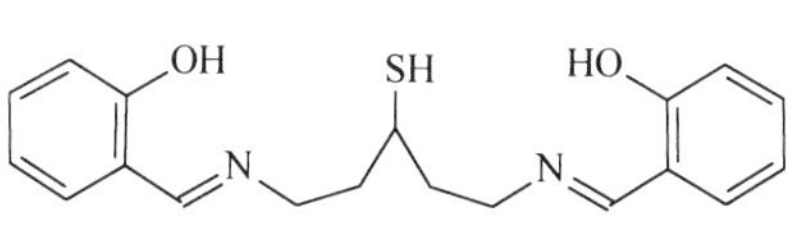

$\mathbf{H_3L^{11}}$

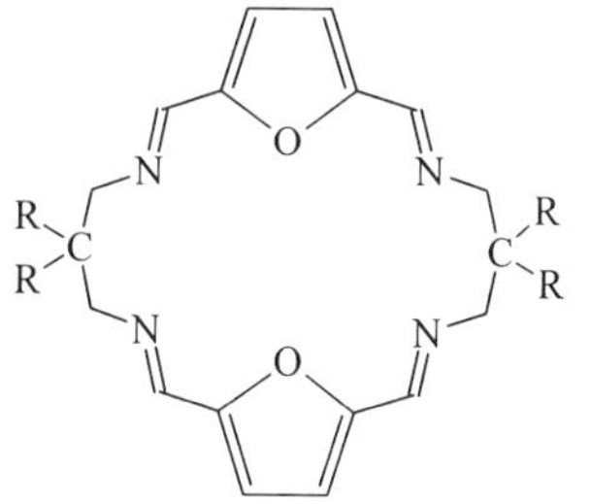

$R = H$, $\mathbf{L^A}$        $R = Me$, $\mathbf{L^B}$

Figure 10.  Drawing of the dinucleating ligands utilized in the magnetostructural correlations on copper(II) pyrazolato complexes.

TABLE X
Pyrazolate-Bridged Dinuclear Copper(II) Complexes Containing Binucleating Ligands

| Complex | Formula | Binucleating Ligand | References |
|---|---|---|---|
| 183 | $[Cu_2(L^3(pz)]$ | $H_3L^3$ | 140 |
| 184 | $[Cu_2(L^4)(pz)](ClO_4)_2$ | $HL^4$ | 141 |
| 185 | $[Cu_2(L^5(pz)](ClO_4)_2$ | $HL^5$ | 141 |
| 186 | $[Cu_2(L^1)(pz)]$ | $H_3L^1$ | 141 |
| 187 | $[Cu_2(L^2)(pz)]$ | $H_3L^2$ | 141, 142 |
| 188 | $[Cu_2(L^6)(pz)]$ | $H_3L^6$ | 141 |
| 189 | $[Cu_2(L^6)(dmpz)]$ | $H_3L^6$ | 141 |
| 190 | $[Cu_2(L^7)(pz)]$ | $H_3L^7$ | 141 |
| 191 | $[Cu_2(L^8)(pz)]$ | $H_3L^8$ | 145 |
| 192 | $[Cu_2(L^9)(pz)]$ | $H_3L^9$ | 145 |
| 193 | $[Cu_2(L^{10})(pz)]$ | $H_3L^{10}$ | 145 |
| 194 | $[Cu_2(L^{11})(pz)]$ | $H_3L^{11}$ | 147 |
| 195 | $[Cu_2(L^A)(pz)_2](ClO_4)_2$ | $L^A$ | 146 |
| 196 | $[Cu_2(L^A)(dmpz)_2](ClO_4)_2$ | $L^A$ | 146 |
| 197 | $[Cu_2(L^B)(dmpz)_2](ClO_4)_2$ | $L^B$ | 146 |

$$H_3L^{13}$$

Air-stable solids of formula $[\{Ir(C_5Me_5)\}_2(\mu\text{-}H)_2(\mu\text{-}pz^*)](BF_4)$ (Hpz* = Hpz, **198**; Hmpz, **199**; Hdmpz, **200**) were isolated by treating $[\{Ir(C_5Me_5)\}_2(\mu\text{-}OH)_3](BF_4)$ with equimolar amounts of Hpz* in isopropanol (150). The related cationic rhodium complex $[\{Rh(C_5Me_5)\}_2(\mu\text{-}H)_2(\mu\text{-}pz)](BF_4)$, **201**, was analogously obtained (128). Complex **201**, reacted with neutral ligands (L) giving products of the formula $[\{Rh(C_5Me_5)\}_2(\mu\text{-}pz)L_2](BF_4)$ (L = $t$-BuNC or CO), with dihydrogen being displaced (128). Spectroscopic studies on the latter two complexes indicated that CO and $t$-BuNC ligands are moving between terminal and bridging positions. The synthetic route reported above also allowed the preparation of mixed ($\mu$-pz*)($\mu$-H)($\mu$-PPh$_2$) complexes (131). Thus, the reaction of [{Rh(C$_5$Me$_5$)}$_2$($\mu$-OH)$_3$](BF$_4$) with equimolar amounts of Hpz and HPPh$_2$ in refluxing propan-2-ol afforded [{Rh(C$_5$Me$_5$)}$_2$($\mu$-H)($\mu$-pz)($\mu$-PPh$_2$)](BF$_4$), **202**, (131).

Treatment of a suspension of [Ru(H)(cod)(NH$_2$NMe$_2$)$_3$](PF$_6$) with Hpz and Et$_3$N gave [{Ru(H)(pz)(cod)}$_2$(Hpz)], **203**, (151). The X-ray structure of **203**

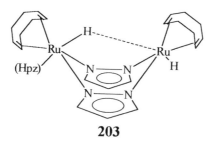

**203**

revealed a semibridging hydride ligand in an unsymmetrical dimer with the ruthenium atoms linked by two pyrazolate groups. Both $^1$H and $^{13}$C NMR spectra showed that this structure persists in solution.

Treatment of **203** with PMe$_3$ affords [{Ru(H)(pz)(cod)}$_2$(PMe$_3$)], **204**, in which the neutral pyrazole ligand is replaced by trimethylphosphine (151). Further examples of ruthenium complexes containing bridging pyrazolate and hydride ligands were prepared (152, 153) from the ruthenium(II) precursor [Ru(H)(cod)(NH$_2$NMe$_2$)$_3$](PF$_6$).

The dinuclear cationic complex [(cod)(Hpz)Ru($\mu$-H)($\mu$-pz)$_2$Ru(Hpz)-(cod)](PF$_6$), **205**, was synthesized by the reaction of the precursor with Hpz in acetone under reflux (152). A related neutral derivative, [(cod)(Hpz)Ru($\mu$-H)($\mu$-pz)$_2$Ru(Cl)(cod)], **206**, formed quantitatively on allowing **203** to stand in CH$_2$Cl$_2$. The X-ray crystal structures of both **205** and **206** revealed their close similarity to **203**. An unexpected dinuclear ruthenium complex containing one terminal and one semibridging hydride ligand, [(cod)$_2$Ru$_2$(H)($\mu$-N=C(Me)pz}($\mu$-pz)($\mu$-H)], **207**, was obtained by reacting [Ru(H)(cod)-(NH$_2$NMe$_2$)$_3$](PF$_6$) with excess K[R$_2$B(pz)$_2$] (R = Et or Ph) in acetonitrile (153). The X-ray crystal structure of **207** confirmed the presence of a semibridging hydride ligand trans to a terminal hydride as well as the presence of a bridging amidine group, formed as the result of the bis(1-pyrazolyl)borate fragmentation followed by a pyrazolate/acetonitrile coupling, and stabilized in an unusual agostic 18-electron, 16-electron diruthenium dihydride complex (153).

The easy dissociation of the acetone molecule of [Ru(H)($\eta^1$-OCMe$_2$)(CO)$_2$\{P(i-Pr)$_3$\}$_2$](BF$_4$) allowed the synthesis of the cationic complex [Ru(H)(CO)$_2$(Hpz)\{P(i-Pr)$_3$\}$_2$](BF$_4$), by reaction with pyrazole. The Hpz li-

**207**

gand in the latter species contains an acidic NH group, which is capable of reacting with $[Rh(\mu\text{-OMe})(\text{diolefin})]_2$ to give the heterodinuclear complexes $[(CO)_2\{P(i\text{-}Pr)_3\}_2Ru(\mu\text{-H})(\mu\text{-pz})Rh(\text{diolefin})]$ (diolefin = tfbb, **208**; cod, **209**). The presence of a bridging hydride ligand in these compounds is substantiated by $^1H$ NMR data (154). The existence of a ruthenium–rhodium interaction in **208** and **209** was supported by the X-ray crystal structure of **209**. Reactions between $Ru_3(CO)_{12}$ and Hpz* gave $[Ru_3(\mu\text{-H})(\mu\text{-pz*})(CO)_{10}]$ (Hpz* = Hpz, **210**; Hdmpz, **211**; Hdfmpz, **212**). These compounds, which resulted from the oxidative addition of Hpz* to the ruthenium cluster with concomitant loss of

TABLE XI

Di- and Trinuclear Complexes Containing the $(\mu\text{-pz*})_a(\mu\text{-H})_b$ Fragment

| Complex | Formula[a] | Notes | References |
|---|---|---|---|
| 198 | $[\{Ir(C_5Me_5)\}_2(\mu\text{-H})_2(\mu\text{-pz})](BF_4)$ | X-ray | 150 |
| 199 | $[\{Ir(C_5Me_5)\}_2(\mu\text{-H})_2(\mu\text{-mpz})](BF_4)$ | | 150 |
| 200 | $[\{Ir(C_5Me_5)\}_2(\mu\text{-H})_2(\mu\text{-dmpz})](BF_4)$ | | 150 |
| 201 | $[\{Rh(C_5Me_5)\}_2(\mu\text{-H})_2(\mu\text{-pz})](BF_4)$ | | 128 |
| 202 | $[\{Rh(C_5Me_5)\}_2(\mu\text{-H})(\mu\text{-pz})(\mu\text{-PPh}_2)](BF_4)$ | | 131 |
| 203 | $[(\text{cod})(Hpz)Ru(\mu\text{-H})(\mu\text{-pz})_2Ru(H)(\text{cod})]$ | X-ray | 151 |
| 204 | $[(\text{cod})(PMe_3)Ru(\mu\text{-H})(\mu\text{-pz})_2Ru(H)(\text{cod})]$ | | 151 |
| 205 | $[\{Ru(\text{cod})(Hpz)\}_2(\mu\text{-H})(\mu\text{-pz})_2](PF_6)$ | X-ray | 152 |
| 206 | $[(\text{cod})(Hpz)Ru(\mu\text{-H})(\mu\text{-pz})_2RuCl(\text{cod})]$ | X-ray | 152 |
| 207 | $[(\text{cod})(H)Ru\{\mu\text{-N}{=}C(Me)pz\}(\mu\text{-pz})(\mu\text{-H})Ru(\text{cod})]$ | | 153 |
| 208 | $[(CO)_2\{P(i\text{-}Pr)_3\}_2Ru(\mu\text{-H})(\mu\text{-pz})Rh(\text{tfbb})]$ | | 154 |
| 209 | $[(CO)_2\{P(i\text{-}Pr)_3\}_2Ru(\mu\text{-H})(\mu\text{-pz})Rh(\text{cod})]$ | X-ray | 154 |
| 210 | $[Ru_3(\mu\text{-H})(\mu\text{-pz})(CO)_{10}]$ | | 155 |
| 211 | $[Ru_3(\mu\text{-H})(\mu\text{-dmpz})(CO)_{10}]$ | | 155 |
| 212 | $[Ru_3(\mu\text{-H})(\mu\text{-dfmpz})(CO)_{10}]$ | X-ray analysis | 155 |
| 213 | $[Os_3(\mu\text{-H})(\mu\text{-pz})(CO)_{10}]$ | | 156 |
| 217 | $[Ru_3(\mu\text{-H})_2(\mu\text{-dmpz})_2(ER_3)_2(CO)_8](\text{cat})^a$ | | 158 |

[a] E = Si, R = Et, Ph, or OMe; E = Sn, R = Bu or Ph; cat = $Et_4N$ or PPN.

two CO molecules, do not show any bands assignable to $\mu$-CO ligands in their IR spectra (155). The X-ray crystal structure of the 3,5-bis(trifluoromethyl) complex, $[Ru_3(\mu\text{-H})(\mu\text{-dfmpz})(CO)_{10}]$, **212**, was also determined (155).

A related osmium complex, $[Os_3(\mu\text{-H})(\mu\text{-pz})(CO)_{10}]$, **213**, was obtained on treatment of $[Os_3(CO)_{10}(Me_3CN)_2]$ with pyrazole (156).

## C. Complexes Containing the $(\mu\text{-pz*})_a(\mu\text{-L})_b$ Fragment (L = Neutral Ligand)

The preparation of an unusual heterobridged mixed-valence rhodium complex bridged by one carbonyl group and two pyrazolate ligands was described (157). The complex $[(C_5Me_5)Rh(\mu\text{-pz})_2(\mu\text{-CO})Rh(dppp)](BPh_4)$, **214**, was obtained by reacting the heterovalent Rh(III)–Rh(I) complex $[(C_5Me_5)(Cl)Rh(\mu\text{-pz})_2Rh(CO)_2]$, **55**, with 1,3-bis-(diphenylphosphino)propane (dppp) in methanol in the presence of NaBPh$_4$. The unambiguous characterization of **214** came from its X-ray crystal structure determination (157).

The anionic cluster $[Ru_3(\mu\text{-dmpz})(\mu\text{-CO})_3(CO)_7]^-$, **215**, was obtained in high yield, by treating $[Ru_3(\mu\text{-H})(\mu\text{-CO})(CO)_{10}]^-$ with Hdmpz (158). Its formulation was proposed on the basis of IR, $^1$H, and $^{13}$C NMR data. The reactivity of **215** with protic acids, alkynes, tertiary silanes, and tertiary stannanes was described. The already known neutral hydrido derivative $[Ru_3(\mu\text{-H})(\mu\text{-dmpz})(CO)_{10}]$, **211**, (155), was obtained from the reaction of **215** with trifluoroacetic acid. The reaction of **215** with diphenylacetylene afforded the anionic complex $[Ru_3(\mu\text{-dmpz})(\mu_3\text{-PhC}_2)(\mu\text{-CO})_2(CO)_6]^-$, **216**, in which the alkyne ligand interacts with three ruthenium centers as revealed by an X-ray crystal structure analysis (158). The reaction of **215** with HSiR$_3$ (R = Et, Ph, or OMe) and HSnR$_3$ (R = Bu or Ph) to give the disilyl or distannyl dihydrido derivatives $[Ru_3(\mu\text{-dmpz})(\mu\text{-H})_2(ER_3)_2(CO)_8]^-$, **217a–217e**, (E = Si or Sn) was also described (158).

Following the synthesis of the first rhodium(I) complex with an A-frame geometry, $[Rh_2(\mu\text{-S})(CO)_2(\mu\text{-dppm})_2]$ (159), several other examples of this type of derivatives in which a $Rh_2(\mu\text{-L})(\mu\text{-dppm})$ fragment is present were published. In 1985, the preparation of a series of new A-frame rhodium(I) and rhodium(II) complexes containing a pyrazolate group acting as a capping ligand have been reported (160). These studies were extended to include the synthesis of several additional complexes, including their full characterization and their chemical properties (161–166). Most of these complexes have the basic structure types I and II. Significant examples are listed in Table XII.

The former type of derivatives (L = L' = CO) are readily obtained by the addition of AgClO$_4$ and Hpz* to trans-$[Rh_2Cl_2(CO)_2(\mu\text{-EE}')_2]$ in the presence

TABLE XII

Pyrazolate-Bridged Dirhodium Compounds of the General Formulas $[Rh_2(\mu\text{-EE}')(\mu\text{-pz*})(L)(L')](A)$ and $[Rh_2(\mu\text{-EE}')(\mu\text{-pz*})(L)(L')(Z)_2](A)$

| Type | E | E' | L | L' | Hpz* | Z | A | Notes | Reference |
|------|---|----|---|----|------|---|---|-------|-----------|
| I | P | P | CO | CO | Hpz | | $ClO_4$ | | 160 |
| I | P | P | CO | CO | Hdmpz | | $ClO_4$ | X-ray | 160 |
| I | P | P | CO | $Me_3CNC$ | Hpz | | $PF_6$ | | 163 |
| I | P | P | $Me_3CNC$ | $Me_3CNC$ | Hpz | | $PF_6$ | | 163 |
| I | P | As | CO | CO | Hdmpz | | $PF_6$ | | 163 |
| I | As | As | CO | CO | Hpz | | $ClO_4$ | | 161 |
| I | As | As | CO | CO | Hdmpz | | $PF_6$ | X-ray | 163 |
| II | P | P | CO | CO | Hpz | I | $ClO_4$ | | 160 |
| II | P | P | CO | CO | Hdmpz | I | $ClO_4$ | X-ray | 160 |
| II | As | As | CO | CO | Hpz | I | $ClO_4$ | | 161 |
| II | P | P | CO | $Me_3CNC$ | Hpz | I | $ClO_4$ | | 161 |
| II | P | P | $Me_3CNC$ | $Me_3CNC$ | Hpz | I | $BF_4$ | X-ray | 161 |

of KOH or on direct treatment of pyrazolate ion, $(pz*)^-$, with $[Rh_2(CO)_2(\mu\text{-}$ $EE')(\mu\text{-Cl})](PF_6)$.

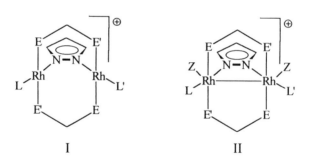

I            II

When 1 equiv of $t$-BuNC is added to $[Rh_2(CO)_2(\mu\text{-EE}')(\mu\text{-pz*})](PF_6)$ a new series of complexes, formulated as containing one CO and one isocyanide ligand, is obtained. The addition of 2 equiv of $t$-BuNC results in the replacement of both CO ligands (163).

Complexes of Type II (Z = Cl, Br, or I; L = L' = CO) were obtained by bicentric oxidative addition of halogens to the dicarbonyl complexes of Type I (160, 163). In these complexes, the carbonyl ligand can also be gradually substituted by isocyanide ligands. The substitution involves the intermediate formation of species containing a bridging carbonyl ligand, as seen in the case of $[Rh_2(CO)_2(\mu\text{-dppm})_2(\mu\text{-pz})(I)_2](ClO_4)$ (Scheme 10) (161).

A related dinuclear heterobridged dirhodium complex triply bridged by two pyrazolate and one dppm ligand $[Rh_2(CO)_2(\mu\text{-pz})_2(\mu\text{-dppm})(I)_2]$, was reported

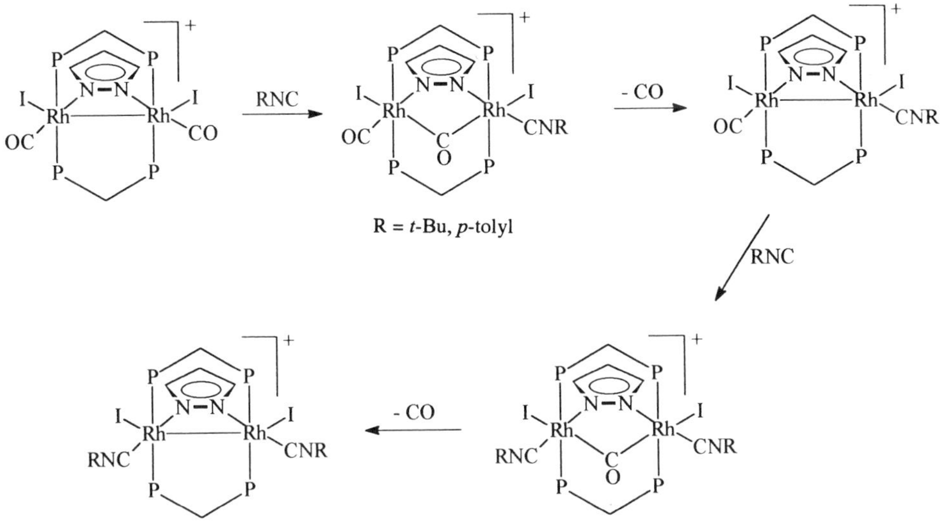

$R = t\text{-Bu}, p\text{-tolyl}$

Scheme 10. The stepwise substitution of carbonyl ligands by RNC (R = $t$-Bu or $p$-tolyl) in the derivative [Rh$_2$($\mu$-pz)($\mu$-dppm)$_2$I$_2$(CO)$_2$]$^+$ is accompanied by the intermediate formation of species containing a bridged carbonyl ligand (see text). [Adapted from (161).]

and its X-ray crystal structure was determined (162). The latter complex was prepared by addition of dppm to [Rh$_2$(CO)$_4$($\mu$-pz)$_2$(I)$_2$].

Extensive electrochemical studies of complexes of Type I were also reported (164–166). It was ascertained that [Rh$_2$]$^{2+}$ di-isocyanide complexes are more susceptible to oxidation than their dicarbonyl analogues, and that one-electron oxidations of the di-isocyanide complexes produce stable and isolable paramagnetic [Rh$_2$]$^{3+}$ species (165). In particular, paramagnetic complexes of the type [Rh$_2$($\mu$-EE')$_2$($t$-BuNC)$_2$($\mu$-pz*)](PF$_6$)$_2$ (EE' = dppm, dapm, or dpam; Hpz* = Hpz, 4-MepzH, Hdmpz, 4-BrdmpzH, or 3,4,5-BrpzH) containing the [Rh$_2$]$^{3+}$ core were obtained via controlled potential electrolysis of the parent [Rh$_2$]$^{2+}$ species [Rh$_2$($\mu$-EE')$_2$($t$-BuNC)$_2$($\mu$-pz*)](PF$_6$). The X-ray crystal structure of [Rh$_2$($\mu$-dppm)$_2$($t$-BuNC)$_2$($\mu$-dmpz)](PF$_6$)$_2$ showed a Rh$\cdots$Rh distance of about 2.83 Å, which is consistent with a bonding order of 0.5 for the [Rh$_2$]$^{3+}$ core (166).

## V. COMPLEXES CONTAINING MONODENTATE PYRAZOLATE GROUPS

The interest for pyrazolate complexes containing the pz* group as a monodentate ligand arises because of their recognized usefulness in the synthesis of dinuclear and trinuclear heterometallic derivatives. The mononuclear complex cis-[(PPh$_3$)$_2$ClPt(dppz)], **218,** was obtained by reacting cis-Pt(PPh$_3$)$_2$Cl$_2$ with Na(dppz) in a 1 : 1 molar ratio (14). The coordination geometry of **218,** which was structurally characterized (14), is very similar to that of the cation in cis-[PdCl(PEt$_3$)$_2$(Hdmpz)](BF$_4$), which was reported in 1981 by Busnell et al. (167). The metalloligand **218** is able to bind to various metals such as Cu, Ag, and Au. Three such heterobimetallic derivatives, cis-[(PPh$_3$)$_2$ClPt($\mu$-dppz)AgCl], **219,** cis-[(PPh$_3$)$_2$BrPt($\mu$-dppz)CuCl], **220,** and cis-[(PPh$_3$)$_2$BrPt($\mu$-4-Brdppz)AuBr], **221,** were characterized by single-crystal X-ray analysis (14).

From Vaska's compound, trans-[Ir(PPh$_3$)$_2$(CO)Cl], the corresponding mononuclear iridium(I) pyrazolate complexes, trans-[Ir(PPh$_3$)$_2$(CO)(pz*)], was synthesized on treatment with Hpz* (Hpz* = Hdmpz, 4-NO$_2$dmpzH, or Hdfmpz) and KOH (168). Evidence for the monodentate nature of the pz* group came from IR and NMR spectra. The presence of a monodentate pyrazolate ligand in trans-[Ir(PPh$_3$)$_2$(CO)(dfmpz)], **222,** was supported by an X-ray crystal structure determination (168). The iridium(I) pyrazolates undergo oxidative addition by dihydrogen, yielding [Ir(H)$_2$(pz*)(CO)(PPh$_3$)$_2$]. The nitrogen atom not involved in coordination was protonated with HBF$_4$ to give the corresponding cationic pyrazole complexes [Ir(CO)(PPh$_3$)$_2$(Hpz*)](BF$_4$) (168). The reaction of iridium(I) and iridium(III) pyrazolate complexes [Ir(PPh$_3$)$_2$(CO)(dmpz)], **223,** and [Ir(H)$_2$(dmpz)(CO)(PPh$_3$)$_2$], **224,** with Au(tht)X (X = Cl or Br) afforded the Ir$^{(I)}$/Au$^{(I)}$ and Ir$^{(III)}$/Au$^{(I)}$ derivatives [Ir(PPh$_3$)$_2$(CO)($\mu$-dmpz)AuX] and [Ir(PPh$_3$)$_2$(CO)(H)$_2$($\mu$-dmpz)AuCl], **225** (169). Complex **225** was structurally characterized (169).

Mention is made here to a member of a series of dimeric terpyridine platinum(II) complexes, [(tpy)Pt($\mu$-pz)Pt(tpy)](ClO$_4$)$_3$, **226.** These complexes were used to study the effect of the Pt$\cdot\cdot\cdot$Pt separation on absorption and emission spectra (170, 171). The X-ray crystal structure of Complex **226** was determined A single pyrazolate ligand bridges two Pt centers. The square planar coordination around each metal is completed by a tridentate terpyridine ligand (170).

In 1979, a series of pyrazolate complexes of the formula [M(pz*)$_2$(L—L)] (M = Pt or Pd; Hpz* = Hpz or Hdmpz; L—L = chelating ligand or diolefin) was described (172). The X-ray crystal structures of [(dppe)Pt(dmpz)$_2$], **227** and [4,4'-Me$_2$bipy)Pt(dmpz)$_2$], **228,** were reported (173, 174). In all these complexes, the pyrazolate groups are monodentate so that all the molecules described can be considered, at least in principle, bidentate metal-containing ligands. Thus, complexes [M(pz*)$_2$(L—L)] were used as ligands toward selected acceptors, affording bimetallic derivatives containing doubly bridged pyrazolate groups (175, 176).

Recently, the synthesis and X-ray crystal structure of [Cu(dcmpz)$_2$(py)$_2$], **229,** were reported (40). Complex **229** was obtained either by reacting [Cu(dcmpz)]$_n$, **12,** (27) with dioxygen in pyridine or by the direct reaction of CuCl$_2$ with Hdcmpz in the presence of pyridine. Complexes having two

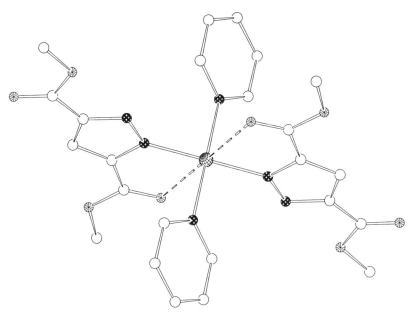

Figure 11. Structure drawing of the mononuclear copper(II) species [Cu(dcmpz)$_2$(py)$_2$], **229**, containing two unidentate 3,5-dicarbomethoxypyrazolate groups in a trans conformation. [Based on data from (40).]

monohapto pyrazolate ligands invariably exhibit a cis-conformation (173). Complex **229** represents the first example of a structurally characterized bis-(pyrazolate) species containing two monodentate pyrazolate groups trans to each other (Fig. 11). Since Complex **229** has two nucleophilic nitrogen atoms not yet involved in coordination with metal centers, this complex has the potential of obtaining homo- and heterotrimetallic systems.

The reaction of [{Rh(C$_5$Me$_5$)}$_2$($\mu$-OH)$_3$](ClO$_4$) with an excess of Hpz or 4-BrpzH, afforded the neutral complexes [{Rh(C$_5$Me$_5$)(pz)}$_2$($\mu$-pz)$_2$], **230,** and [{Rh(C$_5$Me$_5$)(4-Brpz)}$_2$($\mu$-4-Brpz)$_2$], **231** (52) in the presence of KOH. An X-ray structural analysis of Complex **230** confirmed the presence of either exo-bidentate or monodentate pyrazolate groups. Interestingly, the addition of HClO$_4$ to **231** causes the protonation of the monodentate 4-Brpz ligand, with formation of [{Rh(C$_5$M$_5$)(4-BrpzH)}$_2$($\mu$-4-Brpz)$_2$](ClO$_4$)$_2$. The latter easily regenerated **231** on addition of KOH (52). The related mixed-valence rhodium(III)–rhodium(I) complex [(C$_5$Me$_5$)(pz)Rh($\mu$-pz)$_2$Rh(tfbb)], **232**, was obtained by reacting [Rh(tfbb)(Hpz)$_2$](ClO$_4$) with the appropriate acac complex, [Rh(C$_5$Me$_5$)(acac)Cl], in the presence of KOH (79). The $^1$H NMR of **232** confirmed the existence of terminal and bridging pyrazolate groups.

The synthesis and molecular structure of a monomeric vanadium(III) anionic complex, K[{HB(dmpz)$_3$}VCl$_2$(dmpz)], **233**, were reported. Complex **233** was obtained by reacting VCl$_3$ with the tris(3,5-dimethylpyrazolyl)borate ligand in solvents other than DMF. The monodentate dmpz group in this species clearly arises from ligand degradation. Several potential pathways for this decomposition have been advanced. These include enhanced hydrolysis of the B—N bond by traces of water in the presence of a strong Lewis acid such as VCl$_3$ (177).

A trioxo(pyrazolyl)rhenium(VII) complex, O$_3$Re(pz), of unclarified pz-hapticity, was recently reported (178). This complex was obtained by reacting a frozen solution of Re$_2$O$_7$ with solid Na(pz). Its IR spectrum exhibits a strong $\nu$(Re=O) band at 913 cm$^{-1}$, characteristic of a perrhenate anion, which may indicate that the structure of this compound may be more complex than supposed.

## A.   Mononuclear Complexes Containing Neutral Pyrazoles and Monodentate Pyrazolate Ligands

In 1986, the preparation and characterization of new mononuclear (pentamethylcyclopentadienyl) iridium(III) complexes containing monodentate pyrazolate ligands as well as neutral pyrazoles, [(C$_5$Me$_5$)Ir(pz)$_2$(Hpz)], **66,** [(C$_5$Me$_5$)Ir(dmpz)$_2$(Hdmpz)], **234,** and [(C$_5$Me$_5$)Ir(dmpz)(Hdmpz)$_2$](BF$_4$), **235,** were reported (95). Spectroscopic measurements revealed in **66** the equivalence of the pyrazole ligands; this fact has been attributed to an intramolecular proton exchange between the neutral pyrazole and the two anionic pyrazolate ligand. On the contrary, the $^1$H NMR spectrum of **234** showed the presence of the Hdmpz ligand sharing the N—H proton with only one dmpz group, whereas the second pyrazolate ligand was free of interaction. An X-ray crystal structure determination confirmed these observations (95). Treatment of **234** with an equimolar amount of HBF$_4$ afforded the ionic complex **235.** The $^1$H NMR of **235** clearly showed that the second proton is bound to the nonchelated ligand.

By taking advantage of the formal similarity between [RB(pz)$_3$]$^-$ and the deprotonated form of the iridium complex **66,** heterodinuclear complexes, synthesized by using **66** as a building block, were described (96). Ruthenium(II) complexes containing neutral pyrazoles as well as pyrazolate anions as ligands were also described (102, 179). The complex [Ru(H)(pz)(CO)(Hpz)(PPh$_3$)$_2$], **80,** was obtained on treatment of [Ru(H)(Cl)(CO)(Hpz)(PPh$_3$)$_2$] with the stoichiometric amount of KOH and Hpz (102). The reaction between cis-[RuCl$_2$(dmso)$_4$] and K[H$_2$B(pz)$_2$] yielded [Ru(pz)$_2$(Hpz)$_3$(dmso)], **236,** whose X-ray crystal structure was determined (179). This determination provided evidence in support of the rupture of the boron–nitrogen bond in the dihydrobis(1-pyrazolyl)borate ligand.

**234**                                    **235**

Neutral and cationic ruthenium complexes containing monodentate pyrazolate groups were obtained by using the $\beta$-diketonato complex [Ru($p$-cymene)(acac)Cl], or the dinuclear complex [{Ru($p$-cymene)Cl}$_2$($\mu$-Cl)] as the starting material (7) (Scheme 11).

Complex **237** (Scheme 11) could be further deprotonated by addition of a second equivalent of KOH, which yielded [Ru($p$-cymene)(pz)$_2$(Hpz)], **238**. These complexes were characterized by analytical and IR measurements, NMR studies, and by determination of the X-ray crystal structure of **237**. In the latter, there is an intramolecular hydrogen bond between the NH of one of the pyrazoles (Hpz) and the pyrazolate group. The NH of the second Hpz ligand is engaged in an intermolecular hydrogen bond with the BF$_4$ anion (7). Deprotonation by NBu$_4$OH of the pyrazole complexes [R$_2$Pd(Hpz)$_2$] (R = C$_6$F$_5$ or C$_6$Cl$_5$) or [R$_2$Pt(Hpz*)$_2$] (R = C$_6$F$_5$, Hpz* = Hpz or Hdmpz) afforded pyrazole–pyrazolate complexes of the formula [R$_2$M(pz*)(Hpz')](NBu$_4$) (180).

The IR and NMR data for the latter species showed that both pyrazolyl ligands were identical. The X-ray structure determinations of [(C$_6$F$_5$)$_2$Pd(pz)(Hpz)](NBu$_4$), **242**, and [(C$_6$F$_5$)$_2$Pt(pz)(Hpz)](NBu$_4$), **243**, established the existence of an intramolecular hydrogen bond between the two pyrazolyl ligands, the coordination at palladium, and platinum centers being essentially square planar (180).

The synthesis and crystal structures of dimeric platinum(II) and palladium(II) complexes of the formula [M(pz*)$_2$(Hpz*)$_2$]$_2$ (M = Pt, Hpz* = Hpz, **244**; M = Pd, Hpz* = Hpz, **245**; Hdmpz, **246**) were reported (44, 45). Consistent with their IR spectra, the structures of **244** and **246** revealed the presence of two monomeric [M(pz*)$_2$(Hpz*)$_2$] units linked by four symmetrical N—H· · ·N bridges between the Hpz* and the pz* ligands (Fig. 12).

Complex **246** exhibits high rigidity even in solution, as indicated by the absence of any fluxional behavior (evidence from $^1$H NMR measurements at

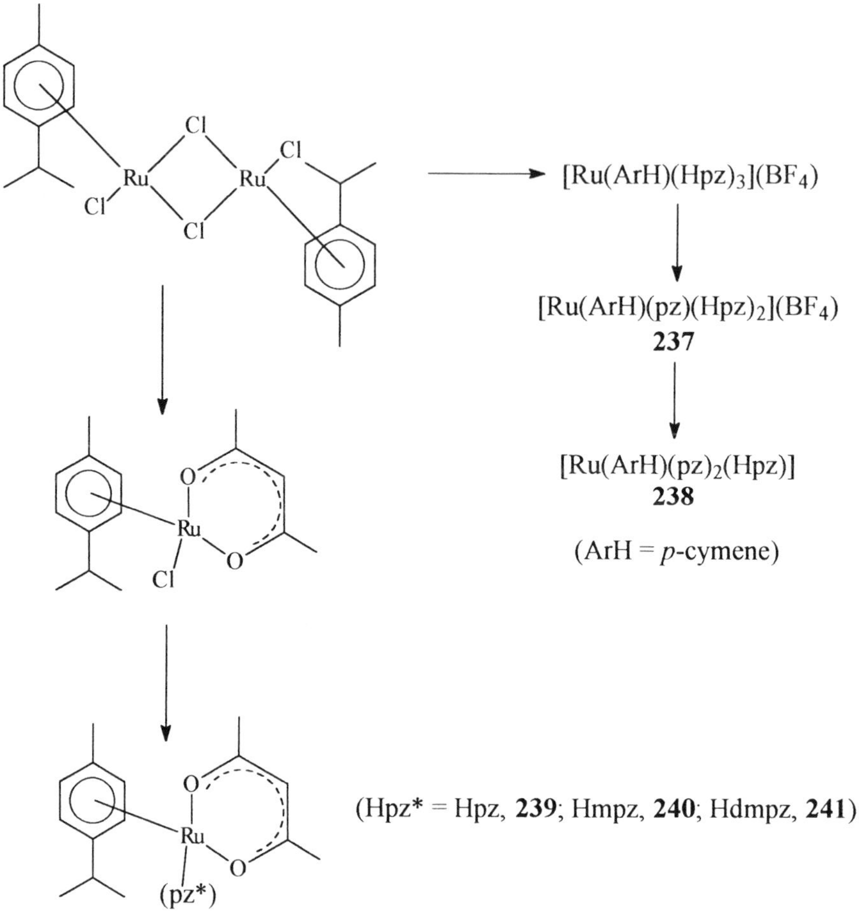

$[Ru(ArH)(Hpz)_3](BF_4)$

$[Ru(ArH)(pz)(Hpz)_2](BF_4)$
**237**

$[Ru(ArH)(pz)_2(Hpz)]$
**238**

$(ArH = p\text{-cymene})$

$(Hpz^* = Hpz, \textbf{239}; Hmpz, \textbf{240}; Hdmpz, \textbf{241})$

Scheme 11. Neutral and cationic ruthenium complexes containing unidentate pyrazolate groups have been obtained by using the dinuclear complex $[\{Ru(p\text{-cymeme})Cl\}_2(\mu\text{-Cl})]$ or the $\beta$-diketonato complex $[Ru(p\text{-cymene})(acac)Cl]$ as starting materials. [Adapted from (7).]

variable temperature) (45). It was verified that **246** reacts rapidly, in $CH_2Cl_2$ solution at room temperature, with a number of heterocumulenes, such as $CS_2$, COS, and $p$-tolylisocyanate (181). The reaction with COS gave $[Pd\{SC(O)dmpz\}_2]$, **247**. The X-ray crystal structure of **247** showed that monomeric, centrosymmetric molecules, containing a square planar trans-$PdN_2S_2$ chromophore, are formed (Fig. 13) (181).

Interestingly, Complex **246**, reacts, in the solid state, with liquid COS at $-78°C$, yielding a product formulated as $[Pd(dmpz)_2(Hdmpz)_2]_2 \cdot COS$, **248**. The latter compound exhibits a strong, sharp $\nu_{as}(SCO)$ IR absorption at 2045 $cm^{-1}$ $[\nu_{as}(SCO)$ of free COS $= 2062$ $cm^{-1}]$. This absorption was initially at-

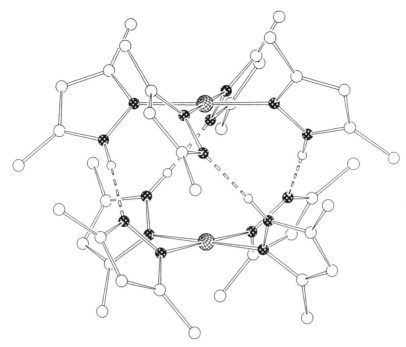

Figure 12. Structure drawing of [Pd(dmpz)$_2$(Hdmpz)$_2$]$_2$, **246.** [Based on data from (45).]

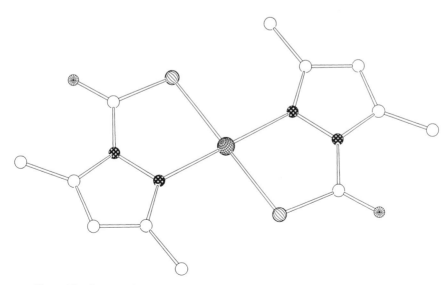

Figure 13. Structure drawing of [Pd{SC(O)dmpz}$_2$], **247.** [Based on data from (181).]

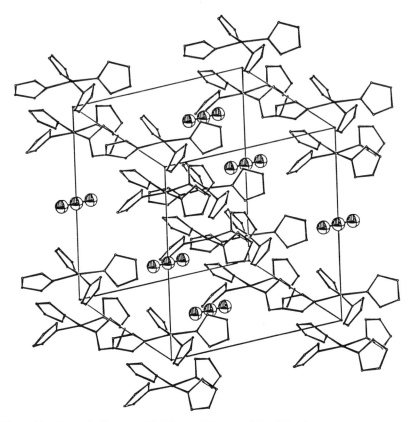

Figure 14.  Unit cell diagram of [Pd(dmpz)₂(Hdmpz)₂]·COS, **248,** showing clathrated, uncoordinated, COS molecules. [Based on data from (181).]

tributed to the still elusive $\eta^1$-S coordination mode of the COS molecule. A structural characterization of **248** was performed by X-ray powder diffraction (XRPD) methods (181). This study showed that in **248** the identity of the [Pd(dmpz)₂(Hdmpz)₂]₂ molecules is essentially retained, while, unexpectedly, COS molecules are hosted in the crystal lattice as nonbonded fragments (Fig. 14).

## VI.  COMPLEXES CONTAINING THE M(μ-pz*)₃M OR M(μ-pz*)₄M CORE

Only a few examples of these unusual species have been reported. Recently, the synthesis and structural characterization of the hydroperoxo complex [(C₅Me₅)Ir(μ-pz)₃Rh(OOH)(dppe)](BF₄), **249,** were described (182). This complex was obtained by protonation of the heterodinuclear complex [(C₅Me₅)(pz)Ir(μ-pz)₂Rh(dppe)] (96) in the presence of dioxygen. An interest-

TABLE XIII
List of Structurally Characterized Transition Metal Complexes Containing Monodentate pz*
Ligands

| Complex | Formula | Reference |
|---------|---------|-----------|
| **31** | $[Zn(dmpz)_2(Hdmpz)]_2$ | 42 |
| **44** | $[Co(dmpz)_2(Hdmpz)]_2$ | 72 |
| **78** | $[(C_5Me_5)(pz)Ir(\mu\text{-}pz)_2Ag(PPh_3)]$ | 96 |
| **218** | $cis\text{-}[Pt(dppz)(PPh_3)_2Cl]$ | 14 |
| **222** | $trans\text{-}[Ir((dfmpz)(CO)(PPh_3)_2$ | 168 |
| **227** | $[(dppe)Pt(dmpz)_2]$ | 173 |
| **229** | $[Cu(dcmpz)_2(py)_2]$ | 40 |
| **230** | $[\{Rh(C_5Me_5)(pz)\}_2(\mu\text{-}pz)_2]$ | 52 |
| **233** | $K[\{HB(dmpz)_3\}VCl_2(dmpz)]$ | 177 |
| **234** | $[(C_5Me_5)Ir(dmpz)_2(Hdmpz)]$ | 95 |
| **237** | $[Ru(p\text{-}cymene)(pz)(Hpz)_2]$ | 7 |
| **242** | $[Pd(C_6F_5)_2(pz)(Hpz)](NBu_4)$ | 180 |
| **243** | $[Pt(C_6F_5)_2(pz)(Hpz)](NBu_4)$ | 180 |
| **244** | $[Pt(pz)_2(Hpz)_2]_2$ | 44 |
| **246** | $[Pd(dmpz)_2(Hdmpz)_2]_2$ | 45 |

ing feature of Complex **249** is the existence of a triply pyrazolate bridge connecting the two transition metal atoms. This structural feature is usual in the chemistry of tris(pyrazolyl)borato transition metal complexes, but extremely rare in bi- or polynuclear pyrazolate bridged transition metal compounds (4, 5, 12).

In 1969 Trofimenko (183) prepared the triply bridged anionic complex $[(CO)_3Mn(\mu\text{-}pz)_3Mn(CO)_3]^-$, from $BrMn(CO)_5$ and $Na(pz)$. The structure was assigned on the basis of IR and NMR data. The nickel nitrosyl complex $[(NO)Ni(\mu\text{-}dmpz)_3Ni(NO)]^-$ had been reported 10 years later (184). The two Ni(NO) moieties are bridged by three dmpz ligands, with the nickel atom existing in a distorted tetrahedral coordination geometry.

In 1985, it was reported that the direct reaction of $Rh_2(O_2CCH_3)_4$ with $Na(dmpz)$ in MeCN afforded the quadruply bridged pyrazolate complex $[Rh_2(\mu\text{-}dmpz)_4(MeCN)_2]$, **250** (185). The formulation of this unprecedented compound was supported by an X-ray crystal structure analysis (185). The MeCN ligands can be removed by heating under vacuum to give $[Rh_2(dmpz)_4]$, which in turn formed adducts with different neutral monodentate ligands yielding species of the general formula $[Rh_2(dmpz)_4(L)_2]$ (L = py or Hdmpz).

## VII.  COMPLEXES CONTAINING endo-BIDENTATE PYRAZOLATE LIGANDS

In 1981, the synthesis and the first structure analysis of a compound with an endo-bidentate pyrazolate anion, $[U(C_5H_5)_3(\eta^2\text{-}pz)]$, **251,** were reported (186).

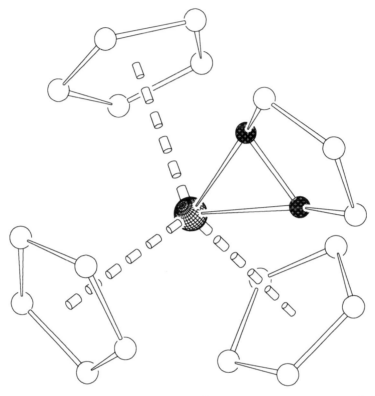

Figure 15.    Structure drawing of [U(C$_5$H$_5$)$_3$($\eta^2$-pz)], **251**. [Based on data from (186).]

This compound was readily prepared from [U(C$_5$H$_5$)$_3$Cl] and Na(pz). The U—N distances are 2.40 and 2.36 Å. The reason for such a 0.04-Å difference appeared unclear. The molecular structure consists of discrete U(C$_5$H$_5$)$_3$(pz) molecules in which the U$^{(IV)}$ ion is coordinated by three $\eta^5$-(C$_5$H$_5$) rings and by the two nitrogens of the pyrazolate group (Fig. 15).

In 1982, the same authors reported on other examples of the endo-bidentate pz* ligand. The reaction between [U(C$_5$Me$_5$)$_2$Cl$_2$] and stoichiometric amounts of Na(pz) afforded [U(C$_5$Me$_5$)$_2$Cl($\eta^2$-pz)], **252**, or [U(C$_5$Me$_5$)$_2$($\eta^2$-pz)$_2$], **253** (187). These compounds were characterized by IR, $^1$H NMR, vis and mass spectra, and by single-crystal X-ray diffraction.

Recently, the synthesis and X-ray crystal structure analysis of a cationic zirconocene complex containing an endo-bidentate pyrazolate, were reported (188) (Fig. 16). The complex [(C$_5$H$_5$)$_2$Zr(thf)($\eta^2$-pz)](BF$_4$), **254**, was obtained by treating dimethylzirconocene with one molar equivalent of Hpz and, subsequently, with (Bu$_3$NH)(BPh$_4$). Complex **254** contains a $\eta^2$-pyrazolate($N,N'$) li-

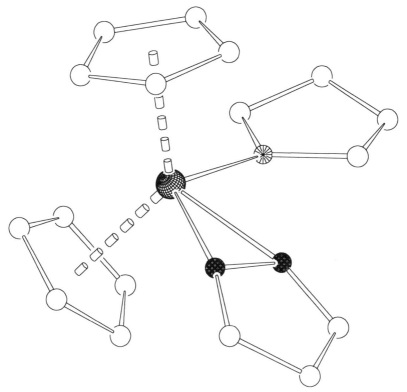

Figure 16. Structure drawing of the $[(C_5H_5)_2Zr(thf)(\eta^2\text{-}pz)]^+$ cation **255**. [Based on data from (188).]

gand bonded to zirconium. Both nitrogen centers are nearly equidistant from zirconium. The preparation of a related zirconium endo-bidentate pyrazolate complex, $[(C_5H_5)_2Zr(Hpz)(\eta^2\text{-}pz)](BPh_4)$, **255,** was also reported. Its structure was proposed by means of $^1H$ NMR evidences (188).

The reaction of $[O(CH_2CH_2C_5H_4)_2MCl]$ (M = Y or Lu) with Na(dmpz) gave the species $[O(CH_2CH_2C_5H_4)_2M(dmpz)]$ (M = Y, **256**; Lu, **257**) probably containing the dmpz group as an endo-bidentate ligand (189). Complexes **256** and **257** were shown to be highly sensitive toward air and water. Their partial hydrolysis gave the mixed-bridged complexes **258** and **259,** whose X-ray crystal structures were determined.

Complexes of the formula $Nd(pz)_3$, or $[Nd(dmpz)_3(thf)]_2$, were obtained in poor yields by the reaction of neodynium metal with $Hg(C_6F_5)_2$ and either Hpz or Hdmpz in THF at room temperature. The X-ray crystal structure of the dmpz derivative showed that the compound is a centrosymmetric dimer with two

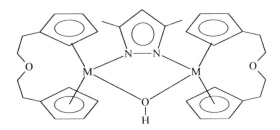

M = Y, **256**; Lu, **257**                    M = Y, **258**; Lu, **259**

bridging dmpz ligands and two symmetrically bridging thf ligands. Each neo-dymium atom is eight-coordinate with four nitrogen atoms from terminal $\eta^2$-dmpz ligands, one nitrogen from each bridging dmpz ligand and two oxygens from bridging thf molecules (190).

Monomeric tris-($\eta^2$-pyrazolate)-lanthanoid complexes with the bulky 3,5-di(*t*-butyl)pyrazolate ligand, of general formula $[M(t\text{-}Bu_2pz)_3(thf)_2]$ (M = La, Nd, Gd, or Er), were obtained by reaction of $Hg(C_6F_5)_2$ and 3,5-*t*-BupzH with an excess of lanthanoid metal in THF at room temperature. The X-ray crystal structure of the neodymium derivative was also reported (191).

Reaction of lanthanoid trichlorides, $MCl_3$ (M = Y, Ho, Yb, or Lu), with Na(dmpz) yielded complexes of the formula $[M_3(\mu\text{-}dmpz)_6(\eta^2\text{-}dmpz)_3(\mu_3\text{-}O)Na_2(thf)_2]$ and $[M_3(\mu\text{-}dmpz)_6(\eta^2\text{-}dmpz)_3(\mu_3\text{-}O)Na_2(Hdmpz)_2]$. The X-ray crystal structures of $[Yb_3(\mu\text{-}dmpz)_6(\eta^2\text{-}dmpz)_3(\mu_3\text{-}O)Na_2(thf)_2]$ and $[Ho_3(\mu\text{-}dmpz)_6(\eta^2\text{-}dmpz)_3(\mu_3\text{-}O)Na_2(Hdmpz)_2]$ revealed trigonal bipyramids containing an oxygen atom at the center, three lanthanoid atoms at the equatorial positions, and two sodium atoms at the apical vertices. Holmium and ytterbium each are heptacoordinated by four nitrogen atoms of the bridging pyrazolate ligands, two nitrogen of the endo-bidentate pyrazolate ligand, and the central oxygen atom (192).

Transition metal complexes containing the pyrazolate group as anionic non-coordinating ligand, although rare, have been reported (5). They are undoubt-edly of lesser importance and will not be further discussed.

## VIII.  COMPLEXES CONTAINING DINUCLEATING FUNCTIONALIZED PYRAZOLATE LIGANDS

The synthesis, characterization, and properties of a large series of metal complexes built by pyrazole based dinucleating ligands possessing two chelat-ing arms attached to the 3- and 5-positions of the pyrazole ring were reported. Such ligands were designed with the aim of preparing dinuclear complexes with an appropriate metal–metal separation that is very important in studies on functional models for some bimetallic biosites (Fig. 17).

The ability of the trianion of pyrazole-3,5-dicarboxylic acid ($H_3PZ^1$) to form dinuclear complexes was utilized in the preparation of a family of anionic complexes of rhodium(I) and iridium(I), using cod, CO, and $PPh_3$ as ancillary ligands (193). The X-ray crystal structure of the complex $[Rh_2(CO)_4-(PZ^1)](NBu_4)$, **260**, was determined. Stacking interactions in some of the rhodium and iridium anionic carbonyl complexes in the solid state have been inferred from the dramatic change of color with changes in the countercation. The electrochemical oxidation of **260** produced a red solid analyzing as the mixed-valence compound $[Rh_2(CO)_4(PZ^1)]$. Bulk magnetic measurements on the red solid showed that it is diamagnetic, indicating either a strong spin–spin coupling or a metal–metal bond between the formally Rh(II) centers from two different molecules.

The same electrochemical oxidation, carried out on the iridium complexes $[Ir_2(CO)_4(PZ^1)](R_4N)$ (R = Bu or Pr), gave dark conducting materials that analyzed as $[Ir_2(CO)_4(PZ^1)](R_4N)_{0.5}$ (193). Conductivity measurements for the latter species gave values that are 1000 times higher than the values found for the most conducting unoxidized precursor (the $Me_4N^+$ salt of the iridium complex).

Divalent metal complexes of the general formula $(Bu_4N)_2[M_2(PZ^1)_2]$ (M = Pt, Pd, Ni, and Cu) were prepared by reacting $H_2PtCl_4$, $PdCl_2$, $Ni(NO_3)_2$, and $Cu(NO_3)_2$ with $H_3PZ^1$ in the presence of $Bu_4NOH$ (194). The structure of both copper and palladium complexes were determined as tetrabutylammonium salts (194) (Fig. 18).

Dicopper(II) complexes of the general formula $[Cu_2(PZ)(N_3)(H_2O)] \cdot H_2O$ were obtained from all the ligands $H_3PZ^1$–$H_3PZ^4$, with deprotonation of both the pyrazole and amide taking place (195). It is presumed that both the pyrazolate and azide groups function as bridges to two copper ions. The end-to-end mode of coordination seems the most plausible for the bridging azide ion in the present complexes. The corresponding complex with an acetate group was isolated only with $H_3PZ^4$, $[Cu_2(PZ^4)(O_2CMe)] \cdot 2MeOH$. Infrared data support the presence of $MeCOO^-$ ion as a bridging group (195).

Two types of dinuclear complexes of the general formula $[M_2(PZ)(X)](BPh_4)_2$ (Type A) and $[M_2(PZ)_2](BPh_4)_2$ (Type B) were obtained by using ligands $HPZ^5$–$HPZ^9$ (196–198) (Table XIV).

X-ray crystal structures of the manganese(II) derivatives ($PZ^7$ or $PZ^9$) and of the copper(II)–$PZ^5$ complex were determined (196–198). All manganese(II) compounds showed catalytic activity toward disproportionation of $H_2O_2$ in DMF at 0°C. A probable mechanism for this reaction has also been proposed.

The reaction of $H_3PZ^{10}$ with $CuCl_2 \cdot 2H_2O$ gave, according to the temperature, two different isomeric dinuclear compounds of the formula $[Cu_2(H_2PZ^{10})_2Cl_2] \cdot 2H_2O$. The molecular structures of the two isomers are closely related, differing only in hydrogen bonding (199).

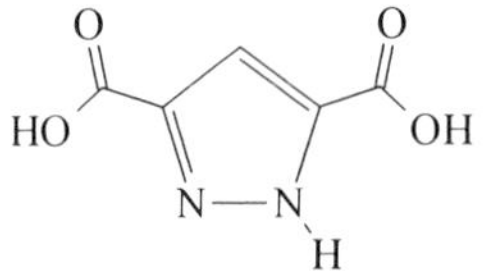

H₃PZ¹

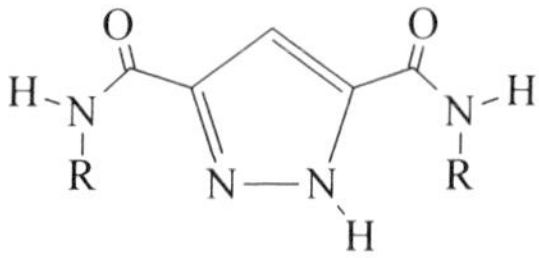

R = (CH$_2$)$_2$NEt$_2$, H₃PZ²
R = (CH$_2$)$_3$NMe$_2$, H₃PZ³
R = CH$_2$(2-py), H₃PZ⁴

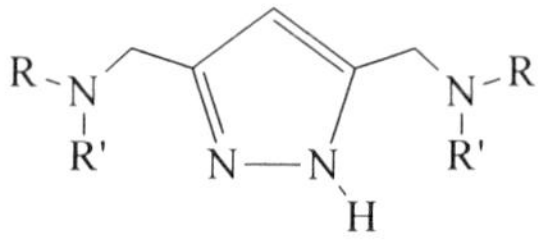

R = (CH$_2$)$_2$NEt$_2$, R' = H, HPZ⁵
R = (CH$_2$)$_3$NMe$_2$, R' = H, HPZ⁶
R = CH$_2$(2-py), R' = H, HPZ⁷
R = R' = (CH$_2$)$_2$NEt$_2$, HPZ⁸
R = R' = CH$_2$(2-py), HPZ⁹

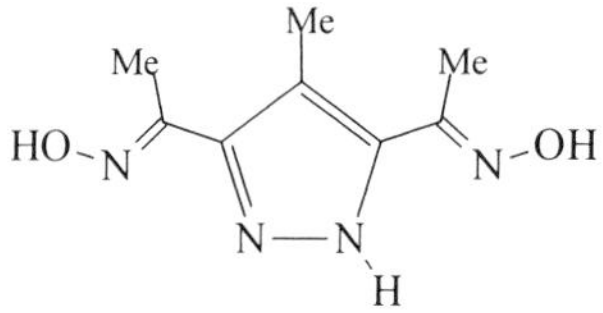

H₃PZ¹⁰

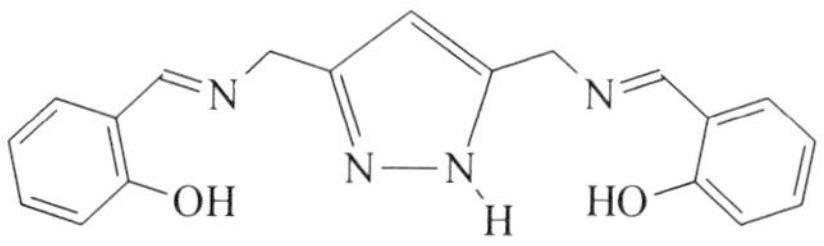

H₃PZ¹¹

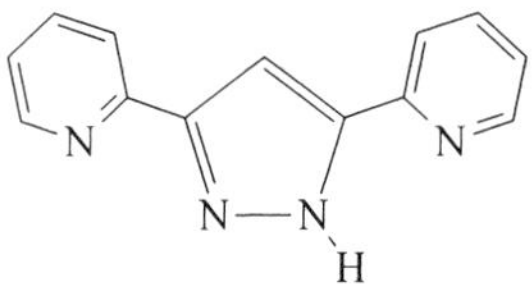

HPZ¹²

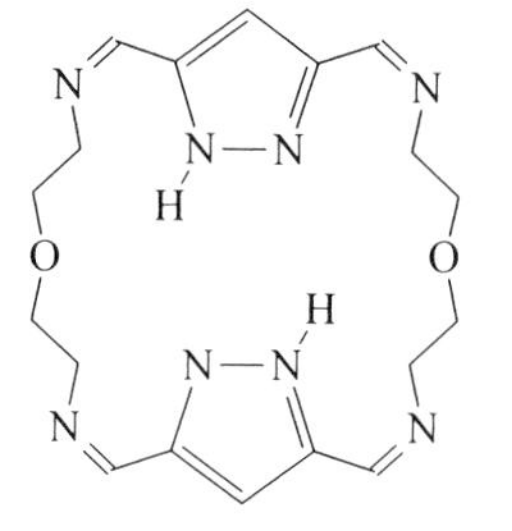

H₂PZ¹³

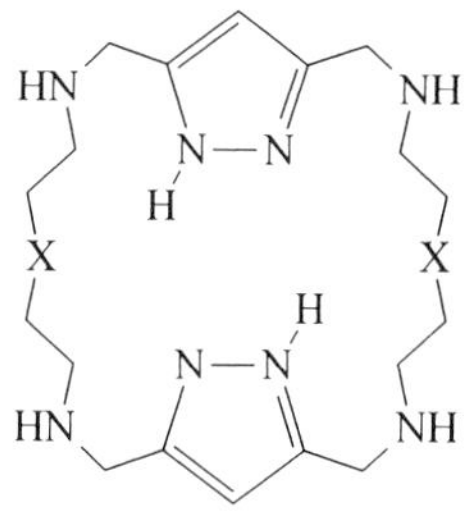

X = NH, H₂PZ¹⁴     X = O, H₂PZ¹⁵

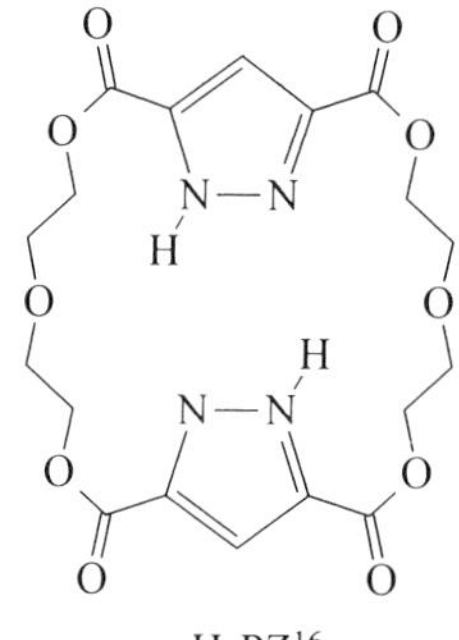

H₂PZ¹⁶

Figure 17. Drawings of the dinucleating functionalized pyrazolate ligands discussed in this chapter.

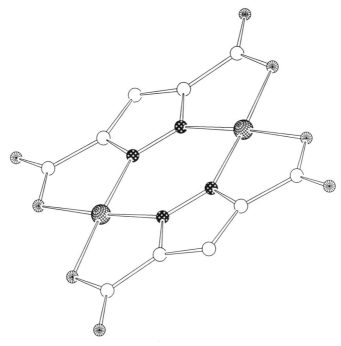

Figure 18.   Structure drawing of the $[Pd_2(PZ^1)_2]^{2-}$ anion. [Based on data from (194).]

A manganese(III) complex, $[Mn_4(PZ^{11})_2(MeO)_4(MeOH)_4](ClO_4)_2$, has been obtained from the dinucleating ligand $H_3PZ^{11}$ and its crystal structure has been reported (200). The ligand $H_3PZ^{11}$ may incorporate a pair of manganese ions by using its two tridentate coordination sites, that is, pyrazolate N and salicylideneaminate N and O. The bridge by the endogeneous pyrazolate group provides the Mn · · · Mn separation of about 3.48 Å. In DMF solutions, the complex exhibits a high catalase-like activity. The absorption spectrum in this solvent suggests that the complex structure differs from the tetranuclear structure found in the solid state.

TABLE XIV
Dinuclear Complexes of the General Formula $[M_2(PZ)(X)(BPh_4)_2$ (Type A) and
$[M_2(PZ)_2](BPh_4)_2$ (Type B)

| Type A | | | Type B | |
|--------|-----|-----|--------|-----|
| Metal | PZ | X | Metal | PZ |
| Mn | $PZ^8$ | $MeCOO^-$ | Cu | $PZ^5$ |
| | $PZ^9$ | $PhCOO^-$ | | $PZ^6$ |
| Co, Cu | $PZ^8$ | $N_3^-$ | Mn | $PZ^7$ |

Reaction of 3,5-bis(pyridin-2-yl)pyrazole ($HPZ^{22}$) with metal nitrates, perchlorate, chlorides and bromides, yielded compounds of the general formula $[M(PZ^{12})]X \cdot nH_2O$ (X = $NO_3$, M = Co, Cd, Ni, Zn, Cu; X = $ClO_4$, M = Cu; X = Cl, M = Mn, Ni, Co, Zn; X = Br, M = Ni, Co, Zn) (201, 202). The nickel(II) derivative containing the $NO_3^-$ anion has been known since 1969 and on the basis of spectroscopic and magnetic data was described as possessing a dinuclear structure (203). A tentative dinuclear structure was assigned to the other complexes of the type $[M(PZ^{12})]X \cdot nH_2O$ as well.

The crystal and molecular structure of $[Ni_2(PZ^{12})_2(MeOH)_4]Cl_2$ were determined (201), which confirms the previous assignment. More recently (204), the X-ray crystal structure of the $Cu^{(II)}/NO_3^-$ complex was solved; it revealed a tetranuclear arrangement, that is, $[Cu_4(PZ^{12})_4(H_2O)_4](NO_3)_4$ (204). However, this compound can be considered as composed of two weakly associated pairs of dimers, with $\pi(\eta^5)$ interdinuclear interactions.

Dinuclear copper(II) complexes, $[Cu_2(PZ^{13})](ClO_4)_2$ and $[Cu(PZ^{14})](ClO_4)_2$, were obtained from the 26-membered macrocycles containing two pyrazole rings, $H_2PZ^{13}$ and $H_2PZ^{14}$ (205). The X-ray crystallographic analysis of the $PZ^{14}$ derivative demonstrates that inside the macrocyclic cavity, the two pyrazolate rings are simultaneously acting as exo-bidentate ligands linking both metal cations, with the Cu$\cdots$Cu separation being 3.9 Å.

With the use of $^{13}$C NMR techniques, the formation of the related zinc(II) dinuclear complexes of the formula $[Zn_2(PZ)]^{2+}$, has been recently evidenced by treatment of $ZnCl_2$ with the ligands $H_2PZ^{14}$, $H_2PZ^{15}$, and $H_2PZ^{16}$ (206, 207).

The PNNHP molecule, when deprotonated, behaves as a planar tetradentate ligand consisting of a central pyrazolate unit with symmetrically disposed methylenediphenylphosphine arms (PNNP). Its geometry provides for two metals to reside within a cooperative distance but does not allow for metal–metal bond formation (208).

PNNHP

Two distinct classes of planar bimetallic complexes containing the PNNP ligand were isolated and characterized: the neutral $[M_2(PNNP)(\mu\text{-}X)L_2]$ and the cationic $[M_2(PNNP)(L)_4]^+$ (208).

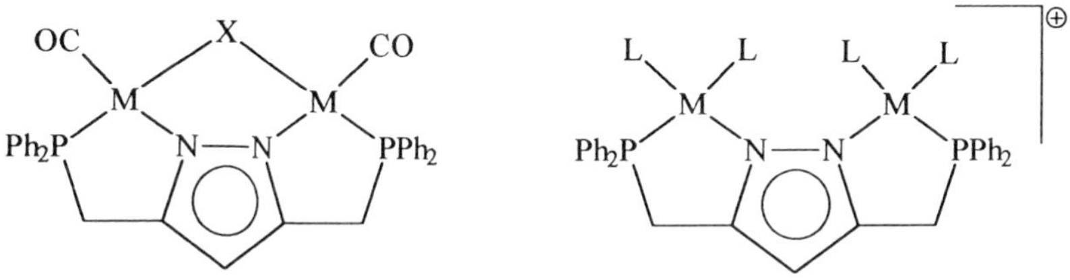

Particular attention has been devoted to oxidative addition and reductive elimination reactions of $[M_2(PNNP)(\mu\text{-}X)L_2]$ with acyl and alkyl halides. Depending on the electron richness of the metals, a complete spectrum of possibilities was observed: from reversible single oxidative addition on one of the metals to irreversible double oxidative addition on both metals (2) (Scheme 12).

Oxidative addition to one metal leads to deactivation of the other one despite the fact that no metal–metal bonds are formed. The crystal structure of the

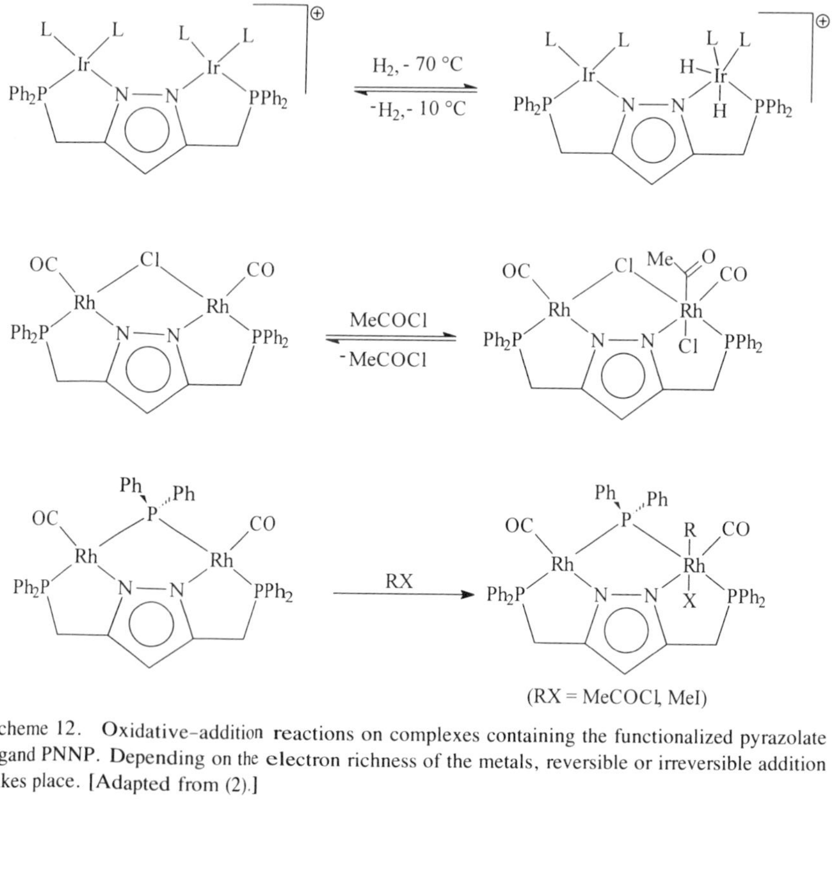

Scheme 12. Oxidative-addition reactions on complexes containing the functionalized pyrazolate ligand PNNP. Depending on the electron richness of the metals, reversible or irreversible addition takes place. [Adapted from (2).]

rhodium(I)/rhodium(III) complex [Rh$_2$($\mu$-PPh$_2$(CO)$_2$(PNNP)(Me)(I)] was also determined (2). The bis(diene)species of rhodium and iridium are precursors for catalytic hydrogenation of alkenes and alkynes.

## IX.  CATALYTIC ACTIVITY OF HOMO- AND HETEROBRIDGED PYRAZOLATE COMPLEXES

Several studies were carried out in the last decade to test the catalytic activity of mononuclear pyrazole complexes as well as dinuclear and polynuclear species containing bridging pyrazolate groups. Particularly investigated were the pyrazolate complexes in which the M($\mu$-pz*)$_2$M (Type A) or the ''M($\mu$-pz*)($\mu$-X)M'' (Type B) framework is present.

Type A                Type B

Some evidence has been presented that the dinuclear structure is retained during the catalytic cycle in the case of homobridged Type A complexes. Indeed, studies on ($\mu$-pz*)$_2$ complexes demonstrated the stability and flexibility of the six-membered cyclic core M—[N—N]$_2$—M. Such compounds generally exist in boat conformations, which thereby facilitate the interactions between the two metal centers, a factor that has important consequences on catalytic activity. In fact, pz* is an extremely versatile ligand in that it is able to bridge an unusually wide range of intermetallic separations. It was suggested that the shorter metal–metal distances generally lead to higher reactivity (209). Accordingly, [Ir($\mu$-dmpz)(CO)$_2$]$_2$ evidenced catalytic activity in hydrogenation reactions. The related compounds [Ir($\mu$-mpz)(CO)$_2$]$_2$ and [Ir($\mu$-pz)(CO)$_2$]$_2$, which possess a longer metal–metal separation, were shown to be inert for the same catalytic reactions (210). However, [Ir($\mu$-dfmpz)(cod)]$_2$, which exhibits a metal–metal distance much shorter than the related dmpz derivative, [Ir($\mu$-dmpz)(cod)]$_2$, was found to be inert as a hydrogenation catalyst (211). Photoelectron spectroscopy studies carried out on a series of representative iridium dimer complexes, having a variety of metal–metal distances, led to believe that electronic interactions are of primary importance in determining the $Ir_2$ chemistry (211). The catalytic reactions that have been investigated to date can be referred to as (a) hydroformylation reactions, (b) hydrogen-transfer reactions, and (c) oxidation or oxygenation reactions. Moreover, it was found that some rhodium–pyrazolate complexes are able to promote the polymerization of monoaryl acetylenes.

## A. Hydroformylation Reactions

The catalytic activity of the mixed-bridged dinuclear complex [$Rh_2(\mu$-pz)($\mu$-S-$t$-Bu)(CO)$_2${P(OMe)$_3$}$_2$], **111,** in the selective hydroformylation of 1-hexene to afford the corresponding aldehydes was explored and the results were compared with those observed for the symmetrical pyrazolate-free precursor [$Rh_2(CO)_2(\mu$-S-$t$-Bu)$_2${P(OMe)$_3$}$_2$] (117, 212). Very mild conditions (5 bar, 80°C) were chosen in both cases. Complex **111** exhibits lower activity; however, the selectivity is about the same as for the symmetrical precursor ($\sim$80% heptanal and 20% 2-methylhexanal). Complex **111** does not give rise to a detectable dismutation and was found to be practically unchanged after reaction. A catalytic cycle in which all the intermediate species remain dinuclear was proposed for the hydroformylation of 1-hexene catalyzed by the symmetrical complexes [$Rh_2(CO)_2(\mu$-S-$t$-Bu)$_2$L$_2$] [L = P(OMe)$_3$, PPh$_3$, P(OPh)$_3$] (213). This proposal is based mainly on the results obtained by varying the phosphorus ligands and on extended Hückel calculations carried out for several proposed intermediates. It is possible that Complex **111** could follow such a cycle (117).

It was also reported (214) that the same hydroformylation reaction can be achieved by various rhodium(I) double bridged pyrazolate complexes of the general formula [$Rh_2(\mu$-pz*)$_2$(CO)$_2$L$_2$] (L = P(OMe)$_3$, P(OPh)$_3$, PPh$_3$; Hpz* = Hpz, Hmpz, or Hdmpz). Good catalytic activity was found. The latter decreases in the sequence P(OPh)$_3$ > P(OMe)$_3$ > PPh$_3$ (214). The effect of the bridging pyrazolate ligand was also studied (215, 216).

Recently, the hydroformylation of 1-dodecene and cyclohexene by the dinuclear doubly bridged pyrazolate complexes [(PR$_3$)(CO)Rh($\mu$-pz*)]$_2$, **46–49,** was reported (75). Thus, at 120°C and at a CO and H$_2$ pressure of 28 atm, these two olefins were transformed into the corresponding aldehydes with greater than 95% yield. All four complexes catalyze the transformation of 1-dodecene to 1-undecane-carboxaldheyde in preference to the iso compound, although this regioselectivity is rather poor (75).

Recently, it has been shown that Complexes **47** and **49** are able to catalyze the polymerization of Ph—C≡CH and the substituted arylacetylenes 4-HC$_6$H$_4$—C≡CH (X = MeO, Me, or Cl) at 25°C in a stereoregular manner

(74). The cis-oriented poly(arylacetylenes) so formed, were found to depoly-merize selectively at 200–225°C to the corresponding 1,3,5-triaryl-benzenes derivatives.


## B.  Hydrogen-Transfer Reactions

Some pyrazolate complexes were found to possess high catalytic activity for various hydrogen-transfer processes and to be of considerable value in synthesis. The heterodinuclear complexes $[(CO)(H)(PPh_3)_2Ru(\mu\text{-pz})(\mu\text{-Cl})M(\text{diolefin})]$ (M = Rh or Ir; diolefin = cod or tfbb) catalyze the hydrogen transfer from isopropanol to cyclohexanone (101). The heterodinuclear complexes were shown to be more active catalysts than the mononuclear pyrazole compounds $[Ru(H)(Cl)(Hpz)(PPh_3)_2]$ and $[M(H)(Cl)(Hpz)(\text{diolefin})]$ (101).

Various hydrogenation catalysts based on $(C_5Me_5)Rh$ complexes with pyrazole-type ligands have been studied (217). Olefins such as cyclohexene and 1-hexene are hydrogenated under ambient conditions with $[\{Rh(C_5Me_5)Cl\}_2(\mu\text{-Cl})_2]$ in the presence of Hpz* and $NEt_3$ (Hpz* = Hpz, Hmpz, or Hdmpz). The catalytic activities are affected mainly by the pyrazole-type ligand. The activity decreases in the order Hdmpz > Hmpz > Hpz, and is highest for the combination $[\{Rh(C_5Me_5)Cl\}_2(\mu\text{-Cl})_2] + 2\ Hdmpz + 2Et_3N$.


## C.  Oxidation or Oxygenation Reactions

Recently, it has been found that the copper(I) homoleptic pyrazolate complexes $[Cu(dmpz)]_3$, **6**, and $[Cu(dppz)]_4$, **8**, are catalytically active in the oxidative coupling of primary aromatic amines to give the corresponding azobenzenes in the presence of $O_2$ at atmospheric pressure (17, 137).

The reaction proceeds with a selectivity of 100%. In the case of $[Cu(dppz)]_4$, the oxidative coupling of a series of different para-substituted aromatic amines was investigated in order to study the influence of the para substitution on the activity of the catalytic system (137).

Complex **6** also exhibits catalytic activity in the oxidation reaction of various other substrates such as $PPh_3$, dibenzylamine, and CO, with formation of triphenylphosphine oxide, N-benzylidenebemzylamine, and $CO_2$, respectively.

The octanuclear copper(II) hydroxo complex $[Cu_8(dmpz)_8(OH)_8]$, **182**, has been identified as the active intermediate species when $[Cu(dmpz)]_3$ is used (137). In the case of $[Cu(dppz)]_4$, the intermediacy of a copper(II) imido species (17) has been proposed on the basis of chemical and spectroscopic evidences.

## X. CONCLUSIONS

The pyrazolate group has been recognized as a particularly useful ligand that is capable of holding two or more metal ions in close proximity with formation of multimetallic transition metal systems. The increasing interest in this area is due to the high relevance of such complexes for multimetal-centered catalysis, biological mimicry, multielectron-transfer reactions, and metal–metal interactions. The extent of such interactions can be controlled by varying the metal–metal distance and the charge and $\pi$-donor/acceptor properties of the ligand used to bridge the metals.

The purpose of this review was to emphasize the remarkable progress in the coordination chemistry of the pyrazolate ligand in the last decade. Quite a lot of interesting and unhoped-for results were attained. These results are summarized below.

1. The metal–pyrazolate saga has been known since the pioneering work of Büchner in the nineteenth century. In the last few years, a definitive characterization of several binary complexes has been possible due to the advent of the recent method of ab initio structure determination from X-ray powder diffraction data. In this respect, it is likely that a number of new molecules and/or crystalline phases, which have so far escaped a complete characterization, will be structurally analyzed, thus confirming the variability of the coordination modes and stoichiometries attributable to "simple" pyrazolato complexes.

2. It is now widely accepted that the nuclearity of binary pyrazolate complexes, $[M(pz^*)_n]_m$, is dependent not only on the nature of the substituents present in the heterocyclic ring, but also on the synthetic strategy used. This important observation allows these systems to be prepared in a selective way.

3. As suggested by Trofimenko in his review articles, controlled extension of the "$M(pz^*)_2M$" core to form polynuclear compounds may constitute an interesting area of research. In these last few years, many elegant examples of homo- and heteropolynuclear systems containing such fragment as well as mixed-bridging $(\mu$-$pz^*)(\mu$-$X)$ groups have been described. The synthesis and chemical and physical behavior of such compounds in larger sequences may represent a fascinating objective in the future. In particular, heteronuclear complexes are of crucial interest because of the special features of the reactivity, which might result from the presence of adjacent metals having different chemical properties (complexes containing electron-rich and electron-deficient metal centers).

4. The monodentate mode of coordination of the pz* group has been now
found in many structurally characterized transition metal compounds.
Thus, complexes of the type M(pz*), *cis*-M(pz*)$_2$, and *trans*-M(pz*)$_2$
have been described. The utility of some of these species as building
blocks for the preparation of heteronuclear complexes is well documented, although further studies are necessary in order to gain more insight into the properties of these systems.

## ABBREVIATIONS

| | |
|---|---|
| acac | Acetylacetonate |
| cod | Cycloocta-1,5-diene |
| cyclop | (1*S*,2*S*)-(+)-1,2-Bis(diphebylphosphinomethyl)cyclohexane |
| *p*-cymene | *p*-Isopropyl-methylbenzene |
| dapm | (Diphenylarsino)(diphenylphosphino)methane |
| debd | 1,6-Bis-(3,5-dimethylpyrazol-1-yl)-2,5-dimethyl-2,5-diazahexane |
| DMF | *N*,*N*-Dimethylformamide |
| dmso | Dimethyl sulfoxide (ligand) |
| DMSO | Dimethyl sulfoxide (solvent) |
| dpae | 1,2-Bis(diphenylarsino)ethane |
| dpam | Bis(diphenylarsino)methane |
| dppb | 1,4-Bis(diphenylphosphino)butane |
| dppe | 1,2-Bis(diphenylphosphino)ethane |
| dppm | Bis(diphenylphosphino)methane |
| dppp | 1,3-Bis(diphenylphosphino)propane |
| ET | Electron transfer |
| Hdbpz | 3,5-Di-*tert*-butylpyrazole |
| Hdcmpz | 3,5-Dicarbomethoxypyrazole |
| Hdfmpz | 3,5-Bis(trifluoromethyl)pyrazole |
| Hdmepz | 3(5),4-Dimethyl-5(3)-ethylpyrazole |
| Hdmpz | 3,5-Dimethylpyrazole |
| Hdppz | 3,5-Diphenylpyrazole |
| Hmpz | 3(5)-Methylpyrazole |
| Hpz | Pyrazole |
| Hpz* | A general pyrazole |
| Htmpz | 3,4,5-Trimethylpyrazole |
| IR | Infrared spectroscopy |
| nbd | Norbornadiene (bicyclo-[2.2.1.]-heptadiene) |
| NMR | Nuclear magnetic resonance |
| phen | 1,10-Phenanthroline |

| py | Pyridine |
|---|---|
| py$^+$ | Pyridinium or substituted pyridinium |
| pz* | A general pyrazolate anion |
| tfbb | Tetrafluorobenzobarrelene (5,6,7,8-tetrafluoro-1,4-Dihydro-1,4-ethenonaphthalene) |
| tht | Tetrahydrothiophene |
| tmpz | Trimethylpyrazolate |
| XPS | X-ray photoelectron spectroscopy |
| XRPD | X-ray powder diffraction |
| 4-XdmpzH | 4-Substituted 3,5-dimethylpyrazole |
| 4-XdppzH | 4-Substituted 3,5-diphenylpyrazole |
| 4-XpzH | 4-Substituted pyrazole |

## ACKNOWLEDGMENTS

We are indebted to Dr. Norberto Masciocchi, Dr. Massimo Moret, and Professor Angelo Sironi (Dipartimento di Chimica Strutturale e Stereochimica Inorganica) and to Professor Sergio Cenini (Dipartimento di Chimica Inorganica, Metallorganica e Analitica) for the invaluable contribution to the pyrazolate chemistry carried out in our laboratories over the past years.

We also thank the Italian Consiglio Nazionale delle Ricerche (C.N.R.) and Ministero dell'Universitá e della Ricerca Scientifica e Tecnologica (MURST) for the financial support.

## REFERENCES

1. See, for example, K. D. Karlin and J. Zubieta, Eds., *Biological and Inorganic Copper Chemistry*, Adenine Press, New York, 1986.

2. T. G. Schenck, C. R. C. Milne, J. F. Sawyer, and B. Bosnich, *Inorg. Chem.*, *24*, 2338 (1985).

3. J. Catalan, J. L. M. Abboud, and J. Elguero, *Adv. Heterocycl. Chem.*, *41*, 187 (1987).

4. J. G. Vos and W. L. Groeneveld, *Inorg. Chim. Acta*, *26*, 71 (1978).

5. S. Trofimenko, *Progress in Inorganic Chemistry*, Wiley-Interscience, New York, 1986, Vol. 34, p. 115.

6. S. Trofimenko, *Chem. Rev.*, *72*, 497 (1972).

7. D. Carmona, J. Ferrer, L. A. Oro, M. C. Apreda, C. Foces-Foces, F. H. Cano, J. Elguero, and M. L. Jimeno, *J. Chem. Soc. Dalton Trans.*, 1463 (1990).

8. B. Bovio, G. Banditelli, and A. L. Bandini, *Inorg. Chim. Acta*, *96*, 213 (1985).

9. M. Mohan, M. R. Bond, T. Otieno, and C. J. Carrano, *Inorg. Chem.*, *34*, 1233 (1995).

10. W. Hückel and H. Bretschneider, *Chem. Ber.*, *70*, 2024 (1937).

11. Q. Mingoia, *Gazz. Chim. Ital.*, *61*, 449 (1931).

12. S. Trofimenko, *Chem. Rev.*, *93*, 943 (1993).

13. N. Masciocchi, M. Moret, P. Cairati, A. Sironi, G. A. Ardizzoia, and G. La Monica, *J. Am. Chem. Soc.*, *116*, 7668 (1994).

14. J. P. Fackler, Jr., R. G. Raptis, and H. H. Murray, *Inorg. Chim. Acta*, *193*, 173 (1992).

15. H. H. Murray, R. G. Raptis, and J. P. Fackler, Jr., *Inorg. Chem.*, *27*, 26 (1988).

16. R. G. Raptis, and J. P. Fackler, Jr., *Inorg. Chem.*, *27*, 4179 (1988).

17. G. A. Ardizzoia, S. Cenini, G. La Monica, N. Masciocchi, and M. Moret, *Inorg. Chem.*, *33*, 1458 (1994).

18. E. Büchner, *Chem. Ber.*, *22*, 842 (1889).

19. H. Okkersen, W. L. Groeneveld, and J. Reedijk, *Recl. Trav. Chim. Pays-Bas*, *92*, 945 (1973).

20. E. Keller, SCHAKAL 92: a computer program for the graphical representation of crystallographic models, University of Freiburg, Germany, 1992.

21. G. A. Ardizzoia, G. La Monica, N. Masciocchi, and M. Moret, unpublished results.

22. C. B. Singh, S. Satpathy, and B. Sahoo, *J. Inorg. Nucl. Chem.*, *33*, 1313 (1971).

23. M. K. Ehlert, S. J. Rettig, A. Storr, R. C. Thompson, and J. Trotter, *Can. J. Chem.*, *68*, 1444 (1990).

24. G. A. Ardizzoia, and G. La Monica, *Inorg. Synth.*, *31*, 299 (1997).

25. M. K. Ehlert, S. J. Rettig, A. Storr, R. C. Thompson, and J. Trotter, *Can. J. Chem.*, *70*, 2161 (1992).

26. M. K. Ehlert, A. Storr, and R. C. Thompson, *Can. J. Chem.*, *70*, 1121 (1992).

27. G. A. Ardizzoia, E. M. Beccalli, G. La Monica, N. Masciocchi, and M. Moret, *Inorg. Chem.*, *31*, 2706 (1992).

28. F. Bonati, A. Burini, B. R. Pietroni, and R. Galassi, *Gazz. Chim. Ital.*, *123*, 691 (1993).

29. J. Berthou, J. Elguero, and C. Rerat, *Acta Crystallogr.*, *B26*, 1880 (1970).

30. F. Krebs-Larsen, M. S. Lehmann, I. Sotofte, and S. E. Rasmussen, *Acta Chem. Scand.*, *24*, 3248 (1970).

31. J. A. S. Smith, B. Wehrle, F. Aguilar-Parrilla, H. H. Limbach, C. Foces-Foces, F. H. Cano, J. Elguero, A. Baldy, M. Pierrot, M. M. T. Khurshid, and J. B. Larcombe-McDouall, *J. Am. Chem. Soc.*, *111*, 7304 (1989).

32. A. L. Llamez-Saiz, C. Foces-Foces, F. H. Cano, P. Jiménez, J. Laynez, W. Meutermans, J. Elguero, H. H. Limbach, and F. Aguilar-Parrilla, *Acta Crystallogr.*, *B50*, 746 (1994).

33. F. Aguilar-Parrilla, G. Scherer, H. H. Limbach, C. Foces-Foces, F. H. Cano, J. A. S. Smith, C. Toiron, and J. Elguero, *J. Am. Chem. Soc.*, *114*, 9657 (1992).

34. R. G. Raptis, R. J. Staples, C. King, and J. P. Fackler, Jr., *Acta Crystallogr.*, *C49*, 1716 (1993).

35. M. K. Ehlert, S. J. Rettig, A. Storr, R. C. Thompson, and J. Trotter, *Can. J. Chem.*, *67*, 1970 (1989).

36. M. K. Ehlert, S. J. Rettig, A. Storr, R. C. Thompson, and J. Trotter, *Can. J. Chem.*, *69*, 432 (1991).

37. M. K. Ehlert, A. Storr, R. C. Thompson, F. W. B. Einstein, and R. J. Batchelor, *Can. J. Chem.*, *71*, 331 (1993).

38. J. G. Vos and W. L. Groeneveld, *Inorg. Chim. Acta*, *24*, 123 (1977).

39. S. Emori and K. Sadakata, *Bull. Chem. Soc. Jpn.*, *67*, 1743 (1994).

40. M. Angaroni, G. A. Ardizzoia, G. La Monica, E. M. Beccalli, N. Masciocchi, and M. Moret, *J. Chem. Soc. Dalton Trans.*, *2715 (1992)*.

41. M. K. Ehlert, A. Storr, and R. C. Thompson, *Can. J. Chem.*, *71*, 1412 (1993).

42. M. K. Ehlert, S. J. Rettig, A. Storr, R. C. Thompson, and J. Trotter, *Can. J. Chem.*, *68*, 1494 (1990).

43. W. Burger and J. Strähle, *Z. Anorg. Allg. Chem.*, *529*, 111 (1985).

44. W. Burger and J. Stähle, *Z. Anorg. Allg. Chem.*, *539*, 27 (1986).

45. G. A. Ardizzoia, G. La Monica, S. Cenini, M. Moret, and N. Masciocchi, *J. Chem. Soc. Dalton Trans.*, 1351 (1996).

46. G. Minghetti, G. Banditelli, and F. Bonati, *J. Chem. Soc. Dalton Trans.*, 1851 (1979).

47. B. Bovio, F. Bonati, and G. Banditelli, *Inorg. Chim. Acta*, *87*, 25 (1984).

48. W. L. Driessen, F. Paap, and J. Reedijk, *Recl. Trav. Chim. Pays-Bas*, *114*, 317 (1995).

49. G. A. Ardizzoia, G. La Monica, M. A. Angaroni, and F. Cariati, *Inorg. Chim. Acta*, *158*, 159 (1989).

50. G. A. Ardizzoia, M. A. Angaroni, G. La Monica, N. Masciocchi, and M. Moret, *J. Chem. Soc. Dalton Trans.*, 2277 (1990).

51. B. F. Fieselmann and G. D. Stucky, *Inorg. Chem.*, *17*, 2074 (1978).

52. L. A. Oro, D. Carmona, M. P. Lamata, C. Foces-Foces, and F. H. Cano, *Inorg. Chim. Acta*, *97*, 19 (1985).

53. J. A. Bailey, S. L. Grundy, and S. R. Stobart, *Organometallics*, *9*, 536 (1990).

54. S. Trofimenko, *J. Org. Chem.*, *33*, 890 (1968).

55. G. A. Ardizzoia, M. Angaroni, G. La Monica, M. Moret, and N. Masciocchi, *Inorg. Chim. Acta*, *185*, 63 (1991).

56. T. N. Sorrell, *Tetrahedron*, *45*, 3 (1989).

57. A. L. Bandini, G. Banditelli, F. Bonati, G. Minghetti, and M. T. Pinillos, *Inorg. Chim. Acta*, *99*, 165 (1985).

58. R. G. Raptis and J. P. Fackler, Jr., *Inorg. Chem.*, *29*, 5003 (1990).

59. R. G. Raptis, H. H. Murray, and J. P. Fackler, Jr., *Acta Crystallogr.*, *C44*, 970 (1988).

60. G. Lopez, G. Garcia, J. Ruiz, G. Sanchez, J. Garcia, and C. Vicente, *J. Chem. Soc. Chem. Commun.*, 1045 (1989).

61. G. Lopez, G. Garcia, G. Sanchez, J. Garcia, J. Ruiz, J. A. Hermoso, A. Vegas, and M. Martinez-Ripoll, *Inorg. Chem.*, *31*, 1518 (1992).

62. G. Lopez, J. Ruiz, G. Garcia, J. M. Marti, G. Sanchez, and J. Garcia, *J. Organomet. Chem.*, *412*, 435 (1991).

63. G. López, G. Garcia, G. Sánchez, M. D. Santana, J. Ruiz, and J. Garcia, *Inorg. Chim. Acta*, *188*, 195 (1991).

64. G. López, J. Ruiz, G. Garcia, C. Vicente, J. Casabó, E. Molins, and C. Miravitlles, *Inorg. Chem.*, *30*, 2605 (1991).

65. G. Lopez, J. Ruiz, G. Garcia, C. Vicente, J. M. Martì, J. A. Hermoso, A. Vegas, and M. Martinez-Ripoll, *J. Chem. Soc. Dalton Trans.*, 53 (1992).

66. G. Lopez, J. Ruiz, G. Garcia, C. Vicente, V. Rodriguez, G. Sanchez, J. A. Hermoso, and M. Martinez-Ripoll, *J. Chem. Soc. Dalton Trans.*, 1681 (1992).

67. V. K. Jain, S. Kannan, and E. R. T. Tiekink, *J. Chem. Soc. Dalton Trans.*, 2231 (1992).

68. V. K. Jain and S. Kannan, *Polyhedron*, *11*, 27 (1992).

69. F. Bonati, H. C. Clark, and C. S. Wong, *Can. J. Chem.*, *58*, 1435 (1980).

70. V. K. Jain, S. Kannan, and E. R. T. Tiekink, *J. Chem. Soc. Dalton Trans.*, 3625 (1993).

71. M. Maekawa, M. Munakata, T. Kuroda, and Y. Nozaka, *Inorg. Chim. Acta*, *208*, 243 (1993).

72. M. K. Ehlert, S. J. Rettig, A. Storr, R. C. Thompson, and J. Trotter, *Can. J. Chem. 71*, 1425 (1993).

73. R. Usón, L. A. Oro, M. A. Ciriano, M. R. Pinillos, A. Tiripicchio, and M. Tiripicchio-Camellini, *J. Organomet. Chem.*, *205*, 247 (1981).

74. I. Amer, H. Schumann, V. Ravindar, W. Baidossi, N. Goren, and J. Blum, *J. Mol. Cat.*, *85*, 163 (1993).

75. H. Schumann, H. Hemling, V. Ravindar, Y. Badrieh, and J. Blum, *J. Organomet. Chem.*, *469*, 213 (1994).

76. F. Barceló, P. Lahuerta, M. A. Ubeda, C. Foces-Foces, F. H. Cano, and M. Martinez-Ripoll, *J. Chem. Soc. Chem. Commun.*, 43 (1985).

77. F. Barceló, P. Lahuerta, M. A. Ubeda, C. Foces-Foces, F. H. Cano, and M. Martinez-Ripoll, *Organometallics*, *7*, 584 (1988).

78. G. W. Bushnell, D. O. K. Fjeldsted, S. R. Stobart, M. J. Zaworotko, S. A. R. Knox, and K. A. Macpherson, *Organometallics*, *4*, 1107 (1985).

79. L. A. Oro, D. Carmona, J. Reyes, C. Foces-Foces, and F. H. Cano, *J. Chem. Soc. Dalton Trans.*, 31 (1986).

80. L. A. Oro, D. Carmona, and J. Reyes, *J. Organomet. Chem.*, *302*, 417 (1986).

81. L. A. Oro, D. Carmona, J. Reyes, C. Foces-Foces, and F. H. Cano, *Inorg. Chim. Acta*, *112*, 35 (1986).

82. D. Carmona, F. J. Lahoz, L. A. Oro, J. Reyes, and M. P. Lamata, *J. Chem. Soc. Dalton Trans.*, 3551 (1990).

83. K. A. Beveridge, G. W. Bushnell, K. R. Dixon, D. T. Eadie, S. R. Stobart, J. L. Atwood, and M. J. Zaworotko, *J. Am. Chem. Soc.*, *104*, 920 (1982).

84. J. Powell, A. Kuksis, S. C. Nyburg, and W. W. Ng, *Inorg. Chim. Acta*, *64*, L211 (1982).

85. J. L. Atwood, K. A. Beveridge, G. W. Bushnell, K. R. Dixon, D. T. Eadie, S. R. Stobart, and M. J. Zaworotko, *Inorg. Chem.*, *23*, 4050 (1984).

86. D. G. Harrison and S. R. Stobart, *J. Chem. Soc. Chem. Commun.*, 285 (1986).

87. R. D. Brost, and S. R. Stobart, *J. Chem. Soc. Chem. Commun.*, 498 (1989).

88. M. A. Ciriano, F. Viguri, L. A. Oro, A. Tiripicchio, and M. Tiripicchio-Camellini, *Angew. Chem. Int. Ed. Engl.*, *26*, 444 (1987).

89. R. D. Brost, D. O. K. Fjeldsted, and S. R. Stobart, *J. Chem. Soc. Chem. Commun.*, 488 (1989).

90. D. O. K. Fjeldsted, S. R. Stobart, and M. J. Zaworotko, *J. Am. Chem. Soc.*, *107*, 8258 (1985).

91. S. Trofimenko, *Inorg. Chem.*, *10*, 1372 (1971).

92. K. A. Beveridge, G. W. Bushnell, S. R. Stobart, J. L. Atwood, and M. J. Zaworotko, *Organometallics*, *2*, 1447 (1983).

93. M. A. Ciriano, M. A. Tena, and L. A. Oro, *J. Chem. Soc. Dalton Trans.*, *2123* *(1992)*.

94. C. Tejel, M. A. Ciriano, L. A. Oro, A. Tiripicchio, and F. Ugozzoli, *Organometallics*, *13*, 4153 (1994).

95. D. Carmona, L. A. Oro, M. P. Lamata, J. Elguero, M. C. Apreda, C. Foces-Foces, and F. H. Cano, *Angew. Chem. Int. Ed. Engl.*, *25*, 1114 (1986).

96. D. Carmona, F. J. Lahoz, L. A. Oro, M. P. Lamata, and S. Buzarra, *Organometallics*, *10*, 3123 (1991).

97. D. Carmona, L. A. Oro, M. P. Lamata, M. L. Jimeno, J. Elguero, A. Belguise, and P. Lux, *Inorg. Chem.*, *33*, 2196 (1994).

98. R. Seeber, G. Minghetti, M. I. Pilo, G. Banditelli, and S. Zamponi, *J. Organomet. Chem.*, *402*, 413 (1991).

99. R. S. Farid, I. J. Chang, J. R. Winkler, and H. B. Gray, *J. Phys. Chem.*, *98*, *5176* (1994).

100. R. S. Farid, L. M. Henling, and H. B. Gray, *Acta Crystallogr.*, *C49*, 1363 (1993).

101. M. P. Garcia, A. M. Lopez, M. A. Esteruelas, F. J. Lahoz, and L. A. Oro, *J. Organomet. Chem.*, *388*, 365 (1990).

102. M. P. Garcia, A. M. Lopez, M. A. Esteruelas, F. J. Lahoz, and L. A. Oro, *J. Chem. Soc. Dalton Trans.*, 3465 (1990).

103. J. A. Cabeza, C. Lanzaduri, L. A. Oro, A. Tiripicchio, and M. Tiripicchio-Camellini, *J. Organomet. Chem.*, *322*, C16 (1987).

104. J. A. Cabeza, C. Lanzaduri, L. A. Oro, D. Belletti, A. Tiripicchio, and M. Tiripicchio-Camellini, *J. Chem. Soc. Dalton Trans.*, 1093 (1989).

105. F. Neumann and G. Süss-Fink, *J. Organomet. Chem.*, *367*, 175 (1989).

106. F. Neumann, H. Stoeckli-Evans, and G. Süss-Fink, *J. Organomet. Chem.*, *379*, 151 (1989).

107. S. J. Sherlock, M. Cowie, E. Singleton, and M. M. de V. Steyn, *J. Organomet. Chem.*, *361*, 353 (1989).

108. D. Carmona, J. Ferrer, R. Atencio, F. J. Lahoz, L. A. Oro, and M. P. Lamata, *Organometallics*, *14*, 2057 (1995).

109. M. T. Pinillos, A. Elduque, and L. A. Oro, *Polyhedron*, *11*, 1007, 1992.

110. M. T. Pinillos, A. Elduque, and L. A. Oro, *Inorg. Chim. Acta*, *178*, 179 (1990).

111. B. M. Louie, S. J. Rettig, A. Storr, and J. Trotter, *Can. J. Chem.*, *63*, 688 (1985).

112. M. T. Pinillos, A. Elduque, J. A. Lopez, F. J. Lahoz, and L. A. Oro, *J. Chem. Soc. Dalton Trans.*, 1391 (1991).

113. F. H. Cano, C. Foces-Foces, L. A. Oro, M. T. Pinillos, and C. Tejel, *Inorg. Chim. Acta*, *128*, 75 (1987).

114. M. T. Pinillos, A. Elduque, J. A. Lopez, F. J. Lahoz, L. A. Oro, and B. E. Mann, *J. Chem. Soc. Dalton Trans.*, 2389 (1992).

115. P. Kalck, J. M. Frances, P. M. Pfister, T. G. Southern, and A. Thorez, *J. Chem. Soc. Chem. Commun.*, 510 (1983).

116. A. Tiripicchio, F. J. Lahoz, L. A. Oro, and M. T. Pinillos, *J. Chem. Soc. Chem. Commun.*, 936 (1984).

117. C. Claver, P. Kalck, M. Ridmy, A. Thores, L. A. Oro, M. T. Pinillos, M. C. Apreda, F. H. Cano, and C. Foces-Foces, *J. Chem. Soc. Dalton Trans.*, 1523 (1988).

118. M. T. Pinillos, A. Elduque, L. A. Oro, F. J. Lahoz, F. Bonati, A. Tiripicchio, and M. Tiripicchio-Camellini, *J. Chem. Soc. Dalton Trans.*, 989 (1990).

119. L. A. Oro, D. Carmona, M. P. Garcia, F. J. Lahoz, J. Reyes, C. Foces-Foces, and F. H. Cano, *J. Organomet. Chem.*, *296*, C43 (1985).

120. M. P. Garcia, A. Portilla, L. A. Oro, C. Foces-Foces, and F. H. Cano, *J. Organomet. Chem.*, *322*, 111 (1987).

121. M. T. Pinillos, A. Elduque, E. Martin, N. Navarro, F. J. Lahoz, J. A. Lopez, and L. A. Oro, *Inorg. Chem.*, *34*, 111 (1995).

122. M. T. Pinillos, A. Elduque, E. Martin, N. Navarro, L. A. Oro, A. Tiripicchio, and F. Ugozzoli, *Inorg. Chem.*, *34*, 3105 (1995).

123. M. T. Pinillos, C. Tejel, L. A. Oro, M. C. Apreda, C. Foces-Foces, and F. H. Cano, *J. Chem. Soc. Dalton Trans.*, 1133 (1989).

124. A. Elduque, L. A. Oro, M. T. Pinillos, C. Tejel, A. Titipicchio, and F. Ugozzoli, *J. Chem. Soc. Dalton Soc. Trans.*, 2807 (1991).

125. G. Lopez, J. Ruiz, G. Garcia, C. Vicente, J. M. Marti, and M. D. Santana, *J. Organomet. Chem.*, *393*, C53 (1990).

126. D. Carmona, A. Mendoza, J. Ferrer, F. J. Lahoz, and L. A. Oro, *J. Organomet. Chem.*, *431*, 87 (1992).

127. L. A. Oro, D. Carmona, M. P. Lamata, M. C. Apreda, C. Foces-Foces, F. H. Cano, and P. M. Maitlis, *J. Chem. Soc. Dalton Trans.*, 1823 (1984), and references cited therein.

128. D. Carmona, L. A. Oro, M. P. Lamata, M. P. Puebla, J. Ruiz, and P. M. Maitlis, *J. Chem. Soc. Dalton Trans.*, 639 (1987).

129. J. A. Bailey, S. L. Grundy, and S. R. Stobart, *Organometallics*, *9*, 536 (1990).

130. M. P. Lamata, D. Carmona, L. A. Oro, M. C. Apreda, C. Foces-Foces, and F. H. Cano, *Inorg. Chim. Acta*, *158*, 131 (1989).

131. D. Carmona, M. P. Lamata, M. Esteban, F. J. Lahoz, L. A. Oro, M. C. Apreda, C. Foces-Foces, and F. H. Cano, *J. Chem. Soc. Dalton Trans.*, 159 (1994).

132. M. Valderrama, M. Scotti, J. Cuevas, D. Carmona, M. P. Lamata, J. Reyes, F. J. Lahoz, E. Onate, and L. A. Oro, *J. Chem. Soc. Dalton Trans.*, 2735 (1992).

133. M. K. Ehlert, S. J. Rettig, A. Storr, R. C. Thompson, and J. Trotter, *Inorg. Chem.*, *32*, 5176 (1993).

134. M. K. Ehlert, S. J. Rettig, A. Storr, A. C. Thompson, and J. Trotter, *Acta Crystallogr.*, *C50*, 1023 (1994).

135. M. Angaroni, G. A. Ardizzoia, T. Beringhelli, G. La Monica, D. Gatteschi, N. Masciocchi, and M. Moret, *J. Chem. Soc. Dalton Trans.*, 3305 (1990).

136. F. B. Hulsbergen, R. W. M. ten Hoedt, G. C. Verschoor, J. Reedijk, and A. L. Spek, *J. Chem. Soc. Dalton Trans.*, 539 (1983).

137. G. A. Ardizzoia, M. A. Angaroni, G. La Monica, F. Cariati, S. Cenini, M. Moret, and M. Masciocchi, *Inorg. Chem.*, *30*, 4347 (1991).

138. R. Robson, *Inorg. Nucl. Chem. Lett.*, *6*, 125 (1970).

139. P. A. Vigato, S. Tamburini, and D. E. Fenton, *Coord. Chem. Rev.*, *106*, 25 (1990) and references cited therein.

140. T. N. Doman, D. E. Williams, J. F. Banks, R. M. Buchanan, H. R. Chang, R. J. Webb, and D. N. Hendrickson, *Inorg. Chem.*, *29*, 1058 (1990). See also references cited therein.

141. W. Mazurek, B. J. Kennedy, K. S. Murray, M. J. O'Connor, J. R. Rodgers, M. R. Snow, A. G. Wedd, and P. R. Zwack, *Inorg. Chem.*, *24*, 3258 (1985).

142. Y. Nishida and S. Kida, *Inorg. Chem.*, *27*, 447 (1988).

143. D. Ajó, A. Bencini, and F. Mani, *Inorg. Chem.*, *27*, 2437 (1988).

144. K. Matsumoto, S. Ooi, W. Mori, and Y. Nakao, *Bull. Chem. Soc. Jpn.*, *60*, 4477 (1987).

145. P. Iliopoulos, K. S. Murray, R. Robson, J. Wilson, and G. A. Williams, *J. Chem. Soc. Dalton Trans.*, 1585 (1987).

146. M. G. B. Drew, P. C. Yates, F. S. Esho, J. Trocha-Grimshaw, A. Lavery, K. P. McKillop, S. M. Nelson, and J. Nelson, *J. Chem. Soc. Dalton Trans.*, 2995 (1988).

147. P. Iliopoulos, G. D. Fallon, and K. S. Murray, *J. Chem. Soc. Dalton Trans.*, 1823 (1988).

148. C. J. McKenzie and R. Robson, *Inorg. Chem.*, *26*, 3615 (1987).

149. M. Mikuriya, K. Nakadera, and T. Kotera, *Chem. Lett.*, 637 (1993).

150. L. A. Oro, D. Carmona, M. P. Puebla, M. P. Lamata, C. Foces-Foces, and F. H. Cano, *Inorg. Chim. Acta*, *112*, L11 (1986).

151. T. V. Ashworth, D. C. Liles, and E. Singleton, *J. Chem. Soc. Chem. Commun.*, 1317 (1984).

152. T. V. Ashworth, D. C. Liles, and E. Singleton, *Inorg. Chim. Acta*, *98*, L65 (1985).

153. M. O. Albers, S. F. A. Crosby, D. C. Liles, D. J. Robinson, A. Shaver, and E. Singleton, *Organometallics*, *6*, 2014 (1987).

154. R. Atencio, C. Bohanna, M. A. Esteruelas, F. J. Lahoz, and L. A. Oro, *J. Chem. Soc. Dalton Trans.*, 2171 (1995).

155. M. I. Bruce, M. G. Humphrey, M. R. Snow, E. R. T. Tiekink, and R. C. Wallis, *J. Organomet. Chem.*, *314*, 311 (1986).

156. J. R. Shapley, D. E. Samkoff, C. Bueno, and M. R. Churchill, *Inorg. Chem.*, *21*, 634 (1982).

157. L. A. Oro, D. Carmona, J. Reyes, C. Foces-Foces, and F. H. Cano, *Inorg. Chim. Acta*, *112*, 35 (1986).

158. J. A. Cabeza, R. J. Franco, V. Riera, S. Garcia-Granda, and J. F. Van der Maelen, *Organometallics*, *14*, 3342 (1995).

159. C. P. Kubiak and R. Eisenberg, *J. Am. Chem. Soc.*, *99*, 6129 (1977).

160. L. A. Oro, D. Carmona, P. L. Perez, M. Esteban, A. Tiripicchio, and M. Tiripicchio-Camellini, *J. Chem. Soc. Dalton Trans.*, 973 (1985).

161. D. Carmona, L. A. Oro, P. L. Peréz, A. Tiripicchio, and M. Tiripicchio-Camellini, *J. Chem. Soc. Dalton Trans.*, 1427 (1989).

162. L. A. Oro, M. T. Pinillos, A. Tiripicchio, and M. Tiripicchio-Camellini, *Inorg. Chim. Acta*, *99*, L13 (1985).

163. C. J. Janke, L. J. Tortorelli, J. L. E. Burn, C. A. Tucker, and C. Woods, *Inorg. Chem.*, *25*, 4597 (1986).

164. C. Woods, L. J. Tortorelli, D. P. Rillema, J. L. E. Burn, and J. C. DePriest, *Inorg. Chem.*, *28*, 1673 (1989).

165. L. J. Tortorelli, C. Woods, and A. T. McPhaie, *Inorg. Chem.*, *29*, 2726 (1990).

166. L. J. Tortorelli, C. Woods, and C. F. Campana, *Acta Crystallogr.*, *C48*, 1311 (1992).

167. G. W. Bushnell, K. R. Dixon, D. T. Eadie, and S. R. Stobart, *Inorg. Chem.*, *20*, 1545 (1981).

168. A. L. Bandini, G. Banditelli, F. Bonati, G. Minghetti, F. Demartin, and M. Manassero, *J. Organomet. Chem.*, *269*, 91 (1984).

169. A. L. Bandini, G. Banditelli, F. Bonati, G. Minghetti, F. Demartin, and M. Manassero, *Inorg. Chem.*, *26*, 1351 (1987).

170. J. A. Bailey and H. B. Gray, *Acta Crystallogr.*, *C48*, 1420 (1992).

171. J. A. Bailey, V. M. Miskowski, and H. B. Gray, *Inorg. Chem.*, *32*, 369 (1993).

172. G. Minghetti, G. Banditelli, and F. Bonati, *J. Chem. Soc. Dalton Trans.*, 1851 (1979).

173. B. Bovio, F. Bonati, and G. Banditelli, *Gazz. Chim. Ital.*, *115*, 613 (1985).

174. W. P. Schaefer, W. B. Connick, V. M. Miskowski, and H. B. Gray, *Acta Crystallogr.*, *C48*, 1776 (1992).

175. F. Bonati and H. C. Clark, *Can. J. Chem.*, *56*, 2513 (1978).

176. A. L. Bandini, G. Banditelli, G. Minghetti, and F. Bonati, *Can. J. Chem.*, *57*, 3237 (1979).

177. E. Kime-Hunt, K. Spartalian, M. DeRusha, C. M. Nunn, and C. J. Carrano, *Inorg. Chem.*, *28*, 4392 (1989).

178. I. A. Degnan, W. A. Herrmann, and E. Herdtweck, *Chem. Ber.*, *123*, 1347 (1990).

179. M. M. Taqui Khan, P. S. Roy, K. Venkatasubramanian, and N. H. Khan, *Inorg. Chim. Acta*, *176*, 49 (1990).

180. G. Lopez, J. Ruiz, C. Vicente, J. M. Marti, G. Garcia, P. A. Chaloner, P. B. Hitchcock, and R. M. Harrison, *Organometallics*, *11*, 4090 (1992).

181. N. Masciocchi, M. Moret, A. Sironi, G. A. Ardizzoia, S. Cenini, and G. La Monica, *J. Chem. Soc. Chem. Commun.*, 1955 (1995).

182. D. Carmona, M. P. Lamata, J. Ferrer, J. Modrego, M. Perales, F. J. Lahoz, R. Atencio, and L. A. Oro, *J. Chem. Soc. Chem. Commun.*, 575 (1994).

183. S. Trofimenko, *J. Am. Chem. Soc.*, *91*, 5410 (1969).

184. K. S. Chong, S. J. Rettig, A. Storr, and J. Trotter, *Can. J. Chem.*, *57*, 3099 (1979).

185. A. R. Barron, G. Wilkinson, M. Motevalli, and M. B. Hursthouse, *Polyhedron*, *4*, 1131 (1985).

186. C. W. Eigenbrot, Jr., and K. N. Raymond, *Inorg. Chem.*, *20*, 1553 (1981).

187. C. W. Eigenbrot, Jr., and K. N. Raymond, *Inorg. Chem.*, *21*, 2653 (1982).

188. D. Röttger, G. Erker, M. Grehl, and R. Fröhlich, *Organometallics*, *13*, 3897 (1994).

189. H. Schuman, J. Loebel, J. Pickardt, C. Qian, and Z. Xie, *Organometallics*, *10*, 215 (1991).

190. G. B. Deacon, B. M. Gatehouse, S. Nickel, and S. N. Platts, *Austr. J. Chem.*, *44*, 613 (1991).

191. J. E. Cosgriff, G. B. Deacon, B. M. Gatehouse, H. Hemling, and H. Schumann, *Angew. Chem. Int. Ed. Engl.*, *32*, 874 (1993).

192. H. Schumann, P. R. Lee, and J. Loebel, *Angew. Chem. Int. Ed. Engl.*, *28*, 1033 (1989).

193. J. C. Bayon, C. Net, P. Esteban, P. G. Rasmussen, and D. F. Bergstrom, *Inorg. Chem.*, *30*, 4771 (1991).

194. J. C. Bayon, P. Esteban, G. Net, P. G. Rasmussen, K. N. Baker, C. W. Hahn, and M. M. Gumz, *Inorg. Chem.*, *30*, 2572 (1991).

195. T. Kamiusuki, H. Okawa, E. Kitaura, M. Koikawa, N. Matsumoto, S. Kida, and H. Oshio, *J. Chem. Soc. Dalton Trans.*, 2077 (1989).

196. T. Kamiusuki, H. Okawa, N. Matsumoto, and S. Kida, *J. Chem. Soc. Dalton Trans.*, 195 (1990).

197. T. Kamiusuki, H. Okawa, E. Kitaura, K. Inoue, and S. Kida, *Inorg. Chim. Acta*, *179*, 139 (1991).

198. M. Itoh, K. Motoda, K. Shindo, T. Kamiusuki, H. Sakiyama, N. Matsumoto, and H. Okawa, *J. Chem. Soc. Dalton Trans.*, 3635 (1995).

199. B. Mernari, F. Abraham, M. Lagrenee, M. Drillon, and P. Legoll, *J. Chem. Soc. Dalton Trans.*, 1707 (1993).

200. K. Shindo, Y. Mori, K. Motoda, H. Sakiyama, N. Matsumoto, and H. Okawa, *Inorg. Chem.*, *31*, 4987 (1992).

201. J. Casabo', J. Pons, K. S. Siddiqi, F. Teixidor, E. Molins, and C. Miravitlles, *J. Chem. Soc. Dalton Trans.*, 1401 (1989).

202. M. Munakata, L. P. Wu, M. Yamamoto, T. Kuroda-Sowa, M. Maekawa, and S. Kitagawa, *J. Chem. Soc. Dalton Trans.*, 4099 (1995).

203. P. W. Ball and A. B. Blake, *J. Chem. Soc. (A)*, 1415 (1969).

204. J. Pons, X. Lopez, J. Casabo', F. Teixidor, A. Caubet, J. Rius, and C. Miravitlles, *Inorg. Chim. Acta*, *195*, 61 (1992).

205. M. Kumar, V. J. Aran, P. Navarro, A. Ramos-Gallardo, and A. Vegas, *Tetrahedron Lett.*, *35*, 5723 (1994).

206. M. Kumar, V. J. Aran, and P. Navarro, *Tetrahedron Lett.*, *34*, 3159 (1993).

207. C. Acerete, J. M. Bueno, L. Campayo, P. Navarro, M. I. Rodriguez-Franco, and A. Samat, *Tetrahedron Lett.*, *50*, 4765 (1994).

208. T. G. Schenck, J. M. Downes, C. R. C. Milne, P. B. Mackenzie, H. Boucher, J. Whelan, and B. Bosnich, *Inorg. Chem.*, *24*, 2334 (1985).

209. K. A. Beveridge, G. W. Bushnell, S. R. Stobart, J. L. Atwood, and M. J. Zaworotko, *Organometallics*, *2*, 1447 (1983).

210. S. Nussbaum, S. J. Rettig, A. Storr, and J. Trotter, *Can. J. Chem.*, *63*, 692 (1985).

211. D. L. Lichtenberger, A. S. Copenhaver, H. B. Gray, J. L. Marshall, and M. D. Hopkins, *Inorg. Chem.*, *27*, 4488 (1988).

212. P. Kalck, J. M. Frances, P. M. Pfister, T. G. Southern, and A. Thorez, *J. Chem. Soc. Chem. Commun.*, 510 (1983).

213. A. Dedieu, P. Escaffre, J. M. Frances, P. Kalck, and A. Thorez, *Nouv. J. Chim.*, *10*, 631 (1986).

214. P. Kalck, A. Thorez, M. T. Pinillos, and L. A. Oro, *J. Mol. Catal.*, *31*, 311 (1985).

215. C. Claver, P. Kalck, L. A. Oro, M. T. Pinillos, and C. Tejel, *J. Mol. Catal.*, *43*, 1 (1987).

216. R. Uson, L. A. Oro, M. T. Pinillos, M. Royo, and E. Pastor, *J. Mol. Catal.*, *14*, 375 (1982).

217. L. A. Oro, M. Campo, and D. Carmona, *J. Mol. Catal.*, *39*, 341 (1987).

# Recent Trends in Metal Alkoxide Chemistry

**RAM C. MEHROTRA** *and* **ANIRUDH SINGH**

*Department of Chemistry*
*University of Rajasthan*
*Jaipur, India*

## CONTENTS

*Progress in Inorganic Chemistry, Vol. 46*, Edited by Kenneth D. Karlin.
ISBN 0-471-17992-2 © 1997 John Wiley & Sons, Inc.

# I. INTRODUCTION

Metal alkoxides can be represented by the general formula $M(OR)_n$, where M is a metal, or metalloid, of valency $n$, and R is an alkyl, alkene, or substituted alkyl group. In spite of the significantly polarized nature of the $M^{\delta-}-O^{\delta-}-R$ bond which is due to the highly electronegative nature of oxygen, most of the metal alkoxides display markedly covalent characteristics; they are volatile and soluble in organic solvents. These characteristics have been explained on the basis of the inductive effect of the alkyl group, which increases with branching. Although the name metal alkoxides is generally used in this chapter, alternative terms (e.g., TEOS and trimethylborate or trimethoxyborane) are still prevalent for alkoxide derivatives [e.g., $Si(OEt)_4$ and $B(OMe)_3$, respectively] of electronegative elements.

Metal alkoxides tend to oligomerize with the formation of $M(\mu\text{-}OR)M$ bridges, leading to the attainment of a higher coordination state of the central metal, M. The extent of this type of oligomerization tends to be lowered by the enhanced steric effect of the increased branching of the alkyl group, which renders them more volatile. For example, the $n$-, $sec$-, and $tert$-butoxides of zirconium were reported to exhibit average degrees of oligomerization with values of 3.6, 2.5, and 1.0, respectively, in refluxing benzene. These compounds can be distilled at 243, 164, and 46°C under 0.1-mm pressure (1, 2).

In addition to the homometallic alkoxides (i.e., binary alkoxides of single metals), many heterometallic alkoxides have also been recently described (3, 4a). Following the detection of alkoxo salts in the titrations of strongly basic alkali alkoxides with alkoxides of less electropositive metals (e.g., aluminum) by Meerween et al. in 1929 (4b), a large number of so-called double alkoxides were described (5, 6). These alkoxides have the general formula $ML_n$, where L is a bidentate (and sometimes a tri- or tetradentate) ligand [e.g., $\{Al(OR)_4\}^-$ ($L_{Al}$), $\{Nb(OR)_6\}^-/\{Ta(OR)_6\}^-$ ($L_{Nb}/L_{Ta}$), or $\{Zr_2(OR)_9\}^-$ ($L_{Zr}$)] that coordinates the central metal M (almost any metal or metalloid of valency $n$). These "double," later termed "bimetallic" (7), alkoxides also involve alkoxy bridges between different metals [e.g., $M(\mu\text{-}OR)M'$]. In fact, it is interesting that instead of behaving as alkoxo salts the bimetallic alkoxides of even strongly electropositive alkali metals such as [$KZr_2(O\text{-}i\text{-}Pr)_9$] show apparently covalent characteristics, for example their volatility, their nonconducting nature, and their solubility in organic solvents. Despite the extraordinary stability of such bimetallic alkoxides (7), the first alkoxide species with three different metals in the same molecular entity, for example [$(i\text{-}Pr\text{-}O)_2Al(\mu\text{-}O\text{-}i\text{-}Pr)_2Be(\mu\text{-}O\text{-}i\text{-}Pr)_2Be(\mu\text{-}O\text{-}i\text{-}Pr)_2Zr(O\text{-}i\text{-}Pr)_3$] was described in 1985 and its stability was ascribed to the small size of the beryllium metal (8). Since 1988, however, many tri- and tetrametallic alkoxide derivatives with the general formula

$M(L_{Al})_x(L_{Nb})_y(L_{Zr})_z(OR)_a X_{(n-x-y-z-a)}$ (where X is a ligand such as Cl and $x$, $y$, $z$ have integral values including zero), have been described, adding a new dimension to heterometal coordination chemistry.

Ebelman (9) synthesized ethyl orthosilicate for the first time in 1846. Although he observed that on standing at room temperature their compound slowly converted into a glassy gel of silica because of its hydrolysis by atmospheric moisture; it is only since the 1950s that the potential of metal alkoxides as precursors for the low-temperature synthesis of oxide-ceramic materials has been gradually realized (10–12). From the extraordinary homogeneity of the final ceramic material obtained by hydrolysis together with the mixtures of alkoxides of many different metals, Dislich (13) conjectured that such ultrahomogeneity could not arise merely from a much more intimate physical mixing of metal alkoxides at the molecular level in solution (compared to that obtained by the grind and bake technique of fusing the oxides of the metal at high temperatures) and consequently suggested the possibility of in situ formation of some sort of complex species among alkoxides of different metals. Although this interesting conjecture had been made by Dislich (13), as early as 1971, he was unaware of the actual isolation of a large number of bi- (5, 7) and triheterometallic (8) alkoxide complexes until he heard Mehrota's lecture in 1987 on the chemistry of metal alkoxides at the IVth International Workshop on Glasses and Ceramics at Kyoto. These compounds were described in hundreds of articles and reviews (3–7, 14–17) as well as several invited lectures, including those at the International Conferences on Coordination Chemistry (18–20). The advantages of using presynthesized bimetallic species, for example, the corresponding alkali hexaethoxyniobates: $(Li/K)Nb(OEt)_6)$ (rather than mixtures of component alkali and niobium ethoxides) were demonstrated in the preparation of more homogeneous crystalline lithium niobate fibers (21) and potassium niobate disks (22). The rapid progress made in these directions has been presented in a number of review articles. Some examples follow:

1. Heterometallic alkoxides as Precursors for Ceramic Materials' (3).
2. Heterometallic alkoxides and oxoalkoxides as intermediates in chemical routes to mixed-metal oxides (23).
3. Volatile metallorganic precursors for depositing inorganic electronic materials (24).

Interestingly, the synthesis of the heterotrimetallic alkoxides of beryllium, which began in 1985 (8), was extended in 1988 to include heterotri- and tetra-metallic alkoxides of copper (25) as well as other $3d$ metals (26). These last two pioneering publications were included as invited papers in two reputed international journals, leading to almost universal acceptance of the formation of the stable species called ''heterometallic alkoxides'' (3, 4, 27–33). These alkoxides were earlier designated by the use of loose terms such as alkoxo-salts (4b), double alkoxides (2, 5, 6), mixed alkoxides (34), and polymetallic alkoxides (35). Both from the points of view of their extraordinary stability without any auxiliary carbonyl-type ligands (31, 36) and the possibilities of synthesizing single-source precursors (33, 37) that were suited to the composition of the targeted ceramic materials by the sol–gel process, these new species have led to a rather novel dimension in our understanding of heterometallic coordination chemistry. Their chemistry was recently reviewed in two articles (38, 39).

Although many heterobimetallic alkoxides were characterized structurally (cf. Section III.A), the crystal structure of the first homoleptic heterotrimetallic isopropoxide $[Ba\{Zr_2(O\text{-}i\text{-}Pr)_9\}$ $(\mu\text{-}O\text{-}i\text{-}Pr)_2Cd(\mu\text{-}O\text{-}i\text{-}Pr)]_2$ was recently elucidated (39a).

In the following periodic table (Table I), the elements for which the homometallic and heterometallic alkoxide chemistry have been explored in detail are circled and put within a square, respectively. The metals of later $4d$ and $5d$ groups are underlined by a semicircle to indicate that not much is known about their alkoxide chemistry (40, 41) due to their instability, which is brought about by facile decomposition pathways (e.g., $\beta$-hydrogen elimination). The elements whose alkoxides have been utilized as synthons for ceramic and other materials are marked with an asterisk (*).

This brief introduction depicts four phases in the development of metal alkoxide chemistry.

1. From 1846–1950, during which time the alkoxo chemistry of only a few (e.g., Na, K, Mg, B, Al, Si, Ti, and As) metals and metalloids was investigated.

2. From 1950–1970, during which time the homometallic chemistry of almost all the elements in the periodic table was illuminated.

3. From 1971–1985, during which time the chemistry of bimetallic alkoxides was looked into in some detail (except X-ray crystal structure elucidation).

4. From 1985 to the present which is now witnessing an unprecedented spurt in the chemistry of heterometal alkoxides and oxometal alkoxides (42) in general, along with alkoxide derivatives of special (substituted, chelating, or unsaturated) alcohols, as well as the rapidly developing applications of metal alkoxides for ceramic and catalytic materials coupled with an increasing emphasis on their actual structural elucidation.

This chapter gives a brief account of these newly emerging dimensions of alkoxide chemistry since 1985.

In an excellent comprehensive review (34) on alkoxides, and aryloxides which was published in 1987, the description of heterometallic alkoxides, called

## TABLE I
### Elements for Which Homo- and Heterometallic Alkoxides Are Known

| | | | | | | | | | | | | | | | | | |
|---|---|---|---|---|---|---|---|---|---|---|---|---|---|---|---|---|---|
| H | | | | | | | | | | | | | | | | | He |
| Li• | Be | | | | | | | | | | | B | C | N | O | F | Ne |
| Na• | Mg• | | | | | | | | | | | Al• | Si | P | S | Cl | Ar |
| K• | Ca• | Sc | Ti• | V• | Cr• | Mn | Fe• | Co• | Ni• | Cu• | Zn• | Ga• | Ge | As | Se | Br | Kr |
| Rb• | Sr | Y | Zr• | Nb• | Mo | Tc | Ru | Rh | Pd | Ag | Cd | In | Sn• | Sb | Te | I | Xe |
| Cs | Ba• | La• [Δ] | Hf | Ta• | W | Re | Os | Ir | Pt | Au | Hg | Tl | Pb | Bi | Po | At | Rn |
| Fr | Ra | Ac [ΔΔ] | | | | | | | | | | | | | | | |

| Δ | La | Ce | Pr | Nd | Pm | Sm | Eu | Gd | Tb | Dy | Ho | Er | Tm | Yb | Lu |
|---|---|---|---|---|---|---|---|---|---|---|---|---|---|---|---|
| ΔΔ | Ac | Th | Pa | U | Np | Pu | Am | Cm | Bk | Cf | Es | Fm | Md | No | Lw |

mixed-metal alkoxides, was limited to about one-half of a page. The formation of heterometal alkoxides involves $M(\mu\text{-OR})M'$ bonds similar to $M(\mu\text{-OR})M$ bonds in the oligomerization of homometal or binary alkoxides. Both types of compounds have essentially analogous chemistry and have been grouped together for the first time in this chapter. Synthetic procedures for both have been described (Section II) together, laying emphasis on the post-1985 literature, followed by a discussion of their general properties and structural features (Section III). These findings indicate wherever possible, the correlation between the two groups. Sections IV and V deal with the chemistry of alkoxides involving some special alcohols and of oxo-alkoxides. These areas are where most of the work has been carried out since 1985. The emerging chemistry of polynuclear metal hydride alkoxides as well as that of oxovanadium and oxomolybdenum alkoxide clusters is, however, not being included in this chapter because of the excellent reviews that were published in 1995 on these topics by well-known authorities in these fields [i.e., Chisholm et al. (43a) and Zubieta et al. (43b), respectively]. Finally, the Sections VI and VII deal with the applications of metal alkoxides in the synthesis of a variety of metalloorganic derivatives as precursors for ceramic materials, followed by a brief section (Section VIII) on their future potentials.

## II. SYNTHETIC PROCEDURES

As the methods of synthesis employed for homo- and heterometallic alkoxides are quite similar, they are all grouped together. The procedures already covered in detail in early literature are generally indicated by representative equations with details given only for more recently illustrative examples.

### A. Reactions of Metals with Alcohols

#### 1. Synthesis of Homometallic Alkoxides

Electropositive metals M, for example Groups 1 (IA) (alkali), 2 (IIA) (alkaline earths), 3 (IIIB) (lanthanides, scandium, and yttrium), and 13 (IIIA) (aluminum), react directly with alcohols to yield (1, 2, 6) the corresponding metal alkoxides.

$$M + n\text{ROH} \longrightarrow M(\text{OR})_n + n/2H_2 \uparrow \tag{1}$$

The facility of these reactions is influenced by the nature of the metal as well as the alcohol. For example, the dissolution of less electropositive bi- and trivalent metals requires catalysts such as $I_2$ or/and $HgCl_2$, whose role is not yet fully understood except for cleansing the surface of the metal or formation of iodide, which might be expected to be more reactive.

In a recent publication (44), the dissolution of metallic uranium in 2-propanol in the presence of stoichiometric amounts of iodine has been shown to yield a mixed iodide–isopropoxide of uranium(IV).

$$U + \geq I_2 + 2i\text{-PrOH} \xrightarrow{i\text{-PrOH}} UI_2(O\text{-}i\text{-Pr})_2(i\text{-PrOH})_2 + H_2 \uparrow \qquad (2)$$

The reactions of some simple as well as chelating alcohols (e.g., 2-alkoxy-alkanols) with metals, resulting in the formation of oxo-alkoxides (in place of simple binary alkoxides) will be dealt with in Sections IV and V.

### 2. Synthesis of Heterometallic Alkoxides

The reactions of bivalent metals (e.g., Mg, Ca, Sr, or Ba) with alcohols are extremely slow even in the presence of catalysts such as $I_2$ or $HgCl_2$. This could be due to the rather poor solubility of alkoxides of bivalent metals (M′), the superficial layers of which on the free metal might hinder direct reactivity further. It was shown (6, 14–16) that alkoxides of a number of metals such as Al, Zr, Nb, and Ta, not only facilitate the reactions but also result in the formation of soluble volatile heterobimetallic alkoxides, for example,

$$M' + 2ROH + 2Al(OR)_3 \longrightarrow [M'\{Al(OR)_4\}_2] + H_2 \uparrow \qquad (3)$$

The role of these added metal alkoxides might be due to the higher activity of the proton of the alcohol molecule, which forms an adduct of the type, $\begin{matrix} R \\ \diagdown \\ \phantom{x}O \\ \diagup \\ H \end{matrix} \longrightarrow Al(OR)_3$. In a recent publication, Vaarstra et al. (45) suggested a rational mechanism for the formation of the bimetallic alkoxides $Ba_2Zr_4(OR)_{20}$ and $LiZr_2(OR)_9 \cdot ROH$, through the acidic protons of the adduct alcohol molecules in $Zr_2(OR)_8 \cdot (ROH)_2$.

In addition to the complexes of $[Mg\{Al(O\text{-}i\text{-Pr})_4\}_2]$ (6) and $[Mg\{Al(O\text{-}sec\text{-Bu})_4\}_2]$ (46), which were synthesized according to Eq. 3. Another interesting derivative, $Mg_2\{Al_3(O\text{-}i\text{-Pr})_{13}\}$, was synthesized (47) by taking the reactants in the requisite molar ratios and characterized by X-ray crystallography.

$$2Mg + 4i\text{-Pr-OH} + 3Al(O\text{-}i\text{-Pr})_3 \longrightarrow$$

$$Mg_2Al_3(O\text{-}i\text{-Pr})_{13} + 2H_2 \uparrow \qquad (4)$$

The reactions of Ca and Ba(M) with 2-propanol in the 2:3 and 1:3 molar ratios represented by the presence of different molar ratios of Al(O-$i$-Pr)$_3$} were found to yield (48) similar bimetallic isopropoxides in different M/Al ratios

$$2M + 4i\text{-Pr-OH} + 3Al(O\text{-}i\text{-Pr})_3 \xrightarrow{i\text{-PrOH}}$$

$$M_2Al_3(O\text{-}i\text{-Pr})_{13} + 2H_2 \uparrow \tag{5}$$

$$M + 2i\text{-Pr-OH} + 3Al(O\text{-}i\text{-Pr})_3 \xrightarrow{i\text{-PrOH}}$$

$$MAl_3(O\text{-}i\text{-Pr})_{11} + H_2 \uparrow \tag{6}$$

### 3. Synthesis by Electrolytic and Metal Deposition Methods

The feasibility of synthesizing methoxides of copper and lead by electrolyzing a solution of NaOMe in MeOH by using copper/lead as sacrificial anode was suggested (34) as early as 1906. This technique has been recently exploited by Lehmkuhl and Eisenbach (49) for the dialkoxides of a number of metals, $M(OR)_2$ (where M = Fe, Co, and Ni; R = Me, Et, $t$- and $n$-Bu), by using LiCl as a conducting electrolyte, by Turova et al. (50) for alkoxides of other metals, $M'(OR)_n$ (where M' = Al, Sc, Y, Ga, Ti, Zr, Hf, Nb, Ta, Fe, Co, Ni, or Ln; R = an alkyl group with 2–5 carbon atoms), and by electrochemical dissolution of the metals in an electrolyte, by using a special cylindrical glass cell without separation of the cathode and the anode spaces. Following their investigations of the synthesis of copper alkoxides (51), Banait et al. (52) extended this technique to the synthesis of mercuric alkoxides by using tetrabutylammonium chloride as an electrolyte. The method was reported (53) to be exploited commercially in Russia.

It may not be out of place to mention that metal vapor deposition techniques that yielded highly promising results in organometallics (34, 54) may emerge as one of the exciting routes for the synthesis of metal alkoxide derivatives as well. To date, neither of these methods has been extended to the synthesis of heterometal alkoxides.

### B. Reactions of Metal Hydroxides/Oxides with Alcohols

Similar to the conventional method for the preparation of organic esters by esterification reactions, the alkoxides (orthoesters) of mainly metalloids as well as organometallic moieties (e.g., $R_xB$, $R_xGe$, $R_xSn$, or PhHg) can be synthesized (6, 34) by the following types of reactions. These reactions can be pushed to the right by removal of water, which is formed, azeotropically with an organic solvent (e.g., benzene/toluene).

$$M(OH)_n + nROH \longrightarrow M(OR)_n + nH_2O \uparrow \qquad (7)$$

$$MO_{n/2} + nROH \longrightarrow M(OR)_n + \frac{n}{2} H_2O \uparrow \qquad (8)$$

In view of the importance of alkoxysilanes and alkoxysiloxanes as precursors for glasses and ceramic materials, a process of obtaining these from portland cement and silicate minerals appears to be of industrial importance (55). If mild conditions are employed, the specific silicon–oxygen framework in the original mineral is often retained in the final alkoxysilane or alkoxysiloxane obtained, for example,

$$\underset{\substack{\text{Tricalcium silicate} \\ \text{(with SiO}_4\text{ framework)}}}{Ca_3(SiO_4)O} \xrightarrow[\substack{+H^+}]{+EtOH} (HO)_4Si.xEtOH$$

$$\Big\downarrow \longleftarrow +H^+ + EtOH$$

$$Si(OEt)_4 \qquad (9)$$

$$\underset{\text{(Wollastonite)}}{CaSiO_3} + CaCl_2.2H_2O \xrightarrow[\Delta]{N_2} Ca(SiO_3)_4Cl_8$$

$$+ H^+ \downarrow + EtOH$$

$$[(HO)_2SiO]_4.xEtOH \qquad (10)$$

The complex $Mo_2O_5(Me)_2 \cdot 2MeOH$ was detected (56) as a condensation product of $MoO_3$ vapor with methanol, water, and THF at $-196°C$. Such co-condensation reactions in the vapor phase again appear to have considerable potential as a novel synthetic route.

## C.  Reactions of Metal Salts

### 1.  Synthesis of Binary Alkoxides

While the chlorides of strongly electropositive elements such as thorium, lanthanides, and some later $3d$ metals form adducts with alcohols (57), anhydrous chlorides of boron (6) and silicon (9) form the corresponding alkoxides or alkyl orthoesters, $B(OR)_3$ and $Si(OR)_4$, with primary and secondary alcohols. The reactions of tertiary alcohols are complicated by the facile side formation of tertiary alkyl chloride and water by the reaction between tertiary alcohols and hydrogen chloride produced in the initial reaction. This hydrogen chloride can be more readily absorbed by a base-like pyridine, which results again in the formation of simple orthoesters (6).

The chlorides of metals with an intermediate electronegativity tend to form chloride alkoxides, but these reactions can be brought to completion in the presence of proton acceptors such as $NH_3$ or alkali metals (K/Na/Li).

$$TiCl_4 + \underset{(excess)}{3EtOH} \longrightarrow TiCl_2(OEt)_2 \cdot EtOH + 2HCl \uparrow \qquad (11)$$

$$TiCl_4 + \underset{(excess)}{4EtOH} \xrightarrow[+4NH_3]{benzene} Ti(OEt)_4 + 4NH_4Cl \downarrow \qquad (12)$$

$$TiCl_4 + 4NaOEt \xrightarrow[ethanol]{benzene} Ti(OEt)_4 + 4NaCl \downarrow \qquad (13)$$

$$TiCl_4 + 4LiOMe \xrightarrow[methanol]{benzene} Ti(OMe)_4 \downarrow + 4LiCl \qquad (14)$$

Precaution has to be taken while using alkali metals as proton acceptors not to use them in excess, in order to avoid the formation of bimetallic alkoxides. For example,

$$TiCl_4 + 9NaOEt \xrightarrow{EtOH} NaTi_2(OEt)_9 + 8NaCl \downarrow \qquad (15)$$

A recent application of the Eq. 15 is the preparation of *tert*- butoxides of arsenic, antimony, and bismuth (58), for example,

$$ECl_3 + 3(t\text{-}BuOH) + 3Et_3N \longrightarrow$$

$$E(O\text{-}t\text{-}Bu)_3 + 3Et_3N \cdot NCl \downarrow \qquad (16)$$

E = As or Sb

$$BiCl_3 + 3KO\text{-}t\text{-}Bu \longrightarrow Bi(O\text{-}t\text{-}Bu)_3 + 3KCl \downarrow \qquad (17)$$

Recently (33, 59), soluble monomeric alkoxides (aryloxides) of calcium and barium have been synthesized by the reactions of their iodides with the potassium salts of sterically hindered alcohols (phenols).

$$MI_2 + 2KOR \xrightarrow{THF} M(OR)_2(thf)_n + 2KI \downarrow \qquad (18)$$

M = Ca; R = $C(Ph)_2CH_2C_6H_4Cl$-4; M = Ba; R = $C_6H_2t$-$Bu_2$-2,6-Me-4)

Similarly, the reaction of freshly prepared $PbF_2$ with KO-$i$-Pr yielded lead(II) isopropoxide (39, 60).

$$PbF_2 + 2KO\text{-}i\text{-}Pr \longrightarrow Pb(O\text{-}i\text{-}Pr)_2 + K_2F_2 \downarrow \qquad (19)$$

The straightforward reaction between well-crystallized $LnCl_3 \cdot 3Pr\text{-}i\text{-}OH$ with 3 mol of KO-$i$-Pr to yield simple triisopropoxides, $Ln(O\text{-}i\text{-}Pr)_3$, has been reported to be complicated in a few cases. For example, the reaction of $YCl_3$ with 3 mol of NaO-$t$-Bu in THF can be represented as

$$3YCl_3 + 8NaO\text{-}t\text{-}Bu \xrightarrow{\text{THF}}$$

$$Y_3(\mu\text{-}O\text{-}t\text{-}Bu)(\mu_3\text{-}Cl)(\mu\text{-}O\text{-}t\text{-}Bu)_3(O\text{-}t\text{-}Bu)_4(thf)_2 + 8NaCl \downarrow \quad (\text{Ref. } 61)$$

$$(20)$$

However, reaction 20 in $1:2$ molar ratio yields (62) a product with a composition corresponding to $Y_3(O\text{-}t\text{-}Bu)_7Cl_2(thf)$. This product was characterized by X-ray crystallography as $Y_3(\mu_3\text{-}O\text{-}t\text{-}Bu)(\mu_3\text{-}Cl)(\mu\text{-}O\text{-}t\text{-}Bu)_3(O\text{-}t\text{-}Bu)_4(thf)_2$. By contrast, the reaction of $LaCl_3$ with 3 equiv of NaO-$t$-Bu was reported (61) to be straightforward, yielding $\{La(O\text{-}t\text{-}Bu)_3\}_3 \cdot 2thf$, as represented in Eq. 21.

$$3LaCl_3 + 9NaO\text{-}t\text{-}Bu \longrightarrow$$

$$La_3(\mu_3\text{-}O\text{-}t\text{-}Bu)_2(\mu\text{-}O\text{-}t\text{-}Bu)_3(O\text{-}t\text{-}Bu)_4(thf)_2 + 9NaCl \downarrow \qquad (21)$$

Although cerous isopropoxide, $Ce(O\text{-}i\text{-}Pr)_3$ can be prepared from the reaction between $CeCl_3 \cdot 3i\text{-}PrOH$ and 3 mol of KO-$i$-Pr, its isolation in the pure state is rather difficult due to its comparative nonvolatility and tendency to be oxidized by atmospheric oxygen. However, volatile $[Ce\{Al(O\text{-}i\text{-}Pr)_4\}_3]$ could be distilled (18) at about 200°C 1 mm$^{-1}$ from a reaction mixture of $CeCl_3$ with 3 mol of $K\{Al(O\text{-}i\text{-}Pr)_4\}$ in an inert atmosphere.

Ceric isopropoxide, $Ce(O\text{-}i\text{-}Pr)_4$ could not be prepared by a similar route because of the instability of $CeCl_4$, but the same compound could be isolated (6) as an isopropanol adduct, $Ce(O\text{-}i\text{-}Pr)_4 \cdot Pr\text{-}i\text{-}OH$, by the reaction between the more stable $(pyH)_2CeCl_6$ with isopropanol in the presence of excess anhydrous ammonia.

$$(pyH)_2CeCl_6 + 5Pr\text{-}i\text{-}OH + 6NH_3 \longrightarrow$$

$$Ce(O\text{-}i\text{-}Pr)_4 \cdot (Pr\text{-}i\text{-}OH) + 6NH_4Cl \downarrow + 2py \qquad (22)$$

Ceric alkoxide, $Ce(OR)_4$ (where R = Me, Et, or $i$-Pr), can be more conveniently prepared (63) by the reaction between commercially available CAN and the corresponding alkanol in the presence of anhydrous ammonia.

$$(NH_4)_2Ce(NO_3)_6 + 4ROH + 4NH_3 \longrightarrow$$

$$Ce(OR)_4 + 6NH_4NO_3 \downarrow \qquad (23)$$

The reaction of CAN with NaO-$t$-Bu can be utilized (64) for the preparation of binary as well as sodium ceric tertiary butoxides.

$$(NH_4)_2Ce(NO_3)_6 + 6NaO\text{-}t\text{-}Bu \xrightarrow{\text{THF}}$$

$$Ce(O\text{-}t\text{-}Bu)_4(thf)_2 + 2NH_3 + 6NaNO_3 \downarrow + 2t\text{-BuOH} \qquad (24)$$

$$2(NH_4)_2Ce(NO_3)_6 + 13NaO\text{-}t\text{-}Bu \xrightarrow{\text{THF}}$$

$$NaCe_2(O\text{-}t\text{-}Bu)_9 + 4NH_3 + 12NaNO_3 \downarrow + 4t\text{-BuOH} \qquad (25)$$

$$(NH_4)_2Ce(NO_3)_6 + 8NaO\text{-}t\text{-}Bu \xrightarrow{\text{THF}}$$

$$Na_2Ce(O\text{-}t\text{-}Bu)_6(thf)_4 + 2NH_3 + 6NaNO_3 \downarrow + 2t\text{-BuOH} \qquad (26)$$

Although so far utilized (64a) only for the preparation of aryloxide derivatives, ether adducts of the lanthanide nitrates, such as $[Ln(NO_3)_3 \cdot MeO(CH_2CH_2O)_4Me]$, may also prove to be convenient starting materials for their alkoxide analogues by reactions with alkali alkoxides.

The stability of ceric alkoxides $Ce(OR)_4$ [in spite of the oxidizing nature of Ce(IV)] appears to be similar to that of distillable $Cr(O\text{-}t\text{-}Bu)_4$, obtained from the disproportionation of $Cr(O\text{-}t\text{-}Bu)_3$, which can be prepared by the reaction of $CrCl_3 \cdot 3thf$ with 3 mol of LiO-$t$-Bu in $t$-BuOH free THF, or by a one-pot procedure (65).

$$CrCl_3(thf)_3 + 4KO\text{-}t\text{-}Bu + CuCl \xrightarrow{\text{THF}}$$

$$Cr(O\text{-}t\text{-}Bu)_4 + Cu \downarrow + 4KCl \downarrow \qquad (27)$$

## 2. Synthesis of Heterobimetallic Alkoxides from Metal Chlorides

The reactions of metal chlorides with alkali metal alkoxometalates have been extensively employed (4a, 6, 7, 12, 17, 19) for the synthesis of soluble heterobimetallic alkoxides of many metals. The following equations illustrate this point:

$$LnCl_3 + 3[KAl(O\text{-}i\text{-}Pr)_4] \longrightarrow [Ln\{Al(O\text{-}i\text{-}Pr)_4)\}_3] + 3KCl \downarrow \quad (28)$$

(Ref. 18)

$$CdCl_2 + 2[KZr_2(O\text{-}i\text{-}Pr)_9] \longrightarrow [Cd\{Zr_2(O\text{-}i\text{-}Pr)_9\}_2] + 2KCl \downarrow \quad (29)$$

(Ref. 66)

$$NiCl_2 + 2[KM(O\text{-}i\text{-}Pr)_6] \longrightarrow [Ni\{M(O\text{-}i\text{-}Pr)_6\}_2] + 2KCl \downarrow \quad (30)$$

(M = Nb/Ta) (Ref. 24)

A wide variety of heterobimetallic alkoxides, which were synthesized according to Eqs. 28–30, are represented by the following examples (3) for elements "E" with valency "$n$":

$$[E\{M(O\text{-}i\text{-}Pr)_4\}_n] \qquad [E\{M'(O\text{-}i\text{-}Pr)_6\}_n] \qquad [E\{M''_2(O\text{-}i\text{-}Pr)_9\}_n]$$

$$M = \text{Al or Ga} \qquad M' = \text{Nb or Ta} \qquad M'' = \text{Zr, Hf, or } Sn^{IV}$$

$n = 1$: E = Li, Na, K, Rb, Cs, RMg, $R_3Sn$, $R_2Sb^{III}$, $RSb^{IV}$, and so on

$n = 2$: E = Be, Mg, Ca, Ba, $Cr^{II}$, $Mn^{II}$, $Fe^{II}$, $Co^{II}$, $Ni^{II}$, $Cu^{II}$, Zn, Cd,

$\qquad\qquad Hg^{II}$, $R_2Sn^{IV}$, $R_2Pb^{IV}$, and so on

$n = 3$: E = Al, Ga, In, $Tl^{III}$, Sc, Y, Ln, $Cr^{III}$, $Mn^{III}$, $Fe^{III}$, $RSi^{IV}$, and so on

$n = \qquad$ E = $Sn^{IV}$, $Ce^{IV}$, $Th^{IV}$, $U^{IV}$, and so on

The chelating alkoxometallate ligands such as $\{M(OR)_4\}^-$, $\{M'(OR)_6\}^-$, and $\{M''_2(OR)_9\}^-$ may be represented as a group by a common symbol, L, and the most common ones may be represented as indicated in parentheses after each specific ligand: $\{Al(O\text{-}i\text{-}Pr)_4\}^-$ ($L_{Al}$), $\{M'(O\text{-}i\text{-}Pr)_6\}^-$ ($L_M$), and $\{Zr_2(O\text{-}i\text{-}Pr)_9\}^-$ ($L_{Zr}$).

For metals such as Zr, Hf, Nb, and Ta, only partial substitution appears to be possible (7), which results in the formation of heteroleptic derivatives.

$$ZrCl_4 + 2KL_{Al} \longrightarrow ZrCl_2(L_{Al})_2 + 2KCl \downarrow \quad (31)$$

$$NbCl_5 + 2KL_{Al} \longrightarrow NbCl_3(L_{Al})_2 + 2KCl \downarrow \quad (32)$$

Partially substituted alkoxometalate chlorides, $ECl_xL_{n-x}$ [e.g., $ClBeL_{Al}$, $ClMgL_{Al}$, $ClZnL_{Al}$, $ClCdL_{Al}$ (67), $ClCoL_{Al}$, $ClCoL_M$, and $ClCoL_{Zr}$ (68);

ClFeL$_{M'}$ (69), ClCuL$_{Al}$, and ClCuL$_{Ta}$; and ClCuL$_{Zr}$ (70), ClLa(L$_{Zr}$)$_2$, and Cl$_2$LaL$_{Zr}$ (71) were synthesized even in the cases of central metals that can form the homoleptic derivatives EL$_n$. In spite of inherent difficulties in the X-ray structural elucidation of metal alkoxides (72), it was possible to determine X-ray structures of the chloride bridged dimers [Cd($\mu$-Cl)L$_{Zr}$]$_2$ (73) and [Pr(L$_{Al}$)$_2$($\mu$-Cl)Pr-$i$-OH]$_2$ (74).

These alkoxometalate chloride derivatives, ECl$_x$L$_{n-x}$, can be converted into their alkoxometalate alkoxide or $\beta$-diketonate analogues by their reaction with alkali alkoxide or $\beta$-diketonate in the appropriate molar ratios.

$$Cl_x M(L)_{n-x} + xKOR \longrightarrow (RO)_x M(L)_{n-x} + xKCl \downarrow \qquad (33)$$

Heterobimetallic isopropoxides of two lanthanide metals (Sm and Yb) in the divalent state were prepared recently (75) by the following reaction:

$$LnI_2 + 2KAl(O\text{-}i\text{-}Pr)_4 \xrightarrow[-2KI]{THF} Ln\{Al(O\text{-}i\text{-}Pr)_4\}_2 \qquad (34)$$

[Ln = Sm(dark purple); Yb(yellow)]

Interestingly, the main molecular ions at $m/z$ 882 (66%) for Ln = Sm and 994 (74%) for Ln = Yb were observed in the mass spectra of the derivatives. These results correspond to the composition Ln$\{Al_3(O\text{-}i\text{-}Pr)_{11}\}$ for which a plausible structure of the type with hexacoordinated lanthanides was suggested.

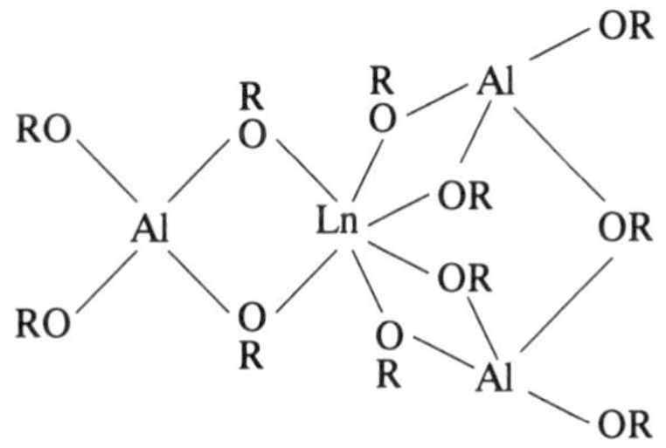

By contrast, the reaction between an ether or THF soluble ZnCl$_2$ with alkali metal $t$-butoxides, is straightforward (76) and yields volatile dimeric alkali tri-*tert*-butoxyzincates.

$$ZnCl_2 + 2MZn(OCMe_3)_3 \longrightarrow \tfrac{1}{2}[MZn(OCMe_3)_3]_2 + 2MCl \qquad (35)$$

M = Na or K

It is worth mentioning that the reactions of $NiCl_2$ (77) and $MnCl_2$ (78) with $KSb(OEt)_4$ were reported to yield heterooxometallic alkoxides (cf. Section V).

### 3.  Synthesis of Tri- and Higher Meterometallic Alkoxides

In spite of significant success in the synthesis of a large variety of stable heterobimetallic alkoxides containing several ligands such as $\{Al(O$-$i$-$Pr)_4\}^-(L_{Al})$, $\{Nb(O$-$i$-$Pr)_6\}^-(L_{Nb})$, $\{Ta(O$-$i$-$Pr)_6\}^-(L_{Ta})$, and $Zr_2(O$-$i$-$Pr)_9]^-(L_{Zr})$ by the chloride method, the preparation of derivatives with more than two different metals in the same molecular species was not even attempted because of the expected instability of such heterometallic derivatives.

The first clue to the formation of a heterotrimetallic alkoxide was, however, detected in 1985 in the case of beryllium. Its stability was initially ascribed to the small size of beryllium, which hindered the disproportionation of the species. The heterobimetallic isopropoxide $(Pr$-$i$-$O)Be(\mu$-$O$-$i$-$Pr)_2Al(O$-$i$-$Pr)_2$ was found on the basis of NMR studies to dimerize on ageing (8) into a product that could be reported as $[(O$-$i$-$Pr)_2Al(\mu$-$O$-$i$-$Pr)_2Al(O$-$i$-$Pr)_2]$. This observation led to the feasibility of reactions of the following types, which resulted for the first time in 1985 in the synthesis and characterization of heterotrimetallic isopropoxides.

$$(Pr\text{-}i\text{-}O)_2Al(\mu\text{-}O\text{-}i\text{-}Pr)_2Be(O\text{-}i\text{-}Pr) + Zr(O\text{-}i\text{-}Pr)_4 \longrightarrow$$

$$[(Pr\text{-}i\text{-}O)_2Al(\mu\text{-}O\text{-}i\text{-}Pr)_2Be(\mu\text{-}O\text{-}i\text{-}Pr)_2Zr(O\text{-}i\text{-}Pr)_3] \qquad (36)$$

This type of synthesis was, therefore, repeated by the chloride method as illustrated by the following equation:

$$(Pr\text{-}i\text{-}O)_2Al(\mu\text{-}O\text{-}i\text{-}Pr)_2BeCl + KNb(O\text{-}i\text{-}Pr)_6 \xrightarrow[-KCl]{}$$

$$[(Pr\text{-}i\text{-}O)_2Al(\mu\text{-}O\text{-}i\text{-}Pr)_2Be(\mu\text{-}O\text{-}i\text{-}Pr)_2Nb(O\text{-}i\text{-}Pr)_4] \qquad (37)$$

Many hetero- (tri- or tetra-) metallic alkoxides of lanthanons (28, 79), zinc and cadmium (38, 80, 81), tin(II) and tin(IV) (82, 83), manganese(II) (84), iron(II) (85), iron(III) (86, 87), cobalt(II) (88, 89), nickel(II) (90), copper(II) (91, 92) and magnesium (93) were since isolated by the research group Mehrotra and Singh. These consist of a central metal atom ligated to one, two, or more chelating alkoxometalate ligands such as $L_{Al}$, $L_{Zr}$, $L_{Nb}$, or $L_{Ta}$, in addition to other similar ligands such as $\{Zr(O$-$i$-$Pr)_5\}^-$ and $\{Al(O$-$i$-$Bu)_4^-$. These ligands appear to confer greater stability to a number of heterometal alkoxide systems (28, 88). The general synthetic procedure could be represented as follows:

$$MCl_n + xKL_{Al} \xrightarrow[-xKCl]{} MCl_{n-x}(L_{Al})_x$$

$$MCl_{n-x}(L_{Al})_x + yKI_{Zr} \xrightarrow[-yKCl]{} MCl_{n-x-y}(L_{Al})_x(L_{Zr})_y$$

$$MCl_{n-x-y}(L_{Al})_x(L_{Zr})_y + zKL_{Nb} \xrightarrow[-zKCl]{} MCl_{n-x-y-z}(L_{Al})_x(L_{Zr})_y(L_{Nb})_z$$

$$(40)$$

and so on (where $x$, $y$, $z$ can have any integral value including zero).

The intermediate chloride derivatives, for example, $MCl_{n-x-y-x}(L_{Al})_x(L_{Zr})_y(L_{Ta})_z$, were also isolated in a few cases and can be converted into alkoxometalate alkoxide or other (e.g., $\beta$-diketonate) derivatives by the following types of reactions:

$$MCl_{n-x-y-z}(L_{Al})_x(L_{Zr})_y(L_{Ta})_z + (n - x - y - z)KOR \longrightarrow$$

$$M(OR)_{n-x-y-z}(L_{Al})_x(L_{Zr})_y(L_{Ta})_z + (n - x - y - z)KCl \downarrow \quad (41)$$

The replacement of alkoxo by chelating $\beta$-diketonato ligands selectively modulates their hydrolyzability, which proved to be of considerable value in the sol–gel process for ceramic materials using alkoxides as precursors (3, 31, 33, 35).

The first crystallographically characterized heterotrimetallic isopropoxide was synthesized (39a) by the following reactions:

$$CdI_2 + KZr_2(O\text{-}i\text{-}Pr)_9 \xrightarrow{toluene} ICd\{Zr_2(O\text{-}i\text{-}Pr)_9\} + KI \downarrow \quad (42)$$

$$ICd\{Zr_2(O\text{-}i\text{-}Pr)_9\} + KBa(O\text{-}i\text{-}Pr)_3 \longrightarrow$$

$$\tfrac{1}{2}[\{Ba\{Zr_2(O\text{-}i\text{-}Pr)_9\}(\mu\text{-}O\text{-}i\text{-}Pr)_2Cd(\mu\text{-}O\text{-}i\text{-}Pr)\}_2] + KI \downarrow \quad (43)$$

A distinctive feature, unprecedented in heterometallic alkoxide chemistry, is the exchange of the central metal atoms between two chelating ligands; this rearrangement is possibly favored by the greater oxophilicity of barium and its tendency to attain higher coordination states (39a).

## D. Alcoholysis and Transesterification Reactions of Metal Alkoxides

Metal alkoxides (homo as well as hetero) undergo facile and reversible alcoholysis and transesterification reactions that can be represented by the following equations:

$$M(O\text{-}i\text{-}Pr)_n + xROH \underset{}{\overset{benzene}{\rightleftharpoons}}$$

$$M(O\text{-}i\text{-}Pr)_{n-x}(OR)_x + xPr\text{-}i\text{-}OH \uparrow \quad (44)$$

$$M(O\text{-}i\text{-}Pr)_n + x\text{MeCOOR} \underset{}{\overset{\text{cyclohexane}}{\rightleftharpoons}}$$

$$M(O\text{-}i\text{-}Pr)_{n-x}(OR)_x + x\text{MeCOO-}i\text{-Pr} \uparrow \qquad (45)$$

$$M(O\text{-}i\text{-}Pr)_n + x\text{MeCOOSiMe}_3 \underset{}{\overset{\text{cyclohexane}}{\rightleftharpoons}}$$

$$M(O\text{-}i\text{-}Pr)_{n-x}(OSiMe_3)_x + x\text{MeCOO-}i\text{-Pr} \uparrow \qquad (46)$$

$$M\{Al(O\text{-}i\text{-}Pr)_4\}_n + nx\text{ROH} \underset{}{\overset{\text{benzene}}{\rightleftharpoons}}$$

$$M\{Al(OR)_x(O\text{-}i\text{-}Pr)_{4-x}\}_n + nx\text{Pr-}i\text{-OH} \uparrow \qquad (47)$$

$$M\{Al(O\text{-}i\text{-}Pr)_4\}_n + nx\text{MeCOOR} \underset{}{\overset{\text{cyclohexane}}{\rightleftharpoons}}$$

$$M\{Al(OR)_x(O\text{-}i\text{-}Pr)_{4-x}\}_3 + nx\text{MeCOOPr-}i\text{-OH} \uparrow \qquad (48)$$

With most of the R groups, the above reactions can be pushed to completion if the liberated isopropanol or acetate is removed azeotropically with a solvent such as benzene or cyclohexane. Alternatively, the reactions lead to the isolation of mixed alkoxides, many of which can be distilled without apparent disproportionation (6). For the replacement of isopropoxide by *tert*-butoxide groups, transesterification with *tert*-butyl acetate is generally preferred (6) because of a larger difference in the boiling points of *tert*-butyl and isopropyl acetate.

It may be interesting to mention that the reacting molecule ROH may be bidentate (e.g., an alkoxy- or aminoalkanol, cf. Section IV) or the enol form of a $\beta$-diketonate ($\beta$-dikH) generally functioning as a bidentate or a carboxylic acid (RCOOH) that may be either mono- or bidentate. The alcoholysis/transesterification reactions thus provide convenient routes for the synthesis of other alkoxide derivatives or complexes with other bidentate or even higher chelating ligands such as Schiff bases (6).

These alcoholysis reactions probably proceed by an $S_N2$ mechanism involving a four-membered cyclic transition state, for example,

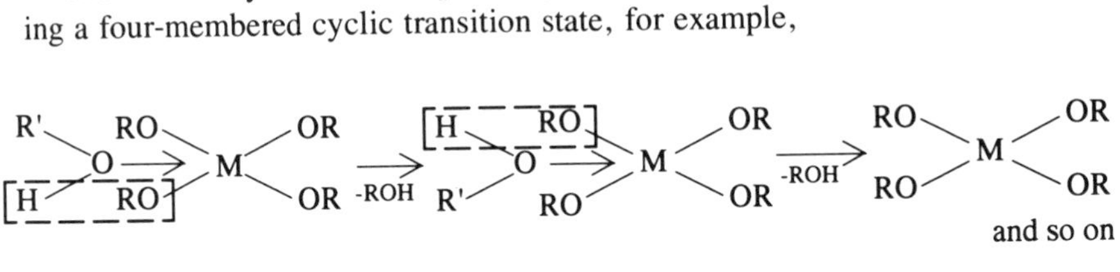

As expected, such reactions are susceptible to steric factors, as exemplified by the formation of $\{Al(O\text{-}i\text{-}Pr)(O\text{-}t\text{-}Bu)_2\}_2$ and $Ln\{Al(O\text{-}i\text{-}Pr)(O\text{-}t\text{-}Bu)_3\}_3$ as the final products in the alcoholysis/transesterification reactions of relevant binary as well as heterometallic isopropoxides.

Note that the primary alkoxides (e.g., ethoxides) of some later $3d$ metals such as $Mn^{II}$, $Fe^{II}$, $Co^{II}$, and $Ni^{II}$ do not undergo alcoholysis (7, 19) or transesterification with *tert*-butanol or butyl acetate, whereas their *tert*-butoxides M(O-*t*-Bu)$_2$ are easily converted into ethoxides with ethanol.

Because of a renewed interest in the hydrolysis reactions of metal alkoxides, which arises from their use as precursors in the sol–gel process, methanalysis reactions (which might be closest to the hydrolysis processes) have received pointed attention with some interesting results. For example, stagewise methanolysis of Ln{Al(O-*i*-Pr)$_4$}$_3$ appears to follow (18, 31) the pattern indicated below

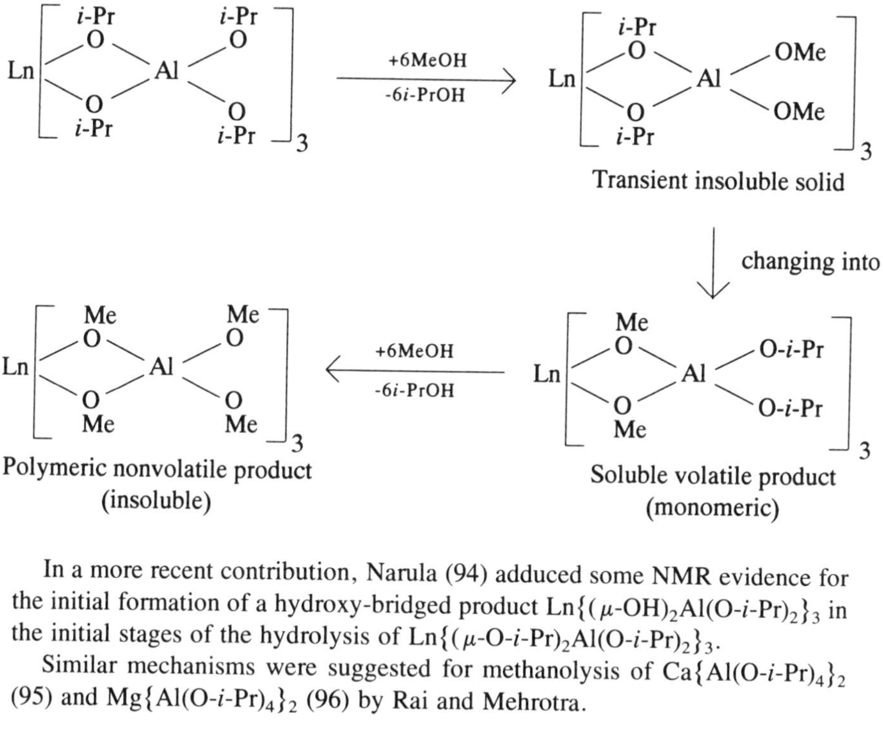

Transient insoluble solid

changing into

Polymeric nonvolatile product
(insoluble)

Soluble volatile product
(monomeric)

In a more recent contribution, Narula (94) adduced some NMR evidence for the initial formation of a hydroxy-bridged product Ln{($\mu$-OH)$_2$Al(O-*i*-Pr)$_2$}$_3$ in the initial stages of the hydrolysis of Ln{($\mu$-O-*i*-Pr)$_2$Al(O-*i*-Pr)$_2$}$_3$.

Similar mechanisms were suggested for methanolysis of Ca{Al(O-*i*-Pr)$_4$}$_2$ (95) and Mg{Al(O-*i*-Pr)$_4$}$_2$ (96) by Rai and Mehrotra.

## E. Reactions of Metal Dialkylamides and Bis(trimethylsilyl)amides with Alcohols

This method is particularly successful for those metals that have a greater affinity for oxygen compared to nitrogen and is generally rendered more facile due to the higher volatility of liberated dialkylamines.

$$M(NR_2)_n + nR'OH \longrightarrow M(OR)_n + nR_2NH \uparrow \qquad (49)$$

Introduced in 1956 by Jones et al. (97) for $U(OR)_4$, the method was utilized in 1961 for synthesis of alkoxides of $V^{IV}$, $Cr^{IV}$, $Nb^{IV}$, $Ta^{IV}$, and $Sn^{IV}$.

In spite of the difficulties involved (98) in the synthesis of the dialkylamide and bis(trimethylsilyl) amide reagents, the procedure was increasingly employed for the synthesis of metal alkoxides, particularly in their lower oxidation states [e.g., $Cr^{II}$ and $Mn^{II}$ (99)]. This method proved to be specially useful for the preparation of alkoxides of molybdenum and tungsten in their 3- and 4-oxidation states (7) and more recently of $U_2(O\text{-}t\text{-Bu})_8 \cdot (t\text{-BuOH})$ (100):

$$Mo_2(NMe_2)_6 + 6ROH \xrightarrow{\text{alkane}} Mo_2(OR)_6 + 6NMe_2H \uparrow \qquad (50)$$

$$(R = Me, Et, CH_2CH_2Me, CHMe_2, CH_2CMe_3,$$

$$CMe_2Ph, C(CF_3)_3, SiMe_3, SiEt_3, \text{etc.})$$

$$Mo(NMe_2)_4 + 4ROH \longrightarrow Mo(OR)_4 + 4NMe_2H \uparrow \qquad (51)$$

$$(R = Me, Et, CHMe, CMe_3, \text{and } CH_2CMe_3)$$

$$U(NEt_2)_4 + 9t\text{-BuOH} \xrightarrow[25°C]{\text{toluene}} U_2(O\text{-}t\text{-Bu})_8 \cdot (t\text{-BuOH}) \qquad (52)$$

Out of the above, $Mo_2(OR)_6$ derivatives with sterically hindered groups have a particularly significant importance as the dimerization occurs in their cases, not through the alkoxy bridges, but by $Mo \equiv Mo$ triple bonds.

This route has been recently utilized for the synthesis of soluble volatile alkoxides of trivalent bismuth (101) and lanthanides (28, 102, 103) in low-coordination states as well as three-coordinate aryloxides of lanthanides (104) and actinides (105), for example:

$$Bi(NMe_2)_3 + 3ROH \longrightarrow Bi(OR)_3 + 3NMe_2H \uparrow \qquad (53)$$

where $R = CMe_2Et$ (Ref. 101)

$$Ln\{N(SiMe_3)_2\}_3 + 3ROH \longrightarrow$$

$$Ln(OR)_3 + 3NH(SiMe_3)_2 \uparrow \qquad (54)$$

[Ln = Y, R = $CEt_3$ (Ref. 104); Ce(III), R = $C\text{-}t\text{-Bu}_3$ (Ref. 102),
    $OC_6H_3\text{-}t\text{-Bu}_2\text{-}2,6$ (Ref. 104);
    = An (an actinide): Pu, R = $OC_6H_3\text{-}t\text{-Bu}_2\text{-}2,6$ (Ref. 105)]

$$3Ln\{N(SiMe_3)_2\}_3 + 9ROH \longrightarrow$$

$$[Ln_3(OR)_9(ROH)_2] + 9HN(SiMe_3)_2 \qquad (55)$$

[Ln = Y, La; R = $t$-Bu, $t$-Am (Ref. 103)]

$$Zn\{N(SiMe_3)_2\}_2 + 2ROH \longrightarrow$$

$$Zn(OR)_2 + 9HN(SiMe_3)_2 \uparrow \qquad (56)$$

(R = CEt$_3$ or CMeEt$_2$) (Ref. 106)

This method has been utilized for the synthesis of alkoxides of zinc and cadmium as well as the lanthanides with some specially hindered alcohols (cf. Section IV).

The reaction of Al$_2$(NMe$_2$)$_6$ with $t$-BuOH is slow due to steric factors. However, the end product Al(O-$t$-Bu)$_3$}$_2$ is found instead of a mixed product of the type (Bu—$t$-O)$_2$Al($\mu$-O-$i$-Pr)($\mu$-O-$t$-Bu)Al(O-$t$-Bu)$_2$, which is finally obtained in the reaction of aluminium isopropoxide with excess $t$-BuOH.

$$Al_2(NMe_2)_6 + 5t\text{-BuOH} \xrightarrow[\text{(6 h)}]{-5HNMe_2}$$
(20-fold excess)

$$Al_2(O^-t\text{-Bu})_4(\mu\text{-O-}t\text{-Bu})(\mu\text{-NMe}_2)$$

(12 h) $\quad$ $t$-BuOH (excess)

$$(Bu^-t\text{-O})_2Al(\mu\text{-O-}t\text{-Bu})_2Al(O\text{-}t\text{-Bu})_2 \quad (57)$$

(Ref. 107)

The amide method was found to be convenient and straightforward in many cases. An example is the synthesis of soluble barium *tert*-butoxide, which has a cubane structure (106b). The preparation of this compound by other routes (e.g, direct action of metals with alcohols or chloride–alkoxide interchange), sometimes tends to be complicated due to the incorporation of chloride or oxide moieties within the structural framework of the final alkoxide product.

## F. Miscellaneous Methods

In this section, we describe a few recent examples of some other methods employed (6) for the preparation of metal alkoxides.

### 1. Redox Reactions

This route was described (108) as early as 1956 for the preparation of some uranium alkoxides in the +5 and +6 oxidation states, respectively.

$$2U(OEt)_4 + Br_2 \longrightarrow 2U(OEt)_4Br$$

$$\downarrow +2NaOEt$$

$$2U(OEt)_5 + 2NaBr \tag{58}$$

$$2NaU(OEt)_6 + (PhCO)_2O_2 \longrightarrow$$

$$2U(OEt)_6 + 2PhCO_2Na \tag{59}$$

Recent applications may be exemplified by the following reactions:

$$2(C_5Me_5)_2Yb(NH_3) + 2t\text{-BuOOBu-}t \xrightarrow[-NH_3]{}$$

$$2(C_5Me)_2Yb(O\text{-}t\text{-Bu}) \tag{60}$$

(Ref. 109)

$$Ce(OC\text{-}t\text{-Bu}_3)_3 + 2t\text{-BuOOBu-}t \longrightarrow$$

$$Ce(OC\text{-}t\text{-Bu}_3)_2(O\text{-}t\text{-Bu})_2 + t\text{-Bu}_3OOC\text{-}t\text{-Bu}_3 \tag{61}$$

(Ref. 110)

## 2. From Metal Alkyls/Hydrides

Since the preparation of RBeOR' from the reaction of $BeR_2$ with R'OH by Coates et al. (6), it was shown that reactions of metal alkoxides with organo-lithium or other main group metal alkyls, such as Grignard reagents or aluminium alkyls, can yield (4a) either substitution (Eq. 59) or addition products (Eqs. 60 and 61); the latter often results in formation of heterometallic alkoxides, for example,

$$M(OR)_n + M'R' \longrightarrow M(OR)_{n-1}R' + M'OR \tag{62}$$

$$M(OR)_n + M'R' \longrightarrow M'MR'(OR)_n \tag{63}$$

$$U(OCH\text{-}t\text{-Bu}_2)_4 + LiMe \longrightarrow LiUMe(OCH\text{-}t\text{-Bu}_2)_4 \tag{64}$$

Ref. (111)

Similar to the earlier observations of Coates et al. (6) on the insertion of MeCHO and $(Me)_2CO$ into metal–alkyl bonds of $Be(Me)_2$, which result in the formation of $MeBeOCH(Me)_2$ and $MeBeOC(Me)_3$, a reaction was reported between $(C_6Me_5)_2ThCl(Me)$ and $(Me)_2CO$ (112).

$$(Me)_2Be + MeCHO \longrightarrow MeBeOCH(Me)_2 \tag{65}$$

$$(Me)_2Be + MeCO \longrightarrow MeBeOC(Me)_3 \tag{66}$$

$$(C_5Me_5)_2ThClMe + (Me)_2CO \longrightarrow (C_5Me_5)_2TlCl[OC(Me)_2] \tag{67}$$

Many low-valent heterometallic titanium–aluminium alkoxides that activate dinitrogen were reported in the reactions between titanium alkoxides and trialkylaluminium (113).

The reaction of aluminium trihydride with a number of alcohols was studied in 1968 by Nöth and Suchy (114).

$$AlH_3 + nROH \longrightarrow Al(OR)_nH_{3-n} + n/2H_2 \uparrow \tag{68}$$

The disproportionation of mixed derivatives of $Al(OR)_nH_{3-n}$ into $AlH_3$ and $HAl(OR)_2$ was shown to decrease with the branching of the alkoxide groups involved.

In addition to Eq. (68), a few interesting preparations are illustrated below to reflect the accessibility of novel products by such preparative procedures.

$$[(C_5H_5)_2Y(\mu\text{-}H)(thf)_2]_2 + 2MeOH \longrightarrow$$

$$[\{[(C_5H_5)_2Y(\mu\text{-}OMe)\}_2] + 2H_2 \uparrow \tag{69}$$

(Refs. 28, 115)

$$\{Cp_2Y(\mu\text{-}H)\}_3(\mu_3\text{-}H) + 3MeOH \longrightarrow$$

$$\{Cp_2Y(\mu\text{-}OCH)\}(\mu_3\text{-}H) + 3H_2 \uparrow \tag{70}$$

(Refs. 28, 116)

$$Zr_2(O\text{-}i\text{-}Pr)_8(i\text{-}PrOH)_2 \left\langle \begin{array}{l} \xrightarrow{+KH} KZr_2(O\text{-}i\text{-}Pr)_9 + i\text{-}PrOH \\ \\ \xrightarrow{+2KH} K_4Zr_2O(O\text{-}i\text{-}Pr)_{10} + \cdots \end{array} \right.$$

(Ref. 117)

$$\tag{71}$$

$$(\text{bpy}) \, \text{NiR}_2 \xrightarrow{\text{R'OH}} \text{No reaction}$$

$$\text{R'} = \text{nonfluorinated alkyl group}$$

$$(\text{bpy}) \, \text{NiR}_2 \xrightarrow{\text{HOCH(CF}_3)_2} (\text{bpy})\text{Ni}\{\text{OCH(CF}_3)_2\}_2$$

(Ref. 118)                                                                                        (72)

$$(\text{bpy}) \, \text{NiR}_2 \xrightarrow[-\text{N}_2]{+\text{N}_2\text{O}} (\text{bpy})\text{Ni}(\text{OR})(\text{R}) \qquad (73)$$

(Ref. 119)                              (R = Et or $i$-Bu)

### 3. Redistribution and Exchange Reactions

The exchange reactions between titanium tetrachloride and titanium/silicon alkoxides at low temperatures lead to the deposition of less soluble chloride alkoxides of titanium (6).

$$3\text{TiCl}_4 + \underset{(\text{excess})}{\text{Ti}(\text{OEt})_4} \longrightarrow 4\text{TiCl}_3(\text{OEt}) \qquad (74)$$

$$\text{TiCl}_4 + \text{SiCl}_2(\text{OEt})_2 \longrightarrow \text{TiCl}_3(\text{OEt}) + \text{SiCl}_3(\text{OEt}) \qquad (75)$$

Alkyltin halides and isopropoxides similarly exchange their groups readily as illustrated by the following equations (6).

$$2\text{RSnCl}_3 + \text{RSn}(\text{O-}i\text{-Pr})_3 \longrightarrow 3\text{RSnCl}_2(\text{O-}i\text{-Pr}) \qquad (76)$$

$$\text{RSnCl}_3 + 2\text{RSn}(\text{O-}i\text{-Pr})_3 \longrightarrow 3\text{RSnCl}(\text{O-}i\text{-Pr})_2 \qquad (77)$$

In two recent publications (120, 121), it was shown that in place of the expected elimination of esters in the reactions of $\text{Sn}(\text{O-}t\text{-Bu})_4$ with $\text{Sn}(\text{OAc})_4$, as well as with $\text{Me}_3\text{Si}(\text{OAc})$ occurring in a refluxing hydrocarbon (e.g., toluene) solvent, only ligand exchanges of the following types were found to occur when the above reactions were carried out in coordinating solvents such as pyridine:

$$3\text{Sn}(\text{O-}t\text{-Bu})_4 + 3\text{Sn}(\text{OAc})_4 \xrightarrow{\text{pyridine}} 2\text{Sn}(\text{O-}t\text{-Bu})(\text{OAc})_3$$

$$+ \, 2\text{Sn}(\text{O-}t\text{-Bu})_2(\text{OAc})_2 + 2\text{Sn}(\text{O-}t\text{-Bu})(\text{OAc})_3 \qquad (78)$$

$$\text{Sn(O-}t\text{-Bu)}_4 + \text{Me}_3\text{Sn(OAc)} \xrightarrow{\text{pyridine}}$$

$$\text{Sn(O-}t\text{-Bu)}_3\text{(OAc)} \cdot \text{py} + \text{Me}_3\text{Si(O-}t\text{-Bu)} \qquad (79)$$

### 4. Synthesis of Heterobimetallic Alkoxides from the Component Metal Alkoxides

Lewis acid–base reactions between component alkoxides have been used primarily for the synthesis of bimetallic alkoxides involving (1) alkali alkoxides and less basic alkoxides, and (2) between binary alkoxides of other metals.

1. Preparation of alkali alkoxometalates was reported by the reactions of alkali alkoxides (Lewis bases) with alkoxides of less electropositive metals and metalloids, for example, beryllium (122), zinc (123), boron (124), aluminium (125, 126), tin(II) (127), tin(IV) (128, 129), antimony(III) (58, 130), bismuth (58), titanium (131), zirconium (133, 134), thorium (125), tantalum (125, 134), niobium (125, 135), and copper (136), as represented by the following illustrative equations:

$$\text{M(OR)}_n + \text{M}'\text{OR} \longrightarrow \text{M}'\text{M(OR)}_{n+1} \qquad (80)$$

$[n = 2 \quad M = \text{Be}; M' = \text{Na}, K; R = i\text{-Pr}$

$\qquad M = \text{Sn(II)}; M' = \text{Li}, \text{Na}, K, \text{Rb}, \text{Cs}; R = t\text{-Bu}$

$n = 3 \quad M = \text{Al}, \text{Sb(III)}, \text{Bi(III)}; M' = \text{Li}, \text{Na}, K, \text{Rb}, \text{Cs};$

$\qquad R = \text{Et}, i\text{-Pr}, t\text{-Bu}$

$n = 4 \quad M = \text{Ti}; M' = \text{Li}, \text{Na}, K; R = i\text{-Pr}$

$\qquad M = \text{Zr}, \text{Sn(IV)}; M' = K, \text{Rb}, \text{Cs}; R = t\text{-Bu}$

$\qquad M = \text{Nb(IV)}; M' = \text{Na}, R = i\text{-Pr}$

$n = 5 \quad M = \text{Nb}, \text{Ta}; M' = \text{Li}, \text{Na}, K, \text{Rb}, \text{Cs}; R = \text{Et}, i\text{-Pr}, t\text{-Bu}]$

$$2\text{M(OR)}_4 + \text{M}'\text{OR} \longrightarrow \text{M}'\text{M}_2\text{(OR)}_9$$

$[M = \text{Zr}, \text{Hf}, \text{Sn}^{IV}, \text{Th}; M' = \text{Li}, \text{Na}, K; R = \text{Et}, i\text{-Pr}; \qquad (81)$

$M = \text{Nb}^{IV}; M' = \text{Na}; R = \text{Me}]$

$$4\text{Cu(O-}t\text{-Bu)} + 4\text{M}'\text{(O-}t\text{-Bu)} \longrightarrow \text{M}'_4\text{Cu}_4\text{(O-}t\text{-Bu)}_8 \qquad (82)$$

(M′ = Na or K)

$$2Al(O\text{-}i\text{-}Pr)_3 + 3M'O\text{-}i\text{-}Pr \longrightarrow M'_3Al_2(O\text{-}i\text{-}Pr)_9 \tag{83}$$

(M′ = Li, Na K)  (Refs. 6, 125)

$$3Zr(O\text{-}i\text{-}Pr)_4 + 3M'(O\text{-}i\text{-}Pr) \xrightarrow[\text{molar ratio}]{1:1\,\text{or higher}}$$

$$M'_2Zr_3(O\text{-}i\text{-}Pr)_{14} + MO\text{-}i\text{-}Pr$$

(Refs. 6, 125) $\hspace{8cm}$ (84)

(M′ = Na or K)

As indicated above, the nature of alkoxyl groups, alkali metals (6, 126), or molar ratios of the reactants were shown to influence the nature and composition of the bimetallic alkoxide formed.

2. Preparation of heterometallic alkoxides between pairs of metals as similar to each other as aluminum and gallium (137) or even niobium and tantalum (138) were described. However, the formation constants of the latter derivative were found to be rather low, precluding its isolation in view of the dynamism of the equilibrium.

$$2[(MeO)_4Nb(\mu\text{-}OMe)_2Ta(OMe)_4] \longrightarrow$$

$$[Nb(OMe)_5]_2 + [Ta(OMe)_5]_2 \tag{85}$$

Note that the equilibrium of the following type of reaction was postulated by Dislich (13) in 1971 to explain the ultrahomogeneity of the final ceramic product obtained in the sol–gel process that employed a mixture of alkoxides with different metals as precursors:

$$mSi(OR)_4 + nB(OR)_3 + pAl(OR)_3 + qNaOR$$

$$\xrightarrow[\text{molar ratio}]{\text{complexation in alcoholic}} (Si_m B_n Al_p Na_q)(OR)_{4m+3n+3p+p} \tag{86}$$

Obviously, the equilibria in the above types of complexation reactions would be very labile and require further investigations. However, numerous examples of such complexation reactions are illustrated below:

$$Ln(O\text{-}i\text{-}Pr)_3 + 3Al(O\text{-}i\text{-}Pr)_3 \longrightarrow [Ln\{Al(O\text{-}i\text{-}Pr)_4\}_3] \tag{87}$$

(Ref. 18)

$$Th(O\text{-}i\text{-}Pr)_4 + 4Al(O\text{-}i\text{-}Pr)_3 \longrightarrow [Th\{Al(O\text{-}i\text{-}Pr)_4\}_4] \tag{88}$$

(Ref. 139)

$$Zr(O\text{-}i\text{-}Pr)_4 \cdot Pr\text{-}i\text{-}OH + 2Al(O\text{-}i\text{-}Pr)_3 \longrightarrow \tag{89}$$

$$(i\text{-}PrO)_2Al(\mu\text{-}O\text{-}i\text{-}Pr)_2Zr(O\text{-}i\text{-}Pr)_2(\mu\text{-}O\text{-}i\text{-}Pr)_2Al(O\text{-}i\text{-}Pr)_2$$

$$M(O\text{-}i\text{-}Pr)_5 + 2Al(O\text{-}i\text{-}Pr)_3 \longrightarrow \tag{90}$$

$$(Pr\text{-}i\text{-}O)_2Al(\mu\text{-}O\text{-}i\text{-}Pr)_2M(O\text{-}i\text{-}Pr)_3(\mu\text{-}O\text{-}i\text{-}Pr)_2Al(O\text{-}i\text{-}Pr)_2$$

(Ref. 140)

$$M(O\text{-}t\text{-}Bu)_2 + M'(O\text{-}t\text{-}Bu) \longrightarrow M(O\text{-}t\text{-}Bu)_3M' \tag{91}$$

(M = Ge, Sn, Pb; M' = In or Tl)

(Ref. 141).

$$2Sn(O\text{-}t\text{-}Bu)_2 + M(O\text{-}t\text{-}Bu)_2 \longrightarrow M\{Sn(O\text{-}t\text{-}Bu)_3\}_2 \tag{92}$$

(M = Sr or Ca)

(Ref. 142)

$$2Tl(OR) + M(OR)_4 \longrightarrow Tl_2M(OR)_6 \tag{93}$$

$[M = Sn^{IV}, R = Et$ (Ref. 127); $R = CH(CF_3)_2$ (Ref. 143)$]$

$$M(OEt)_2 + Sb(OEt)_3 \longrightarrow M\{Sb(OEt)_5\} \tag{94}$$

(M = Mn, Fe, Co, Ni)

(Ref. 144)

It is interesting that the first example of a heterometallic alkoxide, $(O\text{-}i\text{-}Pr)_2Al(\mu\text{-}O\text{-}i\text{-}Pr)_2Be(\mu\text{-}O\text{-}i\text{-}Pr)_2Be(\mu\text{-}O\text{-}i\text{-}Pr)_2Zr(O\text{-}i\text{-}Pr)_3$ was synthesized by refluxing $Be(O\text{-}i\text{-}Pr)_2$ and $Al(O\text{-}i\text{-}Pr)_3$ together in 1:2 molar ratio and by adding to it 1 mol of $Zr(O\text{-}i\text{-}Pr)_4$ (8).

## III. PROPERTIES AND STRUCTURAL FEATURES

### A. General Properties

#### 1. Introduction

Of the three main classes of compounds with M—O—C functionality (32), that is, carboxylates (145), $\beta$-diketonates (146), and alkoxides (6), the chemistry of metal alkoxides has developed at a fast pace only since the 1950s. The organic moiety R attached to oxygen in metal alkoxides may be alkyl, fluoro-substituted alkyl (cf. Section V.4.B), chelating alkyl (cf. Section IV.C), or alkenyl (simple or substituted) (cf. Section IV.D). Since the 1970s, heterometallic alkoxides have received considerable attention. Although efforts were continuously made to shed light on the essential bonding features of alkoxide derivatives by other physicochemical techniques, unusual difficulties were encountered in their structural elucidation by X-ray crystallography until the late 1980s (72). However, our more recent expertise in crystal structure determination has helped us to confirm the basic framework generally indicated by earlier investigations, which is a corroboration of the essentials of the two earlier conclusions. These conclusions are (a) oligomeric homo- and heterometallic alkoxides involve simple $M(\mu\text{-}OR)M$ or $M(\mu\text{-}OR)M'$ bridges, respectively and (b) the degree of association $n$ of a homometallic alkoxide $[M(OR)_x]_n$ appears to be the smallest (147) to enable the metal atom to attain the preferred coordination state of the metal. A similar conclusion appears to be discernable (148) in heterobimetallic alkoxides in that the value of $(n_x + m_y)/(x + y)$ for $M_n M'_m(OR)_{nx + my}$ appears to be the lowest (where $x$ and $y$ are the valencies of M and M', respectively), so that all the M and M' atoms can be accommodated in a closed polyhedron by $\mu_2$- and $\mu_3$-OR functionalities, as required for a compact structural unit.

Furthermore, metal atoms tend to accommodate themselves in convenient geometries, for example, the metal atoms of $[KZr_2(O\text{-}i\text{-}Pr)_9]$ (117) and $[BaZr_2(O\text{-}i\text{-}Pr)_{10}]_2$ (149), arranging themselves in a triangular configuration, whereas the atoms of $[K_2Zr_2(O\text{-}t\text{-}Bu)_{10}]$ (133) and $[K_4Zr_2O(O\text{-}i\text{-}Pr)_{10}]$ (117) are arranged by depicting tetrahedral and octahedral geometries, respectively. In developing the above argument further, Caulton and co-workers (148) explained the serpentine structures of $Pb_4Zr_2(O\text{-}i\text{-}Pr)_{16}$ and $Pb_2Zr_4(O\text{-}i\text{-}Pr)_{20}$, which are obtained by the interaction of $Pb(O\text{-}i\text{-}Pr)_2$ and $Zr(O\text{-}i\text{-}Pr)_4$ in an equimolar ratio, as represented by Eq. 95.

$$6Pb(O\text{-}i\text{-}Pr)_2 + 6Zr(O\text{-}i\text{-}Pr)_4 \longrightarrow$$

$$Pb_4Zr_2(O\text{-}i\text{-}Pr)_{16} + Pb_2Zr_4(O\text{-}i\text{-}Pr)_{20} \qquad (95)$$

The close relationships in the structural features of homo- and heterometallic alkoxides can be exemplified by those of $Al\{Al(O\text{-}i\text{-}Pr)_4\}_3$ (150) and $Ln\{Al(O\text{-}i\text{-}Pr)_4\}_3$ (18). The molecular weights of the latter in benzene correspond to their empirical formulas. All of these compounds can be distilled in the range of 200–180°C/0.1 mm, with a lowering of the boiling point as was expected from increasing the covalent character which results from lanthanide contraction in the series. The tetrameric aluminum isopropoxide $Al\{Al(O\text{-}i\text{-}Pr)_4\}_3$ or $\{Al(O\text{-}i\text{-}Pr)_3\}_4$, however, disproportionates and distills as a dimeric vapor around

100°C under reduced pressure. The freshly distilled liquid is trimeric, but it changes on ageing (allowing to stand) into a more stable crystalline tetrameric form. This ageing phenomenon, which was described (151) as early as 1953, appears albeit to a lesser extent in other metal alkoxy derivatives and its prominence in aluminum isopropoxide may be ascribed to the change of coordination number 4 for aluminum in the dimer to 6 for the central aluminum in the tetrameric form. A similar observation about the dimerization on ageing of heterobimetallic $(i\text{-Pr-O})_2\text{Al}(\mu\text{-O-}i\text{-Pr})_2\text{Be(O-}i\text{-Pr})$ into $[(i\text{-Pr-O})_2\text{Al}(\mu\text{-O-}i\text{-Pr})_2\text{Be}(\mu\text{-O-}i\text{-Pr})_2\text{Al(O-}i\text{-Pr})_2]$ because of the tendency of beryllium to attain a higher tetracoordinate state led to the synthesis of the first heterotrimetallic species, for example, $(i\text{-Pr-O})_2\text{Al}(\mu\text{-O-Pr})_2\text{Be}(\mu\text{-O-}i\text{-Pr})_2\text{Nb(O-}i\text{-Pr})_4$.

The lower stability of the $\text{Al}\{\text{Al(O-}i\text{-Pr})_4\}_4$ could be ascribed to the steric strain on the central aluminum atom compared to that on the larger central lanthanide atoms in $\text{Ln}\{\text{Al(O-}i\text{-Pr})_4\}_3$.

The lack of an associative tendency and the higher solubility of heterobimetallic alkoxides appear to be general features. These features are prominently exhibited in the monomeric, volatile, and soluble tetraalkoxoaluminate derivatives, $\text{M}\{\text{Al(OR)}_4\}_2$, of bivalent metals (7) [e.g., Be, Mg, Ca, Sr, Ba, Zn, Cd, $\text{Mn}^{\text{II}}$, $\text{Fe}^{\text{II}}$, $\text{Co}^{\text{II}}$, $\text{Ni}^{\text{II}}$, and $\text{Cu}^{\text{II}}$], whereas the corresponding homometallic alkoxides $[\text{M(OR)}_2]_n$ are generally polymeric, nonvolatile, and insoluble. This salient difference in the properties of homo- and heterometallic alkoxides could easily be understood by considering the chelating (bidentate) nature of ligands

such as $\text{Al}\begin{smallmatrix}&\nearrow\text{O}\\ &\searrow\text{O}\end{smallmatrix}^{R}_{R}$ . It is interesting to point out that although generally biden-

tate, sterically less demanding ligands such as $\{\text{Al(OMe)}_4\}^-$ appear to bind central metal atoms (e.g., nickel) in a tridentate manner, for example, $\{(\text{MeO})\text{Al}(\mu_2\text{-OMe})_3\}_2\text{Ni}$, due to the tendency of the central nickel atom to attain a hexacoordination state (7).

The general properties of the significantly covalent behavior of metal alkoxides (both homo- and heterometallic), in spite of the polar character of the $\text{M}^{\delta+}-\text{O}^{\delta-}$ bond, were already dealt with in some detail (6). The effect of steric and inductive factors on the extent of polarization of the $\text{M}^{\delta+}-\text{O}^{\delta-}$ bond as well as the consequent degree of association and volatility can be exemplified by the boiling points (under $\sim 1$-mm pressure) and the observed degrees of their association (given in parentheses) by the three isomeric butoxides of zirconium: $\text{Zr(O-}n\text{-Bu})_4$ ($\sim 250°C$; 3.5); $\text{Zr(O-sec-Bu})_4$ ($\sim 150°C$; 2.0; $\text{Zr(O-}t\text{-Bu})_4$ ($\sim 50°C$; 1.0). However, the similarities in the molecular association of the neopentyloxides of Ti, Zr, and Al to the secondary rather than primary amyloxides have been adduced to indicate the higher predominance of steric rather than inductive factors in the above directions. Similarly, the insolubility and nonvolatility of the methoxides of most of the metals, in contrast to their higher alkoxides, appear to be mainly due to extensive association arising from the low steric effects of the methyl groups.

## 2.  Nature of $M^{\delta+} - O^{\delta-}$ Bonds

In addition to the normal covalent bond, the oxygen atom of the alkoxo group can form single or double $p\pi$-$p\pi$ or $p\pi$-$d\pi$ donor bonds with the metal atoms:

$$M\text{-}\overset{..}{\underset{..}{O}}\text{-}R \quad \longleftrightarrow \quad \overset{-}{M} \Longleftarrow \overset{+}{O}R \quad \longleftrightarrow \quad \underset{-}{M} \rightleftharpoons \underset{+}{O}R$$

Furthermore, the alkoxo group can bind the metals (similar or different) in $\mu_2$ or $\mu_3$ fashion:

In view of the higher electronegativity of oxygen, the $M^{\delta+} - O^{\delta-} - R$ bond is polarized in the direction shown. The extent of this polarization should be attenuated by the increasing $+I$ inductive effect of the more ramified alkyl group. In addition to the inductive effect, association of the molecule through coordination bridges of the type

$$\begin{array}{c} \delta^- \\ \diagdown \; \overset{\delta^+}{M} \overset{O}{\diagup} \overset{R}{\underset{\delta^-}{\diagdown}} \overset{\delta^+}{M} \diagup \\ \diagup \qquad \underset{R}{\overset{O}{}} \qquad \diagdown \end{array}$$

should also tend to reduce the negative charge(s) on the oxygen atom(s) and the positive charge(s) on the metal atoms.

The latter effect should be enhanced at the more electropositive metal, say $M'$, compared to that at M, making the overall bridged species more covalent. However, it was difficult to understand the salient covalent characteristics (e.g., volatility and solubility in organic solvents) coupled with the low conductivity (6) of the derivatives of even strongly electropositive metals (e.g., alkali and alkaline earth). This understanding was particularly difficult for heterobimetallic isopropoxides such as $[KZr_2(O\text{-}i\text{-}Pr)_9]$, as depicted by its nonconducting nature (6). The role of the polydentate bonding mode of the potassium complex which was postulated (6) on $^1$H NMR evidence, was confirmed by the X-ray crystallography of $[KU_2(O\text{-}t\text{-}Bu)_9]$ (152).

In contrast to the above, the somewhat (albeit lower) covalent characteristics of MAl(O-$i$-Pr)$_4$ derivatives and the variations observed when M was changed from Li to Cs, led to the conjecture (6) of a formulation such as $i$-Pr-OH → K($\mu_2$-O-$i$-Pr)$_2$Al(O-$i$-Pr)$_2$. This formulation exhibited more facile solubility in propane-2-ol compared to analogous derivatives of other alkali metals. Even though the existence of this basic unit ($i$-Pr-OH)$_2$K($\mu$-O-$i$-Pr)$_2$Al(O-$i$-Pr)$_2$ was confirmed by X-ray crystallography (126), it was shown to be an associated species [($i$-Pr-OH)$_2$K($\mu$-O-$i$-Pr)$_2$Al(O-$i$-Pr)$_2$]$_n$ (Fig. 35; Section III.B), in which potassium atoms are actually hexacoordinated. The difference(s) in the behavior of analogous MAl(O-$i$-Pr)$_4$ derivatives of other alkali metals await further structural elucidation.

The nature of alkoxide → metal (RO → M) $\pi$ bonding, which has been conceptualized (6) since the early 1950s, has begun to receive more attention. This attention is exemplified by investigations on the nature of $\pi$ bonding in alkoxides of the formula M$_2$(OR)$_6$, which are known to adopt one of the two structural types, **I** or **II**. In **I**, alkoxide bridge formation allows maximum M—O bonding, for example, in Al$_2$(O-$t$-Bu)$_6$ (153), while the preference for the unbridged structure reflects the predominating role of metal–metal bonding for $d^3$ Mo$^{3+}$ and W$^{3+}$ (154):

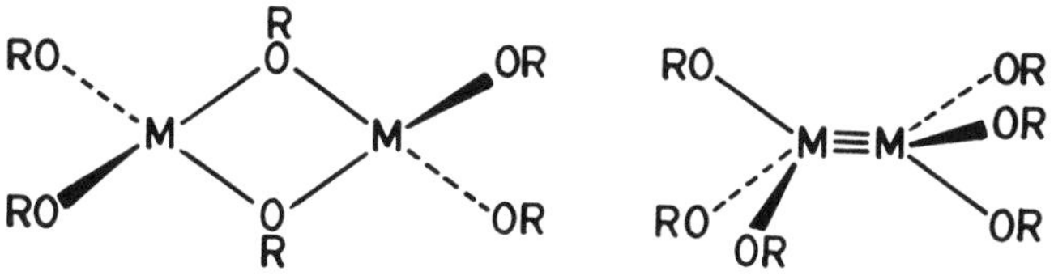

Both ab initio and Fenske-H type molecular orbital (MO) calculations were performed on the model M$_2$(OH)$_6$ (M = B, Al, or Mo) in order to compare the M—O bonding in Group 13 (IIIA) (B or Al) and Group 6 (VIB) (Mo) alkoxides. The Al$_2$(O-$t$-Bu)$_6$ (**I**) exhibits a pseudotetrahedral geometry, with the Al—O bridging distances being about 0.15 Å longer than the terminal Al—O bond distances. The $\pi$-bond orders for terminal M—OR groups were found to be only 0.26 for M = Al and 0.37 Å for M = B. This result indicated a significant degree of ionic character for the terminal Al—O bond, which was reflected in its observed shortening by about 0.15 Å compared to the sum of Al and O atomic radii. The B—O bonds have been found to be significantly more covalent than Al—O bonds, which is consistent with their relative electronegativities. For M = Mo, the short Mo—O bond lengths can be attributed to significant O$_{p\pi}$–Mo$_{d\pi}$ interaction, with less of an ionic contribution.

While dealing with the role of the alkoxide ligand in alkene and acetylene metathesis reactions, Schrock (155) emphasized that a combination of $\sigma$ (as estimated by the $pK_a$ value of the relevant alcohol and the ionization potential of the metal) and steric effects could qualitatively lead to the same type of conclusions as those attributed more recently to "the degree of $\pi$-bonding" alone. These conclusions were derived from the actually observed M—O length(s) and M—O-R bond angles, measured in the X-ray structures of the derivatives. The marked differences [to the extent of $\sim 10^7$–$10^9$ among EtOH, $(Me)_2CHOH$, and $(Me)_3COH$ and $\sim 10^{14}$ between $(Me)_3COH$, and $(CF_3)_3COH$] in $pK_a$ values of alcohols, mentioned against the formula of each alcohol [e.g., $MeCH_2OH$, 9.89; $(Me)_2CHOH$, 17.1; $(Me)_3COH$, 19.1; $(CF_3)_2CHOH$, 9.3; $(CF_3)_2(Me)COH$, 9.6; $(CF_3)_3COH$, 5.4] should obviously affect the nature of the $\sigma$ bond. This result could be utilized along with individual steric effects in predicting–understanding the nature of M—O-R bonds in the respective alkoxide derivatives.

Although dealing primarily with Al—O bond interaction in four-coordinate aluminum aryloxide systems, Barron (156) tried to define a normal Al—O bond (157) as a considerably polar covalent $\sigma$ bond with an additional $\pi$ interaction between the $O_p$ and Al–X $\sigma^*$ antibonding orbitals, weakening the latter in derivatives such as $AlR_n(OAr)_{3-n}(L)_x$ ($n = 0, 1,$ or 2 and $x = 0$ or 1).

Following some interesting studies on the structural and reactivity manifestations of O → M $\pi$ donations in later transition metal alkoxide derivatives— $IrH_2(OCH_2CF_3)(PCy_3)_2$ (158), $(\eta^5\text{-}C_5Me_5)Ru(OCH_2CF_3)$ $(PCy_3)$ (159), and $RuH(OSiPh_3)(CO)(P\text{-}t\text{-}Bu_2Me)_2$ (160)—Caulton and co-workers (161) have more recently attempted a rationalization of the observed trans/cis stereochemistry of $W(OCH_2CF_3)_2Cl_2(PMe_2Ph)_2$. This complex was synthesized by the reaction of $WCl_4(PMePh)_2$ with 2 equiv of $TlOCH_2CF_3$. The finding that this derivative was unreactive toward water was explained on the basis of its nearly saturated ($\sim 18$) valence electron structure, thus resisting the coordination of a water molecule on the central tungsten atom and making oxygen atoms of the $OCH_2CF_3$ groups inaccessible for electrophilic attack. Both conditions result from the strong O → W $\pi$ donation. The low-spin electronic state of this molecule was compared with that of $W(OC_6H_4Br)_2Cl_2(PMe_2Ph)_2$ (162) in contrast with the high-spin form, which is preferred by the analogue $W(OC_6H_3Ph_2\text{-}2,6)_2Cl_2(PMe_2Ph_2)_2$ (163). Reference may also be made to a recent comprehensive review (164) on the chemistry of the C—O bond in alkoxide and related ligands.

### 3. Molecular Association and Adduct Formation

Although the observation of Bradley (1) that "a remarkable feature of metal alkoxides is that they form very few coordination compounds with donor molecules" remains valid for homometal alkoxides (except for halogenoalkoxide derivatives; cf. Section IV.B), the newer species of oxo-alkoxides tend to form adducts with donor molecules such as thf in consonance with their geometrical features (Section V). The effect of steric factors (size of the central metal atom and ramification of the alkyl group) on the nuclearity of the homometal alkoxides of group 4 (IVB) metals was demonstrated in the early 1950s on the basis of their molecular weight measurements in solution (165). These results remain valid even for alkoxides derived from special alcohols (Section IV), but very few detailed physicochemical investigations were carried out to date.

In a 1992 study (166), a number of dimethylaluminum alkoxides of the type $\{Me_2Al(\mu\text{-}OR)\}_x$ (R = Me, Et, Pr, $i$-Pr, Bu, $i$-Bu, $sec$-Bu, $t$-Bu, $C_5H_{11}$, $CH_2CH_2\text{-}i\text{-}Pr$, $CH_2\text{-}t\text{-}Bu$, $C_6H_{13}$, $C_8H_{17}$, $C_{10}H_{21}$, or $C_{12}H_{25}$) were synthesized and studied by $^1H$, $^{13}C$, $^{17}O$, and $^{27}Al$ NMR, IR, and mass spectrometry. With the exception of R = $CH_2CH_2\text{-}i\text{-}Pr$, all the compounds of branched-chain alkoxide groups were purely dimeric (i.e., $x = 2$), while the former, as well as the $n$-alkyl derivatives, depict in solution an equilibrium between their dimeric and trimeric ($x = 2$ or 3) forms. The kinetics of the conversion of $\{Me_2Al(O\text{-}Pr)\}_2$ into $\{MeAl(O\text{-}Pr)_2\}_2$ was investigated, $\Delta H$ and $\Delta S$ were determined, and the identity of intermediates was probed by NMR and mass spectrometry.

In addition to the commonly operative conventional bridging mode through alkoxy oxygen, the dimeric lanthanum aryloxide derivative $La_2(OC_6H_3\text{-}i\text{-}Pr_2\text{-}2,6)_6$ was shown (167) to be bridged by intermolecular $\eta^6\text{-}\pi\text{-}$arene interactions of the bulky aryloxide ligands. These derivatives were suggested to be too demanding sterically to allow the bridging through their oxygen atoms. By contrast, the crystal structure of the diammonia adduct of $La_2(OC_6H_3\text{-}i\text{-}Pr\text{-}2,6))_6\cdot(NH_3)_2$ curiously shows the bridging to occur through aryloxy oxygen atoms with the two lanthanum atoms that depict a distorted (more open) trigonal bipyramid geometry in the unit $LaO_4N$ (N depicting an $NH_3$ molecule coordinated to each La atom).

Characterization (168) of $Zr_2(O\text{-}i\text{-}Pr)_8\cdot(HO\text{-}i\text{-}Pr)_2$ by $^1H$ and $^{13}C$ NMR spectroscopy and single-crystal X-ray diffraction has in general confirmed the edge-shared octahedral structure (Fig. 7; Section III.B) suggested by Bradley et al. (6). This actual structural elucidation has also clearly brought out the role of an unsymmetric hydrogen bonding. This novel type of hydrogen bonding has been demonstrated to be a general feature in not only adducts, such as isomorphous $Hf_2(O\text{-}i\text{-}Pr)_8\cdot(HO\text{-}i\text{-}Pr)_2$, and analogous $Ce_2(O\text{-}i\text{-}Pr)_8\cdot(HO\text{-}i\text{-}Pr)_2$ (168), but also in $Mo_4(OCH_2\text{-}c\text{-}Bu)_{12}(Bu\text{-}c\text{-}CH_2OH)$ (169) and $Nb_2Cl_5(O\text{-}i\text{-}Pr)(HO\text{-}i\text{-}Pr)_4$ (170).

An even more fascinating type of adduct formation involving hydrogen bonding may be exemplified (171) by $cis$-$PtMe(OCH(CF_3)_2)(PMe_3)_2\cdot(HOCH(CF_3)_2)_2$ (cf. Fig. (49)).

A detailed study of the central metal atom Th, which attains a higher coordination state through dimerization of $\{Th(OCH\text{-}i\text{-}Pr_2)_4\}$ or alternatively by formation of an adduct such as $Th(OCH\text{-}i\text{-}Pr_2)_4\cdot$(quinoline), was published (172) recently.

An example of a more recent study (173) of this commonly observed phenomenon of adduct formation of heterometallic alkoxides may be cited in the isolation and molecular structure elucidation of the adduct $[Ba_2Cu(\mu_3, \eta^2\text{-} OCHMeCH_2NMe_2)_2(\mu, \eta^2\text{-thd})_2(\eta^2\text{-thd})_2(HO\text{-}i\text{-}Pr)_2]$, where thdH is the $\beta$-diketone (2,2,6,6-tetramethylheptane-3,5-dione) ligand.

## 4. Physicochemical Studies

A description of the various physicochemical properties (e.g., volatility, dipole moments, refractivity, density, viscosity, magnetic susceptibilities, IR, ESR spectra, mass spectra, and NMR spectra) for elucidation of the structural features of homo- and heterometallic alkoxides was already presented in a number of publications [e.g., (3, 4, 6, 19, 34)].

With the growing sophistication of X-ray crystallographic techniques, the definitive bonding modes of more and more alkoxide derivatives in the solid state are being determined. But with the rapidly advancing uses of metal alkoxides as precursors in the sol–gel process, a clearer understanding of their solution structures has become essential to understanding of the mechanistic aspects of hydrolysis and condensation as key steps in this process.

A brief description of a few physicochemical techniques (e.g., NMR; EXAFS and XANES; vibrational, electronic, and ESR spectroscopy; magnetic studies, Mössbauer as well as mass spectrometry) that have been utilized extensively during the last decade to shed light on the structural features of metal alkoxides, is presented in Sections, III.A.a–f.

a. **NMR Studies $^1$H and $^{13}$C NMR.** These studies have become almost routine measurements on new homo- and heterometal alkoxides, as these techniques sometimes provide valuable information on the terminal and bridging ($\mu_2/\mu_3$) alkoxide groups (4a, 6, 7, 28, 34, 38, 39).

In many cases, however, derivation of useful structural information becomes somewhat difficult from their NMR spectra due to exchange phenomena of the following types: (a) intermolecular bridge/terminal or terminal/terminal site exchange; (b) the Brønsted basicity of the alkoxide groups makes them vulnerable to proton-catalyzed site exchange with the free alcohol molecules present in the system from adventitious hydrolysis; (c) the overlapping of the resonances; and (d) the broadening of the signals due to quadrupolar effects. In spite of such handicaps, a correlation between the NMR chemical shifts and the bonding mode of the alkoxide groups can be made based on the principle that terminal alkoxide groups give signals at the highest field, followed by doubly ($\mu_2$-OR) and then triply ($\mu_2$-OR) bridging groups more and more downfield.

For example, the molecular structure of $Y_3(\mu_3\text{-OR})(\mu_3\text{-Cl})(\mu_2\text{-OR})_4(thf)_2$ (Fig. 17) (61), where R = $t$-Bu, contains eight $t$-Bu groups in six distinct environments. This finding is in excellent agreement with the $1:2:1:2:1:1$ pattern: $\delta$ 1.88 ($\mu_3$-OR); 1.50 1.43 (2:1) ($\mu_2$-OR), and 1.31, 1.24, and 1.22 (2:1:1) (three types of terminal groups) in its $^1$H NMR spectrum (161). Similarly the $^{13}$C {$^1$H} NMR spectrum of the complex in THF-$d_8$ also showed six chemically nonequivalent O-$t$-Bu groups with the quaternary carbon signals in the range 76.9–72.1 ppm and the methyl carbon resonances in the 36.4–35.2 range; although only five of the six expected methyl resonances could be resolved.

Another recent example (174) of the elucidation of molecular structure in solution by NMR spectroscopy is that of [Sn(O-$i$-Bu)$_4$·HO-$i$-Bu]$_2$. The $^1$H and $^{13}$C{H} NMR data for the product were consistent with the formula [Sn(O-$i$-Bu)$_4$·HO-$i$-Bu]$_2$ and the $^{119}$Sn{H} NMR spectrum exhibited $^2J\,^{119}$Sn–$^{117}$Sn consistent with a dimeric unit. The data are qualitatively analogous to those for [Sn(O-$i$-Pr)$_4$·HO-$i$-Pr]$_2$ (175). The solution dynamic behavior is represented below.

The solution structure of [Sn(O-$i$-Bu)$_4$·HO-$i$-Bu]$_2$ at 273 K is represented below with a symmetrically bridging hydrogen from the coordinated alcohol making this pair of O-$i$-Bu and HO-$i$-Bu groups indistinguishable by $^1$H and

$^{13}$C{$^1$H} NMR. The intramolecular proton-transfer process is fast, with a lower activation barrier than that of the isopropoxide analogue, in which case it can be observed on the NMR time scale. It finally appears that [Sn(O-$i$-Bu)$_4$·HO-$i$-Bu]$_2$ retains its coordinated alcohol more tenaciously than [Sn(O-$i$-Pr)$_4$·HO-$i$-Pr]$_2$.

The kinetic as well as the thermodynamic parameters for the equilibrium

$$Th_2(OCH\text{-}i\text{-}Pr_2)_8 \rightleftharpoons 2Th(OCH\text{-}i\text{-}Pr_2)_4$$

were evaluated (172) by measuring the relative concentrations of the monomer and dimer over a temperature range of 273–353 K from their $^1$H NMR spectra in benzene-$d_8$ or toluene-$d_8$. The OCH-$i$-Pr$_2$ well-resolved methine protons provided a convenient resonance for integration purposes. A plot of $\ln(K_{eq})$ versus $1/T$ led to the following parameters: $\Delta H^0 = 17$ kcal mol$^{-1}$, $\Delta G^0_{298} = 5$ kcal mol$^{-1}$, and $\Delta S^0 = 40$ cal mol$^{-1}$ K$^{-1}$. However, NMR was not able to evaluate the activation parameters due to the very fast rate of equilibrium attainment.

In addition to $^1$H and $^{13}$C NMR spectroscopy, $^{17}$O (176, 177) and $^{19}$F (143) NMR studies were employed and have a bright future in the elucidation of reaction mechanisms in solution.

Furthermore, in the attempt to better characterize homo- as well as hetero-metal (4) alkoxide derivatives, studies with NMR active (178, 179) metal nuclei [e.g., $^{11}$B, $^{27}$Al, $^{29}$Si, $^{49}$Ti, $^{51}$V (180), $^{89}$Y (181), $^{205}$Tl (143), $^{91}$Zr, $^{119}$Sn (174, 182, 120), $^{207}$Pb (183), $^{95}$Mo, and $^{183}$W] could effectively supplement the information derived from other physicochemical measurements.

In a recent publication (184), alkali metal ($^7$Li and $^{23}$Na) NMR spectroscopy confirmed that in solution as well as in the crystalline state, [LiTi(O-$i$-Pr)$_5$]$_2$ exists as a dimer with a rapid isopropoxide exchange, whereas a tight ion pair exists between Na$^+$ and the Ti(O-$i$-Pr)$_5^-$ anion.

Even more recently, Veith et al. (39a) carried out $^{113}$Cd NMR studies on a novel heterometallic alkoxide derivative [{Cd(O-$i$-Pr)$_3$Ba{Zr$_2$(O-$i$-Pr)$_9$}]$_2$ (39a).

In view of the difficulties encountered (72) in the X-ray crystallography of metal alkoxides, solid state NMR (MAS) could complement X-ray structural data or/and provide an alternative/primary source of solid state structural data (185).

**b. EXAFS and XANES Techniques.** Extended X-ray absorption fine structure (EXAFS) and X-ray absorption near edge structure (XANES) techniques (186–188) often provide useful complementary information regarding the local environment around preselected metal atoms for solids (irrespective of their crystallinity or dimensionality) as well as their solutions. These techniques are thus able to provide some structural insight when classical X-ray techniques fail. In addition, these techniques could be of exceptional importance regarding structural integrity in solution. Generally, EXAFS gives average interatomic distances with an accuracy of $\pm 0.02$ Å and indicates the type and number of nearest-neighbor atoms up to a distance of about 4 Å around a central metal atom.

The elucidation of structural details for four titanium alkoxides, $Ti(OR)_4$ (where OR = O-$t$-Am, O-$i$-Pr, OEt, and O-$n$-Bu) may be cited as a practical illustration (189) of the application of these techniques. Both $Ti(O$-$t$-Am$)_4$ and $Ti(O$-$i$-Pr$)_4$ were shown to be tetrahedral monomeric molecules, whereas $\{Ti(OEt)_4\}_n$ and $\{Ti(O$-$n$-Bu$)_4\}_n$ were shown to be oligomers (probably trimeric, $n = 3$) with the Ti—O bond distances of 1.80 (terminal) and 2.05 Å (bridging), as well as Ti—Ti interactions (observed for the first time) of 3.1 Å. Based on the information available from XANES studies, the most probable coordination number of titanium is 5 in a threefold oligomer.

In a recent publication (190), these techniques were applied to elucidate the crystal structure of low-dimensional compounds such as iron oxychloride, FeOCl, and its alkoxy substituents, $FeOCl_{l-x}(OR)_x$, where R is Me or Et. It was shown that the local symmetry around the iron atom becomes higher for methoxy substituents compared to that for FeOCl. In fact, it is similar to that for $\gamma$-FeOOH. Similarity of the one-dimensional electron density map along the $c$ axis, obtained separately from X-ray crystallographic studies and from EXAFS fitting, reflects the reliability of the EXAFS fit and indicates its applicability to other two-dimensional systems with poor crystallinity.

**c. Vibrational, Electronic, and Electron Paramagnetic Resonance Spectroscopy.** Although in principle the IR spectra should be able to provide both fingerprint and structurally diagnostic information, to date this technique has been employed in a large number of cases mainly for the former purpose (6). Only in a few cases, such as that of $U(OMe)_6$, will the IR as well as Raman spectroscopic studies of the $^{16}O$ and $^{18}O$ species could be employed (191) to give more reliable assignments of the $U-{}^{16}O$ stretching frequencies as 505.0 cm$^{-1}$ ($A_{1g}$), 464.8 cm$^{-1}$ ($T_{1u}$), and 414 cm$^{-1}$ ($E_g$).

**d. Electronic Spectra and EPR Studies.** These techniques were used (7, 19) for alkoxides and tetraalkoxyaluminates of later $3d$ metals (e.g., Cr, Mn, Fe, Co, Ni, and Cu) to elucidate their stereochemistry and other structural features. Some interesting conclusions resulted. For example, the electronic spectrum of $Ni\{Al(OR)_4\}_2$ indicates an octahedral geometry for nickel when R = Me, which leads to the plausible structure of the type $[Ni\{(\mu_3\text{-}OMe)_3AlOMe\}_2]$. With a change of R group from Me to Et, $i$-Pr and $t$-Bu, the molecule tends to be tetrahedral, with both forms in equilibrium when R = Et or $i$-Pr. However, the role of steric factors becomes predominant in the tetrahedral form of $[Ni\{(\mu\text{-}O\text{-}i\text{-}Pr)(\mu\text{-}O\text{-}t\text{-}Bu)Al(O\text{-}t\text{-}Bu)_2\}_2]$ and $[Ni\{(\mu\text{-}O\text{-}t\text{-}Bu)_2Al(O\text{-}t\text{-}Bu)_2\}_2]$ (90).

**e. Magnetic Susceptibility Studies and Mössbauer Spectroscopy.** The detailed investigations on magnetic susceptibilities of the methoxides of first-row di- and trivalent metals were summarized already (6, 7).

Observed magnetic susceptibilities of homo- and heterometallic alkoxides of latter $3d$ metals (Cr, Mn, Fe, Co, Ni, or Cu) are in consonance with their structures, which were derived on the basis of electronic spectra (7, 19). For example, chromium(III) alkoxides as well as their substituted derivatives, $Cr(OEt)(acac)_2$ and $Cr(OEt)_2(acac)$, all depict a magnetic susceptibility of $3.8\mu_B$, indicative of an octahedral geometry for chromium with three unpaired electrons.

However, the determination (65) of the magnetic susceptibility of $Cr(O\text{-}t\text{-}Bu)_4$ in $C_6D_6$ at $23°C$ gave a value of $2.86\ \mu_B$, which conforms with the presence two singly occupied lower $e_g$ shells in a tetrahedral environment.

Among the more recent studies, reference may be made to the measurements (192) of magnetic susceptibilities of $Cu_4(\mu\text{-}OCMe_3)_6[OC(CF_3)_3]_2$ and $Cu_3(\mu\text{-}OCMe_3)_4[OC(CF_3)_3]_2$ in the temperature range of 77–300 K. These measurements show that both compounds are antiferromagnetically coupled. The magnetic data for the two fit, respectively, with the theoretical equations for a linear centrosymmetric tetranuclear system $[J = -131.6\ (0.6)\ \text{cm}^{-1}]$ and a corresponding trinuclear equation $[J = -114.4\ (1.5)\ \text{cm}^{-1}]$.

The significance of $[Fe_4S_4]^{n+}$ (193) and $[Mn_4X_4]^{n+}$ (194) (where X = oxo, chloro, or alkoxo ligands) in biological systems, has made the synthesis and properties of iron and manganese alkoxide cubes (195), $[Fe(OMe)(MeOH)(DPM)]_4$, $[Fe(OMe)(MeOH)(DMB)]_4$, $Fe^{III}Fe_3^{II}(OMe)_5(MeOH)_3(OOCC_6H_5)_4$, and $[Mn_4(OEt)_4(EtOH)_2(DPM)_4]$ (where HDPM and HDBM are dipivaloylmethane and dibenzoylmethane, respectively), very interesting. The four metal ions and the bridging alkoxide ligands are located at alternating vertices of a cube, with either alcohol or alkoxide and $\beta$-diketonate or benzoate ligands on the exterior of the core. Spectral studies, including $^1H$ NMR and solution Mössbauer studies, indicate that the cubes remain intact in solvent mixtures containing alcohol. Magnetic measurements, for example, have shown that iron atoms in the $[Fe_4(OMe)_4]^{4+}$ of the first species are ferromagnetically exchange coupled with a coupling constant $J = -1.88$ $\text{cm}^{-1}$, $g = 2.29$, and contributions from zero-field splitting; its predicted ground state ($S_T = 8$) was confirmed by its saturation moment ($15.7\mu_B$ in a 19.7T field at 1.2 K).

Mössbauer spectroscopy is of special significance for species containing paramagnetic $^{57}Fe$ nuclei, which render the NMR technique not as useful. This technique could also be utilized for homo- and heterometallic alkoxides of Mössbauer active nuclei such as $^{119}Sn$ and $^{121}Sb$.

**f. Mass Spectral Studies.** Mass spectral studies (6) of alkoxides such as $(LiO\text{-}t\text{-}Bu)_6$, $Cr(O\text{-}t\text{-}Bu)_4$, $M(O\text{-}t\text{-}Am)_4$ (where M = Ti, Zr, or Hf),

M'[OCH(CF$_3$)$_2$]$_4$ (where M' = Si, Ge, Ti, Zr, or Hf), [Al(O-$i$-Pr)$_3$]$_4$ and [(Bu-$t$-O)$_2$Al($\mu$-O-$i$-Pr)$_2$Al(O-$t$-Bu)$_2$] have provided evidence for their structural patterns in the vapor phase. A parallelism was observed (196) in the fragmentation pattern of a number of [Ln{($\mu$-O-$i$-Pr)$_2$Al(O-$i$-Pr)$_2$}$_3$] complexes, with the pattersn of [Al(O-$i$-Pr)$_3$]$_4$ or Al{$\mu$-O-$i$-Pr)$_2$Al(O-$i$-Pr)$_2$}$_3$. These patterns provided the strongest corroboratory evidence for the stability of the [Ln{Al(O-$i$-Pr)$_4$}$_3$] species in this early stage of development of heterometallic alkoxide chemistry.

Similarly, in addition to the $^1$H NMR and molecular weight measurements in solution, mass spectral confirmation (8) of the stability of alkoxide species such as [($i$-Pr-O)$_3$Zr($\mu$-O-$i$-Pr)$_2$Be($\mu$-O-$i$-Pr)$_2$Al(O-$i$-Pr)$_2$] and [($i$-Pr—O)$_4$Nb($\mu$-O-$i$-Pr)$_3$Be($\mu$-O-$i$-Pr)$_2$Al(O-$i$-Pr)$_2$], provided the first evidence for the formation of heterometallic alkoxides containing more than two different metals in the same molecular species. This result opened up a new dimension in heterometallic alkoxide chemistry.

Detailed mass spectral studies of Zr(OR)$_4$ (where R = Et, $i$-Pr, or $n$-Bu) and some hafnium as well as tin(IV) analogues indicated (197) the presence of M$_3$O(OR)$_{10}$ and M$_4$O(OR)$_{14}$, respectively, along with M(OR)$_4$ oligomers of different molecular complexity.

Because of the growing importance of homo- and heterometallic alkoxides (3, 24) as precursors for ceramic materials by the MOCVD technique (cf. Section VIII), their mass spectroscopic studies should be very useful in ascertaining their applicability.

## B. X-Ray Crystallographic Studies

### 1. Introduction

As mentioned earlier, fairly successful conjectures (147, 148) about the structural features of homo- as well as heterometallic alkoxides were made earlier on the basis of physicochemical studies in solution. Even as late as 1989, Bradley (72) pointed out the special difficulties in the X-ray structural elucidation of metal alkoxide complexes. Such difficulties were reemphasized by Caulton and Hubert-Pfalzgraf (4), particularly the heterometallic alkoxides, which in many cases are obtained in a highly plastic (deformable) state even when crystallizable, which renders them unsuitable for single-crystal X-ray analysis. Therefore, the process of X-ray crystallographic structural characterization was gathering considerable momentum (4, 23, 28, 38, 39) and even with the availability of other more sophisticated techniques, publications were being limited mainly to derivatives characterized crystallographically.

The role of steric factor(s) in structural preferences is well illustrated in the dimeric tetrahedral aluminum $tert$-butoxide, [($t$-BuO)$_2$Al($\mu$-O-$t$-Bu)$_2$Al(O-$t$-Bu)$_2$] (153) (Fig. 1), in contrast to the octa-cum-tetrahedral form of tetrameric aluminum isopropoxide (150, 198), [Al{($\mu$-O-$i$-Pr)$_2$Al(O-$i$-Pr)$_2$}$_3$] (Fig. 2).

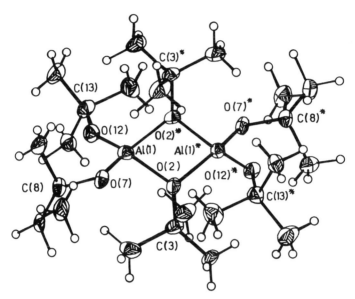

Figure 1. The crystal structure of $(t\text{-Bu-O})_2\text{Al}(\mu\text{-O-}t\text{-Bu})_2\text{Al(O-}t\text{-Bu})_2$. [Adapted from (153).]

The latter type of structural framework (where the only difference is that the six-coordinate central aluminum atom is replaced by a lanthanide) with bidentate ligation of tetraisopropoxoaluminate ligands was suggested (196) mainly on the basis of NMR and mass spectral studies for tris(tetraisopropoxoaluminates) of Group 3 (IIIB) and lanthanide metals, $[\text{Ln}\{(\mu\text{-O-}i\text{-Pr})_2\text{Al(O-}i\text{-Pr})_2\}_3]$ and for derivatives such as $[\text{M}\{\text{Al(O-}i\text{-Pr})_4\}_2]$ [where M = Ni (7), Co (7), and Cu (7)] and $[\text{Cr}\{(\mu\text{-O-}i\text{-Pr})_2\text{Al(O-}i\text{-Pr})_2\}_3]$ (7) on the basis of their UV–vis spectral and magnetic susceptibility data. However, X-ray structural confirmation for the bidentate ligational behavior of the $[\text{Al(O-}i\text{-Pr})_4]^-$ ligand was made possible in 1992 for the praseodynium complex, $[\{\text{Pr}[(\mu\text{-O-}i\text{-Pr})_2\text{Al(O-}i\text{-Pr})_2]_2(\mu\text{-Cl})(i\text{-PrOH})\}_2]$ (Fig. 29) (74), which happens to be the first heteroleptic tetraisopropoxoaluminate derivative characterized by single-crystal X-ray crystallographic studies.

The effect of size and the tendency to attain coordinative saturation of central alkali metals in analogous heterobimetallic isopropoxides of the type $\text{MTi(O-}i\text{-Pr})_5$ (M = Li, Na, or K) has been currently (131, 192) shown by X-ray crystallographic studies (Fig. 23, 36) to result in derivatives with different geometries. Furthermore, the isolation of a heterotrimetallic isopropoxide containing barium, cadmium, and zirconium of the type $[\{\text{Zr}_2(\text{O-}i\text{-Pr})_9\}\text{Ba}(\mu\text{-O-}i\text{-Pr})_2\text{Cd}(\mu\text{-O-}i\text{-Pr})]_2$ (39a), which was obtained by the reaction of $\{\text{Zr}_2(\text{O-}i\text{-Pr})_9\}\text{CdI}$ with $\text{KBa(O-}i\text{-Pr})_3$, was expected to form $[\{\text{Zr}_2(\text{O-}i\text{-Pr})_9\}\text{Cd}(\mu\text{-O-}i\text{-}$

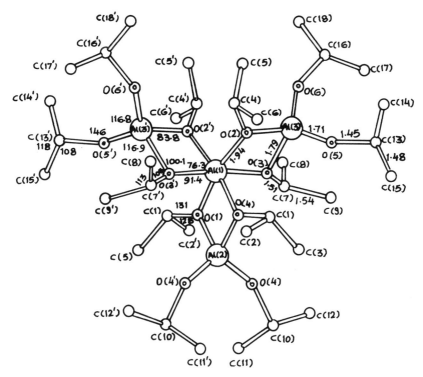

Figure 2.  The crystal structure of tetrameric aluminum isopropoxide, [Al(O-*i*-Pr)$_3$]$_4$ (150 and 198).

Pr)$_2$Ba($\mu$-O-*i*-Pr)$_2$]$_2$. This observation is intriguing in heterometallic alkoxide chemistry. The analytical as well as other physicochemical data could not have distinguished between the expected and the rather unusual actual formulation, which was established by X-ray crystallography (Fig. 37). This structure is consistent with the preferential characteristics of the two metals involved (barium and cadmium) for attaining coordination numbers 6 and 4, respectively.

In addition, incorporation of unexpected ions such as Li$^+$, Cl$^-$, and O$^{2-}$ during the synthesis of homo- and heterometallic alkoxides by conventional methods was observed in some cases (4a, 28, 38, 39, 42).

X-ray crystallography has played a prominent role in establishing the inclusion of heteroinorganic ions. More specifically, in the case of oxo-alkoxides (Section V), where the presence of the O$^{2-}$ ion is undetectable analytically.

Recent successes [cf. (4a, 23, 28, 38, 39, 42, 199)] in X-ray structural elucidation of an increasing number of homo- and heterometallic alkoxides revealed novel features in addition to general quantification of their basic framework, which was postulated on the basis of spectroscopic (mainly NMR) studies described in Section III.A.4.a.

Next, an effort was made to depict the crystal structures of a few typical derivatives, which were arranged according to the nuclearity of the metal alkoxo species. Structures of metal complexes of special (e.g., fluoro/sterically hindered/chelated) alcohols would be dealt with separately in Section IV.

## 2. Mononuclear Derivatives

Because of the strong preference of metal alkoxides to form associated species involving $\mu_2$- and/or $\mu_3$-alkoxo bridges, examples of mononuclear metal alkoxides still continue to be rather scarce. However, such derivatives are gradually becoming accessible either by the use of sterically demanding alkoxo (e.g., $^-$O-$t$-Bu or OC-$t$-Bu$_3$) ligands (39) as well as 2,6-di-tertiarybutylphenoxo ligands (28, 39, 200–202), which do not fall directly under the purview of this chapter. Such mononuclear species can also be obtained by the formation of molecular adducts with suitable donor ligands. The partial substitution and side reactions (see Section IV.A) that lead to the formation of di- and higher nuclear species are noteworthy drawbacks regarding the use of more sterically demanding alkoxo ligands such as $^-$OC-$t$-Bu$_3$.

Alkoxides of Ca, Sr, and Ba containing sterically compact groups (e.g., OMe, or OEt) are highly associated and exhibit low solubility and volatility. It may be wise to mention some more sterically demanding aryloxide derivatives of metals as they provide a clue to the potential synthesis of similar alkoxide derivatives of a number of metal(loid)s that use sterically bulky aryloxide groups. This led in recent years to the isolation of a variety of X-ray crystallographically characterized mononuclear derivatives, such as [M(OC$_6$H$_2$-$t$-Bu$_2$-2,6-Me-4)$_2$(thf)$_3$]. THF (M = Ca or Ba) (59, 203), [Sr(OC$_6$H$_2$-$t$-Bu$_3$-2,4,6)$_2$(thf)$_3$] (204), and hydrogen bonded [Ba(OC$_6$H$_2$-$t$-Bu$_2$-2,6)$_2$(HOCH$_2$CH$_2$NMe$_2$)$_4$] (205).

Since beryllium is smaller in size, mononuclear two-coordinate alkoxide derivatives are more likely to form. The best known examples include only aryloxide derivatives, [Be(OC$_6$H$_3$-$t$-Bu$_2$-2,6)$_2$] (206) [Be(OC$_6$H$_2$Me$_3$-2,4,6)$_2$(OEt$_2$)] and [Be(OC$_6$H$_2$-$t$-Bu$_3$-2,4,6)$_2$(OEt$_2$)] (207), linear (two-coordinate) and planar (three-coordinate) beryllium atoms, respectively.

Furthermore, these bulky aryloxide ligands played a crucial role in making the crystallographically characterized less-familiar three-coordinate compounds of scandium, yttrium, and the lanthanides (Ln), such as Ln(OC$_6$H$_2$-$t$-Bu$_2$-2,6-R-4)$_3$ (28) more accessible.

It may be worthwhile to mention the mononuclear nature of Cr(O-$t$-Bu)$_4$, the structure of which was elucidated by the allied technique of gas-phase electron diffraction (65).

The X-ray crystallographically authenticated mononuclear metal alkoxide

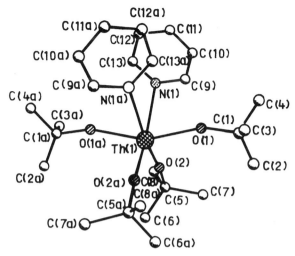

Figure 3. The crystal structure of Th(O-*t*-Bu)₄(py)₂. [Adapted from (208).]

derivatives include Th(O-*t*-Bu)$_4$(py)$_2$ (208), Sn(O-*t*-Bu)$_4$ (209), and Cr(OCH-*t*-Bu$_2$)$_4$ (see Section IV.A).

The derivative Th(O-*t*-Bu)$_4$(py)$_2$ crystallizes in the monoclinic space group $C_{2/c}$ with the central thorium atom lying on a twofold axis. The central thorium atom is bonded to four oxygen atoms of the ⁻O-*t*-Bu groups and two nitrogen atoms of the pyridine ligands in a cis-pseudooctahedral environment as shown in Fig. 3.

Two different Th—O bond distances are observed in the molecule, 2.161(6) Å for *tert*-butoxide groups trans to pyridine and 2.204(6) Å for the *tert*-butoxide ligands trans to one another. The Th—N(py) distance is 2.752(7) Å.

Very little structural data were available for tin(IV) alkoxide derivatives until now (209). Tin(IV) *tert*-butoxide is monomeric in the solid state (Fig. 4), consistent with molecular weight measurements in benzene solution (6).

The environment around the tin(IV) center is a distorted tetrahedral; O(2)—Sn—O(2') and O(7)—Sn—O(7') angles are enlarged ($\sim 115°$), while the O-Sn—O angles (105° and 107°) are reduced compared to the tetrahedral angle. The Sn—O bond lengths are 1.946 and 1.949 Å, slightly smaller than the sum of the Sn and O covalent radii. However, the Sn—O—C angles are 125.0(2)° and 124.1(2)° for O(2) and O(7), respectively, in the region expected for the absence of $\pi$ donation from the alkoxide oxygen to the metal center.

The gas-phase electron diffraction studies on Cr(O-*t*-Bu)$_4$ are consistent with a model of $S_4$ symmetry (65) drawn in Fig. 5.

The arsenic(III) *tert*-butoxide, As(O-*t*-Bu)$_3$, is monomeric even in the solid

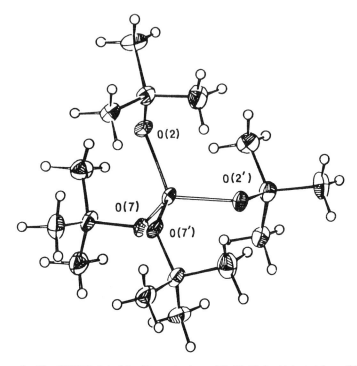

Figure 4.   The ORTEP plot of the X-ray structure of Sn(O-*t*-Bu)$_4$. [Adapted from (209).]

state (58) and displays a three-coordinate pyramidal arsenic atom (As—O = 1.74(2) Å, O-As—O 91.5°), as expected.

### 3.   *Dinuclear Derivatives*

The earlier account of the structural features of di- and polynuclear homo- and heterometallic alkoxides have already been comprehensively reviewed (4a, 23, 28, 38, 39, 42, 199, 202). However, with the current resurgence of interest in the structural characteristics of associated homometal alkoxides and monomeric heterobimetallic alkoxides, we present current status of the solid state structures (as determined by X-ray crystallographic studies) of such derivatives.

**a.   Homometallic Alkoxides.**   Alkali metal derivatives of sterically less demanding alkoxo (e.g., OMe, OEt, or O-*i*-Pr) ligands and even their *tert*-butoxo analogues adopt highly associated structures (38) by forming $\mu_2$- or $\mu_3$-alkoxo bridged bonds in the solid state as well as in solution. However, more sterically hindered ligands such as $^-$OC$_6$H$_2$-*t*-Bu$_2$-2,6-Me-4 (210) and $^-$OC-*t*-

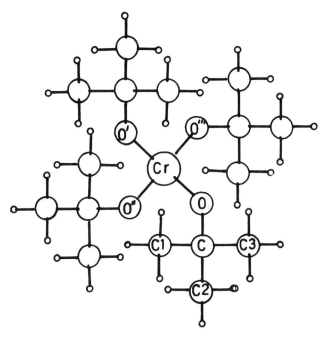

Figure 5.   The structure of Cr(O-t-Bu)₄ as determined by gas-phase electron diffraction. [Adapted from (65).]

Bu₃ (see Section IV.A) make accessibility of dinuclear species with the two- and three-coordinate environment around lithium atoms.

Being large in size, heavier alkaline earth metals usually tend to form larger molecular aggregates (38) with higher (6–8) coordination numbers of the metals; X-ray structures of only two dinuclear alkoxides of heavier alkaline earth metals such as [{Ca($\mu$-OR)(OR)(thf)₃}₂] [R = C(Ph)CH₂C₆H₄-Cl-4 (59)] and [Ba₂(OCPh₃)₄(thf)₃] (211) appear to be known so far.

Due to the large size of the lanthanides and actinides, their alkoxide complexes containing less bulky alkoxo ligands such as ⁻OMe, ⁻OCH₂-t-Bu, or ⁻O-i-Pr exhibit a strong tendency toward oligomerization, which results in the isolation of dinuclear (168) or tetranuclear (see Section IV.B.4) complexes and higher oligomers (see Section IV.B.7). Even the use of bulky tertiary alkoxide [e.g., O-t-Bu, O-t-Am, OC(i-Pr)Me₂, OC(i-Pr)(Me)Et, OCEt₃] ligands restrict the oligomerization to trimers or dimers [Y, La (103), and U (152)].

The X-ray diffraction study of the triphenylmethoxide derivative of lanthanum (212) is shown in Fig. 6.

Tetraisopropoxides with an empirical formula M(O-i-Pr)₄ (where M = Ce,

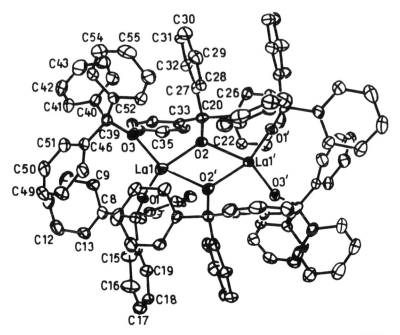

Figure 6.   The ORTEP diagram of [La(OCPh$_3$)$_2$($\mu$-OCPh$_3$)]$_2$. [Adapted from (212).]

Zr, Hf, Th, and Sn) are coordinatively unsaturated and are potential Lewis acids for ligating alcohol molecules during their syntheses (see Section III.A). The dimeric nature of the molecular adduct with an empirical formula Zr(O-$i$-Pr)$_4$(HO-$i$-Pr) in benzene solution led Bradley et al. (213) to suggest an edge-shared octahedral structural (**III**) as early as 1953,

$$
\begin{array}{c}
\text{III}
\end{array}
$$

without mention of hydrogen bonding, a feature that was recently well established (168) by X-ray crystallography studies (Fig. 7).

This structure (Fig. 7) can be described as edge-shared bioctahedral in which the two octahedral zirconium fragments are connected by two doubly ($\mu_2$-)

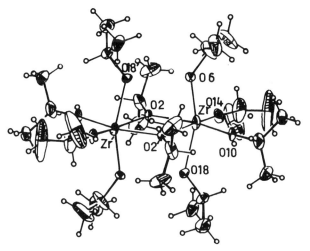

Figure 7.   The crystal structure of $Zr_2(O\text{-}i\text{-}Pr)_8(HO\text{-}i\text{-}Pr)_2$. [Adapted from (168)].

bridging isopropoxo ligands. The structural framework represented in Fig. 7 contains two terminal isopropoxo ligands at each zirconium atom in the same plane as the doubly bridging isopropoxo ligands and the terminal isopropoxo group are perpendicular to the $(Pr\text{-}i\text{-}O)_2Zr(\mu\text{-}O\text{-}i\text{-}Pr)_2Zr(O\text{-}i\text{-}Pr)_2$ plane. In addition, perpendicular to the bridges, there is one isopropanol molecule. This molecule is coordinated to each zirconium atom and is on opposite sides of this plane, which is involved in asymmetric intramolecular hydrogen bonding to an isopropoxo group of the adjacent zirconium atom. The asymmetry in the hydrogen bridge is reflected from the difference in the coordination of isopropanol and isopropoxo ligands to the zirconium centers. Similar structures were recently elucidated by X-ray crystallographic studies for complexes of other metals that include $M_2(O\text{-}i\text{-}Pr)_8(HO\text{-}i\text{-}Pr)_2$ [where M = Ce (168), Hf (168), Sn (214), and $[Sn(O\text{-}i\text{-}Bu)_4 \cdot HO\text{-}i\text{-}Bu]_2$ (174)]. This result is consistent with the similarities in the covalent radii of Ce, Zr, Hf, and Sn (215). By contrast, while titanium(IV) is only slightly smaller than tin(IV), the size difference is sufficient to cause a distinct difference in its metal alkoxide chemistry. However, it is interesting to note that a qualitatively similar pattern of intramolecular hydrogen bonding to that in $Zr_2(O\text{-}i\text{-}Pr)_8(HO\text{-}i\text{-}Pr)_2$ exists in $Ti_2(OPh)_8(PhOH)_2$ (216).

Extensive structural studies (217) on complexes of the type $W_2Cl_4(\mu\text{-}OR)_2(OR)_2(ROH)_2$ reveal intramolecular hydrogen bonding between syn OR and ROH groups (linked by a W=W bond), which are very similar to that in $Zr_2(O\text{-}i\text{-}Pr)_8(HO\text{-}i\text{-}Pr)_2$.

Other examples of hydrogen bonding to coordinated alkoxide include:

$Mo_4(OCH_2$-$c$-$Bu)_{12}(HOCH_2$-$c$-$Bu)$ (169), $[Mo(O$-$i$-$Pr)_2(HO$-$i$-$Pr)_2]_2$ (218). Furthermore, $Nb_2Cl_5(O$-$i$-$Pr)(HO$-$i$-$Pr)_4$ (175) is an example of hydrogen bonding (**IV**) to other electron-pair donors.

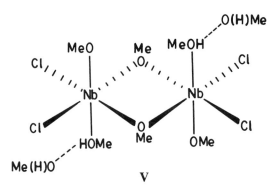

**IV**

In spite of a wide occurrence of ROH· · ·OR hydrogen bonding in dinuclear alkoxide complexes, it can be interrupted by competing effects. For example, the complex $Nb_2Cl_4(OMe)_4(MeOH)$ (175) adopts the structure **V** with a planar

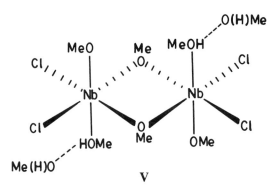

**V**

$Nb_2(OMe)_2(MeOH)_2$ substructure seemingly suited to intramolecular bonding among OMe groups. However, the lattice methanol molecules are more potent electron donors than the Nb—OMe oxygen atoms, and are the favored sites for intermolecular hydrogen bonding. Some other metal alkoxide complexes exhibiting intermolecular hydrogen bonding include the following: $Pd(OCH(CF_3)_2)(OPh)(HOPh)(bpy)$ (219) and $cis$-$PtMe(OCH(CF_3)_2)$-$(HOCH(CF_3)_2)(PMe_3)_2$ (220).

The X-ray crystal structure of $U_2(O$-$i$-$Pr)_{10}$ has confirmed the edged-shared bioctahedral geometry (152) in the solid state (Fig. 8), although $^1H$ NMR studies in solution (6) revealed a monomer–dimer equilibrium.

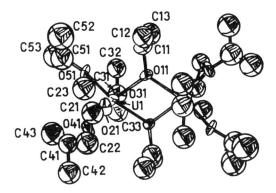

Figure 8.   The ORTEP view of $U_2(O\text{-}i\text{-}Pr)_{10}$. [Adapted from (152).]

The crystal structure of an interesting dinuclear mixed-valence [$U^{IV}$, $U^{V}$] *tert*-butoxide of uranium (152) is shown in Fig. 9. The configuration is face-sharing bioctahedral; the U· · ·U distance is 3.549(1) Å.

Structural studies of bulky Group 14 (IVA) metal(II) alkoxides reveal that (a) gaseous tin(II) *tert*-butoxide (Fig. 10), is a trans dimer (221) with the remarkably small endocyclic OSnO angle of 76(2)°, (b) the crystalline compounds of the type M(O-*t*-Bu)Cl (M = Ge or Sn) (222) is alkoxo-bridge dimers (Fig. 11), and (c) the crystalline tri-*tert*-butylmethoxides of Ge(II) and Sn(II), are isomorphous V-shaped monomers (see Section IV.B).

We would also like to mention the more recently determined single-crystal X-ray structure of [Nb(O-*i*-Pr)$_5$]$_2$ (223), crystals of which are built of dimers formed by sharing an edge of two octahedra (Fig. 12), although the presence of only a monomeric species is indicated in the gaseous phase.

The shape and geometric parameters of the Nb—O core for this compound are similar to those of the earlier values reported for [Nb(OMe)$_5$]$_2$ (224). The

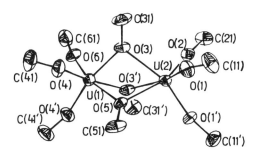

Figure 9.   The crystal structure of $U_2(O\text{-}t\text{-}Bu)_9$. [Adapted from (152).]

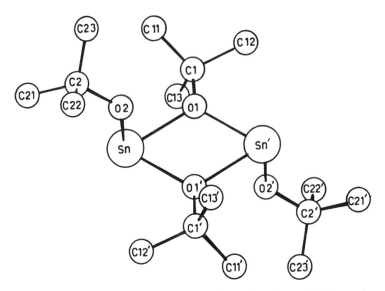

Figure 10.   The molecular structure of [Sn($\mu$-O-$t$-Bu)(O-$t$-Bu)]$_2$ (as determined by gas-phase electron diffraction). [Adapted from (221).]

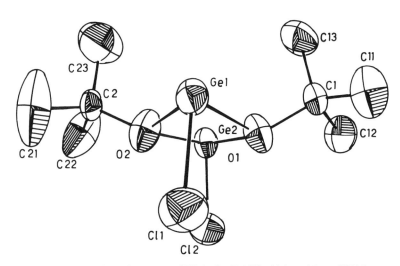

Figure 11.   Crystal structure of [Ge($\mu$-O-$t$-Bu)Cl]$_2$. [Adapted from (222).]

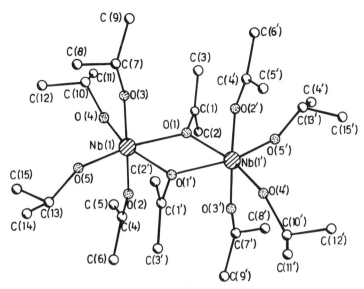

Figure 12. The crystal structure of [Nb(O-$i$-Pr)$_5$]$_2$. [Adapted from (223).]

noteworthy features of these structures are their much smaller sums of axial (3.788 Å) in comparison with the sums of equatorial (4.042–4.035 Å) bond lengths. These results may be interpreted in terms of the different contribution of the $\pi$ component to the interaction of metal and oxygen atomic orbitals (4). This inference is corroborated with the observed larger Nb—O-C bond angles for axial oxygen atoms in comparison with the corresponding angles at the equatorial oxygen atoms (160.2–164.0° and 141.3–142.8°), respectively.

**b. Heterometallic Alkoxides.** Although structures of many heterodinuclear metal alkoxides were established by X-ray methods (Table II) during the last decade, structures of a few typical derivatives are being presented in this section. An account of those structures that are derived from more hindered (e.g., $^-$OCH-$i$-Pr$_2$ or $^-$OC-$t$-Bu$_3$) and fluoro [e.g., $^-$OCH(CF$_3$)$_2$ or $^-$OCMe(CF$_3$)$_2$] alkoxides can be seen in Sections IV.A and V.B.

Structure of a dinuclear heterometallic methoxide derivative containing lithium and boron, (MeOH)$_2$Li($\mu$-OMe)$_2$B(OMe)$_2$ (Fig. 13), was determined by the X-ray diffraction method (124); both lithium and boron atoms were found to be four coordinate.

The derivative LiAl(OCEt$_3$)$_3$Cl(thf) is another interesting crystallographically characterized dinuclear heterometallic alkoxide (225) of Group 13 (IIIA).

Following the principles enunciated mainly by Bradley (147) (Section III.A),

TABLE II

Crystal Data for Some Homo- and Heterometallic Alkoxide Complex Compounds

| Compound | Crystal Class | Space Group | Z | $a$ (Å) $b$ (Å) $c$ (Å) | $\alpha$ (°) $\beta$ (°) $\gamma$ (°) | References |
|---|---|---|---|---|---|---|
| *Homometallic Derivatives* | | | | | | |
| [LiOCMe₂Ph]₆ (Fig. 28) | Monoclinic | $P2_{1/n}$ | 2 | 12.555(2) 12.010(2) 16.405(3) | 90.0 99.44(1) 90.0 | 239 |
| [KO-$t$-Bu]₄ | Cubic | $P\bar{4}3M$ | 4 | 8.372(1) 8.372(1) 8.372(1) | 90 90 90 | 238 |
| [KO-$t$-Bu · Bu-$t$-OH]∞ | Triclinic | $P\bar{1}$ | 2 | 9.862(2) 9.929(4) 6.330(2) | 99.32(2) 90.66(2) 66.80(1) | 238 |
| [RbO-$t$-Bu]₄ | Cubic | $P\bar{4}3M$ | 4 | 8.514(1) 8.514(1) 8.514(1) | 90 90 90 | 238 |
| [RbO-$t$-Bu · Bu-$t$-OH]∞ | Triclinic | $P\bar{1}$ | 2 | 9.886(2) 9.914(2) 6.640(2) | 98.90(1) 90.46(1) 66.76(1) | 238 |
| [Ca($\mu$-OR)(OR)(thf)]₂·2PhMe (R = C(C₆H₅)₂CH₂C₆H₄Cl-4) | Triclinic | $P\bar{1}$ | 1(dimer) | 13.517(2) 15.782(2) 12.129(2) | 111.282(8) 107.205(9) 99.49(1) | 59 |
| [Ba₄(O-$t$-Bu)₈(HO-$t$-Bu)₈] | Cubic | $Pa3$ | 8 | 25.766(4) 25.766(4) 25.766(4) | 90.0 90.0 90.0 | 106a |
| [Ba₂(OCPh₃)₄(thf)₃]·PhMe·THF | Monoclinic | $P2_{1/c}$ | 4 | 18.242(5) 17.337(5) 26.302(7) | 90 97.81(1) 90 | 211 |

| Compound | Crystal system | Space group | Z | a, b, c (Å) | α, β, γ (°) | Ref. |
|---|---|---|---|---|---|---|
| [Y₃(O-t-Bu)₈Cl(thf)₂] (Fig. 17) | Orthorhombic | $Pmcn$ | 4 | 16.976(6) / 13.466(5) / 25.823(8) | 90.0 / 90.0 / 90.0 | 61 |
| [Y₃(O-t-Bu)₇Cl₂(thf)₂] | Monoclinic | $P2_{1/c}$ | 4 | 18.102(7) / 15.871(6) / 21.0844(5) | 90.0 / 104.87(3) / 90.0 | 62 |
| [Y₃(O-t-Bu)₇Cl(thf)₃]⁺[BPh₄]⁻·PhMe | Triclinic | $P\bar{1}$ | 2 | 14.141(3) / 15.035(3) / 18.712(5) | 85.01(2) / 85.02(2) / 81.75(2) | 267 |
| [Y₂(O-t-Bu)₄Cl(thf)₄]⁺[BPh₄]⁻ | Tetragonal | $P4_{2/n}$ | 4 | 18.424(5) / 18.424(5) / 17.738(9) | 90.0 / 90.0 / 90.0 | 267 |
| [Y(O-t-Bu)Cl(thf)₅]⁺[BPh₄]⁻ | Monoclinic | $P2_{1/n}$ | 4 | 13.675(3) / 12.418(3) / 32.803(8) | 90.0 / 90.43(2) / 90.0 | 267 |
| [(C₅H₄SiMe₃)₂Y(OMe)]₂·2PhH | Monoclinic | $C_{2/c}$ | 4 | 16.747(11) / 27.351(14) / 8.404(4) | 90.00(4) / 109.49(4) / 90.05(5) | 268 |
| [La₂(OCPh₃)₆]·2PhMe (Fig. 6) | Triclinic | $P\bar{1}$ | 1 | 12.019(3) / 14.287(2) / 16.375(3) | 73.99(1) / 85.39(2) / 76.07(2) | 212 |
| [La₃(O-t-Bu)₉(HO-t-Bu)₂] (Fig. 16) | Monoclinic | $P2_{1/a}$ | 4 | 28.163(5) / 10.987(2) / 19.974(5) | 90.00 / 90.00 / 99.79 | 103 |
| [La₄(OCH₂-t-Bu)₁₂] | Orthorhombic | $P2_12_12$ | 4 | 17.259(4) / 19.151(4) / 11.771(3) | 90.0 / 90.0 / 90.0 | 249 |
| [Cp₃Ce(O-t-Bu)] | Orthorhombic | $Pnma$ | 4 | 14.351(17) / 13.3467(14) / 8.9677(9) | 90.0 / 90.0 / 90.0 | 269 |
| [Ce(O-t-Bu)₂(NO₃)₂(HO-t-Bu)₂] | Triclinic | $P\bar{1}$ | 2 | 10.7548(56) / 11.0853(49) / 12.4403(56) | 72.760(34) / 88.190(39) / 69.050(35) | 64 |

TABLE II
Crystal Data for Some Homo- and Heterometallic Alkoxide Complex Compounds

| Compound | Crystal Class | Space Group | $Z$ | $a$ (Å) $b$ (Å) $c$ (Å) | $\alpha$ (°) $\beta$ (°) $\gamma$ (°) | References |
|---|---|---|---|---|---|---|
| *Homometallic Derivatives* (Continued) | | | | | | |
| $[Ce_2(O\text{-}i\text{-}Pr)_8(HO\text{-}i\text{-}Pr)_2]$ | Triclinic | $P\bar{1}$ | 1 | 11.385(4) 12.144(5) 9.009(3) | 109.72(1) 111.54(2) 65.70(1) | 168 |
| $[Nd_4(OCH_2\text{-}t\text{-}Bu)_{12}]$ (Fig. 21) | Tetragonal | $P42_{1c}$ | 2 | 20.383(5) 20.383(5) 11.822(3) | 90.0 90.0 90.0 | 249 |
| $[Nd_6(O\text{-}i\text{-}Pr)_{17}Cl]$ (Fig. 29) | Monoclinic | $P2_{1/n}$ | 4 | 24.52(2) 22.60(2) 14.22(1) | 90.0 101.05(5) 90.0 | 256 |
| $[Th(O\text{-}t\text{-}Bu)_4(Py)_2]$ | Monoclinic | $C_{2/c}$ | 4 | 12.656(3) 15.326(3) 16.803(3) | 90.0 96.43(3) 90.0 | 208 |
| $[Th_2(OCHEt_2)_8(Py)_2]$ | Triclinic | $P\bar{1}$ | 1 | 11.302(3) 11.458(3) 14.191(3) | 113.02(2) 92.50(2) 115.82(2) | 250 |
| $[Th_4(O\text{-}i\text{-}Pr)_{16}(Py)_2]$ (Fig. 22) | Triclinic | $P\bar{1}$ | 1 | 11.211(2) 18.689(4) 9.984(2) | 101.65(1) 108.57(1) 75.60(1) | 250 |
| $[U_2(O\text{-}i\text{-}Pr)_{10}]$ (Fig. 8) | Triclinic | $P\bar{1}$ | 1 | 10.974(3) 12.226(3) 10.002(2) | 111.56(2) 110.09(2) 67.87(2) | 152 |
| $[U_2(O\text{-}t\text{-}Bu)_9]$ (Fig. 9) | Orthorhombic | $Pbcm$ | 4 | 13.749(7) 19.977(7) 16.923(8) | 90.0 90.0 90.0 | 152 |

| Compound | Crystal system | Space group | Z | Cell parameters (Å) | Angles (°) | Ref. |
|---|---|---|---|---|---|---|
| $[U_2I_4(O\text{-}i\text{-}Pr)_4(HO\text{-}i\text{-}Pr)_2]$ | Triclinic | $P\bar{1}$ | 2 | 9.643(7)  11.283(8)  9.598(7) | 95.59(3)  117.39(2)  108.59(3) | 44 |
| $[Zr_2(O\text{-}i\text{-}Pr)_8(HO\text{-}i\text{-}Pr)_2]$ (Fig. 7) | Triclinic | $P\bar{1}$ | 4 | 18.262(4)  19.883(5)  12.067(3) | 98.59(1)  96.26(1)  77.49(1) | 168 |
| $[Nb_2(OMe)_{10}]$ | Triclinic | $P\bar{1}$ | 4 | 8.684(3)  9.673(3)  12.543(3) | 69.33(2)  85.84(2)  86.46(3) | 224 |
| $[Mo_2(OCH_2\text{-}t\text{-}Bu)_6]$ | Monoclinic | $P2_{1/n}$ | 2 | 18.160(10)  11.051(7)  9.956(6) | 90.0  104.30(4)  90.0 | 270 |
| $[Mo_2(O\text{-}i\text{-}Pr)_4(HO\text{-}i\text{-}Pr)_4]$ | Tetragonal | $P4/nmm$ | 2 | 12.810(4)  12.810(4)  9.869(5) | 90.0  90.0  90.0 | 271 |
| $[Mo_2(O\text{-}c\text{-}Pen)_4(HO\text{-}c\text{-}Pen)_4]$ | Monoclinic | $C_{2/c}$ | 12 | 62.232(11)  10.442(2)  20.167(3) | 90.0  106.83(1)  90.0 | 271 |
| $[Mo_2(OCH_2\text{-}t\text{-}Bu)_4(HNMe_2)_4]$ | Tetragonal | $I4cm$ | 8 | 22.064(5)  22.064(5)  16.985(4) | 90.0  90.0  90.0 | 271 |
| $[Mo_2(O\text{-}i\text{-}Pr)_4(Py)_4]$ | Triclinic | $P\bar{1}$ | 2 | 18.254(6)  10.327(20)  10.076(2) | 70.92(1)  103.08(1)  104.38(1) | 271 |
| $[Mo_2(OCH_2\text{-}K\text{-}t\text{-}Bu)_4(PMe_3)_4]$ | Monoclinic | $Pa$ | 2 | 19.441(10)  11.619(5)  9.906(4) | 90.0  106.41(2)  90.0 | 271 |
| $[W_4(OEt)_{16}]$ | Triclinic | $P\bar{1}$ | 1 | 12.129(6)  10.985(6)  9.692(5) | 93.28(3)  108.64(2)  105.71(2) | 245 |
| $[Re_3(O\text{-}i\text{-}Pr)_9]$ (severely disordered) | Hexagonal | $P6_{3/m}$ | 6 | 20.335(3)  20.335(3)  16.509(4) | 90.0  90.0  120.0 | 272 |

### TABLE II
#### Crystal Data for Some Homo- and Heterometallic Alkoxide Complex Compounds

| Compound | Crystal Class | Space Group | Z | $a$ (Å) $b$ (Å) $c$ (Å) | $\alpha$ (°) $\beta$ (°) $\gamma$ (°) | References |
|---|---|---|---|---|---|---|
| *Homometallic Derivatives* (Continued) | | | | | | |
| [Re$_3$(H)(O-$i$-Pr)$_8$] | Triclinic | P$\bar{1}$ | 2 | 9.865(2) 11.495(2) 16.537(4) | 88.41(2) 84.99(2) 67.33(2) | 273 |
| [Re$_3$(H)(O-$i$-Pr)$_8$(Py)]$\frac{1}{2}$Py | Monoclinic | P2$_1$/c | 8 | 24.940(6) 15.201(5) 23.664(2) | 90.0 111.84(1) 90.0 | 273 |
| [Fe(OMe)$_2$(O$_2$CCH$_2$Cl)]$_{10}$ (Fig. 34) | Monoclinic | P2$_1$/c | 2 | 11.106(3) 15.958(3) 22.938(3) | 90.0 97.34(1) 90.0 | 260 |
| [Cu(O-$t$-Bu)PPh$_3$]$_2$ | Tetragonal | P4$_1$2$_1$2 | 4 | 10.327(4) 10.327(4) 36.669(20) | 90.0 90.0 90.0 | 274 |
| [Al$_2$(O-$t$-Bu)$_6$] (Fig. 1) | Triclinic | P$\bar{1}$ | 2 | 9.946 9.755 16.332 | 88.89 73.81 88.89 | 153 |
| [Al(O-$i$-Pr)$_3$]$_4$ (Fig. 2) | Tetragonal | P4$_1$2$_1$2 | 4 | 12.317(3) 12.317(3) 31.724(8) | 90.0 90.0 90.0 | 150, 198 |
| [($t$-Bu)$_2$GaOCPh$_3$] | Monoclinic | P2$_1$/n | 4 | 9.533(2) 17.030(4) 14.289(4) | 90.0 90.57(2) 90.0 | 275 |
| [Sn(O-$t$-Bu)$_4$] (Fig. 4) | Monoclinic | C$_{2/c}$ | 4 | | 90.0 | 209 |

294

| Compound | Crystal system | Space group | Z | a, b, c (Å) | α, β, γ (°) | Ref. |
|---|---|---|---|---|---|---|
| [Sn(O-i-Pr)$_4$(HO-i-Pr)]$_2$ | Monoclinic | $P2_{1/c}$ | 4 | 11.742(5) 14.285(3) 12.341(3) | 90.0 95.37(1) 90.0 | 209, 214 |
| [Sn(O-i-Bu)$_4$(HO-i-Bu)]$_2$ | Triclinic | $P\bar{1}$ | 1 | 19.544(4) 11.939(4) 12.615(4) | 111.091(27) 96.652(28) 107.854(28) | 174 |
| [Sn(O-t-Bu)$_3$(OAc)(Py)] | Orthorhombic | $Pnma$ | 4 | 12.638(4) 11.305(5) 16.858(9) | 90.0 90.0 90.0 | 121 |
| *Heterometallic Derivatives* | | | | | | |
| [Li$_5$Sm(O-t-Bu)$_8$] | Tetragonal | $P4_22_12$ | 8 | 17.03(3) 17.03(3) 31.78(1) | 90.0 90.0 90.0 | 276 |
| [LiTi(O-i-Pr)$_5$]$_2$ | Monoclinic | $P2_{1/n}$ | 4 | 11.440(8) 16.396(13) 11.838(8) | 90.0 92.59(5) 90.0 | 131 |
| [LiZr$_2$(O-i-Pr)$_9$(HO-i-Pr)] (Fig. 20) | Monoclinic | $P2_{1/c}$ | 4 | 13.561(2) 16.741(2) 18.332(2) | 90.0 97.98(1) 90.0 | 45 |
| [LiNb(OEt)$_6$]$_6$ | Orthorhombic | $Pbca$ | 16 | 17.704(4) 19.736(5) 21.267(8) | 90.0 90.0 90.0 | 277 |
| [LiAl(OCEt$_3$)$_3$Cl(thf)$_2$] | Monoclinic | $P2_{1/c}$ | 4 | 16.49(3) 10.65(1) 20.05(3) | 90.0 106.97(11) 90.0 | 225 |
| [Na$_2$Ce(O-t-Bu)$_6$(dme)$_2$] | Orthorhombic | $Pna2_1$ | 4 | 20.5337(36) 10.9557(23) 19.4128(38) | 90.0 90.0 90.0 | 64 |
| [NaTi(O-i-Pr)$_5$]$_\infty$ (Fig. 36) | Monoclinic | $P2_{1/a}$ | 4 | 12.616(1) 10.078(4) 17.573(4) | 90.0 103.20 90.0 | 184 |

TABLE II
Crystal Data for Some Homo- and Heterometallic Alkoxide Complex Compounds

| Compound | Crystal Class | Space Group | $Z$ | $a$ (Å) $b$ (Å) $c$ (Å) | $\alpha$ (°) $\beta$ (°) $\gamma$ (°) | References |
|---|---|---|---|---|---|---|
| *Heterometallic Derivatives* (Continued) | | | | | | |
| [NaTh₂(O-t-Bu)₉] | Orthorhombic | *Pnma* | 4 | 11.170(2) 25.408(5) 17.195(3) | 90.0 90.0 90.0 | 208 |
| [KU₂(O-t-Bu)₉]·C₆H₁₄ (Fig. 18) | Monoclinic | *P2₁/c* | 4 | 10.713(3) 25.990(8) 19.480(6) | 90.0 91.13(5) 90.0 | 152 |
| [KTi(O-i-Pr)₅]∞ | Monoclinic | *P2₁/n* | 4 | 12.213(2) 9.792(2) 18.442(3) | 90.0 106.944(2) 90.0 | 184 |
| [K₂Zr₂(O-t-Bu)₁₀] | Monoclinic | *P2₁/c* | 4 | 10.532(1) 18.057(3) 27.618(5) | 90.0 95.54(1) 90.0 | 133 |
| [(i-Pr-OH)₂K{Al(O-i-Pr)₄}]∞ (Fig. 35) | Orthorhombic | *P2₁2₁2₁* | 4 | 11.935(4) 16.513(6) 13.681(4) | 90.0 90.0 90.0 | 126 |
| [KSn(O-t-Bu)₅]∞ | Orthorhombic | *Fddd* | 32 | 25.149(22) 29.658(26) 32.558(43) | 90.0 90.0 90.0 | 129 |
| [KSb(O-t-Bu)₄] | Monoclinic | *C2/c* | 4 | 15.529(8) 12.577(6) 11.510(6) | 90.0 95.09(4) 90.0 | 58 |
| [K₂Sb(O-t-Bu)₅·dioxane] | Monoclinic | *P2₁/m* | 2 | 9.962(5) 17.620(9) 10.316(5) | 90.0 111.34(3) 90.0 | 58 |

296

| Compound | Crystal system | Space group | $Z$ | Cell dimensions (Å) | Angles (°) | Ref. |
|---|---|---|---|---|---|---|
| $[KBi(O\text{-}t\text{-}Bu)_4]$ | Triclinic | $P\bar{1}$ | 2 | 9.942(7)<br>10.357(7)<br>12.097(9) | 102.89(5)<br>91.75(6)<br>104.73(5) | 58 |
| $[Mg\{Ti_2(OEt)_8Cl\}_2]$ | Monoclinic | $P2_{1/n}$ | 2 | 13.900(9)<br>14.624(8)<br>13.982(8) | 90.0<br>95.05(4)<br>90.0 | 255 |
| $[Mg_2Al_3(O\text{-}i\text{-}Pr)_{13}]$ (Fig. 26) | Monoclinic | $P2_{1/c}$ | 4 | 15.927(4)<br>18.239(7)<br>19.914(7) | 90.0<br>91.34(2)<br>90.0 | 47 |
| $[Mg_2Sn_2(O\text{-}t\text{-}Bu)_8]$ | Monoclinic | $P2_{1/c}$ | 2 | 10.100(7)<br>14.951(8)<br>15.009(9) | 90.0<br>97.30(6)<br>90.0 | 278 |
| $[BaZr_2(O\text{-}i\text{-}Pr)_{10}]_2$ | Monoclinic | $P2_{1/c}$ | 2 | 18.066(3)<br>12.549(2)<br>19.409(3) | 90.0<br>94.42<br>90.0 | 149 |
| $[Ba\{Nb(O\text{-}i\text{-}Pr)_6\}_2(HO\text{-}i\text{-}Pr)_2]$ | Triclinic | $P\bar{1}$ | 2 | 10.763(2)<br>14.595(5)<br>19.144(10) | 97.99(4)<br>96.69(3)<br>102.04(2) | 279 |
| $[BaSn_2(O\text{-}t\text{-}Bu)_6]$ | Tetragonal | $R\bar{3}$ | 3 | 10.17(1)<br>10.17(1)<br>29.86(2) | 90.0<br>90.0<br>90.0 | 142 |
| $[\{Pr[Al(O\text{-}i\text{-}Pr)_4]_2(Cl)(HO\text{-}i\text{-}Pr)\}_2]$ (Fig. 30) | Monoclinic | $C_{2/c}$ | 4 | 25.181(6)<br>16.937(3)<br>19.376(3) | 90.0<br>91.60(2)<br>90.0 | 74 |
| $[NbAl_3Me_9(O\text{-}t\text{-}Bu)_3]$ | Tetragonal | $P3\bar{1}_c$ | 2 | 16.226(2)<br>16.226(2)<br>8.219(1) | 90.0<br>90.0<br>90.0 | 280 |
| $[ErAl_3(O\text{-}i\text{-}Pr)_{12}]$ (Fig. 25) | Orthorhombic | $P2_12_12_1$ | 4 | 13.150(1)<br>17.464(2)<br>23.158(3) | 90.0<br>90.0<br>90.0 | 251 |
| $[Cr_2Sn_2(O\text{-}t\text{-}Bu)_8]$ | Monoclinic | $P2_{1/c}$ | 2 | 13.654(9)<br>10.411(8)<br>15.729(9) | 90.0<br>103.27(8)<br>90.0 | 278 |

TABLE II
Crystal Data for Some Homo- and Heterometallic Alkoxide Complex Compounds

| Compound | Crystal Class | Space Group | Z | $a$ (Å) $b$ (Å) $c$ (Å) | $\alpha$ (°) $\beta$ (°) $\gamma$ (°) | References |
|---|---|---|---|---|---|---|
| _Heterometallic Derivatives_ (Continued) | | | | | | |
| [MnGe$_2$(O-$t$-Bu)$_6$] | Orthorhombic | _Pnma_ | 4 | 19.624(9) 16.950(8) 9.910(6) | 90.0 90.0 90.0 | 278 |
| [Co$_2$Ge$_2$(O-$t$-Bu)$_8$] | Monoclinic | _P2$_{1/n}$_ | 2 | 10.054(7) 14.794(8) 14.999(8) | 90.0 97.06(7) 90.0 | 278 |
| [Mn$_2$Sn$_2$(O-$t$-Bu)$_8$] | Monoclinic | _P2$_{1/n}$_ | 2 | 10.072(5) 15.149(8) 15.082(8) | 90.0 97.11(4) 90.0 | 278 |
| [Co$_2$Sn$_2$(O-$t$-Bu)$_8$] | Monoclinic | _P2$_{1/n}$_ | 2 | 10.032(7) 15.052(9) 14.978(9) | 90.0 97.08(7) 90.0 | 278 |
| [Ni$_2$Sn$_2$(O-$t$-Bu)$_8$] | Monoclinic | _P2$_{1/n}$_ | 2 | 9.959(7) 15.083(9) 14.920(9) | 90.0 96.96(7) 90.0 | 278 |
| [CuCl{Zr$_2$(O-$i$-Pr)$_9$}] | Monoclinic | _P2$_{1/n}$_ | 4 | 9.459(2) 24.351(4) 16.654(3) | 90.0 95.06(1) 90.0 | 231 |
| [Cu$_2$Zr$_2$(O-$i$-Pr)$_{10}$] | Triclinic | $P\bar{1}$ | 2 | 12.612(2) 18.196(3) 9.830(1) | 102.88(1) 105.68(1) 92.09(1) | 231 |
| [Na$_2${Zn(O-$t$-Bu)$_3$}$_2$] (Fig. 24) | Triclinic | $P\bar{1}$ | 1 | 9.917(2) 10.066(2) 10.327(2) | 66.51(2) 60.30(2) 81.90(2) | 76 |

| Compound | Crystal system | Space group | Z | a, b, c | α, β, γ | Ref. |
|---|---|---|---|---|---|---|
| [K$_2$\{Zn(O-$t$-Bu)$_3$\}$_2$] | Triclinic | $P\bar{1}$ | 1 | 9.916(3) 10.030(3) 10.270(3) | 62.65(2) 66.01(2) 81.31(2) | 76 |
| [\{Cd[Zr$_2$(O-$i$-Pr)$_9$]Cl\}$_2$] (Fig. 31) | Monoclinic | $P2_{1/n}$ | 2 | 12.5120(10) 19.462(3) 16.4250(10) | 90.0 93.250(10) 90.0 | 73 |
| [CdI\{Zr$_2$(O-$i$-Pr)$_9$\}] | Triclinic | $P\bar{1}$ | 2 | 9.815(10) 11.450(9) 20.69(2) | 95.41(8) 99.81(9) 11.227(7) | 39a |
| [\{Cd(O-$i$-Pr)$_3$\}Ba\{Zr$_2$(O-$i$-Pr)$_9$\}]$_2$ (Fig. 37) | Monoclinic | $P2_{1/c}$ | 2 | 23.20(2) 12.937(11) 19.33(2) | 90.0 97.30(8) 90.0 | 39a |
| [InGe(O-$t$-Bu)$_3$] | Orthorhombic | $Pnma$ | 4 | 12.53(1) 15.51(2) 9.604(8) | 90.0 90.0 90.0 | 199 |
| [InSn(O-$t$-Bu)$_3$] | Tetragonal | $P6_{3/m}$ | 2 | 9.867(9) 9.867(9) 11.21(1) | 90.0 90.0 90.0 | 199, 227 |
| [TlGe(O-$t$-Bu)$_3$] | Orthorhombic | $Pnma$ | 4 | 12.361(9) 15.64(1) 9.604(7) | 90.0 90.0 90.0 | 199 |
| [TlSn(O-$t$-Bu)$_3$] (Fig. 14) | Tetragonal | $P6_{3/m}$ | 2 | 9.944(5) 9.944(5) 11.07(1) | 90.0 90.0 90.0 | 199, 226 |
| [TlPb(O-$t$-Bu)$_3$] | Tetragonal | $P6_{3/m}$ | 2 | 9.94(5) 9.94(5) 10.80(9) | 90.0 90.0 90.0 | 199 |
| [Pb$_2$Zr$_4$(O-$i$-Pr)$_{20}$] (Fig. 32) | Monoclinic | $P2_{1/c}$ | 2 | 12.190(6) 14.701(7) 19.978(13) | 90.0 105.37(3) 90.0 | 148 |
| [Pb$_4$Zr$_2$(O-$i$-Pr)$_{16}$] (Fig. 33) | Monoclinic | $P2_{1/c}$ | 2 | 16.996(6) 10.014(3) 24.924(9) | 90.0 105.86(1) 90.0 | 148 |

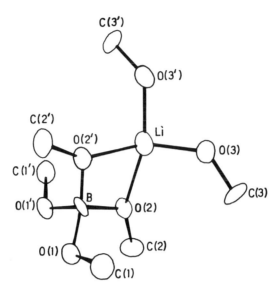

Figure 13.   The crystal structure of $(MeOH)_2Li(\mu\text{-}OMe)_2B(OMe)_2$. [Adapted from (124).]

Veith et al. (142) developed a model with which the formation of cage compounds in metal alkoxides can easily be explained if the following two conditions are satisfied: (1) the metal should be low valent or should bear a low
induced charge; (2) the substituent at the oxygen should be bulky. As expected,
thallium(I) *tert*-butoxide reacts smoothly with tin(II) *tert*-butoxide (141) in a [3
+ 2] cycloaddition (Eq. 88) manner. The distorted trigonal bipyramidal structure (241) of the $SnO_3Tl$ framework of this compound is represented in Fig.
14, which exhibits a one-dimensional cage arrangement in heterobimetallic main
group metal alkoxides.

In the above type of $M(\mu\text{-}O\text{-}t\text{-}Bu)_3M'$ series, the synthesis of $In(O\text{-}t\text{-}Bu)_3Sn$
(141, 227), $Tl(O\text{-}t\text{-}Bu)_3Ge(141, 211)$, $Tl(O\text{-}t\text{-}Bu)_3Pb$ (141, 211), and $In(O\text{-}t\text{-}Bu)_3Ge$ was also achieved (see Eq. 88) and characterized by X-ray crystallographic methods (selected crystal data are given in Table II for brevity).

### 4.   Trinuclear Derivatives

**a.   Homometallic Alkoxides.**   Lead *tert*-butoxide (228), which is prepared
by the amide method (cf. Section II.E), is a trimeric species shown in Fig. 15.

The trimeric structure consists of two three-coordinate lead atoms, a central
six-coordinate lead atom, and six bridging tertiary butoxo ligands. The symmetry of the molecule appears to be $S_6$. However, the absence of a dihedral

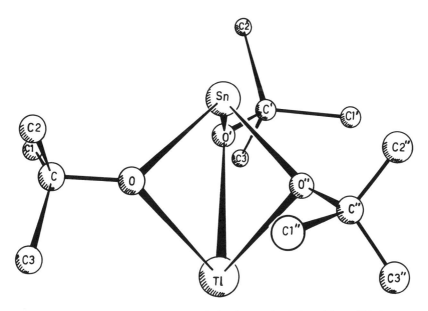

Figure 14. The crystal structure of Tl($\mu$-O-$t$-Bu)$_3$Sn. [Adapted from (226).]

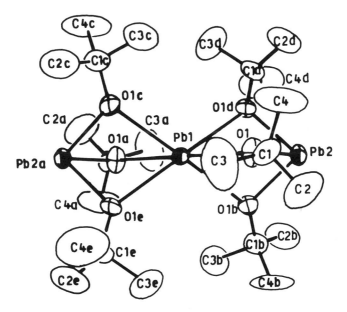

Figure 15. The crystal structure of Pb($\mu$-O-$t$-Bu)$_3$Pb($\mu$-O-$t$-Bu)$_3$Pb. [Adapted from (228).]

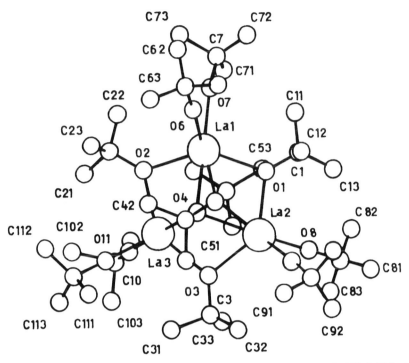

Figure 16.    The crystal structure of [La$_3$($\mu_3$-O-$t$-Bu)$_2$($\mu_2$-O-$t$-Bu)$_3$(O-$t$-Bu)$_4$($t$-BuOH)$_2$]. [Adapted from (103).]

symmetry plane is due to a small torsion angle about the O(1)—C(1) bond. If the methyl groups are omitted, the remaining core has D$_{3d}$ symmetry.

The X-ray crystal structure of [La$_3$(O-$t$-Bu)$_9$($t$-BuOH)$_2$] (Fig. 16) (103) showed that the La$_3$ triangle is capped by two $\mu_3$-O-$t$-Bu groups with three $\mu_2$-O-$t$-Bu bridges. All the lanthanum atoms in Fig. 16 are in distorted octahedral coordination with each metal bonded to two terminal, two $\mu_2$-bridging, and two $\mu_3$-bridging *tert*-butoxide groups. The coordinated alcohol molecules occupy terminal positions on two of the three lanthanum atoms and their presence is revealed by longer La—O bond distances compared with the La—O-$t$-Bu groups.

Another interesting trinuclear derivative is that of the yttrium *tert*-butoxide complex [Y$_3$($\mu_3$-O-$t$-Bu)($\mu_3$-Cl)($\mu_2$-O-$t$-Bu)$_3$(O-$t$-Bu)$_4$(thf)$_2$] (61), as shown in Fig. 17.

The three yttrium atoms consist of a triangle that has $\mu_2$-OCMe$_3$ groups bridging each edge. Above the triangle is a $\mu_3$-Cl ligand and below is a $\mu_3$-

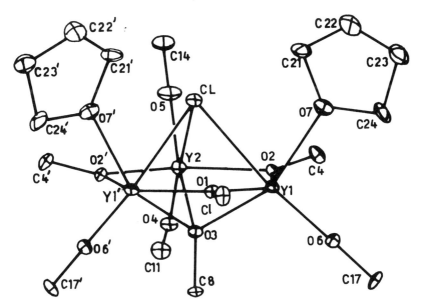

Figure 17.   The crystal structure of $[Y_3(O\text{-}t\text{-Bu})_8Cl(thf)_2]$. [Adapted from (61).]

$OCMe_3$ group. One yttrium atom has two terminal $OCMe_3$ groups; each of the other two yttrium atoms are ligated by one terminal $OCMe_3$ ligand and a molecule of thf. The overall ligand arrangement of this type results in a six-coordinate environment composed of five oxygen-donor atoms and a chloride for each yttrium.

**b. Heterometallic Alkoxides.** These types of derivatives can be divided into two main categories: *closed polyhedra* (triangular units) and *one-dimensional arrangements* (open structures).

*1. Closed Polyhedra.* The crystal and molecular structure of a heterotrinuclear alkoxide $KU_2(O\text{-}t\text{-Bu})_9$ was established by Cotton et al. (152) in 1984 and is shown in Fig. 18, with face-sharing octahedra. The $U_2(O\text{-}t\text{-Bu})_9^-$ anion and the $K^+$ cation form a tight ion pair, with the $K^+$ ion cradled by two bridging and two terminal oxygen atoms at distances of 2.75(2) and 2.93(s) Å.

This structure can be considered in terms of a metal triangle with two $\mu_3$-O-$t$-Bu groups, rather than to create an artificial division into $U_2(O\text{-}t\text{-Bu})_9^-$ and $K^+$ units. The symmetry of this molecule is not threefold, but rather $C_2$; the representation in a triangular form provides a conceptual link to homometallic $M_3(OR)_n$ species.

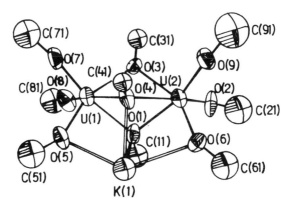

Figure 18.   The crystal structure of $KU_2(O\text{-}t\text{-}Bu)_9$. [Adapted from (152).]

In addition to the above, similar structures for a number of other heterotri-nuclear complexes such as $NaCe_2(O\text{-}t\text{-}Bu)_9$ (64), $NaTh_2(O\text{-}t\text{-}Bu)_9$ (221), and $KZr_2(O\text{-}t\text{-}Bu)_9$ (133) were established by X-ray crystallographic methods.

The single-crystal X-ray structures of $(O\text{-}i\text{-}Pr)_2Al(\mu\text{-}O\text{-}i\text{-}Pr)_2Mg(HO\text{-}i\text{-}Pr)_2(\mu\text{-}O\text{-}i\text{-}Pr)_2Al(O\text{-}i\text{-}Pr)_2$ (229) and $(O\text{-}i\text{-}Pr)_4Nb(\mu\text{-}O\text{-}i\text{-}Pr)_2Ba(HO\text{-}i\text{-}Pr)_2(\mu\text{-}O\text{-}i\text{-}Pr)_2Nb(O\text{-}i\text{-}Pr)_4$ (223) reveal an octahedral environment around Mg, Ba, and Nb atoms, while aluminum is four coordinate.

Some other interesting heterotrinuclear alkoxides with 2 : 1 or 1 : 2 (M : M′) stoichiometry include: $Sr_2Ti(O\text{-}i\text{-}Pr)_8(HO\text{-}i\text{-}Pr)_5$ (4, 230), $(dme)_2Na_2Ce\text{-}(O\text{-}t\text{-}Bu)_6$ (64), and $LaNb_2(O\text{-}i\text{-}Pr)_{13}$ (223). Out of the above, the crystal struc-ture (Fig. 19) of the last compound shows many peculiarities: (a) it consists of

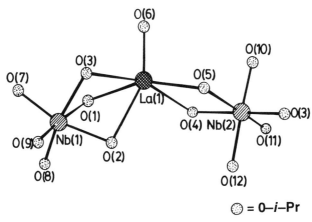

$\circledcirc = O\text{-}i\text{-}Pr$

Figure 19.   The crystal structure of $(O\text{-}i\text{-}Pr)_3Nb(\mu\text{-}O\text{-}i\text{-}Pr)_3La(O\text{-}i\text{-}Pr)(\mu\text{-}O\text{-}i\text{-}Pr)_2Nb(O\text{-}i\text{-}Pr)_4$. [Adapted from (223).]

an asymmetric cyclolinear chain in which the central La atom is in a highly distorted octahedral environment, sharing a face with $Nb(1)O_6$ and an edge with $Nb(2)O_6$ octahedra; (b) the sum of *axial* bond lengths is smaller than for equatorial ones; (c) the bond angles at two axial [O(10) and O(12)] and two equatorial [O(13) and O(11)] oxygen atoms differ significantly [165.1(1) and 135.1(9)°]; [148.8(8) and 166.0(1)°]. The different coordination of bridging O(4) and O(5) atoms, situated in trans positions to the O(13) and O(11) atoms, might be the reason for such a difference in the latter case; and (d) the molecule depicted in Fig. 19 might be considered as one of the possible isomers of the $LaNb_2(O\text{-}i\text{-}Pr)_{13}$ complex and the octahedral coordination of all three metal atoms in this isomer may be a deciding factor for its greater stability.

An interesting tridentate chelation of the $[Zr_2(O\text{-}i\text{-}Pr)_9]^-$ ligand rather than its usual tetradentate ligating behavior (51, 73, 231), which was proposed by us for the complexes of the types $[M\{Zr_2(O\text{-}i\text{-}Pr)_9\}_2]$ (M = Co (232), Cu (233), and $Be\{Zr_2(O\text{-}i\text{-}Pr)_9\}Cl$ (234), was established by X-ray crystallographic studies of $LiZr_2(O\text{-}i\text{-}Pr)_9(HO\text{-}i\text{-}Pr)$ (45) (see Fig. (20)). The lithium is too small to span the distance between O-i-Pr groups terminal on two different zirconium centers; instead, it binds to one terminal and two $\mu_2$-O-i-Pr groups of the [O-i-Pr)$_3$Zr($\mu_2$-O-i-Pr)$_3$Zr(O-i-Pr)$_3$]$^-$ structure. The fourth coordination site on lithium is then occupied by HO-i-Pr, which also hydrogen bonds to the oxygen of a terminal isopropoxo group.

2. *Open Structures (One-Dimensional Arrangement of Metal Atoms).* An interesting "sandwich" type Structure (**VI**) with an $MO_6$ group and trigonal

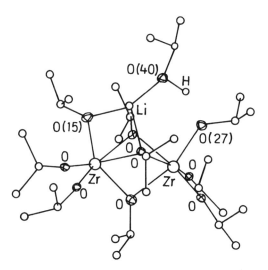

Figure 20.   The crystal structure of $LiZr_2(O\text{-}i\text{-}Pr)_9(HO\text{-}i\text{-}Pr)$. [Adapted from (45).]

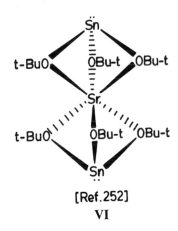

[Ref.252]

**VI**

planar oxygen atoms was elucidated by X-ray diffraction studies (235) for $SrSn_2(O\text{-}t\text{-}Bu)_6$. The barium analogue was also isolated, but calcium is too small to form such a species.

The formation of $Sn(\mu_2\text{-}O\text{-}t\text{-}Bu)_3Sr(\mu_2\text{-}O\text{-}t\text{-}Bu)_3Sn$ can be viewed as two trigonal bipyramids fused together at the common divalent metal corner, which can be either a main group, a transition, or a lanthanide (e.g., Eu) metal.

A similar structure was also established (226) for $MgGe_2(O\text{-}t\text{-}Bu)_6$. The corresponding stannate does not exist; instead a magnesium tin derivative (127) with a dispiro structure (**VII**) was elucidated:

$$
\begin{array}{ccccccc}
 & t\text{-Bu} & & t\text{-Bu} & & t\text{-Bu} & OBu\text{-}t \\
 & O & & O & & O & | \\
\ddot{S}n & & Mg & & Mg & & Sn \cdot\cdot \\
| & & & & & & \\
t\text{-BuO} & O & & O & & O & \\
 & t\text{-Bu} & & t\text{-Bu} & & t\text{-Bu} &
\end{array}
$$

**VII**

In place of Mg(II), other divalent metals such as Ni(II), Co(II), Mn(II), and Cr(II) can be incorporated in this stannate and still maintain the Structure of **VII**.

### 5.  Tetranuclear Derivatives

**a.  Homometallic Alkoxides.** The strong tendency of the monovalent ions of comparable ionic radius [e.g., $K^+(1.44\ \text{Å})$, $Rb^+(1.58\ \text{Å})$, and $Tl^+(1.54\ \text{Å})$] to form alkoxo bridges, is reflected in their preference for triply bridged ($\mu_3$-

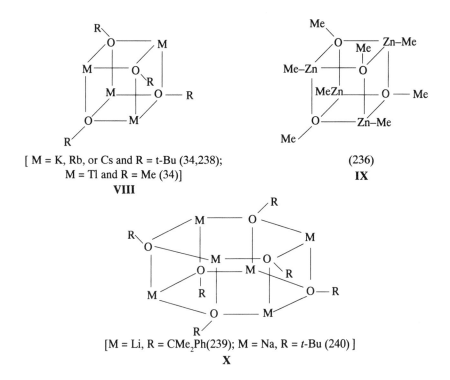

[ M = K, Rb, or Cs and R = t-Bu (34,238);
M = Tl and R = Me (34)]
**VIII**

(236)
**IX**

[M = Li, R = CMe₂Ph(239); M = Na, R = t-Bu (240) ]
**X**

OR) bridges (**VIII** and **IX**) rather than doubly bridged structures (see below). It is also apparent in the structures of $MX^+$ complexes (**X**), where M is a metal or metalloid and X = Cl or an alkyl/alkoxo group (236). Such compact structures are favored by the ability of the metal atoms to utilize more atomic orbitals for bonding purposes (237).

Hexamers (239, 240) and tetramers (34, 238) are common in alkali metal alkoxide chemistry, although with more sterically demanding alkoxo groups, the oligomeric nature becomes lower in the solid state (cf. Section IV.B). Interestingly, [NaO-$t$-Bu]$_n$, crystallizes as two independent molecules in the asymmetric unit: as a hexamer ($n = 6$) and a nonamer ($n = 9$).

Another interesting point is that the barium *tert*-butoxide derivative [Ba(O-$t$-Bu)$_2$(HO-$t$-Bu)$_2$] (106a) adopts a structure quite similar to that of the (alkoxo) alkyl zinc or -magnesium (236). The only difference is that the barium atom being larger in size is additionally coordinated with two molecules of HO-$t$-Bu, in order to satisfy the coordination number 6 for barium atoms. The cubane exhibits (only) intramolecular hydrogen bonding between the two HO-$t$-Bu molecules and the single ⁻O-$t$-Bu ligand on each metal. The remaining *tert*-butoxide groups occupy four corners of the cube.

The arrangement (**VIII** and **IX**) of four metal atoms and four bridging ligands at alternating corners of a cube is a well-precedented unit in inorganic chemistry. More recently, the synthesis and X-ray crystallographic characterization of four cubic alkoxide derivatives of iron and manganese containing the $[M_4(OR)_4]^{n+}$ cores (195), such as $[Fe(OMe)(MeOH)(DPM)]_4$, $[Fe(OMe)(MeOH)(DBM)]_4$, $[Fe^{III}Fe_3^{II}(OMe)_5(MeOH)_3(OBz)_4]$, and $[Mn_4-(OEt)_4(EtOH)_2(DPM)_4]$ are valuable additions to the coordination chemistry of lower valent iron and manganese.

Although copper(I) *tert*-butoxide is a tetramer $[CuO\text{-}t\text{-}Bu]_4$ (241) with alkoxo bridges, it adopts a planar structure, as shown in **XI,** which is presumably favored because it allows copper(I) to have essentially linear O-Cu-O coordination. The Cu· · ·Cu distances lie in the range 2.65–2.77 Å, suggestive of some weak Cu—Cu bonding.

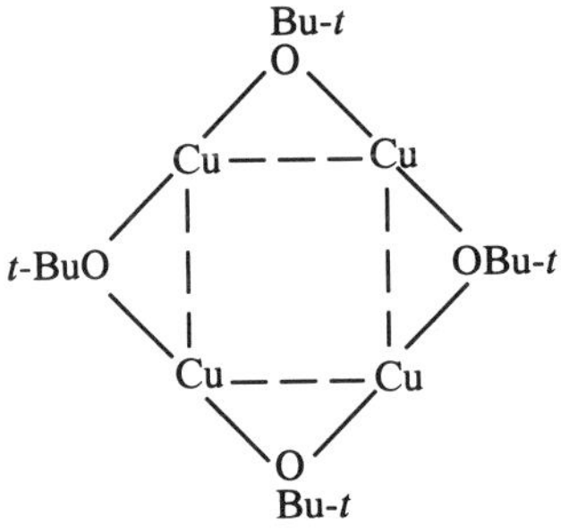

(258)

**XI**

Structure **XII** (M = Ti) for $[Ti(OEt)_4]_4$ was reported by Ibers (242) in 1963 and is of historical significance since it provided the first structural confirmation of Bradley's suggestion for a compound of the general formula $M(OR)_4$. That is, each metal atom achieves an octahedral environment, through the agency of

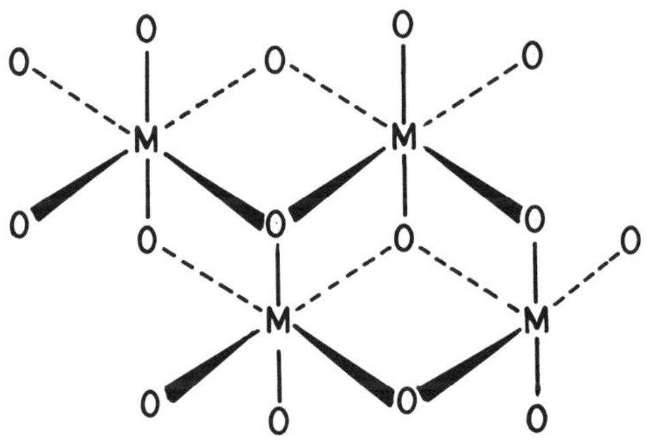

**XII**

four doubly ($\mu_2$) bridging and two triply ($\mu_3$) bridging alkoxo groups. The overall structure involves four fused $MO_6$ octahedral cores, $M_4(\mu_3\text{-}O)_2(\mu_2\text{-}O)_4O_{10}$, wherein the alkyl groups (which are not shown in Structure **XII**) are spread out from the oxygen atoms, resulting in an approximately cylindrical shape.

This type of structure is adopted by a large number of compounds of the formula $M_4(OR)_{16}$ or, in general $M_4X_aY_bZ_c$. In these compounds, the atoms $X = OR$, Y, and Z are anionic (e.g., $O^{2-}$, halide) or neutral (e.g., py or ROH) ligands, and $a + b + c = 16$ (243), such as $[Ti_4(OMe)_{16}]$ (244), $W_4(OEt)_{16}$ (245), and $Mo_4O_8(O\text{-}i\text{-}Pr)_4(py)_4$ (245).

In marked contrast, a phenoxide derivative of strontium, $[Sr(OPh)_2]_4(PhOH)_2(thf)_6$ [another example of the $M_4(OR)_8L_8$ class] (246) adopts the alternative Structure (**XIII**).

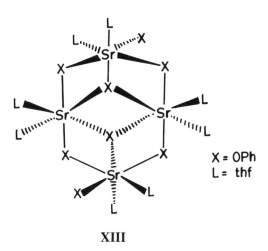

$X = OPh$
$L = thf$

**XIII**

The preference between the cubes **XII** and **XIII** is subtle, since both the structural forms provide coordination number 6 for the concerned metal atoms with approximate octahedral geometry. With the present state of knowledge there appears to be no basis for predicting which one of these two structures would be preferred. However, the compound possesses the generic $M_2M_2'(\mu_3\text{-}X)_2(\mu_2\text{-}X)_4L_n$ stoichiometry and a similar-type structure is also adopted by $Li_2Sn_2(O\text{-}t\text{-}Bu)_6$ [in this case the terminal site on tin(II) is occupied by a lone pair] (247), and $Mo_4O_{10}(OMe)_6^{2-}$ (248). When $n$ (the number of terminal ligands) equals 10, this structure consists of both edge- and face-shared octahedra [cf. Structure **XII**, which is highly suited to a metal such as $Ti^{IV}$ in $Ti_4(OMe)_{16}$ (6)]. The generality of the structural form (4), however, depends in its ability to accommodate a variety of different terminal ligands ($n$ values) and thus metals in oxidation states other than $+4$.

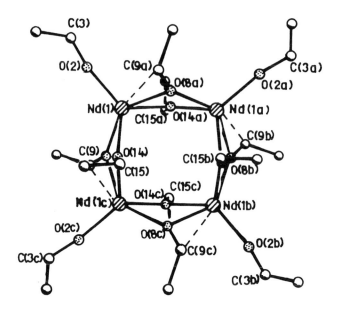

Figure 21.   The crystal structure of $Nd_4(OCH_2\text{-}t\text{-}Bu)_{12}$. [Adapted from (249).]

Another interesting structural form (Fig. 21) has been revealed by X-ray diffraction studies for tetrameric $Ln_4(OCH_2\text{-}t\text{-}Bu)_{12}$ (Ln = La or Nd) (249). Both molecules contain a square of lanthanide metal atoms involving eight $\mu_2$-OR groups: four above and four below the $Ln_4$ plane. However, the neodymium complex contains four molecules of toluene per tetramer within its lattice.

The molecular structure of $Th_4(O\text{-}i\text{-}Pr)_{16}(py)_2$ (Fig. 22) (250) consists of two edge-shared bioctahedra of $[Th_2(O\text{-}i\text{-}Pr)_8(py)]$ units, linked by one axial and one equatorial bridging isopropoxo ligand. The bridging isopropoxo groups that join the halves of the molecule make up the third edge-shared bioctahedral unit. Thorium atoms at each end of the zigzag chain are capped by a pyridine ligand.

**b.   Heterometallic Alkoxides.** The unit **XIII,** with a slight modification as seen in the generic structure **XIV,** is of common occurrence among heterometallic alkoxide derivatives (4) with a 1:1 M/M′ stoichiometry. The unit **XIV** has $C_{2h}$ symmetry, which could not be raised further even if M = M′ (see Structure **XII**), regardless of the values of $p$ and $q$, which represent terminal ligands. The coordination number of M and M′ can be judiciously altered independently. For example, when $p = 3$ and $q = 2$, both metals attain six-coordinate distorted octahedral geometry with the general formula $M_2M_2'(OR)_{16}$ [cf. $M_4(OEt)_{16}$, Ti (242), or W (245)]. However, with $p = 2$, a trigonal bipyramidal geometry can conveniently by adopted by M′, or even a tetrahedron

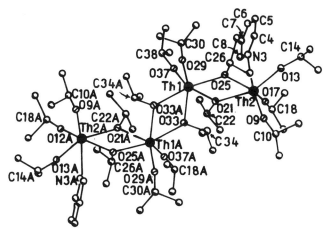

Figure 22. The crystal structure of $Th_4(O\text{-}i\text{-}Pr)_{16}(py)_2$. [Adapted from (250).]

when $p = 1$. The various combinations of $p$ and $q$ can easily conform to the formulas $M_2M_2'(OR)_n$, where $n = 6, 8, 10, 12, 14,$ and $16$.

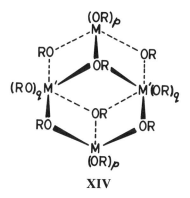

**XIV**

This structural principle is well illustrated by the X-ray crystallographically determined structures of $Li_2Sn_2(O\text{-}t\text{-}Bu)_6$ (247) and $Li_2Ti_2(O\text{-}i\text{-}Pr)_{10}$ (Fig. 23) (131). The lithium atoms are four coordinate, while tin(II) and titanium(IV) centers achieve distorted-pyramidal [with one stereochemically active lone pair for Sn(II)] trigonal bipyramidal coordination geometries, respectively.

Interestingly, the heavier alkali analogues of $LiSn(O\text{-}t\text{-}Bu)_3$ exist as one-dimensional infinite polymers (242, 264) with a common $SnO_3M$ cage (where $M = K$, Rb, and Cs). By contrast, the $M = Tl$ derivative exists as a monomer in the solid state. The polymeric structures for heavier alkali metal derivatives may be attributed to the higher coordination number requirements by the larger

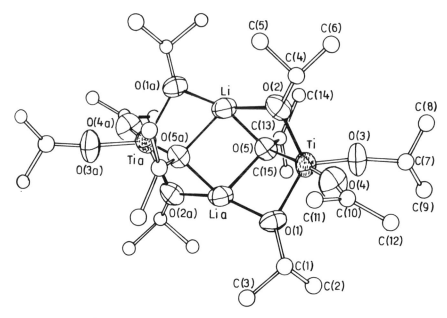

Figure 23.   The crystal structure of [LiTi(O-$i$-Pr)$_5$]$_2$. [Adapted from (131).]

alkali metals, while a lone pair on the Tl(I) appears to be a deciding factor for producing the monomeric derivative Tl($\mu_3$-O-$t$-Bu)$_3$Sn (226).

An interesting heterobimetallic alkoxide derivative containing copper(I) and zirconium, Cu$_2$Zr$_2$(O-$i$-Pr)$_{10}$, was assigned the Structure **XV** crystallographically (248), and consists of a Zr$_2$(O-$i$-Pr)$_9$ face-sharing bioctahedron with two

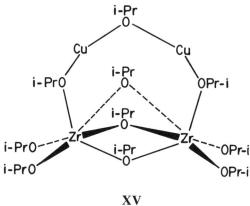

**XV**

$\mu_2$-alkoxide bridging to a $Cu_2^I(O\text{-}i\text{-}Pr)^+$ fragment giving each copper a linear two-coordinate [$\angle\ O-Cu-O = 173.7\%$ (av)] and each zirconium a distorted octahedral environment.

Two tri-*tert*-butoxyzincates, $M_2[Zn(O\text{-}t\text{-}Bu)_3]_2$ (M = Na or K), were prepared (76) and characterized by X-ray crystallography. The centrosymmetric structure (Fig. 24) consists of a $(ZnO\text{-}t\text{-}Bu)_2$ ring with four terminal O-*t*-Bu groups. Sodium metal ions are positioned above·and below the $Zn_2O_2$ plane, each bonded to three oxygen atoms.

The structure of $K_2[Zn(O\text{-}t\text{-}Bu)_3]_2$ is analogous to the sodium derivative shown in Fig. 24 with only small changes occurring as a result of ion size.

The currently determined X-ray structure of a tetranuclear heterobimetallic isopropoxide, $ErAl_3(O\text{-}i\text{-}Pr)_{12}$ (Fig. 25) (251), possesses a noncrystallographically imposed $C_3$ point symmetry, with the rotation axis perpendicular to the plane through the aluminum atoms. The coordination geometry around the erbium atom is a distorted trigonal prism, in contrast to the ideal octahedral coordination for the central aluminum atom in the structure of $AlAl_3(O\text{-}i\text{-}Pr)_{12}$ (see Fig. 2).

Note that the main features of the structure in Fig. 25 are essentially similar to those proposed earlier for most of the tris(tetraisopropoxoaluminates) of lanthanides (6, 196).

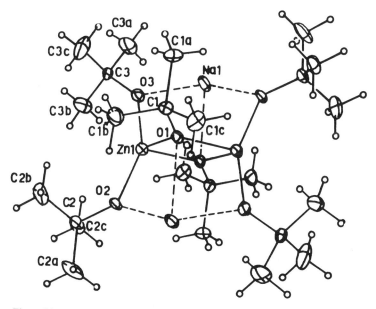

Figure 24. The crystal structure of $Na_2[Zn(O\text{-}t\text{-}Bu)_3]_2$. [Adapted from (76).]

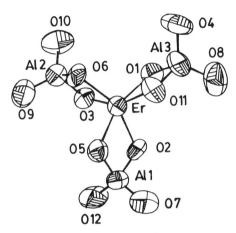

Figure 25.   The crystal structure of $ErAl_3(O\text{-}i\text{-}Pr)_{12}$. [Adapted from (251).]

## 6. Pentanuclear Derivatives

**a. Homometallic Alkoxides.** The X-ray structures of only a few penta-nuclear oxo-alkoxide (see Section V) and aminoalkoxide (see Section IV.C.2) complexes have been reported to date.

**b. Heterometallic Alkoxides.** Note that, despite the excellent progress made during the last decade in the synthesis of various novel types of penta-nuclear derivatives (4a, 28, 38, 39, 252) [e.g., $M\{Zr_2(OR)_9\}_2$ (M = Be, Mg, Ca, Sr, Ba, Zn, Cd, Fe, Co, Ni, or Cu), $M\{Zr_2(OR)_9\}_2Cl$ (M = Y, Ln, Cr, or Fe), and $M_2Al_3(OR)_{13}$] and speculation upon the structure of these complexes, only a few of these predicted structural patterns have been crystallo-graphically established. Bradley (72) reported difficulties in the crystallographic studies of metal alkoxide complexes.

The X-ray structure of $La_2Na_3(OC_6H_4Me\text{-}4)_9(thf)_5$, consists of a trigonal bipyramid of metal atoms (**XVI**) with the lanthanum atoms in the apical positions and the sodium atoms in the equatorial sties (253). A thf ligand is coordinated to each of the five metals. Six of the nine $OC_6H_4Me\text{-}4$ ligands occupy doubly bridging sites and three $OC_6H_4Me\text{-}4$ groups are quadruply bridging. It is rare to find an aryloxide moiety as a $\mu_4$-ligand (4a, 6). The structure of lithium methoxide is one other example in which a $\mu_4$-OR bridge is known (254).

With this arrangement of ligands, each lanthanum is seven coordinate and each sodium is five coordinate. In general, a coordination number of 6 or less is the most common for yttrium, lanthanide alkoxide, and aryloxide complexes (4a, 6). A rare example of an alkoxide complex that contains seven-coordinate yttrium is the cyclic decamer $[Y(OC_2H_4OMe)_3]_{10}$ (see Section V.C.1; Fig. 56).

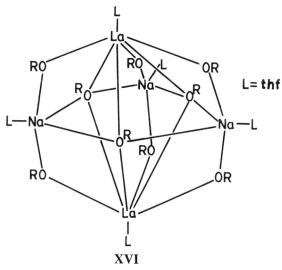

XVI

The molecular structure of an interesting heterobimetallic isopropoxide containing magnesium and aluminum (47), $Mg_2Al_3(O\text{-}i\text{-}Pr)_{13}$, is shown in Fig. 26, and consists of one magnesium in a distorted-trigonal bipyramidal and the other in a tetrahedral environment of isopropoxo groups. The two magnesium atoms are linked by a bridging O-i-Pr group and by an $Al(O\text{-}i\text{-}Pr)_4$ moiety, which is doubly isopropoxo bridged to one magnesium and singly bridged to the second. Figure 26 represents the first crystallographically characterized triply bridging mode for the $Al(O\text{-}i\text{-}Pr)_4^-$ group, which bridges two nonbonded metals.

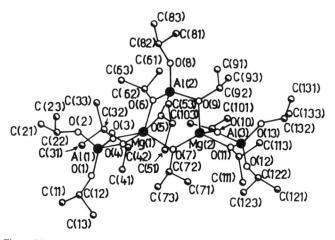

Figure 26.  The crystal structure of $Mg_2Al_3(O\text{-}i\text{-}Pr)_{13}$. [Adapted from (47).]

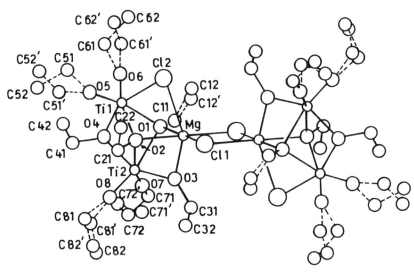

Figure 27. The crystal structure of $[Ti_2(\mu_3\text{-OEt})_2(\mu_2\text{-OEt})_4(\mu\text{-Cl})]_2Mg(\mu_2\text{-Cl})_2$. [Adapted from (255).]

Although X-ray crystallographic data for $Ba\{Zr_2(O\text{-}i\text{-Pr})_9\}_2$ are not acceptably refined (49), it indicate gross skeletal features in which barium is eight coordinate and the two $Zr_2(O\text{-}i\text{-Pr})_9^-$ units are tetradentately bonded to the central barium atom.

An X-ray structural study of an $[Ti_2(OEt)_8Cl]_2Mg_2(\mu\text{-Cl})_2$ complex (255), reveals that the molecule is a centrosymmetric dimer (Fig. 27) with triangular $Ti_2Mg$ units bonded to each other by a double chlorine bridge between the two magnesium atoms. Although all the metal atoms are six coordinate, the titanium atoms are of two different types. The $Ti^1$ atom is bonded to a chlorine atom, while $Ti^2$ is linked only with ethoxo ligands.

### 7. Hexanuclear Derivatives

**a. Homometalic Alkoxides.** Structures of only a few hexanuclear homometallic alkoxide derivatives such as oxo-metal alkoxides (see Section V.B), $[LiOCMe_2Ph]_6$ (239) $[NaO\text{-}t\text{-Bu}]_6$ (240), and $Nd_6(O\text{-}i\text{-Pr})_{17}Cl$ (255) were confirmed by X-ray crystallography.

Structure of the molecule $[LiOCMe_2Ph]_6$ (Fig. 28) has a center of symmetry and exhibits a skeleton related to $[NaO\text{-}t\text{-Bu}]_6$ (240), with the oxygen and lithium atoms alternating at the corners of a hexagonal prism. The oxygen and lithium atoms in the hexameric species are arranged into two near-parallel $Li_3O_3$ planes. There are several features worthy of comment in this structure (Fig.

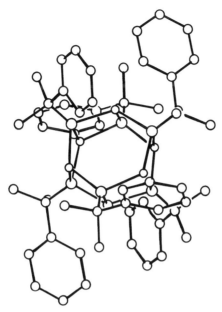

Figure 28.    The crystal structure of [LiOCMe$_2$Ph]$_6$. [Adapted from (239).]

28), for example, the Li—O distances are 1.91(5) Å (av), which is longer compared to the Na—O distances 2.24(1) Å (av) observed in [NaO-$t$-Bu]$_6$ (240). The difference is interpreted for by the larger ionic radii of Na$^+$ with respect to Li$^+$; the lithium atoms in the molecule are three coordinate; the O—Li—O angles within the hexagonal prismatic structure are found to vary over a wide range: 97.75(12)–126.277(12)°; the $\mu_3$-bridging 2-phenyl-2-propoxide ligands are almost equivalent.

An interesting as well as unusual structure (Fig. 29) was reported for the compex Nd$_6$(O-$i$-Pr)$_{17}$Cl (256) in 1978. This structure attested to the complexity possible in lanthanide alkoxide chemistry (28). The structure of the neodymium complex can be regarded as two triangular Nd$_3$($\mu_3$-OR)($\mu_2$-OR)$_3$(OR)$_6$ (where R = O-$i$-Pr) units connected by sharing the $\mu_3$-Cl site (which becomes a $\mu_6$-Cl) and three terminal OR ligands (which become $\mu_2$-OR groups).

**b.   Heterometallic Alkoxides.** X-ray diffraction studies were carried out on a number of hexanuclear heterometallic alkoxides and aryloxides of M$_2$M$_4'$ types, revealing peculiar structural patterns. These include X-ray crystallographic characterization of derivatives as diverse as [Na$_4$Cr$_2$(OPh)$_8$-(thf)$_4$]   (257),   [Na$_2$Gd$_4$O(O-$t$-Bu)$_{12}$]   (258),   [Li$_4$Re$_2$O$_2$Cl$_2$(O-$i$-Pr)$_{10}$(thf)$_2$] (259), [Pr$_2$Al$_4$Cl$_2$(O-$i$-Pr)$_{16}$(HO-$i$-Pr)$_2$] (74), [Ba$_2$Zr$_4$(O-$i$-Pr)$_{20}$] (149), [K$_4$Zr$_2$O-

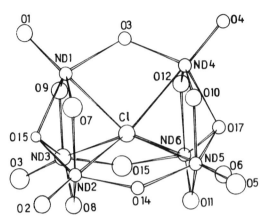

Figure 29. The ORTEP drawing of the Nd, Cl, and O framework in $Nd_6(O\text{-}i\text{-}Pr)_{17}Cl$. [Adapted from (256).]

$(O\text{-}i\text{-}Pr)_{10}]$ (117), $[Pb_4Zr_2(O\text{-}i\text{-}Pr)_{16}]$ (148), and $[Pb_2Zr_4(O\text{-}i\text{-}Pr)_{20}]$ (148). Out of these, structures of oxo-alkoxide complexes will be discussed in Section V.C, whereas the X-ray structures of only a few typical derivatives will be presented below.

For example, a new structural type in the heterometallic alkoxide chemistry of lanthanides was crystallographically established in the case of the complex $[\{Pr[Al(O\text{-}i\text{-}Pr)_4]_2(\mu\text{-}Cl)(HO\text{-}i\text{-}Pr)\}_2]$ (Fig. 30) (74). The molecule is a centro-symmetric dimer composed of two triangular $[Al_2Pr(\mu\text{-}O\text{-}i\text{-}Pr)_4(O\text{-}i\text{-}Pr)_4(HO\text{-}i\text{-}Pr)]^+$ units linked by two chloride bridges, with praseodymium being hepta-coordinated. The geometry around a praseodymium atom may preferably be envisaged as a distorted capped trigonal prims ($C_{2v}$).

A similar chloride bridged structure (Fig. 31) is adopted by $Cd_2Zr_4Cl_2(O\text{-}i\text{-}Pr)_{18}$. The centrosymmetric dimer consists of two triangular $[CdZr_2(O\text{-}i\text{-}Pr)_9]^+$ units. These units are formed by a $[Zr_2(\mu_2\text{-}O\text{-}i\text{-}Pr)_3(O\text{-}i\text{-}Pr)_6]^-$ face-shared bioc-tahedral moiety bonded to $Cd^{2+}$ via its two $\mu_2$-isopropoxo groups and one ter-minal isopropoxo ligand from each zirconium, and is linked by the cadmium atoms via two chlorine bridges (73).

The crystal structures of two heterobimetallic isopropoxides containing lead and zirconium (148), $Pb_2Zr_4(O\text{-}i\text{-}Pr)_{20}$ (Fig. 32) and $Pb_4Zr_2(O\text{-}i\text{-}Pr)_{16}$ (Fig. 33), were determined by X-ray diffraction methods. These molecules adopt rare ser-pentine structures.

The structure of $Pb_2Zr_4(O\text{-}i\text{-}Pr)_{20}$ (Fig. 32) consists of a central (Pr-$i$-O)$Pb(\mu_2\text{-}O\text{-}i\text{-}Pr)_2Pb(O\text{-}i\text{-}Pr)$ fragment each of whose terminal isopropoxo group bridges to a different $Zr_2(O\text{-}i\text{-}Pr)_8$ unit, thereby creating a $Zr_2(O\text{-}i\text{-}Pr)_9^-$ moiety

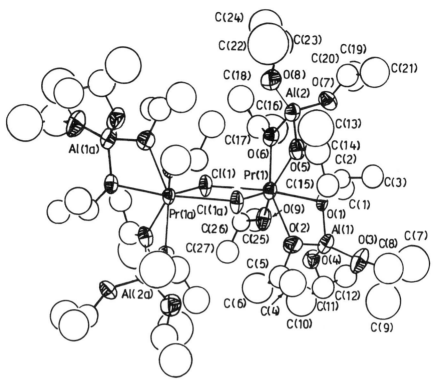

Figure 30. The crystal structure of $[\{Pr[Al(O\text{-}i\text{-}Pr)_4]_2(\mu\text{-}Cl)(HO\text{-}i\text{-}Pr)\}_2]$. [Adapted from (74).]

of face-shared bioctahedral structure. The six metals are slightly distorted from planarity.

Another interesting complex, $Pb_4Zr_2(O\text{-}i\text{-}Pr)_{16}$ (Fig. 33) (148), contains a central $(Pr\text{-}i\text{-}O)Pb(\mu_2\text{-}O\text{-}i\text{-}Pr)_2Pb(O\text{-}i\text{-}Pr)$ unit bonded to two $Pb(\mu_2\text{-}O\text{-}i\text{-}Pr)_3Zr(O\text{-}i\text{-}Pr)_3$ units. The six metals are nearly planar, and each terminal isopropoxo group of the $(Pr\text{-}i\text{-}O)Pb(\mu_2\text{-}O\text{-}i\text{-}Pr)_2Pb(O\text{-}i\text{-}Pr)$ unit binds to a distinct $PbZr(O\text{-}i\text{-}Pr)_6$ unit. The coordination number of $Pb_2$ is 4, and the overall geometry of this lead atom is "sawhorse" or pseudotrigonal bipyramidal, with an (unseen) equatorial lead lone pair.

## 8. Polynuclear Derivatives

In addition to what was already discussed in the preceding sections, great diversification in the structural features of homo- and heterometallic alkoxide complexes were revealed by X-ray crystallography for many poly- (octa-,

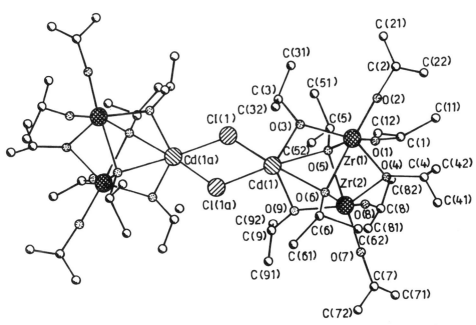

Figure 31.　The crystal structure (ORTEP drawing) of [{Cd[Zr₂(O-*i*-Pr)₉](μ-Cl)}₂]. [Adapted from (73).]

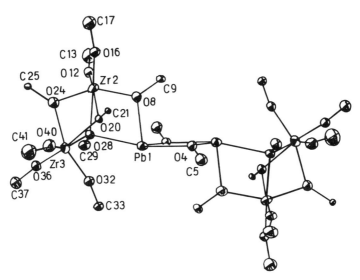

Figure 32.　The crystal structure of Pb₂Zr₄(O-*i*-Pr)₂₀. Methyl groups are omitted for clarity. [Adapted from (148).]

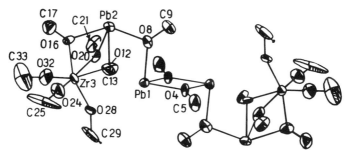

Figure 33. The crystal structure of $Pb_4Zr_2(O\text{-}i\text{-}Pr)_{16}$. Methyl groups are omitted for clarity. [Adapted from (148).]

nona-, deca-, and higher) nuclear complexes. Such unique structural patterns for polynuclear complexes will be dealt with in Sections IV.C.1 and IV.C.2, V. Herein, a few interesting structures are shown in Figs. 34–37, and are briefly discussed.

**a. Homometallic Alkoxides.** The most remarkable among the higher nuclear homometallic alkoxides is that of sodium *tert*-butoxide, which crystallizes with two independent molecules in the asymmetric unit as a hexamer (**IX**) and a nonamer (240).

An interesting circular decanuclear iron(III) methoxide derivative, $[Fe(OMe)_2(O_2CCH_2Cl)]_{10}$, a molecular ferric wheel (Fig. 34) with bridging methoxide and carboxylate ligands, is obtained when basic iron chloroacetate reacts with ferric nitrate in methanol. The X-ray crystallographic study showed this compound to have idealized $D_{5d}$ symmetry and consist of a 20-membered ring comprised of 10 ferric ions linked by 20 bridging methoxide and 10 bridging chloroacetate ligands. The 10 iron atoms are nearly coplanar and attain a distorted octahedral environment of 6 oxygen donor atoms (260).

**b. Heterometallic Alkoxides.** The X-ray structure of a potassium aluminum isopropoxide complex (126) shows a polymer of composition $[(HO\text{-}i\text{-}Pr)_2K(\mu\text{-}O\text{-}i\text{-}Pr)_2Al(\mu\text{-}O\text{-}i\text{-}Pr)_2]_{\infty}$ (Fig. 35) in which chains of alternating K and Al atoms are linked by double isopropoxo bridges, representing a rare example in which all four isopropoxides of an $[Al(O\text{-}i\text{-}Pr)_4]^-$ group are coordinated by second metals. The distorted octahedral coordination sphere about potassium is completed by two molecules of isopropyl alcohol.

The congeners of $[LiTi(O\text{-}i\text{-}Pr)_5]_2$ (cf. Fig. 23) were identified by X-ray crystallographic studies as $[NaTi(O\text{-}i\text{-}Pr)_5]_{\infty}$ and $[KTi(O\text{-}i\text{-}Pr)_5]_{\infty}$ (184). The derivative $[NaTi(O\text{-}i\text{-}Pr)_5]_{\infty}$ (Fig. 36) consists of an infinite linear chain of alternating distorted trigonal bipyramidal titanium isopropoxide moieties and bridging highly distorted tetrahedral sodium atoms $[O(5)-Na-O(4) = 66.9°]$.

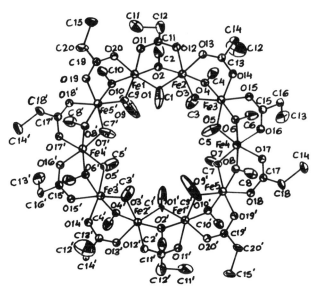

Figure 34.    The crystal structure (an ORTEP view) of $[Fe(OMe)_2(O_2CCH_2Cl)]_{10}$. [Adapted from (260).]

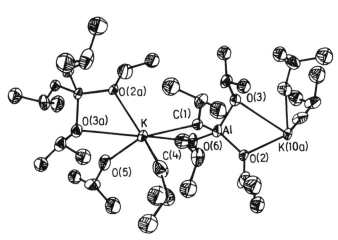

Figure 35.    The crystal structure representing a portion of the $[HO\text{-}i\text{-}Pr)_2K(\mu\text{-}O\text{-}i\text{-}Pr)_2Al(\mu\text{-}O\text{-}i\text{-}Pr)_2]_\infty$ polymer. [Adapted from (126).]

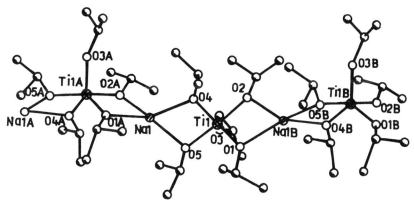

Figure 36. The crystal structure (ball-and-stick drawing) of $[NaTi(O\text{-}i\text{-}Pr)_5]_\infty$. [Adapted from (184).]

The titanium atoms show little distortion from ideal TBP geometry with angles ranging from 87.5(4)° to 174.2(3)° (four $\mu$-O-$i$-Pr and one terminal O-$i$-Pr ligand).

The potassium analogue $[KTi(O\text{-}i\text{-}Pr)_5]_\infty$ is structurally similar to Fig. 36, however, the infinite alternating chain is nonlinear. The potassium atoms are five coordinated and appear to adopt an intermediate geometry between square base prismatic and trigonal bipyramidal. The potassium atoms interact with four of the five isopropoxides of titanium using $\mu$-O-$i$-Pr and $\mu_3$-O-$i$-Pr ligands.

An interesting octanuclear complex, the first X-ray crystallographically established heterotrimetallic alkoxide having the structure $[\{Zr_2(O\text{-}i\text{-}Pr)_9\}Ba(\mu\text{-}O\text{-}i\text{-}Pr)_2Cd(\mu\text{-}O\text{-}i\text{-}Pr)]_2$ possessing a crystallographic center of symmetry (Fig. 37) (39a), is obtained from $ICd\{Zr_2(O\text{-}i\text{-}Pr)_9\}$ and $KBa(O\text{-}i\text{-}Pr)_3$. This molecule consists of two $[Cd(\mu_2\text{-}O\text{-}i\text{-}Pr)_2BaZr_2(\mu_3\text{-}O\text{-}i\text{-}Pr)_2(\mu_2\text{-}O\text{-}i\text{-}Pr)_3(O\text{-}i\text{-}Pr)_4]^+$ units linked together by two isopropoxo groups bridging the cadmium atoms. All barium and zirconium atoms are six coordinated but are in highly distorted octahedral geometry.

## C. Chemical Reactivity

The bond $M^{\delta+}\text{—}O^{\delta+}\text{—}R$ in homo- as well as heterometallic alkoxides is highly reactive. This reactivity is due to the positive charge induced on the metal atoms rendering them highly prone to nucleophilic attack. Almost all of the alkoxides are, therefore, extremely susceptible to hydrolysis by even traces of moisture (including the atmosphere, reaction vessels, etc.) and hence, their synthesis and handling require extraordinarily anhydrous conditions or environment. The facile hydrolyzability of homo- and heterometal alkoxides is now

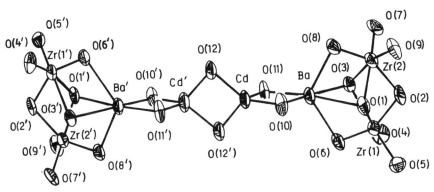

Figure 37.  The crystal structure of $Ba_2Cd_2Zr_4(O\text{-}i\text{-}Pr)_{24}$. [Adapted from (39a).]

being utilized extensively for the preparation of oxide-ceramic materials by the sol–gel process (cf. Section VII).

Metal alkoxides react readily with an excess of hydrogen halides or acyl halides yielding the metal halides. However, by using stoichiometric quantities of these halides the metal alkoxide halide derivatives can be synthesized as desired. The alkoxides also react readily with a large number of protic reagents (R′OH), for example, alcohols, phenols, glycols, carboxylic acids, hydroxy acids, $\beta$-diketones, and alkanolamines, containing reactive hydroxyl groups with the replacement of the alkoxyl groups by the new organic ligand OR′, namely,

$$M(OR)_n + x\text{HOR}' \longrightarrow M(OR')_x(OR)_{n-x} + x\text{ROH} \uparrow \tag{96}$$

Fully substituted derivatives $M(OR')_n$, with an excess of the reagent, can be obtained. However, by using the ethoxide/isopropoxide of the metal in a solvent such as benzene, which forms a lower boiling azeotrope with the liberated alcohol, the reaction can be carried out in any desired stoichiometric ratio by fractioning out the alcohol azeotropically and isolation of a large variety of mixed-alkoxide derivatives, $M(OR)_{n-x}(OR')_x$. Some of these derivatives can be used as synthons for the preparation of a wide variety of interesting derivatives, for example, $M(OR')_x(OR'')_{n-x}$, where R″OH is also another higher alcohol or one of the hydroxy reagents of the type listed as R′OH.

$$M(OR)_{n-x}(OR')_x + (n-x)\text{HOR}'' \longrightarrow$$

$$M(OR')_x(OR'')_{n-x} + (n-x)\text{ROH} \uparrow \tag{97}$$

Depending on the nature of the metal M, the alkoxides are sometimes reactive toward other reagents having -NH or -SH groups. In these cases, the reactions are controlled by thermodynamic factors and appear to be governed by the comparative stability of M—O, M—N, and M—S bonds. For example, the -NH group of alkanolamines or even higher amines are reactive toward alkoxides of a softer (group B) metal such as tin, but not with zirconium analogues, which is a harder (group A) metal.

Metal alkoxides undergo insertion reactions across the M—O bonds (6) when treated with unsaturated substrates such as isocyanates, isothiocyanates, aldehydes, ketones, ketenes as well as with carbon dioxide, carbon disulfide, and sulfur dioxide.

The reactions between alkoxides of different metals between themselves and sometimes with metal halides was found to yield homoleptic or heteroleptic chloro heterobimetallic alkoxide under suitable condition (cf., Section II.C.2).

An extensive account of all of these types of reactions with their possibilities and limitations is already available in a number of books and reviews (1–7, 34, 38, 39). Only a classification of the reactions (6) among different types is, therefore, being reproduced at this stage, followed by a few more recent examples:

1. Hydrolysis reactions.
2. Reactions with hydrogen halides and acyl halides.
3. Reactions with alcohols and silanols.
4. Reactions with organic esters and silyl esters.
5. Reactions with glycols.
6. Reactions with organic acids and acid anhydrides.
7. Reactions with $\beta$-diketones and $\beta$-ketoesters.
8. Reactions with $\beta$-ketoamines and Schiff bases.
9. Reactions with alkanolamines.
10. Reactions with oximes and hydroxylamines.
11. Reactions with thiols.
12. Tischtschenko and Meerwein–Ponndorf–Verley reactions.
13. Adduct formation with coordinating ligands.
14. Insertion reactions with unsaturated substrates.
15. Reactions with other metal alkoxides and chlorides.

For example, the potential of a type 6 reaction (146) was suggested (260, 261) for the preparation of a variety of ferric dialkoxide carboxylates by the facile route of Eq. 98.

$$\text{Fe(OR)}_3 + \text{R'COOH} \longrightarrow \text{Fe(OR)}_2(\text{OOCR'}) + \text{ROH} \uparrow \qquad (98)$$

This reaction followed the highly fascinating "molecular wheel" type structure exhibited (260) by $[\text{Fe(OMe)}_2(\text{OOC}\cdot\text{CH}_2\text{Cl})]_{10}$, which was crystallized from a reaction mixture of basic iron acetate $[Fe_3O(O_2C \cdot CH_2Cl)_6 \cdot H_2O]NO_3$ (0.366 mmol) and $Fe(NO_3)_3 \cdot 9H_2O$ (1.10 mmol) in methanol (65 mL).

Another example of derivatives with interesting structures formed in reactions 7 (147) of aluminum alkoxides with $\beta$-diketones/$\beta$-ketoesters (262) may be cited.

$$Al(OR)_3 + 3H\beta-dik \longrightarrow Al(\beta-dik)_3 + 3ROH \uparrow \qquad (99)$$

$$2Al(OR)_3 + 4H\beta-dik \longrightarrow$$

$$[Al(OR)(\beta-dik)_2]_2 + 4ROH \uparrow \qquad (100)$$

$$2Al(OR)_3 + 2HO\beta-dik \longrightarrow$$

$$[Al(OR)_2(\beta-dik)]_2 + 2ROH \uparrow \qquad (101)$$

It was shown (262) that whereas the tris-$\beta$-diketonates are monomeric with hexacoordinated aluminum, the monoalkoxide bis-$\beta$-diketonates dimerize with alkoxide bridges and can be represented by $[(\beta-dik)_2Al(\mu-OR)_2Al(\beta-dik)_2]$, in which both aluminum atoms are hexacoordinated. The dialkoxide mono-$\beta$-diketonate, instead of adopting a symmetric structure $[(RO)(\beta-dik)Al(\mu-OR)_2Al(\beta-dik)(OR)]$ with both aluminum atoms in the five-coordinate state shows a structure that can be represented by the unsymmetric $[(\beta-dik)_2Al(\mu-OR)_2Al(OR)_2]$, in which one of the aluminum atoms is six coordinate while the other is tetra-coordinate. Although higher (74) coordination for aluminum was reported in a number of glycol derivatives (263), the preferential formation of $[\beta-dik)_2Al(\mu-O-i-Pr)_2Al(O-i-Pr)_2]$ may be considered to arise from the stability of $\{Al(O-i-Pr)_4\}^-$ moiety, a very large number of derivatives of which were already described (4–8, 14–20, 25–30, 38, 39).

In a number of publications (264), it was shown that the terminal isopropoxide groups of $[(acac)_2Al(\mu-O-i-Pr)_2Al(O-i-Pr)_2]$ can be replaced by reactions in requisite molar ratio with a variety of reagents such as different glycols and mercaptoethanol, yielding interesting products, in which the rest of the molecule $[(acac)_2Al(\mu-O-i-Pr)_2Al\diagdown]$ remains intact. This type of interesting reactivity, where the terminal alkoxide group does not disturb the rest of the bridged structure was already reported (18) in the reaction of $[Ln\{(\mu-O-i-Pr)_2Al(O-i-Pr)_2\}_3]$ with acetylacetone in a 1:6 molar ratio, yielding monomeric volatile products, $[Ln(\mu-O-i-Pr)_2Al(acac)_2]$.

In addition to the well-established chemistry of the M—O bond in M—O—C derivatives (4, 6, 7, 14, 28–32, 38, 39), the chemical reactivity of the C—O bond in aryloxide and related derivatives of mainly rhenium and tungsten (43) was recently highlighted by Mayer (164). Although the main focus of this contribution is the physicochemical and structural characteristics as well as applications of homo- and heterometal alkoxides of main group and the earlier transition metals, attention is drawn to the following possible pathways for the formation of oxo-alkoxides (cf. Section V), which are gaining prominence in the sol–gel processes.

$$\overset{\overset{\text{O}}{\|}}{L_n M^{n+}} + R^+ \;\rightleftharpoons\; \overset{\overset{\text{OR}}{|}}{L_m M^{n+}} \tag{102}$$

$$\overset{\overset{\text{O}}{\|}}{L_n M^{n+}} + R^* \;\rightleftharpoons\; \overset{\overset{\text{OR}}{|}}{L_m M^{(n-1)+}} \tag{103}$$

$$\overset{\overset{\text{O}}{\|}}{L_n M} + R' \;\rightleftharpoons\; \overset{\overset{\text{OR}}{|}}{L_m M^{(n-2)+}} \tag{104}$$

Similarly, reference is made to another more recent review (41) of the author's contributions to the chemistry of rhodium, iridium, rhenium, osmium and ruthenium alkoxo, and hydroxo complexes, following a similar earlier publication (40) on the alkoxo derivatives of platinum metals. Reference may also be made to the first X-ray structural study (265, 266) of a siloxy derivative, $[(\text{cod})\text{Rh}(\mu\text{-OSiMe})_3]_2$.

## IV. METAL ALKOXIDES DERIVED FROM SOME SPECIAL ALCOHOLS

The main purpose of synthesizing metal derivatives of special ligands (e.g., fluoro-, alkoxy-, amino-, sterically demanding-, and during the last 10–15 years, alcohols) was to enhance solubility (in organic solvents), volatility, and to reduce the oligomeric nature of the resulting product. These properties make them suitable precursors for oxide-ceramic materials by MOCVD/sol–gel processes (6, 23, 24, 27, 31, 38, 39, 72).

This section will update and present the current status of our knowledge in the title area, with emphasis on the structural features (as elucidated by X-ray crystallography) of the preceding types of metal complexes.

### A. Derivatives of More Sterically Demanding Alcohols

Volatility, solubility, and stereochemistry of metal alkoxides, $[M(OR)_n]_x$, tend to be limited both by their oligomeric nature and the steric demand of the alkoxide moiety, as a consequence of the formation of $\mu_2$-OR/$\mu_3$-OR bridges

(6) to achieve coordinative saturation in the most economic way (147). Tertiary alkoxide ligands (e.g., $^-$O-$t$-Bu, $^-$OC(Et)Me$_2$, and $^-$OCEt$_3$) were often used to prevent oligomerization (6) by steric hindrance, but these become less effective with lower (bi- and tri-) valent metals (6), for which more sterically crowded alkoxo (102, 235, 281–291) and aryloxo (28, 38, 39) ligands are required. Even in cases with divalent metal ions, such as cobalt(II), the more sterically demanding di-$tert$-butylmethoxide ligand, $^-$OCH-$t$-Bu$_2$, yields octahedral cobalt(II) derivative. Obviously, in these cases, the highly sterically hindered alkoxo ligand, $^-$OC-$t$-Bu$_3$, is expected to fulfill both objectives: the enhancement in the volatility and attainment of low-coordination geometry by the concerned metal atom. However, in many cases, very interesting features often arise either from the ease of elimination of isobutylene (102) or the formation of heteroleptic derivatives (281, 287, 289–292), previously unknown in metal alkoxide chemistry, that are primarily due to the bulk of the ligand.

The compounds containing more sterically demanding alkoxide ligands have exhibited an extraordinary variety of structures (see Figs. 38–46, and Table III), with the preparation of not only the new complexes, but of new types of derivatives being a commonplace event during the last decade. Many of the types of compounds discovered contain forms of bonding previously unknown in metal alkoxide (1, 6, 7, 34) and aryloxide (293, 294) chemistry.

## 1. Synthesis and Properties

Out of a number of preparative routes described in Section II for preparing metal alkoxides, those involving (a) the metal–carbon bond cleavage reaction (Section II.F), (b) the chloride–alkoxide exchange reaction (Section II.C.1), (c) the amido–alkoxo exchange reaction (Section II.E), and (d) the alcoholysis reaction (Section II.D) appear to be more convenient and versatile. Schemes 1–6 summarize some of the typical reactions studied for the preparation of metal complexes of alcohol ligands with large steric requirements.

A perusal of the synthetic Schemes 1–6 indicates the vital role played by the steric bulk of an alkoxide and other ancillary ligands in providing access not only to new compounds but to new types of structurally interesting complexes as well (Table III, Figs. 38–46).

$$t\text{-Bu}_2\text{-CHOH} + \text{LiBu} \xrightarrow{\text{hexane}} \text{LiOCH-}t\text{-Bu}_2 \ (\text{Ref.285}) + \text{C}_4\text{H}_{10}$$

$$t\text{-Bu}_3\text{-COH} + \text{LiBu} \xrightarrow{\text{hexane}} \text{LiOC-}t\text{-Bu}_3 \ (\text{Refs.295,296}) + \text{C}_4\text{H}_{10}$$

$$t\text{-Bu}_3\text{-COH} + \text{LiBu} \xrightarrow[-(\text{C}_4\text{H}_{10})]{\text{hexane/THF}} [\{\text{Li}(\mu\text{-OC-}t\text{-Bu}_3)(\text{thf})\}_2] \ (\text{Ref.281})$$

Scheme 1.  Preparation of some lithium alkoxide precursor complexes.

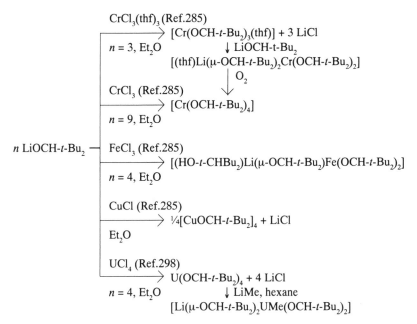

Scheme 2.   Preparation of some di-*tert*-butylmethoxide derivatives of metals via chloride–alkoxide exchange reactions.

Furthermore, the reduced intermolecular associative tendency of more sterically congested metal alkoxides has provided an opportunity for the preparation of Lewis acid–base type molecular addition complexes, such as $Nd_2[OCH-i-Pr_2)_6L_2$ (L = thf or py) (282), $[Nd_2(OCH-i-Pr_2)_6(\mu\text{-dme})]_\infty$ (282), $Nd(OC-t-Bu_3)_3(thf)$ (290), $Th(OCH-i-Pr_2)_4(dme)$ (299), $Th(OCH-i-Pr_2)_4(quinuclidine)$ (299), $Cr(OC-t-Bu_3)_2(thf)_2$ (281), and $Co(OCPh_3)_2(thf)_2$ (300).

The IR (281–300) and NMR (102, 212, 235, 282, 288, 290, 292, 295, 296, 299) spectral data are usually inconclusive for actual structural elucidation, except in providing preliminary information regarding the presence of alkoxide groups and other inorganic and organic moieties and their different environments. However, in some cases, NMR spectra are capable of providing useful information about the behavior of a metal alkoxide in solution. For example, $^1$H NMR spectra (at ambient temperature) of $[Th(OCH-i-Pr_2)_4]_2$ (299) are consistent with the presence of two species (dimer and monomer) in solution.

$n$ LiOC-$t$-Bu$_3$ —

$\xrightarrow[n = 1, Et_2O]{\text{TiCl}_4 \text{ (Ref.295)}}$ (Bu$_3$-$t$-CO)TiCl$_3$ + LiCl

$\xrightarrow[n = 2, Et_2O]{\text{TiCl}_4 \text{ (Ref.295)}}$ (Bu$_3$-$t$-CO)$_2$TiCl$_2$ + 2 LiCl

$\xrightarrow[n = 2, Et_2O]{\text{TiCl}_3 \text{ (Ref.295)}}$ "(Bu$_3$-$t$-CO)$_2$TiCl" + 2 LiCl

$\xrightarrow[n = 2, Et_2O]{\text{ZrCl}_4 \text{ (Ref.295)}}$ [(Bu$_3$-$t$-CO)$_2$ZrCl$_3$.Li(OEt$_2$)$_2$] + LiCl
$\quad$ hexane ↓ -(2Et$_2$O + LiCl)
$\quad\quad\quad$ (Bu$_3$-$t$-CO)$_2$ZrCl$_2$

$\xrightarrow[n = 2, Et_2O]{\text{CrCl}_3 \text{ (Ref.281)}}$ [(thf)$_2$Li($\mu$-Cl)($\mu$-OC-$t$-Bu$_3$)Cr(OC-$t$-Bu$_3$)]

$\xrightarrow[n = 2, \text{hexane/THF/Et}_2O]{\text{CoCl}_2 \text{ (Ref.289)}}$ [(thf)$_3$Li($\mu$-Cl)Co(OC-$t$-Bu$_3$)$_2$]

$\xrightarrow[n = 2, \text{THF}]{\text{UCl}_4 \text{(Ref.297)}}$ (Bu$_3$-$t$-CO)$_2$UCl$_2$(thf)$_2$ + 2 LiCl

Scheme 3.   Preparation of some tri-*tert*-butylmethoxide (tritox) derivatives of metals via chloride–alkoxide exchange reactions.

## 2.  Structures

During the last decade, considerable progress was made in the structural determinations by X-ray crystallography (Table III) of more sterically hindered alkoxide derivatives of metals. This rapid expansion is traced to the realization that (a) the course of many reactions with sterically more demanding alkoxide ligands are often unpredictable, (b) compounds in low-coordination states can be kinetically stabilized with more bulky ligands, and (c) in many cases, unambiguous structural elucidation of the final products are only possible with the help of X-ray diffraction crystallography. Next, a few typical structures will be illustrated in Figs. 38–46, mainly to highlight the potential of such bulky ligands in generating novel structures and interesting ligating modes.

Although alkali metal alkoxides are generally oligomers (see Sections III.B.4

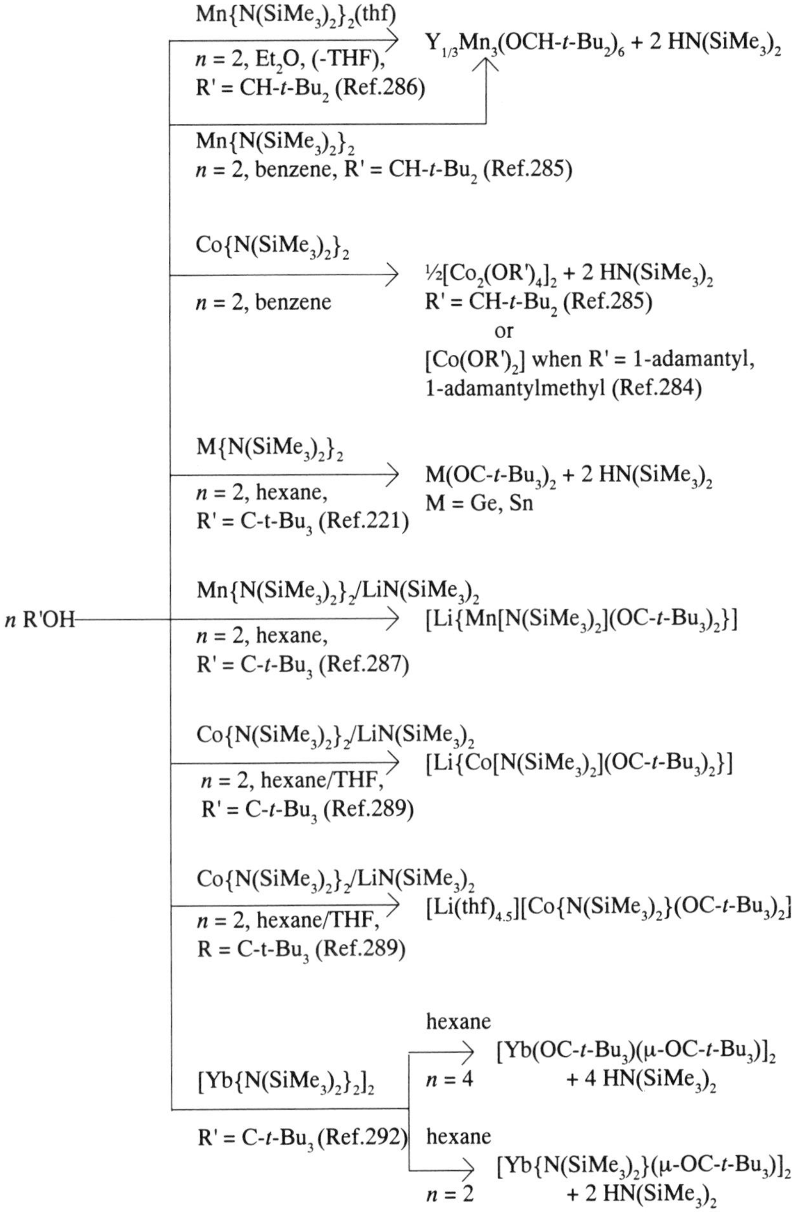

$n$ R'OH ⟶

$\dfrac{Mn\{N(SiMe_3)_2\}_2(thf)}{n = 2, Et_2O, (-THF),\ R' = CH\text{-}t\text{-}Bu_2\ (Ref.286)} \longrightarrow$  Y$_{1/3}$Mn$_3$(OCH-$t$-Bu$_2$)$_6$ + 2 HN(SiMe$_3$)$_2$

$Mn\{N(SiMe_3)_2\}_2$
$n = 2$, benzene, R' = CH-$t$-Bu$_2$ (Ref.285)

$\dfrac{Co\{N(SiMe_3)_2\}_2}{n = 2,\ benzene} \longrightarrow$  ½[Co$_2$(OR')$_4$]$_2$ + 2 HN(SiMe$_3$)$_2$
R' = CH-$t$-Bu$_2$ (Ref.285)
or
[Co(OR')$_2$] when R' = 1-adamantyl, 1-adamantylmethyl (Ref.284)

$\dfrac{M\{N(SiMe_3)_2\}_2}{n = 2,\ hexane,\ R' = C\text{-}t\text{-}Bu_3\ (Ref.221)} \longrightarrow$  M(OC-$t$-Bu$_3$)$_2$ + 2 HN(SiMe$_3$)$_2$
M = Ge, Sn

$\dfrac{Mn\{N(SiMe_3)_2\}_2/LiN(SiMe_3)_2}{n = 2,\ hexane,\ R' = C\text{-}t\text{-}Bu_3\ (Ref.287)} \longrightarrow$  [Li{Mn[N(SiMe$_3$)$_2$](OC-$t$-Bu$_3$)$_2$}]

$\dfrac{Co\{N(SiMe_3)_2\}_2/LiN(SiMe_3)_2}{n = 2,\ hexane/THF,\ R' = C\text{-}t\text{-}Bu_3\ (Ref.289)} \longrightarrow$  [Li{Co[N(SiMe$_3$)$_2$](OC-$t$-Bu$_3$)$_2$}]

$\dfrac{Co\{N(SiMe_3)_2\}_2/LiN(SiMe_3)_2}{n = 2,\ hexane/THF,\ R = C\text{-}t\text{-}Bu_3\ (Ref.289)} \longrightarrow$  [Li(thf)$_{4.5}$][Co{N(SiMe$_3$)$_2$}(OC-$t$-Bu$_3$)$_2$]

[Yb{N(SiMe$_3$)$_2$}$_2$]$_2$
R' = C-$t$-Bu$_3$ (Ref.292)

hexane, $n = 4$ ⟶  [Yb(OC-$t$-Bu$_3$)(µ-OC-$t$-Bu$_3$)]$_2$ + 4 HN(SiMe$_3$)$_2$

hexane, $n = 2$ ⟶  [Yb{N(SiMe$_3$)$_2$}(µ-OC-$t$-Bu$_3$)]$_2$ + 2 HN(SiMe$_3$)$_2$

Scheme 4.   Synthesis of more sterically hindered divalent metal alkoxides by alcoholysis of metal amides.

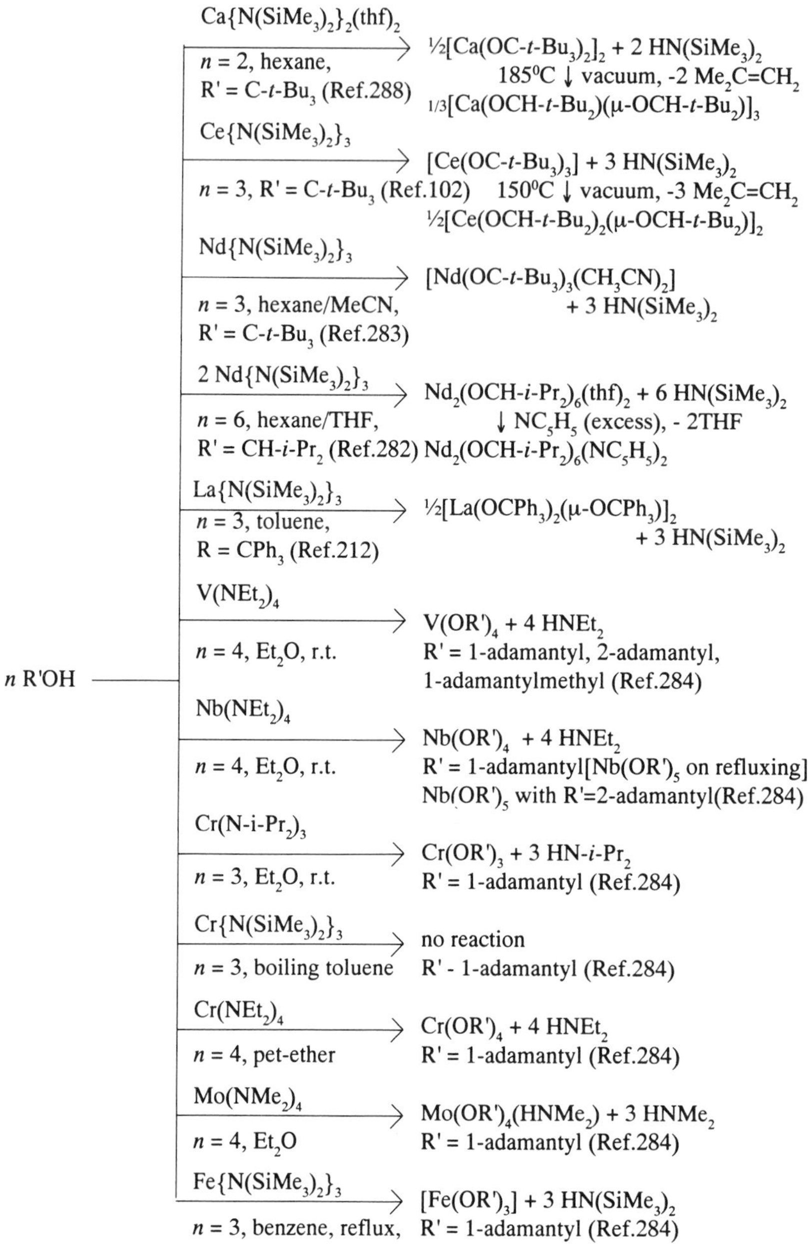

Ca{N(SiMe$_3$)$_2$}$_2$(thf)$_2$

$\xrightarrow{\hspace{2cm}}$ ½[Ca(OC-$t$-Bu$_3$)$_2$]$_2$ + 2 HN(SiMe$_3$)$_2$

$n$ = 2, hexane,                                    185°C ↓ vacuum, -2 Me$_2$C=CH$_2$

R' = C-$t$-Bu$_3$ (Ref.288)   $_{1/3}$[Ca(OCH-$t$-Bu$_2$)(μ-OCH-$t$-Bu$_2$)]$_3$

Ce{N(SiMe$_3$)$_2$}$_3$

$\xrightarrow{\hspace{2cm}}$ [Ce(OC-$t$-Bu$_3$)$_3$] + 3 HN(SiMe$_3$)$_2$

$n$ = 3, R' = C-$t$-Bu$_3$ (Ref.102)   150°C ↓ vacuum, -3 Me$_2$C=CH$_2$

                                                       ½[Ce(OCH-$t$-Bu$_2$)$_2$(μ-OCH-$t$-Bu$_2$)]$_2$

Nd{N(SiMe$_3$)$_2$}$_3$

$\xrightarrow{\hspace{2cm}}$ [Nd(OC-$t$-Bu$_3$)$_3$(CH$_3$CN)$_2$]

$n$ = 3, hexane/MeCN,                         + 3 HN(SiMe$_3$)$_2$

R' = C-$t$-Bu$_3$ (Ref.283)

2 Nd{N(SiMe$_3$)$_2$}$_3$

$\xrightarrow{\hspace{2cm}}$ Nd$_2$(OCH-$i$-Pr$_2$)$_6$(thf)$_2$ + 6 HN(SiMe$_3$)$_2$

$n$ = 6, hexane/THF,                            ↓ NC$_5$H$_5$ (excess), - 2THF

R' = CH-$i$-Pr$_2$ (Ref.282) Nd$_2$(OCH-$i$-Pr$_2$)$_6$(NC$_5$H$_5$)$_2$

La{N(SiMe$_3$)$_2$}$_3$

$\xrightarrow{\hspace{2cm}}$ ½[La(OCPh$_3$)$_2$(μ-OCPh$_3$)]$_2$

$n$ = 3, toluene,                                        + 3 HN(SiMe$_3$)$_2$

R = CPh$_3$ (Ref.212)

V(NEt$_2$)$_4$

$\xrightarrow{\hspace{2cm}}$ V(OR')$_4$ + 4 HNEt$_2$

$n$ = 4, Et$_2$O, r.t.         R' = 1-adamantyl, 2-adamantyl,

                                   1-adamantylmethyl (Ref.284)

Nb(NEt$_2$)$_4$

$\xrightarrow{\hspace{2cm}}$ Nb(OR')$_4$ + 4 HNEt$_2$

$n$ = 4, Et$_2$O, r.t.         R' = 1-adamantyl[Nb(OR')$_5$ on refluxing]

                                   Nb(OR')$_5$ with R'=2-adamantyl(Ref.284)

Cr(N-i-Pr$_2$)$_3$

$\xrightarrow{\hspace{2cm}}$ Cr(OR')$_3$ + 3 HN-$i$-Pr$_2$

$n$ = 3, Et$_2$O, r.t.         R' = 1-adamantyl (Ref.284)

Cr{N(SiMe$_3$)$_2$}$_3$ $\xrightarrow{\hspace{1cm}}$ no reaction

$n$ = 3, boiling toluene   R' - 1-adamantyl (Ref.284)

Cr(NEt$_2$)$_4$ $\xrightarrow{\hspace{1cm}}$ Cr(OR')$_4$ + 4 HNEt$_2$

$n$ = 4, pet-ether           R' = 1-adamantyl (Ref.284)

Mo(NMe$_2$)$_4$

$\xrightarrow{\hspace{2cm}}$ Mo(OR')$_4$(HNMe$_2$) + 3 HNMe$_2$

$n$ = 4, Et$_2$O                R' = 1-adamantyl (Ref.284)

Fe{N(SiMe$_3$)$_2$}$_3$

$\xrightarrow{\hspace{2cm}}$ [Fe(OR')$_3$] + 3 HN(SiMe$_3$)$_2$

$n$ = 3, benzene, reflux,   R' = 1-adamantyl (Ref.284)

$n$ R'OH ───

Scheme 5.   Preparation of sterically congested tri- and tetravalent metal alkoxides by alcoholysis of metal amides.

$$Ti(O\text{-}i\text{-}Pr)_4 \ + \ 4\ R'OH \ \xrightarrow[\text{reflux}]{\text{benzene}} \ Ti(OR')_4 \ + \ 4\ HO\text{-}i\text{-}Pr$$

R' = 1-adamantyl, 2-adamantyl, 1-adamantylmethyl (284)

$$[\{(Me_3Si)_2N\}_2Th(CH_2SiMe_2N\text{-}SiMe_3] \ + \ 4\ HOCH\text{-}i\text{-}Pr_2 \ \longrightarrow$$

$$[Th(OCH\text{-}i\text{-}Pr_2)_4]_2 \ (299) \ + \ 3\ HN(SiMe_3)_2$$

Scheme 6.   Miscellaneous routes for the synthesis of sterically congested metal alkoxides.

and III.B.6; Fig. 28), compounds in rare coordination states of three (Fig. 38) and two (Fig. 39) can be isolated by judicious choice of the sterically hindered alkoxide ligands.

The X-ray structure of a interesting homometallic five-coordinate neodymium alkoxide complex $[(t\text{-}Bu_3\text{-}CO)_2Nd(\mu\text{-}Cl)(thf)]_2$ (290) is shown in Fig. 40, and consists of a chloride-bridged dimer with an $Nd_2Cl_2$ planar ring.

The X-ray crystallographic studies of another interesting neodymium(III) complex of the tri-*tert*-butylmethoxide ligand (291), $[(thf)_3Li(\mu\text{-}Cl)Nd(OC\text{-}t\text{-}Bu_3)_3]$, shows it to have the structure shown in Fig. 41, with the neodymium atom in a slightly distorted tetrahedral geometry. This geometry is formed by three $^-OC\text{-}t\text{-}Bu_3$, ligands and a chlorine atom bridging neodymium and lithium.

A similar type of single chloride-bridged structure for a cobalt(II) heterometallic derivative $[(t\text{-}Bu_3\text{-}CO)_2Co(\mu\text{-}Cl)Li(thf)_3]$ (289) has also been crystallographically authenticated. The cobalt atom is in a rare three-coordinate state.

The crystal structure of $[(t\text{-}Bu_3\text{-}CO)_2ZrCl_3 \cdot Li(OEt_2)_2]$ (295) is shown in Fig. 42, and consists of two chloride groups bridging lithium and zirconium atoms. The overall stereochemistry of the zirconium center is pseudotrigonal bipyramidal and the bulky tri-*tert*-butylmethoxide ligands occupy the least-crowded equatorial sites.

X-ray crystal structures of two five-coordinate Nd(III) and Th(IV) alkoxides containing the diisopropylmethoxide ($^-OCH\text{-}i\text{-}Pr_2$) ligand, $[\{Nd(OCH\text{-}i\text{-}Pr_2)_2(thf)(\mu\text{-}OCH\text{-}i\text{-}Pr)\}_2]$ (282) and $[\{Th(OCH\text{-}i\text{-}Pr_2)_3(\mu\text{-}OCH\text{-}i\text{-}Pr_2)\}_2]$ (299) are shown in Figs. 43 and 44, respectively. The neodymium atom (Fig. 43) adopts a distorted trigonal bipyramidal geometry, in which the centrosymmetric $Nd_2O_8$ core can be viewed as two $NdO_5$ trigonal bipyramids joined along a common axial–equatorial edge with the terminal THF molecule occupying an axial position. The terminal Nd—O bond lengths of the diisopropylmethoxide ligands average 2.153(4) Å and are comparable with the average terminal Nd—O bond lengths of 2.05(2), 2.174(2), and 2.148(16) Å observed in $Nd_6(O\text{-}i\text{-}Pr)_{17}Cl$ (256), $Nd(OC\text{-}t\text{-}Bu_2CH_2PMe_2)_3$ (301), and $Nd_5O(O\text{-}i\text{-}Pr)_{13}(HO\text{-}i\text{-}Pr)_2$ (302), respectively. The terminal Nd—O—C (alkoxide) angles are very obtuse and average 164.9(4)°.

TABLE III
Crystal Data for Some More Sterically Demanding Alkoxides

| Compound | Crystal Class | Space Group | Z | a (Å)<br>b (Å)<br>c (Å) | α (°)<br>β (°)<br>γ (°) | References |
|---|---|---|---|---|---|---|
| *Homometallic Derivatives* | | | | | | |
| [LiOC-*t*-Bu$_3$]$_2$ (Fig. 39) | Cubic | *Pa*3 | 4 | 13.856(7)<br>13.856(7)<br>13.856(7) | 90.0<br>90.0<br>90.0 | 296 |
| [LiOC-*t*-Bu$_3$(thf)]$_2$ (Fig. 38) | Orthorhombic<br>Orthorhombic | *Pbc*2$_1$<br>*Pbc*2$_1$ | 4<br>4 | 10.685(1)<br>10.685(1)<br>16.884(3)<br>19.558(3) | 90.0<br>90.0<br>90.0<br>90.0 | 281<br>281 |
| [Ce(OCH-*t*-Bu$_2$)$_3$]$_2$ | Triclinic | *P*$\bar{1}$ | 1 | 11.611(7)<br>12.497(6)<br>12.611(7) | 63.78(4)<br>70.69(4)<br>79.36(4) | 102 |
| [Nd$_2$(OCH-*i*-Pr$_2$)$_6$(thf)$_2$] (Fig. 43) | Monoclinic | *P*2$_{1/n}$ | 2 | 10.949(1)<br>21.433(3)<br>11.965(1) | 90.0<br>95.63(1)<br>90.0 | 282 |
| [Nd$_2$(OCH-*i*-Pr$_2$)$_6$(Py)]$_2$ | Triclinic | *P*$\bar{1}$ | 1 | 12.003(4)<br>12.243(4)<br>11.454(4) | 109.99(1)<br>108.82(1)<br>98.76(1) | 282 |
| [Nd$_2$(OCH-*i*-Pr$_2$)$_6$($\mu$-dme)]$_\infty$ | Triclinic | *P*$\bar{1}$ | 1 | 10.797(3)<br>10.984(3)<br>12.112(3) | 109.08(3)<br>100.32(2)<br>90.80(2) | 282 |
| [Nd(OC-*t*-Bu$_3$)$_2$(Cl)(thf)]$_2$ (Fig. 40) | Triclinic | *P*$\bar{1}$ | 2 | 12.204(4)<br>13.165(5)<br>16.936(6) | 69.88(2)<br>87.62(2)<br>62.34(4) | 290 |
| [Th(OCH-*i*-Pr$_2$)$_4$]$_2$ | Monoclinic | *P*2$_{1/n}$ | 2 | 12.115(2)<br>20.820(3)<br>13.002(2) | 90.0<br>100.62(1)<br>90.0 | 299 |
| [Yb(OC-*t*-Bu$_3$)(NR$_2$)]$_2$ (R = SiMe$_3$) | Monoclinic | *P*2$_{1/n}$ | 4 | 12.987(2)<br>16.179(4)<br>23.621(4) | 90.0<br>91.03(1)<br>90.0 | 292 |

| Compound | Crystal system | Space group | Z | Cell dimensions (Å) | Angles (°) | Ref. |
|---|---|---|---|---|---|---|
| [Cr(OCH-t-Bu₂)₄] | Monoclinic | $C_{2/c}$ | 4 | 20.249(3) / 10.748(1) / 20.275(2) | 90.0 / 116.68(1) / 90.0 | 285 |
| [Cr(μ-OCH-t-Bu₂)(OC-t-Bu₃]₂ | Triclinic | $P\bar{1}$ | 2(dimers) | 8.151(2) / 16.520(4) / 17.434(4) | 94.46(2) / 93.94(2) / 101.01(2) | 286 |
| [Mo(1-ado)₄(NHMe₂)] (1-ado = 1-adamantoxo) | Triclinic | $P\bar{1}$ | 2 | 13.500(1) / 13.168(2) / 11.646(1) | 91.034(10) / 103.835(8) / 70.842(10) | 284 |
| [Mn(OCH-t-Bu)₂]₃ | Monoclinic | $P2_{1/c}$ | 4 | 12.092(2) / 21.868(4) / 22.748(4) | 90.0 / 96.57(2) / 90.0 | 286 |
| [Co{OC(C₆H₁₁)₃}₂]₂·MeOH·½C₆H₁₂·THF | Triclinic | $P\bar{1}$ | 2 | 13.359(4) / 14.589(5) / 23.054(9) | 72.73(2) / 77.14(3) / 64.86(3) | 300 |
| [Co(OCPh₃)₂]₂·n-C₆H₁₄ | Triclinic | $P\bar{1}$ | 2 | 11.787(4) / 13.336(4) / 20.107(5) | 81.50(2) / 80.07(3) / 89.40(3) | 300 |
| [Co(OCPh₃)₂(thf)₂] | Monoclinic | $C_{2/c}$ | 4 | 18.994(9) / 9.600(8) / 23.421(11) | 90.0 / 119.66(3) / 90.00 | 300 |
| [Ge(OC-t-Bu₃)₂] | Cubic | $Pa3$ | 4 | 14.015(6) / 14.015(6) / 14.015(6) | 90.0 / 90.0 / 90.0 | 221 |
| *Heterometallic Derivatives* | | | | | | |
| [Nd(OC-t-Bu₃)₃LiCl(thf)₃] (Fig. 41) | Monoclinic | $P2_{1/n}$ | 4 | 12.507(1) / 19.639(2) / 22.834(3) | 90.0 / 97.440(10) / 90.0 | 291 |
| [U(OCH-t-Bu₂)₄·LiMe] | Monoclinic | $P2_{1/c}$ | 4 | 11.648(2) / 15.866(2) / 23.405(3) | 90.0 / 105.29(1) / 90.0 | 298 |

335

TABLE III

Crystal Data for Some More Sterically Demanding Alkoxides

| Compound | Crystal Class | Space Group | Z | $a$ (Å) $b$ (Å) $c$ (Å) | $\alpha$ (°) $\beta$ (°) $\gamma$ (°) | References |
|---|---|---|---|---|---|---|
| *Heterometallic Derivatives* (Continued) | | | | | | |
| [Zr(OC-$t$-Bu$_3$)$_2$ZrCl$_3$·Li(OEt$_2$)$_2$] (Fig. 42) | Monoclinic | $P2_{1/c}$ | 4 | 12.946(3) 14.026(3) 26.257(4) | 90.0 121.016(12) 90.0 | 295 |
| [Cr(OC-$t$-Bu$_3$)$_2$·LiCl(thf)$_2$] (Fig. 45) | Triclinic | $P\bar{1}$ | 2 | 12.633(12) 12.898(8) 12.998(18) | 61.47(8) 80.06(9) 81.73(6) | 281 |
| [Li$_2$Cr(OCH-$t$-Bu$_2$)$_4$(thf)] | Triclinic | $P\bar{1}$ | 2 | 19.986(2) 11.618(2) 11.390(1) | 117.61(1) 76.45(1) 93.11(1) | 285 |
| [Li{Mn(N(SiMe$_3$)$_2$)(OC-$t$-Bu$_3$)$_2$}] (Fig. 46) | Monoclinic | $Cc$ | 4 | 21.243(4) 11.814(1) 20.334 | 90.0 133.60(1) 90.0 | 287 |
| [Li$_2${MnBr$_2$(OC-$t$-Bu$_3$)$_2$}(thf)$_2$] | Monoclinic | $P2_{1/c}$ | 4 | 17.868(11) 8.563(6) 26.075(20) | 90.0 98.39(6) 90.0 | 287 |
| [CoCl(OC-$t$-Bu$_3$)$_2$·Li(thf)$_3$] | Triclinic | $P\bar{1}$ | 2 | 12.329(15) 12.614(11) 15.440(18) | 78.46(8) 68.77(8) 64.86(7) | 289 |
| [Li(thf)$_4$][Co{N(SiMe$_3$)$_2$}$_2$(OC-$t$-Bu$_3$)$_2$] | Monoclinic | $P2_{1/n}$ | 4 | 11.491(3) 16.143(6) 31.343(27) | 90.0 99.64(2) 90.0 | 289 |
| [Li{Co(N(SiMe$_3$)$_2$)$_2$(OC-$t$-Bu$_3$)$_2$}] | Monoclinic | $I_c$ | 4 | 16.150(10) 11.485(6) 20.762(11) | 90.0 108.90(4) 90.0 | 289 |

336

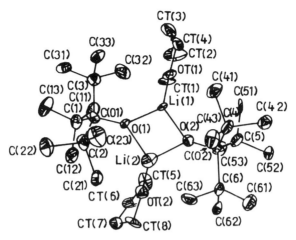

Figure 38. The crystal structure of [Li(OC-*t*-Bu₃)(thf)]₂. [Adapted from (281).]

The X-ray crystal structure of homoleptic thorium alkoxide [Th(OCH-*i*-Pr₂)₄]₂ (299) is shown in Fig. 44. Each thorium atom adopts a distorted trigonal bipyramidal geometry, and the Th₂O₈ core can be viewed as two trigonal bipyramids fused along a common axial–equatorial edge.

The molecular structure of a chromium(II) complex [Cr(OC-*t*-Bu₃)₂·LiCl(thf)₂] (281) is shown in Fig. 45. The structure is noteworthy in several respects, being a rare example of three-coordinate chromium(II) alkoxide. The geometry at chromium with a wide O(1)—Cr—O(2) angle of 157.2(2)° may be considered as T-shaped. All five atoms forming the core O(1), O(2), Cr, Cl, and Li are coplanar. The geometry around the lithium atom is a severely

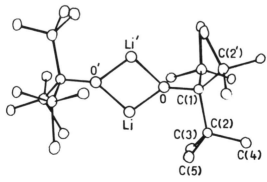

Figure 39. The crystal structure of [Li(μ-OC-*t*-Bu₃)]₂. [Adapted from (296).]

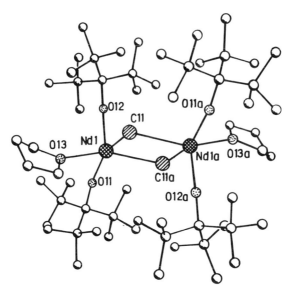

Figure 40.    The crystal structure of $[(t\text{-}Bu_3\text{-}CO)_2Nd(\mu\text{-}Cl)(thf)]_2$. [Adapted from (290).]

distorted tetrahedron as a result of steric crowding. The Li· · ·Cl bond length
[2.417(8) Å] is close to the sum of their ionic radii.

Figure 46 illustrates the molecular structure of a monomeric three-coordinate
manganese alkoxide $\{Li[Mn(N(SiMe_3)_2)(OC\text{-}t\text{-}Bu_3)_2]\}$ (287). The geometry at
manganese is distorted trigonal planar. The two tri-*tert*-butylmethoxide ligands

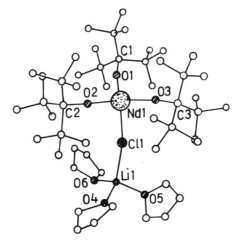

Figure 41.    The crystal structure of $[(t\text{-}Bu_3\text{-}CO)_3Nd(\mu\text{-}Cl)Li(thf)_3]$. [Adapted from (291).]

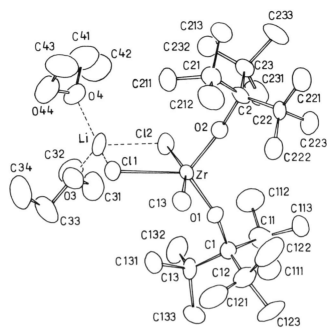

Figure 42. The crystal structure of $[(t\text{-Bu}_3\text{-CO})_2\text{ZrCl}_3\cdot\text{Li}(\text{OEt}_2)_2]$. [Adapted from (295).]

show similar bond angles and bond lengths to those found in $[\text{Cr}(\text{OC-}t\text{-Bu}_3)_2\cdot\text{LiCl}(\text{thf})_2]$ (281) and $[\{\text{Li}(\text{OC-}t\text{-Bu}_3)(\text{thf})\}_2]$ (281). The *tert*-butyl groups in each alkoxide group are staggered with respect to each other, which disrupts the near $C_2$ axis through N, Mn, and Li centers. Interestingly, the lithium atom is two coordinate even when the derivative is prepared in THF solutions. The

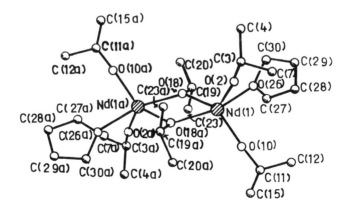

Figure 43. The crystal structure of $[\text{Nd}_2(\text{OCH-}i\text{-Pr}_2)_6(\text{thf})_2]$. [Adapted from (282).]

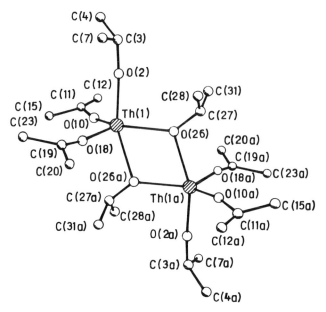

Figure 44.   The crystal structure of [Th(OCH-*i*-Pr$_2$)$_4$]$_2$. [Adapted from (299).]

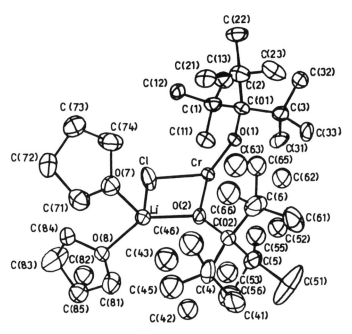

Figure 45.   The crystal structure of [Cr(OC-*t*-Bu$_3$)$_2$.LiCl(thf)$_2$]. [Adapted from (281).]

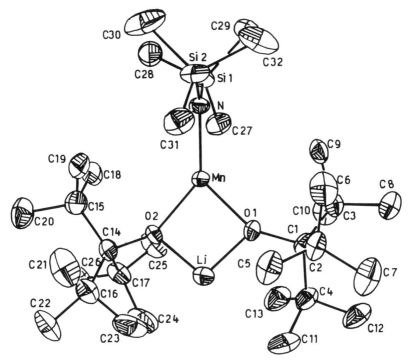

Figure 46.   The crystal structure of [Li{Mn(N(SiMe$_3$)$_2$)(OC-$t$-Bu$_3$)$_2$}]. [Adapted from (287).]

large steric bulk of the $^-$OC-$t$-Bu$_3$ ligand prevents the formally two-coordinate lithium from complexing a THF molecule.

## B.  Derivatives of Haloalcohols

Out of the two groups (fluoro and chloro) of title compounds so far studied, the chemistry of fluoroalkoxides (24, 160, 303–305) and fluoro-$\beta$-diketonates (146, 306–309) has recently attracted considerable attention, mainly due to their enhanced volatility compared with the unfluorinated analogues. Another interesting feature of fluoroalkoxide derivatives in their role as an effective alkene and acetylene metathesis (160).

The replacement of Me by CF$_3$ groups in primary, secondary, and tertiary alcohols could be expected to bring about the following changes: (a) a small increase in the bulkiness of the CF$_3$ compared to the Me group is likely to reduce the oligomerization tendency of the concerned metal alkoxide; (b) the inductive effect of the electron-withdrawing fluoro group makes the alkoxo oxygen ($^-$OR$_F$) less nucleophilic [where R$_F$ is a fluorinated alkyl group, such as CH$_2$CF$_3$, CH(CF$_3$)$_2$, CH(CF$_3$)Ph, C(CF$_3$)$_3$, CMe$_2$(CF$_3$), CMe(CF$_3$)$_2$,

$C(CF_3)(C_6F_5)_2$, or $C(CF_3)_2(C_6F_5)$] and thus a weaker bridging ligand; (c) the introduction of fluoro (or chloro) alkoxide ligands would obviously enhance the electrophilicity of the central metal atom and thus facilitate the attainment of the preferred coordination state either through intermolecular association via alkoxo bridges wherever structurally feasible, or by the expeditious coordination of supplementary neutral ligands available to stabilize the complex; (d) the lower polarizability of the fluorine may reduce the van der Waals attractive forces; and (e) the intermolecular repulsions between the several fluorine lone pairs may lead to increased volatility.

During the past decade, the field of metal haloalkoxides has expanded producing an increasing number of papers each year. The topic of metal complexs of fluorinated alcohols was reviewed in 1988 by Willis (303) covering the literature up to 1986.

The purpose of this section is to update and present the currently known facts in the title area and to indicate the trends of recent investigations. Some interesting features that have emerged over the last 10 years, are highlighted, with an emphasis on the structures determined by X-ray crystallography.

## 1. Fluoroalkoxide Derivatives

Before dealing with the developments of the post decade, it is appropriate to summarize briefly the results accomplished prior to 1986. This summary is merely to indicate and to appreciate the areas in which an extension and exploration of this field has been commonplace.

**a. Developments Prior to 1986.** Numerous preparative routes were devised for the synthesis of homometallic fluoroalkoxide complexes (303) similar to those for hydrogen analogues (see Section II). For example,

1. Reactions of metal chlorides with fluoroalcohols (Section II.C.1).
2. Reactions of metal chlorides with fluoroalcohols in the presence of an amine (Section II.C.1).
3. Salt elimination between the metal chlorides and an alkali metal alkoxide (Section II.C.1).
4. Cleavage of metal–carbon bonds (Section II.F).
5. Amido–alkoxo exchange reactions (Section II.E).

Other synthetic routes (especially when the parent alcohol is unstable) include the reaction of a metal hydride (or fluoride) and a perfluoroketone, the reaction of a fluoroacyl chloride with a metal chloride or a metal, and the reactions of fluorinated epoxides with cesium fluoride.

With the use of one of the aforementioned preparative routes, synthesis of a wide variety of interesting homometallic homoleptic derivatives of many metals and metalloids were already described without X-ray crystallography authentication. Some typical examples include: (a) the synthesis of $M(OCH_2CF_3)_4$ (M = Ti, Zr, or Ge) (310), $M(OCH_2CF_3)_5$ (M = Nb or Ta) (310), $M[OCH(CF_3)_2]_4$ (M = Ti, Zr, or Ge) (311), $M[OCH(CF_3)_2]_5$ (M = Nb or Ta) (311), $M[OCH(CF_3)_2]_3$ (Y and lanthanides) (312), $M[OCH(CF_3)_2]_3 \cdot 2NH_3$ (M = Y and lanthanides) (313), $Zr[OCH(CF_3)_2]_4 \cdot 2NH_3$ (314), $M[(OC(CF_3)_3]_3 \cdot 3NH_3$ (M = La, Gd, Yb, or Y) (315) by the ammonia method (cf. Section II.C.1); (b) preparation of $U(OCH_2CF_3)_5$ (97), $Fe(OCH_2CF_3)_3$ (316), $U[OCH(CF_3)_2]_4(thf)_2$ (317), $U[OC(CF_3)_3]_4(thf)_2$ (317), and $Be[OC(CF_3)_3]_2 \cdot Et_2O$ (318) by the sodium or lithium alkoxide method (cf. Section II.C.1).

In addition to these homoleptic derivatives, $Ti[OC(CF_3)_3]_4$, $Pb[OC(CF_3)_3]_2$, $Bi[OC(CF_3)_3]_3$, and $Te[OC(CF_3)_3]_4$ were also prepared by the reaction of $TiCl_4$ or Pb, Bi, and Te with $(CF_3)_3COCl$ at $0°C$, respectively (319).

Interesting heteroleptic derivatives: such as $TiCl_3(OCH_2CF_3)$ (320, 321), $TiCl_2(OCH_2CF_3)_2$ (320b, 321), and $SnCl_3(OCH_2CF_3)(HOCH_2CF_3)$ (322) (by the direct reaction of the appropriate metal chloride with $CF_3CH_2OH$ in desired molar ratio); $(acac)_2Ti(OCH_2CF_3)_2$ (323) (by alcohol exchange); $VO(OCH_2CF_3)_3$ (324, 325), $VO(O\text{-}i\text{-}Pr)_{3-x}(OCH_2CF_3)_x$ ($x$ = 1, 2, or 3) (325) $MoF_n(OCH_2CF_3)_{6-n}$ ($n$ = 1-6) (326), $MoO(Cl)[OC(CF_3)_3]_3$ (327), $MoO[OC(CF_3)_3]_4$ (327), $WF_5(OR_F)$ [$R_F$ = $CH_2CF_3$, $CH_2CH_2CF_3$, and $CH(CF_3)_2$) (328)] (by the reactions of the corresponding metal chlorides with alkali metal (generally lithium) fluoroalkoxides), were prepared during 1975–1980.

**b. Recent Developments.** During the past decade, some of these preparative routes played an important role in the synthesis of novel derivatives, as shown in Eqs. 105–129. These results exhibited an interesting variation in their structural features [see Section IV.B.1.b.iii].

### i. Homometallic Derivatives

*Metathesis Reactions*

$$VCl_3 + x LiOCH_2CF_3 \xrightarrow[x = 1 \text{ or } 2]{\text{benzene}} VCl_{3-x}(OCH_2CF_3)_x + x LiCl \quad (105)$$

(Ref. 329)

$$CrO_2Cl_2 + x LiOCH(CF_3)_2 \xrightarrow[x = 1 \text{ or } 2]{\text{CCl}_4} Cr_2O_2[OCH(CF_3)_2]_x Cl_{2-x} + 2 LiCl$$

(Refs. 330, 331)                                                                 (106)

$$Mo_2Cl_6(dme)_2 + 6NaOCMe_2CF_3 \xrightarrow{CH_2Cl_2} Mo_2(OCMe_2CF_3)_6 + 6\ NaCl$$

(Ref. 332)                                                                                    (107)

$$WCl_4(PMe_2Ph)_2 + 2TlOCH_2CF_3 \xrightarrow{toluene}$$                   (108)

$$\text{(Ref. 333)} \quad WCl_2(OCH_2CF_3)_2(PMe_2Ph)_2 + 2TlCl$$

$$CuBr_2 + 2NaOR_F + 2Py \xrightarrow{THF} Cu(OR)_F)_2(py)_2 + 2NaBr$$

$$\text{(Ref. 334)} \quad R_F = CH(CF_3)_2,\ C(CF_3)_3$$                    (109)

$$BiCl_3 + 3Na[OCH(CF_3)_2 \xrightarrow{THF} \tfrac{1}{2}[\{Bi[OCH(CF_3)_2]_3(thf)\}_2] + 3NaCl$$

(Refs. 335, 336)                                                                         (110)

*Alcoholysis of Metal Amides*

$$M[N(SiMe_3)_2]_3 + 3R_FOH \xrightarrow{benzene} M(OR_F)_3 + 3HN(SiMe_3)_2 \quad (111)$$

$$M = Y\ (\text{Ref. 337}), \quad R_F = C(CF_3)_2Me$$

$$M = Y\ (\text{Ref. 337}), \quad R_F = CMe_2(CF_3)$$

$$M = La\ (\text{Ref. 338}), \quad R_F = C(CF_3)_2Me$$

$$M = Pr\ (\text{Ref. 338}), \quad R_F = CMe(CF_3)_2$$

$$M = Eu\ (\text{Ref. 338}), \quad R_F = CMe(CF_3)_2$$

$$M[N(SiMe_3)_2]_3 + 3R_FOH \xrightarrow[\text{(excess)}]{benzene} [M(OR_F)_3(NH_3)_x]_n + 3HN(SiMe_3)_2 \quad (112)$$

$$M = Sc\ (\text{Ref. 339}), \quad R_F = CH(CF_3)_2, \quad x = 2, \qquad n = 2$$

$$M = Y\ (\text{Ref. 337}), \quad R_F = CMe(CF_3)_2, \quad x = 3, \qquad n = 0$$

$$M = La\ (\text{Ref. 338}), \quad R_F = CMe(CF_3)_2, \quad x = 1\ or\ 2,\ n = 2$$

$$M = Pr\ (\text{Ref. 338}), \quad R_F = CMe(CF_3)_2, \quad x = 2, \qquad n = 2$$

$$M = Eu\ (\text{Ref. 338}), \quad R_F = CMe(CF_3)_2, \quad x = 1, \qquad n = 2$$

The ammoniate of the products in Eq. 112 originates from the reaction of the acidic fluoroalcohols with the $(Me_3Si)_2NH$ produced during the formation of the metal fluoroalkoxides:

$$(Me_3Si)_2NH + 2R_FOH \longrightarrow NH_3 + 2Me_3SiOR_F \qquad (113)$$

If the reaction is carried out in a donor solvent such as THF or $Et_2O$, the product obtained is sometimes ligated with the donor solvent in preference to ammonia.

$$M[N(SiMe_3)_2]_3 + 3R_FOH \xrightarrow{\text{THF}} [M(OR_F)_3(thf)_x] + 3HN(SiMe_3)_2 \quad (114)$$

$$M = Y \text{ (Ref. 339)}, \quad R_F = CH(CF_3)_2, x = 3$$

$$M = Y \text{ (Ref. 337)}, \quad R_F = CMe(CF_3)_2, x = 3$$

$$M = La \text{ (Ref. 340)}, R_F = CMe(CF_3)_2, x = 3$$

$$M = Eu \text{ (Ref. 340)}, R_F = CMe(CF_3)_2, x = 3$$

$$M[N(SiMe_3)_2]_3 + 3R_FOH \xrightarrow{Et_2O} [M(OR_F)_3(OEt_2)_x] + 3HN(SiMe_3)_2 \quad (115)$$

$$M = Y \text{ (Ref. 337)}, \quad R_F = CMe(CF_3)_2, x = 0.33$$

$$M = La \text{ (Ref. 338)}, \quad R_F = CMe(CF_3)_2, x = 0.33$$

The latter two reactions (Eqs. 114 and 115) reflect the relative donor ability of the solvents THF and $Et_2O$.

*Alcohol-Exchange Reactions*

$$[(CuO\text{-}t\text{-}Bu)_4] + 2R_FOH \longrightarrow Cu_4(O\text{-}t\text{-}Bu)_2(OR_F)_2 + 2Bu{-}t\text{-}OH$$

(Ref. 341) $\qquad\qquad\qquad\qquad\qquad\qquad\qquad\qquad\qquad (116)$

The reaction of a moderate excess of $HOCH(CF_3)_2$ with $[Cu(O\text{-}t\text{-}Bu)_2]_n$ in hydrocarbon solvents affords two main products, $Cu_4(O\text{-}t\text{-}Bu)_6[OCH(CF_3)_2]_2$ and $Cu_3(O\text{-}t\text{-}Bu)_4[OCH(CF_3)_2]_2$ (192), with their relative amounts determined by the reaction stoichiometry.

$$7/n\,[Cu(O\text{-}t\text{-}Bu)_2]_n + 4HOCH(CF_3)_2 \longrightarrow$$

$$\text{(Ref. 192)} \quad Cu_4(O\text{-}t\text{-}Bu)_6[OCH(CF_3)_2]_2$$

$$+ Cu_3(O{-}t\text{-}Bu)_4[OCH(CF_3)_2]_2 + 4Bu\text{-}t\text{-}OH \qquad (117)$$

By contrast, copper(II) methoxide reacts with $HOCH(CF_3)_2$ in refluxing benzene, to afford an insoluble solid of molecular formula $Cu_5(OMe)_6[OCH(CF_3)_2]_4$ (334).

$$5/n\,[Cu(OMe)_2]_n + 4HOCH(CF_3)_2 \longrightarrow Cu_5(OMe)_6[OCH(CF_3)_2]_4 + 4MeOH$$

$$(118)$$

Although the reaction of $[Cu(OMe)_2]_n$ with an excess of $HOCH(CF_3)_2$ does not form $[Cu\{OCH(CF_3)_2\}_2]_n$, the reaction of the mixed-ligand derivative $Cu_5(OMe)_6[OCH(CF_3)_2]_4$ with tmeda in diethyl ether forms the bis(hexafluoroisopropoxo) complex $Cu\{OCH(CF_3)_2\}_2(tmeda)$ via redistribution of coordinated ligands.

$$Cu_5(OMe)_6[OCH(CF_3)_2]_4 + 2tmeda \xrightarrow{\text{Et}_2\text{O}}$$

$$(\text{Ref. 334}) \quad 2[Cu\{OCH(CF_3)_2\}_2(tmeda)] + 3/n\,[Cu(OMe)_2]_n \quad (119)$$

A more convenient general procedure for the synthesis of bis(fluoroalkoxo) copper(II) complex has been designed (334) and is represented by the following reaction:

$$1/n[Cu(OMe)_2]_n + 2R_FOH + L \xrightarrow{\text{Et}_2\text{O}} Cu(OR_F)_2(L) + 2HOMe \quad (120)$$

$$R_F = CH(CF_3)_2, \ L = tmeda, \ teed, \ bpy, \ (py)_2$$

$$R_F = C(CF_3)_3, \ L = tmeda, \ bpy, \ (py)_2$$

Thallium(I) hexafluoroisopropoxide (342) was prepared by the reaction of TlOEt with $HOCH(CF_3)_2$ in ethanol.

*Aryloxo–Alkoxo Exchange Reaction*

The interaction of $[Pd(OPh)_2(bpy)]$ with $HOCH(CF_2)_2$ affords a mixed (aryloxo)(alkoxo) palladium complex (219) according to the reaction shown below [Eq. (121)].

$$[Pd(OPh)_2(bpy)] + HOCH(CF_2)_2 \xrightarrow{\text{CH}_2\text{Cl}_2}$$

$$[Pd\{OCH(CF_2)_2\}(OPh)(HOPh)(bpy)] \quad (121)$$

*Metal–Carbon Bond Cleavage Reactions*

$$cis\text{-}PtMe_2(PMe_3)_2 + HOCH(CF_3)_2 \longrightarrow$$

$$(\text{Ref. 220}) \quad cis\text{-}PtMe\{OCH(CF_3)_2\}(PMe_3)_2 + CH_4 \quad (122)$$

$$cis\text{-PtMe}_2(\text{PMe}_3)_2 + 2\text{HOCH(CF}_3)_2 \longrightarrow$$

$$\text{(Ref. 171)} \quad cis\text{-PtMe}\{\text{OCH(CF}_3)_2\}\{\text{HOCH(CF}_3)_2\}(\text{PMe}_3)_2 + 2\text{CH}_4 \tag{123}$$

$$\text{BiPh}_3 + 3\text{HOC}_6\text{F}_5 \xrightarrow{\text{toluene}} [\text{Bi(OC}_6\text{F}_5)_3(\text{PhMe})]_2 + 3\text{PhH}$$

(Ref. 336)
$$\tag{124}$$

$$2[\text{PPN}][\text{W(CO)}_5\text{CH}_3] + 2\text{HOCH}_2\text{CF}_3 \xrightarrow{\text{THF}}$$

$$\text{(Ref. 343)} \quad [\text{PPN}]_2[\text{W}_2(\text{CO})_8(\text{OCH}_2\text{CF}_3)_2] + 2\text{CH}_4 + 2\text{CO} \tag{125}$$

*Metal–Hydrogen-Bond Cleavage Reactions*

This route was used as early as 1970 by Wales and Weigold (334) for the synthesis of an interesting zirconium derivative $(\text{C}_5\text{H}_5)_6\text{Zr}_3\text{O}_2(\text{OCH}_2\text{CF}_3)_2$ by the reaction of $(\text{C}_5\text{H}_5)_2\text{ZrH}_2$ with $\text{CF}_3\text{CH}_2\text{OH}$. Furthermore, reaction of $\text{Ti(NR}_2)(\text{O-}i\text{-Pr})_3$ with $\text{CF}_3\text{CH}_2\text{OH}$ yielded $\text{Ti(O-}i\text{-Pr})_3(\text{OCH}_2\text{CF}_3)$ (345), indicating facile cleavage of the Ti—N bond. Recently, sodium hexafluoroisopropoxide (342), was prepared by the reaction of sodium hydride with hexafluoroisopropyl alcohol.

*ii. Heterometallic Derivatives.* A variety of new heterobimetallic fluoroalkoxides were synthesized mainly by two routes: (1) metathesis reaction (Eqs. 126–128), and (2) union of two different homometallic alkoxides (Eqs. 129 and 130).

$$\text{CuCl}_2 + x\text{NaOCH(CF}_3)_2 \longrightarrow \text{Na}_{x-2}\text{Cu[OCH(CF}_3)_2]_x + 2\text{NaCl}$$

$$\text{(Ref. 346)} \quad x = 3 \text{ or } 4 \tag{126}$$

$$\text{CuCl}_2 + 2\text{Ba[OR}_\text{F}]_2 \longrightarrow \text{BaCu[OR}_\text{F}]_4 + \text{BaCl}_2$$

$$\text{R}_\text{F} = \text{CH(CF}_3)_2 \text{ (Ref. 346) and CMe(CF}_3)_2 \text{ (Ref. 347)} \tag{127}$$

$$\text{ZrCl}_4 + \text{NaOCH(CF}_3)_2 \xrightarrow{\text{benzene}} \text{Na}_2\text{Zr[OCH(CF}_3)_2]_6 + 4\text{NaCl}$$

(Ref. 342)
$$\tag{128}$$

$$\text{TlOCH(CF}_3)_2 + \text{Zr[OCH(CF}_3)_2]_4 \xrightarrow{\text{benzene}} \text{Tl}_2\text{Zr[OCH(CF}_3)_2]_6$$

(Ref. 342)
$$\tag{129}$$

$$2NaOCH(CF_3)_2 + Zr[OCH(CF_3)_3]_4 \xrightarrow{\text{benzene}} Na_2Zr[OCH(CF_3)_2]_6(C_6H_6)_x$$

(Ref. 342)                                                                    (130)

***iii. Structures.*** In the past few years, there was a surge of interest in the X-ray crystal structures of homo- and heterometallic fluoroalkoxides having secondary $M \cdot \cdot \cdot F—C$ interactions with metals stemming from their use as precursors in CVD studies. For example, Purdy et al. (309) and others (348, 349) showed that the use of fluorinated alkoxide and perfluorinated $\beta$-diketonate systems (which contain $M \cdot \cdot \cdot F—C$ interactions) as CVD precursors not only results in increased volatility but is accompanied by the deposition of metal fluoride in the CVD product.

Another promising result of X-ray structural elucidation of metal alkoxide complexes was the discovery of the presence of inter- and intramolecular $O—H \cdot \cdot \cdot O$ hydrogen bonding (cf. Section III.B.3).

It was mentioned in Section IV.B that fluoroalkoxo groups, as a consequence of steric and electronic factors, favor either monomeric or smaller oligomeric species. Such fluoro compounds exhibit higher Lewis acid behavior compared with their nonfluorinated analogues unless the molecule is more sterically crowded to preclude coordination of donor ligands.

X-ray crystallographic studies also helped to demonstrate that oligomeric species containing both the fluoroalkoxide and nonfluorinated alkoxide components prefer to bridge through the latter group (Fig. 48).

A number of crystal structures of homo- and heterometallic fluoroalkoxides revealed that the above features were determined. Many of these features are included in Table IV and a few typical structures are shown in Figs. 47–51 along with a brief comment on their structural characteristics.

The X-ray analysis of $Ba\{Cu[OCMe(CF_3)_2]_3\}_2$ (347) shows (Fig. 47) barium to be located on a two-fold axis and there are close $Ba \cdot \cdot \cdot F—C$ interactions. 12-coordinate $Ba^{2+}$ cation is surrounded by four oxygen atoms (2.636–2.644 Å) and eight fluorine atoms (2.94–3.14 Å). The $CuO_3$ chromophore is three coordinate, Y shaped, and planar.

There are several other fluoroalkoxide compounds, such as [Na-$(OCH(CF_3)_2)]_4$ (342), $Na_2Cu[OCH(CF_3)_2]_4$ (346), $Na_2Zr(OCH(CF_3)_2)_6$-$(C_6H_6)$ (342), $Na_2Zr(OCH(CF_3)_2)_6(C_6H_6)_2$ (342), and $Tl_2Zr(OCH(CF_3)_2)_6$ (143, 342) in which organofluorine binding to metals was established by X-ray crystallography.

X-ray crystallography of $[Ti\{OCH(CF_3)_2\}_2(OEt)_2(HOEt)]_2$ (Fig. 48), which was prepared by the alcoholysis of $Ti(OEt)_4$ with a large excess of $(CF_3)_2CHOH$ (350), reveals that it adopts a centrosymmetric edge-shared bioctahedral structure (Fig. 48) in which each titanium atom has a coordinated alcohol (6, 168, 351). Although the hydrogen atoms of the alcohols are not located, hydrogen bonding across the dinuclear unit to an alkoxide oxygen atom is indicated by

TABLE IV

Crystal Data for Some Metal Haloalkoxides

### Homometallic Derivatives

| Compound | Crystal Class | Space Group | Z | $a$ (Å) $b$ (Å) $c$ (Å) | $\alpha$ (°) $\beta$ (°) $\gamma$ (°) | References |
|---|---|---|---|---|---|---|
| [NaOCH(CF$_3$)$_2$]$_4$ | Monoclinic | $P2_1$ | 2 | 10.648(1) 9.936(1) 11.315(1) | 90.0 91.40(0) 90.0 | 342 |
| [Sc{OCH(CF$_3$)$_2$}$_3$(NH$_3$)$_2$]$_2$ | Monoclinic | $C2/c$ | 4 | 21.39(3) 9.699(2) 22.488(4) | 90.0 105.70(1) 90.0 | 339 |
| [Y{OCMe(CF$_3$)$_2$}$_3$(thf)$_3$] | Monoclinic | $P2_{1/n}$ | 4 | 10.128(2) 18.065(5) 18.716(4) | 90.0 90.26(1) 90.0 | 339 |
| [Y{OCMe(CF$_3$)$_2$}$_3$(diglyme)] | Orthorhombic | $P2_12_12_1$ | 4 | 9.863(2) 18.322(3) 16.399(3) | 90.0 90.0 90.0 | 337 |
| [La{OCMe(CF$_3$)$_2$}$_3$(thf)$_3$] (Fig. 50) | Monoclinic | $P2_{1/n}$ | 4 | 10.283(3) 18.261(4) 19.116(5) | 90.0 90.35(3) 90.0 | 340 |
| [Pr{OCMe$_2$(CF$_3$)}$_3$]$_3$ | Triclinic | $P\bar{1}$ | 2 | 10.922(2) 11.366(2) 23.404(8) | 75.04(2) 102.86(2) 93.18(1) | 339 |
| [Pr{OCMe(CF$_3$)$_2$}$_3$(NH$_3$)$_2$]$_2$ | Triclinic | $P\bar{1}$ | 1 | 9.459(1) 10.988(1) 13.038(2) | 111.43(2) 86.45(2) 115.59(2) | 338, 339 |
| [Ti{OCH(CF$_3$)$_2$}$_2$(OEt)$_2$(HOEt)]$_2$ (Fig. 48) | Monoclinic | $P2_{1/m}$ | 2 | 10.233(3) 12.702(4) 16.175(7) | 90.0 96.05(3) 90.0 | 350 |

## TABLE IV
### Crystal Data for Some Metal Haloalkoxides

#### Homometallic Derivatives (Continued)

| Compound | Crystal Class | Space Group | Z | $a$ (Å) $b$ (Å) $c$ (Å) | $\alpha$ (°) $\beta$ (°) $\gamma$ (°) | References |
|---|---|---|---|---|---|---|
| [Ti{OCH(CF$_3$)$_2$}$_4$(N≡CMe$_2$)] | Orthorhombic | C222$_1$ | 4 | 10.2872(7) 15.295(1) 18.651(1) | 90.0 90.0 90.0 | 350 |
| [TiCl$_2$(OCH$_2$CH$_2$Cl)$_2$·HOCH$_2$CH$_2$Cl]$_2$ (Fig. 52) | Triclinic | $P\bar{1}$ | 1 | 8.9659(7) 9.2748(8) 10.515(1) | 98.186(7) 111.001(7) 112.435(6) | 351 |
| [Zr(OCMe(CF$_3$)$_2$)$_4$] | Monoclinic | $P2_{1/n}$ | 8 | 16.3216(6) 10.856(3) 31.023(12) | 90.0 104.74(1) 90.0 | 342 |
| [W(OCH$_2$CF$_3$)$_2$Cl$_2$(PMe$_2$Ph)$_2$] | Monoclinic | $P2_{1/c}$ | 4 | 9.583(2) 14.277(3) 19.287(4) | 90.0 102.66(1) 90.0 | 333 |
| [cis-PtMe(OCH(CF$_3$)$_2$)(PMe$_3$)$_2$(HOCH(CF$_3$)$_2$] (Fig. 49) | Trigonal | $P3_1$ | 3 | 9.347(6) 9.347(6) 23.317(3) | | 171 |
| [Cu$_4$(O-$t$-Bu)$_6${OC(CF$_3$)$_3$}$_2$] | Triclinic | $P\bar{1}$ | 1(two half-molecules) | 10.609(3) 10.667(3) 11.151(3) | 100.69(3) 95.37(3) 99.65(3) | 192 |
| [Cu{OCH(CF$_3$)$_3$}$_2$(tmeda)] | Monoclinic | $P2_{1/c}$ | 4 | 9.454(1) 15.045(2) 14.197(6) | 90.0 105.65(2) 90.0 | 334 |
| [Cu{OCMe(CF$_3$)$_2$}$_2$(tmeda)] (Fig. 51) | Monoclinic | $I2/a$ | 4 | 13.601(3) 9.567(2) 16.562(5) | 90.0 106.56(2) 90.0 | 334 |

350

| | | | | a, b, c | angles | Ref. |
|---|---|---|---|---|---|---|
| [Bi{OCH(CF$_3$)$_2$}$_3$(thf)]$_2$ | Monoclinic | $P2_{1/n}$ | 2 | 13.738(7)<br>10.936(7)<br>14.389(8) | 90.0<br>94.528(4)<br>90.0 | 335, 336 |

*Heterometallic Derivatives*

| | | | | | | |
|---|---|---|---|---|---|---|
| [Na$_2$Zr{OCH(CF$_3$)$_2$}$_6$(C$_6$H$_6$)$_2$] | Monoclinic | $C_{2/m}$ | 2 | 16.783(4)<br>14.927(4)<br>9.667(2) | 90.0<br>112.78(1)<br>90.0 | 342 |
| [Na$_2$Zr{OCH(CF$_3$)$_2$}$_6$(C$_6$H$_6$)] | Rhombohedral | $R\overline{3}$ | 1 | 10.440(3)<br>10.440(3)<br>10.440(3) | 106.20(2)<br>112.78(1)<br>106.20(2) | 342 |
| [Ba{Cu[OCMe(CF$_3$)$_2$]$_3$}$_2$]<br>(Fig. 47) | Monoclinic | $C_{2/c}$ | 4 | 23.843(12)<br>18.324(8)<br>10.998(6) | 90.0<br>116.97<br>90.0 | 347 |
| [Na$_2$Cu{OCH(CF$_3$)$_2$}$_4$] | Monoclinic | $C_{2/c}$ | 4 | 15.965(6)<br>12.652(5)<br>11.837(5) | 90.0<br>94.36(3)<br>90.0 | 346 |
| [Tl$_2$Zr{OCH(CF$_3$)$_2$}$_6$] | Monoclinic | $C_{2/c}$ | 4 | 19.879(2)<br>10.182(1)<br>20.456(1) | 90.0<br>122.97(0)<br>90.0 | 143, 342 |

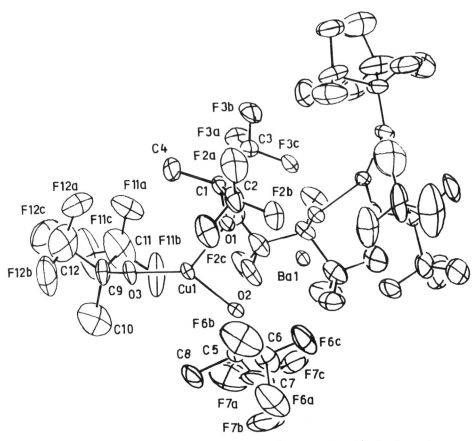

Figure 47. Simplified view of the crystal structure of Ba{Cu[OCMe(CF$_3$)$_2$]$_3$}$_2$ showing environment of barium and copper. Barium interacts with only two [Cu{OCMe(CF$_3$)$_2$}$_3$]$^-$ units. [Adapted from (347).]

the O(2)—Ti(1)—O(5) angle, 167.1(4)°. The alcohol moiety is also readily identifiable by the Ti(1)—O(5) bond length, 2.187(9) Å, which are longer than the corresponding distances for the alkoxides and approximate Ti—O single bond (351). The dinuclear unit of the complex (Fig. 48) is held together by bridging oxygen atoms of ethoxides [Ti(1)—O(1) = 2.001(7) Å and Ti(1)'—O(1) = 2.031(7) Å]. The terminal alkoxides have a relatively short Ti(1)—O(4) average bond length of 1.723(8) Å and a large Ti(1)—O(4)—C(4) angle of 170(1)°, which is normally taken as a reflection of the π donation of both oxygen lone pairs to the titanium atoms (351, 352).

An interesting X-ray crystallographically determined structure of *cis*-PtMe[OCH(CF$_3$)$_2$](PMe$_3$)$_2$[HOCH(CF$_3$)$_2$] (Fig. 49) (171), shows intermolec-

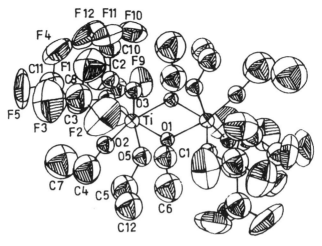

Figure 48.   The crystal structure of [Ti{OCH(CF$_3$)$_2$}$_2$(OEt)$_2$(HOEt)]$_2$. [Adapted from (350).]

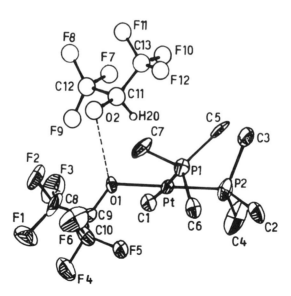

Figure 49.   The crystal structure of *cis*-PtMe[OCH(CF$_3$)$_2$](PMe$_3$)$_2$[HOCH(CF$_3$)$_2$]. [Adapted from (171).]

ular O—H· · ·O hydrogen bonding with a distorted square planar geometry around the platinum atom. The short distance [2.63(5) Å] between two oxygen atoms $O_1$ and $O_2$ supports the presence of O—H· · ·O hydrogen bonding between the alkoxide ligand and the coordinated alcohol molecule.

Single-crystal X-ray analysis showed that [La{OCMe(CF$_3$)$_2$}$_3$(thf)$_3$] (340) is a mononuclear octahedral complex (Fig. 50) with the facial arrangement of ligands. As expected, La—O(hftb) bonds (2.229, 2.237, and 2.222 Å) are significantly shorter than the La—O(thf) bonds (2.635, 2.610, and 2.59 Å). The shorter La—O(hftb) bonds suggest $\pi$ donation from the hftb oxygen to the lanthanum atom and is supported by the wide LaOC(hftb) angles (165.9, 160.9, and 176.1).

The crystallographic analysis of a copper(II) fluoroalkoxide, [Cu{OC-(CH$_3$)(CF$_3$)$_2$}$_2$(tmeda)] (334), reveals a rare example of monomeric tetrahedrally distorted square planar geometry (Fig. 51) with unidentate alkoxide ligands. There are two short intramolecular H· · ·F nonbonded contacts: H(1)· · ·F(21)c = 2.47(5) Å, H1· · ·F(31)c = 2.44(5) Å, and H(1)· · ·F(31)c = 2.44(5) Å. H(7)c· · ·F(3) = 2.55(5) Å and H(7)c· · ·F(3') = 2.51(4) Å

## 2.    Chloroalkoxide Derivatives

Despite the fact that metal fluoroalkoxides of many metals are known (Section IV.B.1), relatively few chloroalkoxide derivatives have been reported (351,

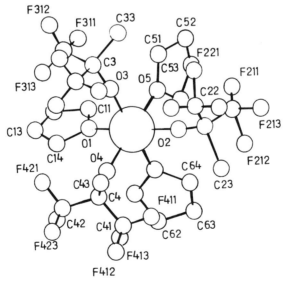

Figure 50.    The crystal structure of [La{OCMe(CF$_3$)$_2$}$_3$(thf)$_3$]. [Adapted from (340).]

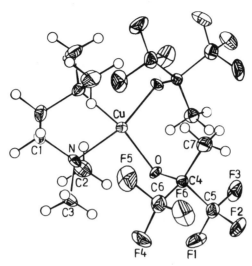

Figure 51. The crystal structure of $[Cu\{OC(CH_3)(CF_3)_2\}_2(tmeda)]$. [Adapted from (334).]

353–365) to date. Similar to fluoro analogues, these compounds also exhibit a tendency to form molecular addition complexes (351, 353–365). A distinct advantage of the chloroalkoxide groups is the enhancement in the steric bulk with reference to their fluoro analogues, and are thus more effective in reducing oligomerization.

**a. Synthesis.** Metal chloroalkoxides can be prepared by a variety of routes, for example, by solvolysis (Eq. 131), metathesis (Eqs. 132–134), alcoholysis (Eq. 135), and metal–carbon bond cleavage (Eq. 136) reactions. Typical reactions are

$$TiCl_4 + \geq 3HOCH_2CH_2X \xrightarrow[-2HCl]{CH_2Cl_2}$$

$$0.5[Cl_2Ti(OCH_2CH_2X)_2 \cdot HOCH_2CH_2X]_2 \tag{131}$$

$$X = Cl, Br, \text{ and } I$$

(Ref. 351)

$$MCl_n + xLiOCH_2CCl_3 \xrightarrow{\text{benzene}} MCl_{n-x}(OCH_2CCl_3)_x + xLiCl \downarrow$$

$$M = Fe, \ n = 3, \ x = 1, 2, \text{ or } 3 \text{ (Refs. 353, 354)} \tag{132}$$

$$MCl_n + xLiOCH_2CCl_3 \xrightarrow{\text{L(donor solvent)}}$$

$$MCl_{n-x}(OCH_2CCl_3)_xL + xLiCL \downarrow \tag{133}$$

M = V, $n = 3$, $x = 3$, L = OEt$_2$ (Ref. 355)

M = V, $n = 3$, $x = 2$, L = 2 OEt$_2$ (Ref. 356)

M = V, $n = 3$, $x = 1$, L = 2OEt$_2$ (Ref. 357)

M = Cr, $n = 3$, $x = 1, 2$, or 3, L = THF (Ref. 358)

M = Co, $n = 2$, $x = 1$ or 2, L = THF (Ref. 359)

$$VCl_3 + xLiOCH_2CH_2Cl \xrightarrow{Et_2O} VCl_{3-x}(OCH_2CH_2Cl)_x + xLiCl \downarrow \qquad (134)$$

$x = 2$ (Ref. 355), 3 (Ref. 360)

$$M(OCH_3)_2 + 2HOCH_2CCl_3 \xrightarrow{benzene} M(OMe)(OCH_2CCl_3) + MeOH \qquad (135)$$

M = Ni (Ref. 361)

M = Cu (Ref. 362)

$$AlEt_3 + 3HOCH_2CCl_3 \xrightarrow{Et_2O} Al(OCH_2CCl_3)_3 + 3C_2H_6$$

(Ref. 363)                                                            (136)


**b.   Coordination Chemistry.** Metal alkoxides in genreal do not form molecular addition complexes with common neutral ligands, because of the preferential intermolecular coordination of alkoxo oxygen leading to coordinately saturated oligomeric or polymeric species. The latter effect is more pronounced with sterically compact ligands such as ⁻OMe and ⁻OEt.

In order to activate a metal alkoxide to behave as an acceptor, the electron-donor ability of the alkoxo oxygen and the degree of oligomerization must be reduced (Section IV.B). Chloroalkoxide (e.g., OCH$_2$CCl$_3$) groups due to the presence of electron-withdrawing bulky chlorine atoms are capable of reducing the oligomeric nature of the metal alkoxide species more effectively. Furthermore, trichloroethoxide forms weak alkoxide bridges due to steric and electronic factors, which could easily be disrupted by hard donor (e.g., oxygen or nitrogen) ligands.

Chadha and co-workers reported the synthesis of a wide range of different types of molecular addition complexes of $3d$ transition metal trichloroethoxides. The principal types are VCl$_2$(OCH$_2$CCl$_3$)·2L (357), V(OCH$_2$CCl$_3$)$_3$·2L (355); CrCl$_2$(OCH$_2$CCl$_3$)·thf, CrCl(OCH$_2$CCl$_3$)$_2$·thf, Cr(OCH$_2$CCl$_3$)$_3$·thf (358); FeCl$_2$(OCH$_2$CCl$_3$)·2L (353), Fe(OCH$_2$CCl$_3$)$_3$·2L (354, 365); CoCl-(OCH$_2$CCl$_3$)·L, Co(OCH$_2$CCl$_3$)$_2$·L (359); Ni(OMe)(OCH$_2$CCl$_3$)·L (364); and Cu(OMe)(OCH$_2$CCl$_3$)·2L (362), where L may be bpy, dma, dmf, dmso, dioxane, HMPA, NMA, phen, ($\alpha$, $\beta$, $\gamma$)pic, py, RNH$_2$ tmu, Ph$_3$AsO, and so on.

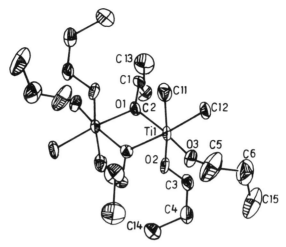

Figure 52.   The crystal structure of [TiCl$_2$(OCH$_2$CH$_2$Cl)$_2$.HOCH$_2$CH$_2$Cl]$_2$. [Adapted from (351).]

These molecular adducts were investigated by IR (351, 353–365), and electronic spectroscopy (353–365), magnetic measurements (353–365), mass spectrometry (in some cases) (357, 358, 361, 364), Mössbauer spectroscopy [for iron(III) complexes] (353, 354, 365). These physicochemical studies led to the proposal of plausible geometries.

**c.   X-Ray Crystallographic Studies.** Out of many metal chloroalkoxide complexes known for over 10 years, the X-ray crystal structure of only the titanium complex, [TiCl$_2$(OCH$_2$CH$_2$Cl)$_2$·HOCH$_2$CH$_2$Cl]$_2$, was determined (351). The molecule (Fig. 52) exists as a dimer in the solid state and lies on a crystallographic inversion center, which makes the halves of the molecule identical. The dimeric structure appears to be strongly favored by intramolecular O—H· · ·Cl hydrogen bonding.

The chlorine atoms on each titanium adopt a cis orientation, with Cl(1) being trans to the coordinated alcohol and Cl(2) being trans to the bridging alkoxide. The central structural feature of the structure shown in Fig. 52 is the planar Ti$_2$O$_2$ core with unsymmetrical Ti(1)—O(1) and Ti(1)′—O(1) bond distances of 1.956 and 2.087 Å, respectively. As expected, the coordinated alcohol Ti(1)—O(2) bond length (2.102 Å) is longer than the titanium—alkoxide Ti(1)—O(3) distance of 1.746 Å.

## C.   Metal Alkoxides Derived from Chelating Alcohols

As mentioned already, the associated nature of metal alkoxides could be ascribed to the tendency of the central metal atom for achieving a higher coordination state. This tendency could be satisifed with a lower degree of association by the use of chelating (e.g., alkoxyl/amino) alcohols.

In view of the oxides of metals such as the alkaline earth, the lanthanides (e.g., Y and La), and copper being constituents of superconductors (366), a more intensive research was initiated for the synthesis of soluble and comparatively (with respect to monodentate) less hydrolyzable alkoxides of such metals as well.

In the following sections a brief account of the synthesis and the structures of metal complexes of chelating alcohols are presented.

### 1. Alkoxyalkoxide Derivatives

**a. Synthesis.** These interesting (homo- and heterometallic) derivatives were prepared by a variety of methods similar to those already described for the derivatives of monodentate alcohols in Section II. These compounds are illustrated by the following reactions (Eqs. 137–154):
(II.A.1)*

$$x\mathrm{M} + (nx + y)\,\mathrm{ROH} \longrightarrow [\mathrm{M(OR)}_n]_x(\mathrm{ROH})_y + nx/2\,\mathrm{H_2} \uparrow$$

[M = Ca (Ref. 367), R = $C_2H_4OMe$, $n = 2$, $x = 9$, $y = 2$;

　　Ba (Ref. 368), R = $O(C_2H_4OMe)_zMe$, $n = 2$, $z = 2$ or 3,

　　$x = 0$, $y = 0$;　　　　　　　　　　　　　　　　　　　　　　(137)

　　Y (Ref. 369), R = $C_2H_4OMe$, $n = 3$, $x = 10$, $y = 0$;

　　Na, Mg, Ca, Sr, Ba, La (Ref. 370), R = $C_2H_4OMe$, $n = 1$(Na),

　　2(Mg, Ca, Sr, Ba), 3(La), $x = 1$, $y = 0$;

　　Al (Ref. 371); R = $C_2H_4OEt$, $n = 3$, $x = 1$, $y = 0$]

By contrast, reaction of barium granules in toluene with 2-methoxyethanol follows a different course, yielding an oxo-alkoxide (Eq. 138) (372):

$$6\mathrm{Ba} + 16\mathrm{HOC_2H_4OMe} \longrightarrow$$

$$\mathrm{H_4Ba_6(\mu_6\text{-}O)(OC_2H_4OMe)_{14}} + O(CH_2CH_2OMe)_2 + 6H_2 \quad (138)$$

(II.A.3)

Anodic dissolution of metals (e.g., Be, Y, Nd, Hf, Co, Ni, Cu, or Zn) in 2-methoxyethanol afford the corresponding alkoxide derivatives (370).
(II.C.1)

---

*The notation in parentheses indicates the section where these synthetic procedures can be found.

$$MCl_n(L)_x + nM'OR \longrightarrow M(OR)_n + nM'Cl \uparrow + xL \tag{139}$$

[M = Fe (Ref. 370), M' = Na, R = $C_2H_4OEt$, $n = 3$, $x = 0$;

    Ni (Ref. 373), M' = K, R = $C_2H_4OEt$, L = $C_5H_5N$, $x = 4$, $n = 2$;

    Cu(I) (Ref. 374), M' = Na, R = $C_2H_4OEt$, $n = 1$, $x = 0$;

    Cu(II) (Ref. 374), M' = Na, R = $C_2H_4OEt$, $n = 2$, $x = 0$;

    Bi(III) (Refs. 370, 375), M' = Na, R = $C_2H_4OEt$ (Ref. 370);

    $C_2H_4OMe$ (Ref. 375), $n = 3$, $x = 0$]

Alcoholysis reactions (Section II.D) were utilized for the synthesis of both types (homo- as well as heterometallic) of derivatives of alkoxyalcohols: (II.D)

$$M(OR)_n + nR'OH \longrightarrow M(OR')_n + nROH \tag{140}$$

[M = Y (Ref. 370), R = $i$-Pr, R' = $C_2H_4OMe$, $n = 3$;

    Ti (Ref. 376), R = $i$-Pr, R' = $C_2H_4OMe$, $n = 4$;

    Zr and Hf (Ref. 370), R = $i$-Pr, R' = $C_2H_4OMe$, $n = 4$;

    Nb (Ref. 370), R = Et or $i$-Pr, R' = $C_2H_4OMe$, $n = 5$;

    Cu (Ref. 377), R = Me, R' = $C_2H_4OMe$, $n = 2$;

    Bi (Ref. 375), R = Et or $i$-Pr, R' = $C_2H_4OMe$, $n = 3$]

$$MoO(OEt)_4 + 4HOC_2H_4OMe \longrightarrow MoO(OC_2H_4OMe)_4 + 4EtOH$$

(Ref. 378) $\tag{141}$

$$Y_5O(O\text{-}i\text{-}Pr)_{13} + 15HOC_2H_4OMe \longrightarrow 5Y(OC_2H_4OMe)_3 + 13i\text{-}PrOH + H_2O$$

(Ref. 369) $\tag{142}$

The last reaction involves the conversion of an oxo-alkoxide $Y_5O(O\text{-}i\text{-}Pr)_{13}$ into a homoleptic yttrium alkoxide $Y(OC_2H_4OMe)_3$, which is probably due to the higher stability of the latter.

$$[(thd)Cu(O\text{-}i\text{-}Pr)]_4 + 4HOC_2H_4OMe \longrightarrow$$

$$[(thd)Cu(OC_2H_4OMe)]_4 + 4Pr\text{-}i\text{-}OH \tag{143}$$

[thd = 2,2,6,6-tetramethyl-3,5-heptanedionato ligand (Ref. 379)]

$(\text{Pr-}i\text{-O})\text{Ba}\{\text{Al(O-}i\text{-Pr})_4 + n\text{ROH} \longrightarrow$

$\qquad \text{BaAl(O-}i\text{-Pr})_{5-n}(\text{OR})_n + n\text{Pr-}i\text{-OH}$ \hfill (144)

[where $n = 3$ or 5, R = $C_2H_4OMe$ (Ref. 380)]

$\text{ClM}\{\text{Al(O-}i\text{-Pr})_4\} + n\text{ROH} \longrightarrow$

$\qquad \text{ClM}\{\text{Al(O-}i\text{-Pr})_{4-n}(\text{OR})_n\} + n\text{Pr-}i\text{-OH}$ \hfill (145)

[M = Zn or Cd (Ref. 381), $n = 1, 2, 3,$ or 4, R = $C_2H_4OMe$]

$\text{M}\{\text{Al(O-}i\text{-Pr})_4\}_2 + 2n\text{ROH} \longrightarrow$

$\qquad \text{M}\{\text{Al(O-}i\text{-Pr})_{4-n}(\text{OR})_n\}_2 + 2n\text{Pr-}i\text{-OH}$ \hfill (146)

[M = Zn or Cd (Ref. 381), $n = 1, 2, 3,$ or 4, R = $C_2H_4OMe$]

Reactions of metal (M = Zn, Cd, Pb, or Bi) amides with 2-methoxyethanol afford structurally interesting derivatives (Eq. 147):
(II.E)

$$x\text{M(NR}_2)_n + (nx + y)\text{HOC}_2\text{H}_4\text{OMe} \longrightarrow$$

$$\cdot\ [x\text{M(OC}_2\text{H}_4\text{OMe})_n(\text{HOC}_2\text{H}_4\text{OMe})_y] + nx/2\text{H}_2 \uparrow \qquad (147)$$

[M = Zn (Ref. 106), R = $SiMe_3$, $n = 2$, $x = 1$, $y = 0$;

Cd (Ref. 382), R = $SiMe_3$, $n = 2$, $x = 9$, $y = 2$;

Pb (Ref. 228), R = $SiMe_3$, $n = 2$, $x = \infty$, $y = 0$;

Bi (Ref. 375b, 383), R = Me, $n = 3$, $x = \infty$, $y = 0$]

New copper(II) derivatives $[(tfd)Cu(OC_2H_4OMe)]_4$ and $[(hfd)Cu(OC_2H_4OMe)]_4$ were prepared by $\beta$-diketonate–alkoxide exchange reactions (379) (Eqs. 148, 149):
(II.F.3)

$$\text{Cu(tfd)}_2 + \text{KOC}_2\text{H}_4\text{OMe} \longrightarrow$$

$$\text{(Ref. 379)}\quad \text{or}$$

$$\text{Cu(hfd)}_2\ \tfrac{1}{4}[(tfd)Cu(OC_2H_4OMe)]_4 + K(tfd)$$

$$\text{or}\ \tfrac{1}{4}[(hfd)Cu(OC_2H_4OMe)]_4 + K(hfd) \qquad (148)$$

The reaction between $[Y(OC_2H_4OMe)_3]_{10}$ and $Cu(acac)_2$ (1 : 3 ratio) was carried out at room temperature over a period of 5 h. This reaction yielded several products, from which extraction by petroleum ether, followed by concentration of the product, produced white crystals of $Y_3(OC_2H_4OMe)_5(acac)_4$ (384) (Eq. 149):

$$[Y(OC_2H_4OMe)_3]_{10} + 3Cu(acac)_2 \longrightarrow$$

$$3Y_3(OC_2H_4OMe)_5(acac)_4 + [Cu(OC_2H_4OMe)(acac)]_x \qquad (149)$$

Novel types of *heterometallic alkoxymethoxide complexes* are accessible by the reactions of two monometallic alkoxide derivatives. (II.F.4)

$$Ba(OC_2H_4OMe)_2 + Ti(OC_2H_4OMe)_4 \longrightarrow BaTi(OC_2H_4OMe)_6 \quad (150)$$

(Ref. 385)

$$Ba(OC_2H_4OMe)_2 + [(acac)Cu(OC_2H_4OMe)]_2 \longrightarrow$$

$$Ba_2Cu_2(OC_2H_4OMe)_4(acac)_4 \cdot 2HOC_2H_4OMe + Cu(OC_2H_4OMe)_2 \downarrow$$

(Ref. 386) $\qquad\qquad\qquad\qquad\qquad\qquad\qquad\qquad\qquad (151)$

$$Ba(OC_2H_4OMe)_2 + [(thd)Cu(OC_2H_4OMe)]_4 \longrightarrow$$

$$\text{(Ref. 379)} \quad BaCu_4(thd)_4(OC_2H_4OMe)_6 \qquad\qquad (152)$$

$$\tfrac{1}{10}[Y(OC_2H_4OMe)_3]_{10} + [(thd)Cu(OC_2H_4OMe)]_4 \longrightarrow$$

$$\text{(Ref. 387)} \quad + (thd)_4Cu_3Y(OC_2H_4OMe)_5 + Cu(OC_2H_4OMe)_2 \downarrow \quad (153)$$

By contrast, the reaction of $[(hfd)Cu(OC_2H_4OMe)]_4$ with $[Y(OC_2H_4OMe)_3]_{10}$ affords a $Y_2Cu_2$ complex.

$$[(hfd)Cu(OC_2H_4OMe)]_4 + [Y(OC_2H_4OMe)_3]_{10} \longrightarrow$$

$$\text{(Ref. 387)} \quad [(hfd)_4YCu(OC_2H_4OMe)_3]_2 + \cdots \qquad (154)$$

**b.  X-Ray Crystallographic Studies.** The alkoxyethoxides such as 2-methoxyethoxide, $MeOCH_2CH_2O^-$, are expected to exhibit a variety of coordination modes, as depicted in Fig. 53, and many of these ligating modes have

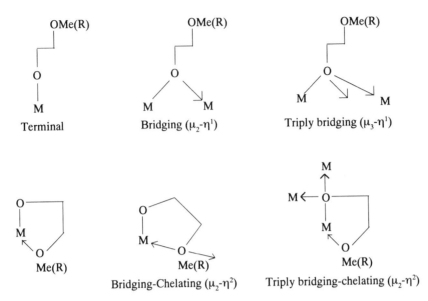

Figure 53.  Coordinating modes of 2-methoxyethoxide and related ligands.

been characterized by single-crystal X-ray crystallographic technique in any one of these forms or in a combination of two or more of these forms in a molecular or polymeric species.

Although chelation was observed for barium (368) and molybdenum (378) derivatives, the main tendency of alkoxyalkoxides is to act as bridging or bridging–chelating ligands. This favors the formation of large (but generally highly soluble) oligomers such as $[M(OC_2H_4OMe)_n]_x$ [M = Ca (Ref. 367), Cd (Ref. 382), $x = 9$, $n = 2$; M = Y (Ref. 369), $x = 10$, $n = 3$) or even infinite polymers $[M(OC_2H_4OMe)_n]_\infty$ (M = Cu (Ref. 377), Pb (Ref. 228), $n = 2$; M = Bi (Refs. 375, 383), $n = 3$].

Salient crystal data for X-ray crystallographically characterized compounds are included in Table V; crystal and molecular structures of only a few typical derivatives are displayed in Figs. 54–61, which is followed by a brief comment on their structural features.

The structural discussions are arranged in the following order: homometallic homoleptic, homometallic heteroleptic (other than oxoalkoxides, see Section V), heterometallic homoleptic, and heterometallic heteroleptic (except oxo-alkoxides, see Section V) derivatives. Further subdivision within each category is in order of increasing nuclearity; a similar pattern is adopted in arranging the derivatives included in Table V.

The X-ray crystal structures are now known for two nonamers of metal 2-methoxyethoxides: $M_9(OC_2H_4OMe)_{18}(HOC_2H_4OMe)_2$, where M = Ca (367)

TABLE V

Crystal Data for Some Selected Metal Complexes of Chelating Alcohols

*Derivatives of Alkoxyalkanols*

*Homometallic Derivatives*

| Compound | Crystal Class | Space Group | Z | $a$ (Å)<br>$b$ (Å)<br>$c$ (Å) | $\alpha$ (°)<br>$\beta$ (°)<br>$\gamma$ (°) | References |
|---|---|---|---|---|---|---|
| [Ca$_9$(OC$_2$H$_4$OMe)$_{18}$(HOC$_2$H$_4$OMe)$_2$]<br>(Fig. 54) | Triclinic | $P\bar{1}$ | 1 | 10.220(4)<br>15.515(5)<br>15.991(4) | 67.29(2)<br>87.17(3)<br>80.98(3) | 367 |
| [Y(OC$_2$H$_4$OMe)$_3$]$_{10}$<br>(Fig. 56) | Monoclinic | $P2_{1/c}$ | 4 | 13.640(6)<br>20.331(8)<br>24.804(9) | 90.0<br>95.45(3)<br>90.0 | 369 |
| [Y$_3$(OC$_2$H$_4$OMe)$_5$(acac)$_4$] | Monoclinic | $P2_{1/n}$ | 4 | 15.981(3)<br>10.637(3)<br>27.427(3) | 90.0<br>98.39(3)<br>90.0 | 384 |
| [(thd)Cu(OC$_2$H$_4$OMe)]$_4$<br>(Fig. 59) | Monoclinic | $P2_{1/n}$ | 2 | 12.450(5)<br>15.573(8)<br>17.324(9) | 90.0<br>93.68(4)<br>90.0 | 379 |
| [(hfd)Cu(OC$_2$H$_4$OMe)]$_4$<br>(Fig. 60) | Monoclinic | $C_{2/c}$ | 4 | 17.046(5)<br>17.211(5)<br>17.634(3) | 90.0<br>93.50(2)<br>90.0 | 379 |
| [Cd$_9$(OC$_2$H$_4$OMe)$_{18}$(HOC$_2$H$_4$OMe)$_2$]<br>(Fig. 55) | Triclinic | $P\bar{1}$ | 1 | 10.342(4)<br>13.666(5)<br>18.838(10) | 98.44(4)<br>113.48(4)<br>102.84(3) | 382 |
| [Pb(OC$_2$H$_4$OMe)$_2$]$_\infty$<br>(Fig. 57) | Monoclinic | $C_{2/c}$ | 4 | 11.411(8)<br>14.641(6)<br>6.278(3) | 90.0<br>108.34(5)<br>90.0 | 228 |
| [Bi$_2$(OC$_2$H$_4$OMe)$_6$]$_\infty$<br>(Fig. 58) | Triclinic | $P\bar{1}$ | 2 | 10.849(8)<br>12.041(9)<br>13.11(1) | 66.57(6)<br>73.54(6)<br>66.71(5) | 101, 375a |

# TABLE V
## Crystal Data for Some Selected Metal Complexes of Chelating Alcohols

| Compound | Crystal Class | Space Group | Z | a (Å) b (Å) c (Å) | α (°) β (°) γ (°) | References |
|---|---|---|---|---|---|---|
| *Heterometallic Derivatives* | | | | | | |
| [Ba$_2$Cu$_2$(OR)$_4$(acac)$_4$ · 2HOR] (where R = CH$_2$CH$_2$OCH$_3$) (Fig. 61) | Triclinic | P$\bar{1}$ | 1 | 10.797(3) 11.269(1) 12.109(1) | 106.18(1) 100.93(2) 102.98(2) | 386 |
| [BaCu$_4$(OC(R)C(H)C(R)O)$_4$(OR')$_2$(HOR')$_4$] (where R = t-Bu, R' = CH$_2$CH$_2$OCH$_3$) | Triclinic | P$\bar{1}$ | 2 | 14.285(5) 16.407(5) 19.409(6) | 77.57(3) 89.60(3) 81.14(3) | 379 |
| [(hfd)$_2$CuY(OCH$_2$CH$_2$OMe)$_3$]$_2$ | Monoclinic | P2$_{1/c}$ | 2 | 10.800(4) 20.067(7) 13.703(5) | 90.0 96.42(3) 90.0 | 387 |
| [(thd)$_4$Cu$_3$Y(OCH$_2$CH$_2$OMe)$_5$] | Monoclinic | C$_C$ | 4 | 33.084(5) 13.980(2) 18.789(3) | 90.0 121.70(10) 90.0 | 387 |
| [(hfd)$_2$(thd)$_2$Cu$_2$Y(OCH$_2$CH$_2$OMe)$_3$] | Triclinic | P$\bar{1}$ | 2 | 14.179(4) 14.425(4) 14.489(4) | 88.95(2) 71.45(2) 77.56(2) | 387 |
| *Aminoalkoxo Derivatives* | | | | | | |
| [Ba$_5$(OH)(OCHMeCH$_2$NMe$_2$)$_4$(thd)$_5$] | Triclinic | P$\bar{1}$ | 2 | 14.59(2) 15.41(2) 27.03(3) | 105.6(1) 96.1(1) 111.7(1) | 401 |
| [Y{(OCH$_2$CH$_2$)$_3$N}$_2$](ClO$_4$)$_3$ · 3C$_5$H$_5$N | Tetragonal | R$\bar{3}$ | 3 | 14.901(2) 14.901(2) 15.255(3) | 90.0 90.0 90.0 | 399a |
| [La{OC$_2$H$_4$)$_3$N}$_2${Nb(O-i-Pr)$_4$}$_3$] | Monoclinic | C$_{2/m}$ | 4 | 26.30(1) 15.183(2) 19.209(2) | 90.0 117.15(2) 90.0 | 399b |

| Compound | Crystal system | Space group | Z | a, b, c | α, β, γ | Ref. |
|---|---|---|---|---|---|---|
| $[Ce_2(O\text{-}i\text{-}Pr)_6(OC_2H_4NMeC_2H_4NMe_2)_2]$ | Monoclinic | $P2_{1/n}$ | 2 | 9.758(3) / 16.314(6) / 12.504(4) | 90.0 / 101.00(3) / 90.0 | 405 |
| $[(Ph_3SiO)Ti(OCH_2CH_2)_3N]$ | Tetragonal | $R\bar{3}$ | 6 | 9.5689(5) / 45.411(4) / 9.5689(5) | 90.0 / 90.0 / 90.0 | 403 |
| $[(MeCO_2)Ti(OCH_2CH_2)_3N]$ | Monoclinic | $P2_{1/n}$ | 2 | 15.563(3) / 6.872(1) / 19.743(4) | 90.0 / 106.65(1) / 90.0 | 403 |
| $[Cu(OC_3H_6NH_2)(HOC_3H_6NH_2)]$ | Monoclinic | $P2_{1/c}$ | 4 | 8.944(2) / 13.689(5) / 10.309(2) | 90.0 / 113.31(2) / 90.0 | 402b |
| $[Cu(OCHMeCH_2NMe_2)_2]$ (Fig. 62) | Monoclinic | $P2_{1/c}$ | 2 | 5.643(3) / 10.357(4) / 11.723(5) | 90.0 / 92.49(5) / 90.0 | 404 |
| $[Cu(OCH_2CH_2N(Me)CH_2CH_2NMe_2)_2]$ (Fig. 63) | Monoclinic | $Cc$ | 4 | 16.405(12) / 10.192(6) / 12.038(10) | 90.0 / 113.60(7) / 90.0 | 404 |
| *Phosphino-Alkoxo Derivatives* | | | | | | |
| $[NaOC\text{-}t\text{-}Bu_2CH_2CH_2PMe_2]_2$ | Orthorhombic | $Pna2_1$ | 4 | 15.530(5) / 8.397(7) / 22.365(8) | 90.0 / 90.0 / 90.0 | 417 |
| $[Y(OC\text{-}t\text{-}Bu_2CH_2CH_2PMe_2)_3]$ | Cubic | $Pa3$ | 8 | 20.521(2) / 20.521(2) / 20.521(2) | 90.0 / 90.0 / 90.0 | 416 |
| $[Nd(OC\text{-}t\text{-}Bu_2CH_2CH_2PMe_2)_3]$ | Rhombohedral | $R\bar{3}$ | 4 | 16.857(3) / 16.857(3) / 16.857(3) | 72.90(1) / 72.90(1) / 72.90(1) | 416 |
| $[TiCl(OC\text{-}t\text{-}Bu_2CH_2CH_2PMe_2)_2]$ | Orthorhombic | $P2_12_12_1$ | 4 | 10.214(1) / 13.964(2) / 21.028(2) | 90.0 / 90.0 / 90.0 | 417 |

## TABLE V
### Crystal Data for Some Selected Metal Complexes of Chelating Alcohols

| Compound | Crystal Class | Space Group | $Z$ | $a$ (Å) $b$ (Å) $c$ (Å) | $\alpha$ (°) $\beta$ (°) $\gamma$ (°) | References |
|---|---|---|---|---|---|---|
| *Phosphino-Alkoxo Derivatives (Continued)* | | | | | | |
| [Pt(OCMe$_2$CH$_2$PPh$_2$)$_2$] · 3.5H$_2$O | Rhombohedral | $R\bar{3}$ | 6 | 17.026(2) 17.026(2) 17.026(2) | 92.76(1) 92.76(1) 92.76(1) | 410 |
| [Pt(OCH$_2$CH$_2$PPh$_2$)$_2$] · H$_2$O | Monoclinic | $C2/c$ | 4 | 14.086(7) 11.217(5) 16.581(4) | 90.0 107.10(3) 90.0 | 419 |
| *Di- and Polyhydroxy Derivatives* | | | | | | |
| Na$_2$[Ti(OCH$_2$CH$_2$O)$_3$] · 4HOC$_2$H$_4$OH | Orthorhombic | $C222_1$ | 8 | 9.061(2) 15.387(8) 33.340(8) | 90.0 90.0 90.0 | 445 |
| K$_2$[Ti(OCH$_2$CH$_2$O)$_3$] · 2.5HOC$_2$H$_4$OH | Monoclinic | $P2_{1/n}$ | 4 | 9.385(2) 16.916(2) 11.837(2) | 90.0 92.31(2) 90.0 | 445 |
| [W(OCMe$_2$CMe$_2$O)$_3$] (Fig. 68) | Monoclinic | $C2/c$ | 4 | 16.841(2) 9.878(1) 13.373(1) | 90.0 109.60(1) 90.0 | 448 |
| Na$_2$[Al(OC$_2$H$_4$O)$_2$(OC$_2$H$_4$OH)] · 4HOC$_2$H$_4$OH | Triclinic | $P\bar{1}$ | 2 | 8.062(2) 8.660(2) 17.257(3) | 99.78(3) 97.61(3) 97.08(3) | 446 |
| Na$_2$[Al(OC$_2$H$_4$O)$_2$(OC$_2$H$_4$OH)] · 5HOC$_2$H$_4$OH | Triclinic | $P\bar{1}$ | 4 | 8.479(3) 19.557(5) 16.641(5) | 90.48(3) 96.68(2) 89.22(2) | 446 |

| Compound | Crystal system | Space group | Z | Cell parameters | Angles | Ref. |
|---|---|---|---|---|---|---|
| Na$_3$[Al$_3$(OC$_2$H$_4$O)$_5$(OC$_2$H$_4$OH)$_2$ · 6HOC$_2$H$_4$OH] | Monoclinic | $P2_{1/c}$ | 4 | 9.531(2)<br>21.490(4)<br>22.298(4) | 90.0<br>99.41(3)<br>90.0 | 446 |
| [Pb(O$_2$CCH$_2$OH)$_2$] | Orthorhombic | $Pbca$ | 8 | 7.1803(11)<br>10.346(2)<br>17.712(4) | 90.0<br>90.0<br>90.0 | 451 |
| [(H$_3$CC(CH$_2$O)$_3$)$_2$Ti$_4$(O-$i$-Pr)$_{10}$] (Fig. 66) | Monoclinic | $P2_{1/n}$ | 2 | 11.9127(10)<br>14.5313(14)<br>15.4676(15) | 90.0<br>106.191(7)<br>90.0 | 441 |
| [(H$_3$CCH$_2$C(CH$_2$O)$_3$)$_2$Ti$_4$(O-$i$-Pr)$_{10}$] | Monoclinic | $P2_{1/n}$ | 2 | 13.989(3)<br>10.806(2)<br>18.468(2) | 90.0<br>99.26(2)<br>90.0 | 441 |
| [(H$_3$CC(CH$_2$O)$_3$)$_2$Zr$_4$(O-$i$-Pr)$_{10}$] (Fig. 67) | Triclinic | $P\bar{1}$ | 4 | 11.5203(11)<br>21.705(2)<br>25.145(2) | 65.520(7)<br>86.015(8)<br>78.304(7) | 441 |
| [BaCu(OC$_2$H$_4$O)$_2$(HOC$_2$H$_4$OH)$_3$] | Monoclinic | $Cc$ | 4 | 12.103(6)<br>13.527(3)<br>17.091(8) | 90.0<br>93.17(2)<br>90.0 | 447 |
| [BaCu(OC$_2$H$_4$O)$_2$(HOC$_2$H$_4$OH)$_6$] | Monoclinic | $P2_{1/n}$ | 4 | 12.127(4)<br>11.913(1)<br>12.540(4) | 90.0<br>102.47(1)<br>90.0 | 447 |

367

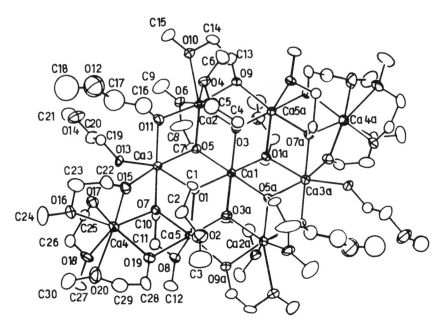

Figure 54.   The crystal structure of $Ca_9(OC_2H_4OMe)_{18}(HOC_2H_2OMe)_2$. [Adapted from (367).]

and Cd (382). The molecular structure of $[Ca_9(\mu_3, \eta^2\text{-}OC_2H_4OMe)_8(\mu_2, \eta^2\text{-}OC_2H_4OMe)_6(\mu_3, \eta^1\text{-}OC_2H_4OMe)_2(\mu_2, \eta^1\text{-}OC_2H_4OMe)_2(HOC_2H_4OMe)_2]$ is shown in Fig. 54, and consists of three six-coordinate and six seven-coordinate calcium atoms for an average coordination number of 6.67 in which all but two of the ether oxygen atoms are chelated to metal centers. In addition, there are two coordinated, but nonchelated, alcohol molecules.

Furthermore, structural features (Table V) of $Cd_9(OC_2H_4OMe)_{18}$-$(HOC_2H_4OMe)_2$ (Fig. 55) reveal it to be related to that of the calcium analogue (Fig. 54), although a lower coordination number is observed for cadmium(II), which is smaller than calcium(II). The molecular structure of the cadmium derivative consists of a centrosymmetric nearly planar arrangement of nine cadmium atoms with the overall structural formula being $[Cd_9(\mu_3, \eta^2\text{-}OC_2H_4OMe)_6(\mu_3, \eta^1\text{-}OC_2H_4OMe)_2(\mu_2, \eta^2\text{-}OC_2H_4OMe)_6(\mu_2, \eta^1\text{-}OC_2H_4OMe)_2(OC_2H_4OMe)_2 \cdot 2HOC_2H_4OMe$. This structure involves cadmium atoms in a six-coordinate environment. The two lattice alcohol molecules (at 4.12 Å from the metal) are hydrogen bonded to the terminal 2-methoxyethoxide ligands.

Despite similarities between both compounds, subtle differences are discerned within the $M_9(OC_2H_4OMe)_{18}$ frameworks: (a) the calcium derivative (M = Ca) consists of 8 triply bridging–chelating 2-methoxyethoxide

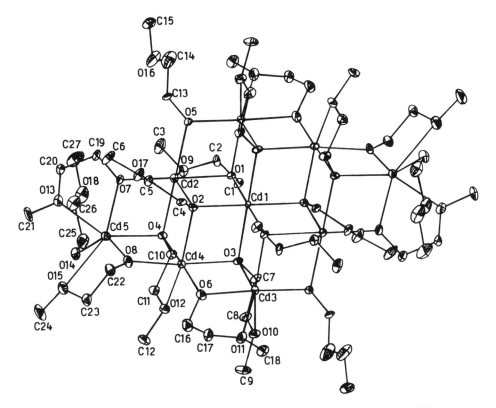

Figure 55.   The crystal structure of $Cd_9(OC_2H_4OMe)_{18}$ moiety. [Adapted from (382).]

groups, whereas in the cadmium analogue (M = Cd) 6 triply bridging–chelating and 2 triply bridging–monodentate alkoxide groups are present, with 10 remaining 2-methoxyethoxide groups in both the complexes present as $(\mu_2,\ \eta^2\text{-}OC_2H_4OMe)_6(\mu_2,\ \eta^1\text{-}OC_2H_4OMe)_2(\eta^1\text{-}OC_2H_4OMe)_2$; (b) two alcohol molecules in the calcium derivative are coordinated to the metal centers in monodentate fashion, whereas in the cadmium derivative, these are hydrogen bonded to the terminal alkoxide ligands; and (c) the average coordination number (C.N.) (6.67) of calcium(II) is higher than that of cadmium(II) (C.N. = 6). These variations are consistent with the differences in ionic radii [for coordination number 6] between calcium(II) and cadmium(II) (1.14 versus 1.09 Å].

Single-crystal X-ray diffraction studies (369) of the derivative $[Y(OC_2H_4OMe)_3]_{10}$ showed a centrosymmetric cyclic arrangement of 10 yttrium atoms, each in a seven-coordinate environment (Fig. 56) attaining a pentagonal bipyramidal arrangement.

Noteworthy features concerning this structure (Fig. 56) are (a) the coordi-

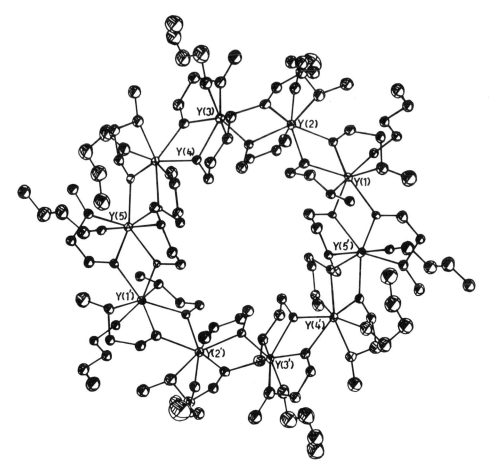

Figure 56.   The crystal structure of $\{Y(OC_2H_4OMe)_3\}_{10}$. [Adapted from (369).]

nation number 7 for yttrium, which is higher than that observed in some phases of $Y_2O_3$ (388), and (b) the structure depicted (Fig. 56) bears no relation to a close-packed structure, which is likely to maximize enthalpy while minimizing entropic considerations.

In addition to the above discrete polynuclear structures (Figs. 55 and 56), which were elucidated by X-ray analysis, only a limited number of polymeric structures were determined for homoleptic 2-alkoxyethoxides (see Table V). For example, $[Pb(OC_2H_4OMe)_2]_\infty$ (228) and $[Bi(OC_2H_4OMe)_3]_\infty$ (375, 383).

The lead(II) derivative $[Pb(\mu, \eta^1\text{-}OC_2H_4OMe)_2]_\infty$ crystallizes as a one-dimensional infinite polymer with four-coordinate lead centers where the potentially chelating 2-methoxyethoxide groups do not coordinate through ether ox-

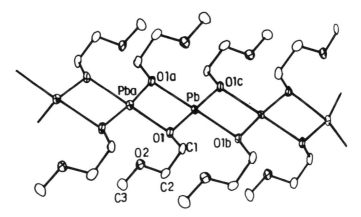

Figure 57.    The crystal structure of $[Pb(\mu, \eta^1\text{-}OC_2H_4OMe)_2]_\infty$. [Adapted from (228).]

ygens but adopt dangling conformations (228) (Fig. 57). The geometry about the lead atom is approximately trigonal bipyramidal, assuming that one of the coordination sites is occupied by a stereochemically active lone pair.

Although the cryoscopic molecular weight determination of $Bi(OC_2H_4OMe)_3$ in benzene depicts a dimeric nature, the X-ray crystallographic studies revealed the presence of one-dimensional (alkoxo-bridged) ribbon-like chains (Fig. 58) (375, 383) with all of the terminal and bridging 2-methoxyethoxide groups with a dangling instead of chelating configuration. The chains are built up from dimeric asymmetric $Bi_2(\mu_2, \eta^1\text{-}OC_2H_4OMe)_4(\eta^1\text{-}OC_2H_4OMe)_2$ units (Fig. 58). These repeat units are linked to each other through bridging alkoxo oxygen atoms, O(13) and O(23).

At the first approximation, all bismuth atoms are pentacoordinated and depict a distorted tetragonal pyramidal geometry. The Bi—O(alkoxo) bond lengths vary from 2.071(16) to 2.573(6) Å. The terminal bond is the shortest and is comparable to the value reported for $Bi(OC_6H_3OMe_2\text{-}2,6)_3$ (389). The O(28)—Bi—O(basal) angles (av 83.7°) are contracted as expected for an $AB_5E$ derivative, where E stands for the lone pair, which is presumably stereochemically active and occupies the sixth octahedral vertex of each bismuth atom. The nonbonding Bi· · ·Bi bond lengths along the chain are alternate: 3.6426(4) Å within the dimeric units [Bi(1)· · ·Bi(2)] and 3.957 Å(av) between two dimeric units.

The molecular structure of the dark blue derivative $[(thd)Cu(OC_2H_4OMe)]_4$ formed by the reaction of $[(thd)Cu(O\text{-}i\text{-}Pr)]_2$ with 2-methoxyethanol is shown in Fig. 59. This structure represents an example of a less common double-chain ladder, such as the centrosymmetrical tetranuclear copper(II) complex (379) with a complete sixfold coordination of the inner copper(1) and copper(1a) atoms.

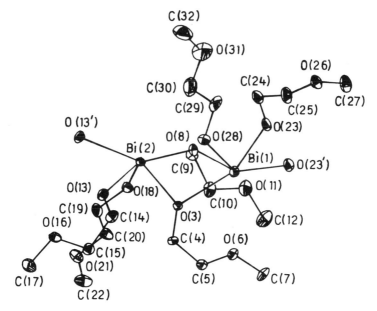

Figure 58. The crystal structure of $[Bi(OC_2H_4OMe)_3]_\infty$. [Adapted from (375, 383).]

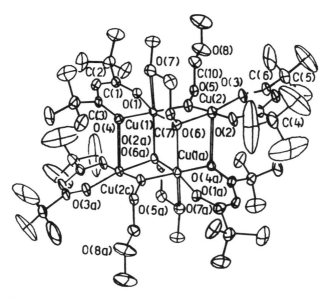

Figure 59. The crystal structure of centrosymmetrical complex $[(thd)Cu(OC_2H_4OMe)]_4$. [Adapted from (379).]

The atoms $Cu^2$ and $Cu^{2a}$ in Fig. 59 clearly adopt a pentacoordination configuration, whereas the inner $Cu^1$ and $Cu^{1a}$ atoms are hexacoordinated. The observed nonbonding $Cu(2) \cdots O(8)$ distance is 4.753(3) Å. The structure depicted above shows four additional $Cu \cdots O$ bonds, which can be described as two strongly interacting planar dimers. The interdimer separations are $Cu(1)-O(6a) = 2.326(4)$ Å; $Cu(2) \cdots O(4a) = 2.605(5)$ Å.

Figure 60 represents a different structural type: The tetrameric cubane system, in which the copper atoms are six coordinate, represents a distorted square pyramidal [4 + 2] environment for copper atoms. The $O(4)$ and $O(6)$ atoms of the 2-methoxyethoxide groups provide additional bonding to the sixth coordination site of the copper atoms. The $Cu(1) \cdots O(4)$ and $Cu(2) \cdots O(6)$ bond distances representing secondary coordinative interaction are 2.672(8) and 2.560(8) Å, respectively. The average $Cu-O$ (basal) bond distance is 1.949 Å.

A particularly interesting copper–barium–2-methoxyethoxide derivative, $Ba_2Cu_2(OC_2H_4OMe)_4(acac)_4 \cdot 2HOC_2H_4OMe$, was recently (386) isolated from the reaction of $(acac)Cu(OC_2H_4OMe)$ with $Ba(OC_2H_4OMe)_2$. The solid state structure is shown in Fig. 61 and indicates that a $\beta$-diketonate moiety was transferred from the copper center to barium.

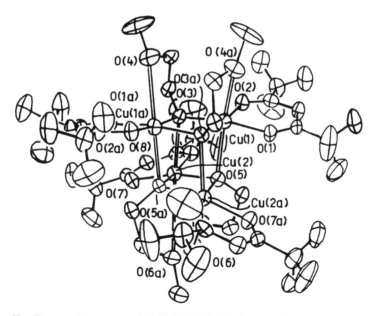

Figure 60. The crystal structure of $[(hfd)Cu(OC_2H_4OMe)]_4$ (crystallographic $C_2$ symmetry). [Adapted from (379).]

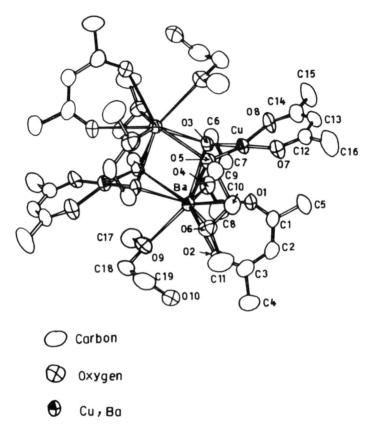

Figure 61.   The crystal structure of $Ba_2Cu_2(OC_2H_4OMe)_4(acac)_4 \cdot 2HOC_2H_4OMe$. [Adapted from (386).]

The metals are arranged in a rhomboid with triply bridging alkoxo oxygens above and below the two triangular planes consisting of two barium atoms and one of the copper atoms for a total of four bridging-alkoxo groups (Fig. 61). Each metal atom is also linked to an acac moiety. The barium is further coordinated by the methoxy oxygen of the triply bridging $OC_2H_4OMe$ ligands (two to each of the barium atoms), and a 2-methoxyethanol, resulting in a coordination number of 9 with a distorted capped rectangular antiprism environment around the barium atoms. The $Cu^{II}d^9$ has a square pyramidal geometry typical of Jahn–Teller distortions for copper(II). It is interesting to note that, in this system, the 2-methoxyethoxide ligands all chelate while the 2-methoxyethanol ligands dangle.

## 2. *Aminoalkoxide Derivatives*

Bidentate chelating aminoalcohols [e.g., $HOCH_2CH_2NH_2$, $HOCH_2$-$CH_2NHMe$, and $HOCH_2CH(R)NH_2$] with active hydrogen atoms on both the nitrogen and oxygen ends of the molecule are versatile mixed-donor (O or N) atom ligands (390). Although nitrogen could be expected to form a dative bond more readily, the higher acidity of the hydroxyl group will tend to make it more reactive. The resulting aminoalkoxide is generally a chelated species of Type **A** though in some cases nonchelating species of Type **B** were also obtained. On chelation, the amino protons tend to become more reactive and may form derivatives of the type $MOCH_2CH_2NH$ under suitable conditions. The latter type derivative may exist either as a mononuclear (**C**), di-, or higher nuclear derivative (**D**).

It is known that molecules with amino groups coordinated to a metal center often require elevated temperatures for activation of the hydrogen atom attached to nitrogen (6), whereas molecules with hydroxyl groups coordinated to the metal atom eliminate hydroxyl protons at room temperature or under mild reaction conditions.

The reactions of aminoalcohols, such as aminoethanol and *N*-methylaminoethanol, with a number of metal alkoxides were investigated in our laboratories (6). These alcohols depict an interesting variation in the reactivity pattern and nature of the final product(s). For example, in reactions of boron, aluminum, silicon, titanium, and zirconium alkoxides with these ligands only the hydroxyl group was found to be reactive. However, with niobium, germanium, and tin (including organometal species), both the hydroxyl as well as the amino groups (e.g., $NH_2$ or NHMe) take part in the replacement reactions yielding cyclic products.

Furthermore, reactions of di- and triethanolaminate derivatives were also investigated earlier in our, as well as other laboratories, which lead to the formation of interesting new compounds (6, 391). Among the atrane derivatives of various elements (392), the silatranes termed as "tryptych," were intensively studied (393, 394) because of their unusual structural (394–397) and physiological (393) properties.

The potential tridentate chelating behavior of diethanolamine (termed as "diptych," was already established by X-ray crystallographic studies on $(C_6H_5)_2Si(OC_2H_4)_2NH$ (398).

More recently (1994), discrete early transition metal (Ti, V, Nb, or Ta) alkoxide complexes containing homochiral trialkanolamine ligands (392) were prepared and their usefulness as highly enantioselective catalysts was demonstrated. It is noteworthy that earlier work on the reactions of tetradentate triethanolamine with transition metal alkoxides was reported to yield insoluble products (6).

Another interesting feature of the triethanolamine ligand was recently demonstrated in the isolation and X-ray crystallographic characterization of a hydrocarbon soluble and sublimable derivative, $[La\{(OC_2H_4)_3N\}_2\{Nb(O\text{-}i\text{-}Pr)_4\}_3]$ (399b).

The aminoalcohols also played a vital role as a modifier in the sol–gel process using metal alkoxides (400).

**a. Synthesis.** Methods used for the preparation of metal aminoalkoxide derivatives are illustrated by the following synthetic routes (Eqs. 155–71): (II.A.)*

$$Ba + HOCHMeCH_2NMe_2 \xrightarrow{\text{THF}}$$

$$\text{(Ref. 401)} \quad \{Ba_5(OH)[OCHMeCH_2NMe_2]_9\}_x + \cdots \qquad (155)$$

$$Ba + HOCHMeCH_2NMe_2 + thdH \xrightarrow{\text{THF}}$$

$$\text{(Ref. 401)} \quad Ba_5(\mu_4\text{-OH})(\mu_3, \eta^2\text{-OCHMeCH}_2NMe)_4(\mu, \eta^1\text{-thd}) + \cdots$$

$$(156)$$

$$Ba + HOCHMeCH_2NMe_2 + thdH \xrightarrow{\text{hexane}}$$

$$\text{(Ref. 401)} \quad [Ba(\mu, \eta^2:\eta^2\text{-thd})(\eta^2\text{-thd})(\eta^2\text{-HOCHMeCH}_2NMe_2)]_2 \qquad (157)$$

(II.C.1)

$$2CuX_2 + 4AOH \longrightarrow [Cu(OA)X]_2 + 2AOH \cdot HX \qquad (158)$$

[X = Cl or Br; AOH = 3-amino-1-propanols,

*N*-methyl-3-amino-1-propanols (Ref. 402a)]

(Ref. 403)  $TiCl_4 + (HOC_2H_4)_3N \longrightarrow ClTi(OC_2H_4)_3N + 3HCl \uparrow$  (159)

(II.D)

$$Cu(OMe)_2 + 2ROH \longrightarrow Cu(OR)_2 + 2MeOH$$

[R = CH_2CH_2NMe_2, CHMeCH_2NMe_2, CH_2CH_2NMeCH_2CH_2NMe_2

(Ref. 404)                                                                      (160)

$$M(OR)_n + N(CH_2CH(OH)R')_3 \xrightarrow{\text{THF}} M(OR)_{n-3}(OCHR'CH_2)_3N + 3ROH$$

[M = Ti, R = *i*-Pr, *n* = 4, R' = Me, *t*-Bu, Ph,                  (161)

cyclohexyl (Ref. 392), H (Ref. 403)

= Nb or Ta, R = Et (Ref. 392)]                              (162)

$$OV(O\text{-}n\text{-}Pr)_3 + N[CH_2CH(OH)R']_3 \longrightarrow OV(OCHR'CH_2)_3N$$

(Ref. 392)  + 3Pr-*n*-OH                                          (163)

$$La(O\text{-}i\text{-}Pr)_3 + 2(HOC_2H_4)_3N \longrightarrow H_3La\{(OC_2H_4)_3N\}_2 \downarrow + 3Pr\text{-}i\text{-}OH$$

(Ref. 399b)                                                            (164)

$$H_3La\{(OC_2H_4)_3N\}_2 + 3Nb(O\text{-}i\text{-}Pr)_5 \longrightarrow$$

(Ref. 399b)  $[La\{(OC_2H_4)_3N\}_2\{Nb(O\text{-}i\text{-}Pr)_4\}_3]$ + 3 Pr-*i*-OH  (165)

$$Ce(O\text{-}i\text{-}Pr)_4 + HOC_2H_4N(Me)C_2H_4NMe_2 \longrightarrow$$

(Ref. (405)  $[Ce(O\text{-}i\text{-}Pr)_3(\mu\text{-}OC_2H_4N(Me)C_2H_4NMe_2)]$ + Pr-*i*-OH

Interestingly, the latter reaction in 2:1 molar ratio afforded $Ce_3O_4(O\text{-}i\text{-}Pr)_2[OC_2H_4N(Me)C_2H_4NMe_2]_2$ (405):

(II.E)

$$Al(NMe_2)_3 + (HOC_2H_4)_3N \xrightarrow{\text{PhMe}}$$

(Ref. 406)  $[Al(OC_2H_4)_3N]_x$ + 3Me_2NH                    (166)

$$Ti(NMe_2)_4 + (HOC_2H_4)_3N \longrightarrow$$

(Ref. 403)   $(Me_2N)\overline{Ti(OC_2H_4)_3N} + 3Me_2NH$        (167)

**(II.F.2)**

$$Ga(CH_3)_3 + HOC_2H_4NH_2 \longrightarrow$$

(Ref. 407)   $(CH_3)_2\overline{GaOC_2H_4NH_2} + CH_4$        (168)

$$Al(C_2H_5)_3 + (HOC_2H_4)_3N \longrightarrow$$

(Ref. 408)   $[\overline{Al(OC_2H_4)_3N}]_x + 3C_2H_6$        (169)

$$R'BH_2 + R-N(C_2H_4OH)_2 \longrightarrow R'\overline{B(OC_2H_4)_2NR} + 2H_2$$   (170)

$[R' = (Me)_2CCH(Me)_2, R = H, Me, Fe (409)]$

**(II.F.3)**

$$Zn(OCH_2CH_2NMeCH_2CH_2NMe_2)_2 \xrightarrow[-2H_2]{170°C(10^{-4} \text{ mm})}$$

(Ref. 106)   $Zn(OCH = CHNMeCH_2CH_2NMe_2)_2$        (171)

Platinum group metal alkoxides are generally kinetically unstable with respect to $\beta$-hydrogen elimination (410–415). Recently, however, phosphino–alkoxide (hybrid) ligands displayed their considerable importance in the stabilization of novel types of platinum group metal alkoxides (Scheme 7). These types of ligands also played an important role in the synthesis of structurally interesting complexes of Li (418), Na (417), Ti$^{III}$ (417), and Zr$^{IV}$ (417) (Scheme 8).

  **b.  X-Ray Structures.** Similar to 2-alkoxyalcohols (see Fig. 53 and Section IV.C.1), aminoalcohols can also form a variety of derivatives with novel structural features. The replacement of oxygen donors in the polyether ligands with more basic amino groups resulted in the isolation of a number of monomeric, volatile copper(II) alkoxides, $Cu(OR')_2$, where $R' = CH_2CH_2NMe_2$, $CHMeCH_2NMe_2$, and $CH_2CH_2N(Me)CH_2CH_2NMe_2$ (404). Single-crystal X-ray crystallographic studies on two of these derivatives corroborate the notion

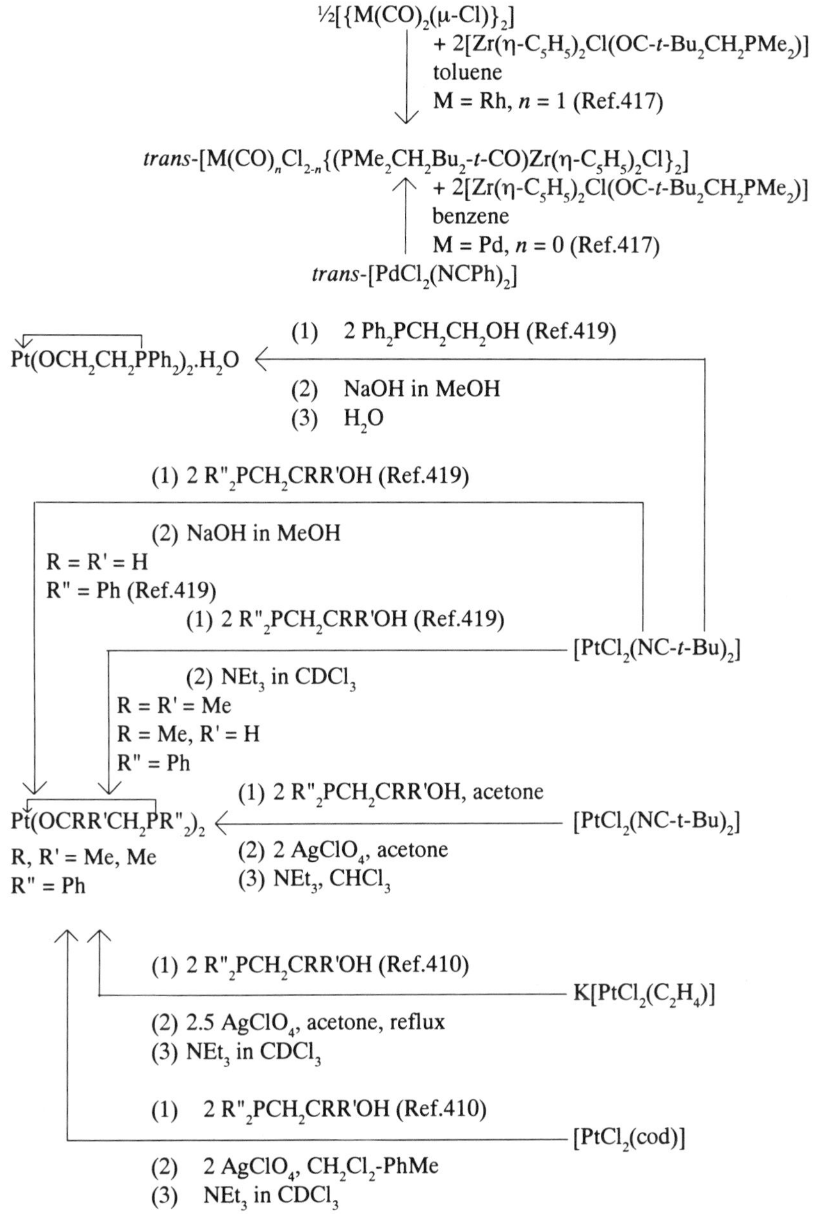

$\frac{1}{2}[\{M(CO)_2(\mu\text{-}Cl)\}_2]$
$+ 2[Zr(\eta\text{-}C_5H_5)_2Cl(OC\text{-}t\text{-}Bu_2CH_2PMe_2)]$
toluene
M = Rh, *n* = 1 (Ref.417)

$\downarrow$

*trans*-$[M(CO)_nCl_{2-n}\{(PMe_2CH_2Bu_2\text{-}t\text{-}CO)Zr(\eta\text{-}C_5H_5)_2Cl\}_2]$
$+ 2[Zr(\eta\text{-}C_5H_5)_2Cl(OC\text{-}t\text{-}Bu_2CH_2PMe_2)]$
benzene
M = Pd, *n* = 0 (Ref.417)

*trans*-$[PdCl_2(NCPh)_2]$

$Pt(OCH_2CH_2PPh_2)_2.H_2O \Longleftarrow$

(1)  2 $Ph_2PCH_2CH_2OH$ (Ref.419)

(2)  NaOH in MeOH
(3)  $H_2O$

(1) 2 $R''_2PCH_2CRR'OH$ (Ref.419)

(2) NaOH in MeOH
R = R' = H
R" = Ph (Ref.419)

(1) 2 $R''_2PCH_2CRR'OH$ (Ref.419)

(2) $NEt_3$ in $CDCl_3$
R = R' = Me
R = Me, R' = H
R" = Ph

$[PtCl_2(NC\text{-}t\text{-}Bu)_2]$

$Pt(OCRR'CH_2PR''_2)_2 \Longleftarrow$

R, R' = Me, Me
R" = Ph

(1) 2 $R''_2PCH_2CRR'OH$, acetone

(2) 2 $AgClO_4$, acetone
(3) $NEt_3$, $CHCl_3$

$[PtCl_2(NC\text{-}t\text{-}Bu)_2]$

(1) 2 $R''_2PCH_2CRR'OH$ (Ref.410)

(2) 2.5 $AgClO_4$, acetone, reflux
(3) $NEt_3$ in $CDCl_3$

$K[PtCl_2(C_2H_4)]$

(1)  2 $R''_2PCH_2CRR'OH$ (Ref.410)

(2)  2 $AgClO_4$, $CH_2Cl_2$-PhMe
(3)  $NEt_3$ in $CDCl_3$

$[PtCl_2(cod)]$

Scheme 7.  Synthetic routes for some homo- and heteroplatinum metal alkoxides.

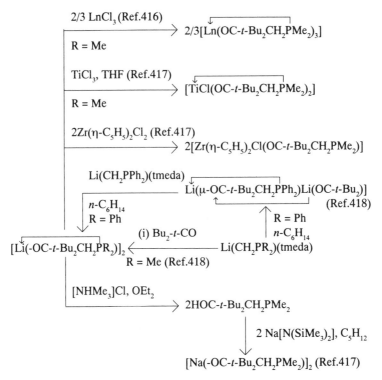

Scheme 8. Synthetic routes for lipophilicmetal [Li, Na, Y, Nd, Ti$^{III}$, and Zr$^{IV}$] derivatives of phosphinoalkoxide ligands.

that the amino group is coordinated to the metal center and results in the observed decrease in oligomerization compared to the ether analogues (Figs. 62 and 63).

The structure in Fig. 62 comprises trans square planar stereochemistry around a copper atom, with average Cu—O and Cu—N bond distances of 1.865(3) and 2.052(3) Å, respectively.

The crystal structure determination of Cu[OCH$_2$CH$_2$N(Me)CH$_2$CH$_2$NMe$_2$]$_2$ reveals a distorted square pyramidal environment around the metal center, which is formed by the tridentate ligation of one of the OCH$_2$CH$_2$N(Me)CH$_2$CH$_2$NMe$_2$ groups and bidentate ligation of the other. The result is a terminal NMe$_2$ group left dangling.

The axial Cu—N(4) bond length of 2.60(1) Å is significantly longer than the basal Cu—N distances of 2.11(1) Å (av), which is typical of Jahn–Teller distorted, five- and six-coordinate derivatives of copper(II) (420, 421). Also note that the Cu—O distances of 1.865(3) Å in Cu(OCHMeCH$_2$NMe$_2$)$_2$ and

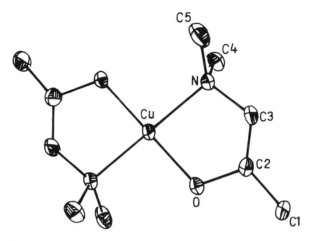

Figure 62. The crystal structure of Cu(OCHMeCH$_2$NMe$_2$)$_2$. [Adapted from (404).]

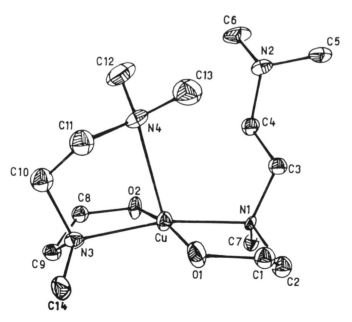

Figure 63. The crystal structure of Cu[OCH$_2$CH$_2$N(Me)CH$_2$CH$_2$NMe$_2$]$_2$. [Adapted from (404).]

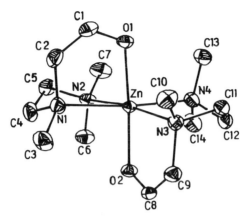

Figure 64.  The crystal structure of [Zn(1,4,5-$\eta^3$-OCH$_2$CHNMeCH$_2$CH$_2$NMe$_2$)$_2$]. [Adapted from (106).]

1.81(1)–1.88(1) Å in [Cu($\mu$-O-$t$-Bu)]$_4$ (422) are comparable to Cu[OCH$_2$CH$_2$N(Me)CH$_2$CH$_2$NMe$_2$]$_2$.

X-ray analysis of the compound Zn(OCH=CHNMeCH$_2$CH$_2$NMe$_2$)$_2$ (106) formed by the sublimation ($10^{-4}$ mmHg, 170°C) of Zn(OCH$_2$CH$_2$-NMeCH$_2$CH$_2$NMe$_2$)$_2$, shows that both the enolate ligands are bound to the zinc center in an $\eta^3$ fashion (Fig. 64) with oxygen atoms in trans positions, with a resulting $C_2$ symmetry for the molecule.

The zinc atom attains a distorted octahedral geometry, with the intraligand L—Zn—L angles compressed below the ideal of 90°. The Zn—O and Zn—N bond distances are 1.99(1) and 2.27(1)–2.35(1) Å.

### 3.   Derivatives of Di- and Polyhydroxy Alcohols

Prior to 1978, the work in this area mainly dealt with the synthesis of tri-, tetra-, and pentavalent metal and metalloid derivatives of aliphatic (glycol) and aromatic (catechol/substituted catechol) diols (6, 423), of a wide variety of structural types. These types depended on (a) the nature of the metal and metalloid and their oxidation states, (b) the type of diol, and (c) the reactants stoichiometries. Out of a large number of structural possibilities, the common types reported earlier are shown in Fig. 65, without X-ray crystallographic authentication.

Out of a number of diols, the catechol ($o$-hydroxybenzene) forms interesting anionic homoleptic metal(loid) complexes, such as eight-coordinate [M(cat)$_4$]$^{n-}$, where cat = catecholate dianion (424), $n$ = 4, M = Th (425), U (425), Ce (426), Hf (426); $n$ = 5, M = Gd (427), and six coordinate [M(cat)$_3$]$^{n-}$ with $n$

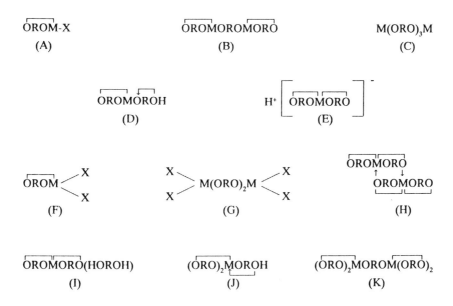

(where R = $-CH_2CH_2-$, $-CMe_2CMe_2-$, $-CHMeCH_2CMe_2-$, (and so on $-C_6H_4-$, being included mainly in the general introduction); M = B, Al, Ga, Sb, Ge, Sn, Ti, Zr, Nd, Ta; X = alkoxide (mainly) as well as amide/alkyl/aryl)

Figure 65. Structural variety in tri-, tetra-, and pentavalent metal(loid) complexes of aliphatic diols.

= 2, M = Mn (428), Sn (429), Si (430); $n$ = 3, M = Cr (431, 432), Fe (431), and V (433). In addition to these complexes, a few neutral derivatives of the type $[M(cat)_3]_x$ [$x$ = 2, M = W (434), Mo (435); $x$ = 1, M = Re (436), Tc (437), Os (438)] with the metal in the formal +6 oxidation state and tetra-chlorocatecholate (434–436) or di-*tert*-butylcatecholate (436–438) ligands are known. The -ate complexes generally were isolated as alkali metal (e.g., Na and K) or trialkylammonium ($NR_4^+$) salts; obviously, these are only soluble in protic solvents and are only slightly soluble or insoluble in aprotic polar solvents (429, 430). These properties inhibited their crystallization and structural elucidation by X-ray crystallography. However, the crystal structure of a ferric complex, $K_3[Fe(Cat)_3]$, was reported in 1976 by Raymond et al. (431).

Extensive work on the reactions of alkoxides of different metal(loid)s (e.g., B, Al, Ga, Si, Ge, Sn, Sb, Se, Te, Ti, Zr, Nb, Ta, lanthanides, and actinides) with a variety of glycols was carried out by Mehrotra and coworkers (6), but the results were not included since they were summarized (6) in 1978. For the last 15 years much research was carried out in this area. The following account deals mainly with their results.

The chemistry of metal derivatives of tri- and polyols is of only recent origin (439–441). The structural elucidation of such derivatives by X-ray crystallography (439–441) has added new dimension to the chemistry of metal alkoxides in general by providing opportunities for better insight into the course of the reaction, which could be much more useful in exploring this field.

Furthermore, the synthesis, spectroscopy, electrochemistry, and magnetic studies of chromium(III) derivatives of hexose recently reported (442) showed the possibility of potential research activity in transition metal saccharide chemistry in the near future.

In this section, we discuss metal derivatives of di- and polyols, excluding the extensive class of polyoxo alkoxide clusters derived from tri- and polyols for which excellent review articles (439, 440) are available.

**a.  Synthesis.** These interesting derivatives were prepared in good yields by a variety of methods (Eqs. 172–186).

$$2\text{THME}-\text{H}_3 + 2[\text{Zr}(\text{O-}i\text{-Pr})_4.(i\text{-Pr-OH})]_2 \xrightarrow[\text{reflux}]{\text{toluene/THF}}$$

$$(\text{Ref. 441})\quad (\text{THME})_2\text{Zr}_4(\text{O-}i\text{-Pr})_{10} + 8i\text{-Pr-OH}$$

$$(\text{THME})\text{-H}_3 = \text{tris(hydroxymethyl)ethane})$$

$$(172)$$

$$2\text{THMR}-\text{H}_3 + 6\text{Ti}(\text{O}-i\text{-Pr})_4 \xrightarrow[\text{reflux}]{\text{toluene/THF}} 6i\text{-Pr-OH}$$

$$(\text{Ref. 441})\quad 2\text{Ti}(\text{O-}i\text{-Pr})_4 + (\text{THMR})_2\text{Ti}_4(\text{O-}i\text{-Pr})_{10}$$

$$-6i\text{-Pr-OH}$$

$$2\text{THMR}-\text{H}_3 + 4\text{Ti}(\text{O-}i\text{-Pr})_4 \xrightarrow[\text{reflux}]{\text{toluene/THF}}$$
$$(\text{Ref. 441})$$

(R = E or P where THME-$\text{H}_3$ = tris(hydroxymethyl)ethane;
THMP-$\text{H}_3$ = tris(hydroxymethyl)propane)

$$(173)$$

$$\text{VO}(\text{OEt})_3 + 2\text{HO}-\text{R}'-\text{OH} \longrightarrow$$

$$(\text{Ref. 443})\quad \overline{\text{OROVO}(\text{OR'OH})} + 3\text{EtOH}$$

[R'(OH)$_2$ = ethylene glycol, *trans*-1,2-cyclohexanediol, (±)-2,3-butanediol, and (2R, 3R)-butanediol]

$$(174)$$

$$2VOCl_3 + 2HOCMe_2-CMe_2OH \xrightarrow{CH_2Cl_2}$$

$$(Ref. 444) \quad [\overline{OCMe_2-CMe_2OVOCl}]_2 + 4HCl \uparrow \qquad (175)$$

$$Ti(O\text{-}i\text{-}Pr)_4 + 7HOC_2H_4OH + 2NaOH \xrightarrow{-4i\text{-}Pr\text{-}OH, -2H_2O}$$

$$(Ref. 445) \quad Na_2Ti(OCH_2CH_2O)_3.4HOC_2H_4OH$$

$$\Big\uparrow -4H_2O$$

$$(Ref. 445) \quad TiO_2 + 7HOC_2H_4OH + 2NaOH \xrightarrow[\text{reflux}]{\text{toluene/THF}} \qquad (176)$$
$$(\text{excess})$$

$$Ti(O\text{-}i\text{-}Pr)_4 + 7HOC_2H_4OH + 2KOH$$

$$(Ref. 445) \xrightarrow[\substack{-4i\text{-}Pr\text{-}OH, \\ -1.5 HOC_2H_4OH, \\ -2 H_2O}]{} K_2Ti(OCH_2CH_2O)_3.2.5HOC_2H_4OH$$

$$(177)$$

$$Al(O\text{-}i\text{-}Pr)_3 + 7HOCH_2CH_2OH + 2NaOH$$

$$(Ref. 446) \xrightarrow[\substack{-3i\text{-}Pr\text{-}OH, \\ -2 H_2O}]{} Na_2Al(OCH_2CH_2O)_2(OC_2H_4OH).4HOC_2H_4OH$$

$$(178)$$

Two crystalline barium–copper–ethylene glycol derivatives were prepared by the reaction of homometallic ethylene glycolates of barium and copper (447) according to the reactions (Eq. 179) illustrated below.

$$Ba(\overline{OCH_2CH_2O})(HOC_2H_4OH)_x + Cu(\overline{OCH_2CH_2O})(HOC_2H_4OH)_y$$

$$(Ref. 447) \xrightarrow{HOC_2H_4OH} BaCu(OC_2H_4O)_2(HOC_2H_4OH)_6$$

$$\Big\downarrow \substack{\text{dilute } (<0.03\,M\,)\text{ solutions layered} \\ \text{with ethyl methyl ketone}}$$

$$BaCu(OCH_2CH_2O)_2(HOC_2H_4OH)_3 \qquad (179)$$

$$W(NMe_2)_6 + 3HOCMe_2\text{--}CMe_2OH \xrightarrow{Et_2O/C_6H_{14}}$$

$$\text{(Ref. 448)} \quad W(OCMe_2\text{--}CMe_2O)_3 + 6Me_2NH \qquad (180)$$

$$2o\text{-}C_6H_4(CH_2OH)_2 + 3AlMe_3 \xrightarrow{Et_2O}$$

$$\text{(Ref. 449)} \quad [o\text{-}C_6H_4(CH_2O)_2]_2Al_3Me_5 + 4CH_4 \qquad (181)$$

$$(182)$$

$$Al_2O_3 + 15HOC_2H_4OH + 4NaOH \longrightarrow$$
$$\text{(excess)}$$

$$\text{(Ref. 446)} \quad Na_2Al(OC_2H_4O)_2(OC_2H_4OH).4HOC_2H_4OH +$$

$$Na_2Al(OC_2H_4O)_2(OC_2H_4OH).5HOC_2H_4OH + 7H_2O$$

$$(183)$$

$$3Al_2O_3 + 13HOC_2H_4OH + 3NaOH \xrightarrow{CH_2Cl_2}$$
$$\text{(excess)}$$

$$\text{(Ref. 446)} \quad Na_2Al_3(OC_2H_4O)_5(OC_2H_4OH)_2.6HOC_2H_4OH +$$

$$7.5H_2O + 1.5Al_2O_3 \qquad (184)$$

$$PbCO_3 + HOCH_2COOH \xrightarrow{H_2O} [Pb(O_2CCH_2OH)_2] + H_2O + CO_2$$

$$\downarrow (\beta\text{-diketonate})M(OR)_2$$

$$\text{(Ref. 451)} \quad [Pb(O_2CCH_2O)_2M(\beta\text{-diketonate})_2] + 2ROH$$

$$(M = Ti, Sn; \beta\text{-diketonate} = acac, dpm; R = i\text{-Pr})$$

$$(185)$$

$$Sb(O\text{-}i\text{-}Pr)_3 + 2HOCMe_2\text{--}CMe_2OH \xrightarrow{\text{benzene}}$$

(Ref. 452)   $H[Sb(OCMe_2\text{--}CMe_2O)_2] + 3i\text{-}Pr\text{-}OH$

$\downarrow$ MOMe, MeOH

$M[Sb(OCMe_2\text{--}CMe_2O)_2] + MeOH$          (186)

(M = Li, Na. or K)

**b. Structures and General Properties.** X-ray crystal structures of many interesting derivatives such as $(H_3CC(CH_2\text{-}\mu_3\text{-}O)(CH_2\text{-}\mu\text{-}O)_2)_2Ti_4(O\text{-}i\text{-}Pr)_{10}$ (Fig. 66) (441), $(H_3CCH_2C(CH_2\text{-}\mu_3\text{-}O)(CH_2\text{-}\mu\text{-}O)_2)_2Ti_4(O\text{-}i\text{-}Pr)_{10}$ (441), and $(H_3CC(CH_2\text{-}\mu\text{-}O)_3)_2Zr_4(\mu\text{-}O\text{-}i\text{-}Pr)_2(O\text{-}i\text{-}Pr)_8$ (Fig. 67) (441), $K_2Ti(C_2H_4O_2)\cdot$ $2.5C_2H_6O_2$ (445), $Na_2Al(OCH_2CH_2O)_2(OCH_2CH_2OH)\cdot 4HOCH_2CH_2OH$ (446), $Na_2Al(OCH_2CH_2O)_2(OCH_2CH_2OH)\cdot 5HOCH_2CH_2OH$ (446), $Na_3Al_3\text{-}$ $(OCH_2CH_2O)_5(OCH_2CH_2OH)_2\cdot 6HOCH_2CH_2OH$ (446), $BaCu(C_2H_6O_2)_6\text{-}$ $(C_2H_4O_2)_2$ (447), $W(OCMe_2CMe_2O)_3$ (Fig. 68) (448), $[Pb(O_2CCH_2OH)_2]$ (451), $[C_{10}H_{20}ClO_7V_2]_2\cdot 4CHCl_3$ (453), $Tl[OC_{12}H_8(OH)](454)$, and $W(OCH_2CH_2O)_3$ (455) were recently determined. The crystallographic data are compiled in Table V. A few typical structures are shown in Figs. 66–68.

The structure shown in Fig. 66, consists of a planar eight-membered, rhom-

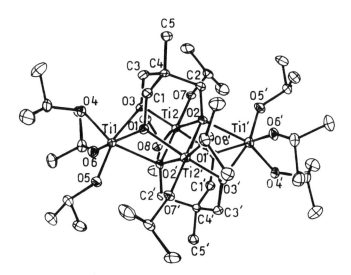

Figure 66.   The crystal structure of $(THME)_2Ti_4(O\text{-}i\text{-}Pr)_{10}$. [Adapted from (441).]

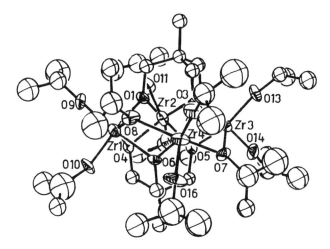

Figure 67.   The crystal structure of $(THME)_2Zr_4(O\text{-}i\text{-}Pr)_{10}$. [Adapted from (441).]

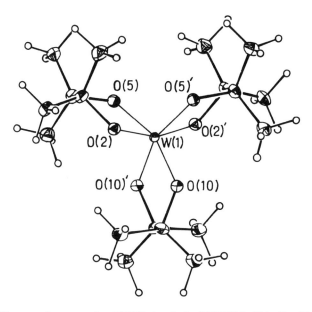

Figure 68.   The crystal structure (an ORTEP drawing) of $W(OCMe_2CMe_2O)_3$. [Adapted from (448).]

bohedrally shapped ring that is typical of a fused-$M_3O_{12}$ structure. In this structure, each titanium is surrounded by six oxygen atoms in a slightly distorted face-shared bioctahedral arrangement and possesses a crystallographic center of symmetry. These are two types (four $\mu$-O and two $\mu_3$-O) of bridging oxygen atoms associated with the THME ligands.

The derivative $(THME)_2Zr_4(\mu$-$O$-$i$-$Pr)_2(O$-$i$-$Pr)_8$ (Fig. 67), is also tetra-nuclear, but there are only $\mu$-O oxygen atoms in the structure. The molecule has an approximate $C_2$ axis passing through $Zr(2)$ and $Zr(4)$. Each zirconium atom is in a highly distorted six-coordinate edge-shared bioctahedral geometry.

Note that the structure shown in Fig. 67, is the first zirconium alkoxide with a $M_4(OR)_{16}$ edge-shared bioctahedral geometry (441).

The crystal structure of $W(OCMe_2CMe_2O)_3$, shown in Fig. 68 is consistent with $d^0$ W(VI) being surrounded by six uninegative O-$CMe_2$ ligands. The derivative has a distorted octahedral $WO_6$ core and a crystallographically imposed twofold axis of symmetry bisecting $O(10)$, $O(10)'$, and $W(1)$. The structure is very similar to that adopted by $W(OCH_2CH_2O)_3$ (448).

Interestingly, metal derivatives of di- and polyhydroxy alcohols are comparatively more hydrolytic resistant than their alkoxide analogues. Some of these compounds are even stable in alkaline solutions. The chelating nature of the alcohols and coordinative saturation achieved by the central metal atoms in the final products appear to be the main factor for their hydrolytic stability.

## D.  Derivatives of Unsaturated Alcohols

In this section, a brief account of the unsaturated alkoxide chemistry of metal(loid)s is presented with the main emphasis on metal enolates. The metal enediolates were also briefly mentioned merely to indicate the future potential for this type of derivatives.

Note that most of the metal enediolate derivatives could be derived from CO activation reactions (456–461); these types of reactions are of considerable importance in reactions catalyzed by $d$- and $f$-block metal compounds (462 a,b). The representative examples of CO activation reactions leads to the formation of diolate derivatives shown below.

$$(C_5Me_5)_2ZrMe_2 \; + \; 2CO \longrightarrow$$

$$\text{(Ref. 456)} \quad (C_5Me_5)_2Zr\{\overline{OC(Me)}{=}\overline{C(Me)O}\}$$

(187)

$$2(C_5Me_5)_2MMe_2 + 4CO \xrightarrow[-78°C]{\text{toluene}}$$

(Refs. 457, 460)   $[(C_5Me_5)_2M(\mu\text{-}O_2C_2Me_2)]_2$   (188)

$$O_2C_2Me_2 = \begin{array}{c} Me \\ {}^{-}O \end{array} C=C \begin{array}{c} Me \\ O^{-} \end{array}$$

$$(C_5Me_5)_2MR_2 + 2CO \xrightarrow{\text{toluene}} (C_5Me_5)_2\overline{MOC(R)}{=}C(R)O$$

M = Th, U; R = CH$_2$SiMe$_3$ (Ref. 457)

M = Th; R = CH$_2$-$t$-Bu (Ref. 457)                    (189)

The above type of dimerization of CO moieties is of potential significance as a model for reductive Fischer–Tropsch catalyzed carbon–carbon coupling (463, 464).

The homoleptic zinc enolate $Zn(1,4,7\text{-}\eta^2\text{-}OCH{=}CHNMeCH_2CH_2NMe_2)_2$, isolated from the dehydrogenation (Eq. 198) reaction of $Zn[OCH_2\text{-}CH_2NMeCH_2CH_2NMe_2)_2$ at $170°C(10^{-4}$ atm) was characterized by X-ray crystallography (106).

The enolate ligands such as $^{-}OCH{=}CH_2$ could conceivably be generated from the common solvent THF by metalation, as shown in Eqs. 190, 191.

$$\overline{OCH_2CH_2CH_2CH_2} + MR \xrightarrow[-RH]{} \overline{HMCHO.CH_2CH_2CH_2}   (190)$$

$$\overline{HMCH.O.CH_2CH_2CH_2} \longrightarrow MOCH{=}CH_2 + CH_2{=}CH_2   (191)$$

Such reactions are readily effected by *tert*-butyllithium.

Another distinct feature of an unsaturated alkoxide moiety (of the type shown above) is the presence of two bridging sites (i.e., oxygen as a hard donor) and $\pi$ electrons of the alkenic bond (i.e., as a soft donor center). This type of combination in an alkoxide ligand might be of considerable potential in binding two metal centers with widely different electronic and chemical characteristics to afford single-source precursors containing an early transition metal and a platinum group metal. Unfortunately, to the best of our knowledge such types of heterometallic alkoxide derivatives were not reported to date.

## 1. Synthesis

Synthesis of the first metal derivative containing an unsaturated alkoxide group appears to have been reported by Haslam (465) in 1955 according to Eq. 192.

$$Ti(O\text{-}i\text{-}Pr)_4 + 2MeCHO \longrightarrow Ti(OCH{=}CH_2)_2(O\text{-}i\text{-}Pr)_2 + 2i\text{-}Pr\text{-}OH \quad (192)$$

This reaction was suggested to proceed by alcohol interchange involving the vinyl alcohol formed by enolization of the acetaldehyde. Naturally, the instability of vinylic alcohols relative to tautomeric aldehydes or ketones precluded synthesis of their metal derivatives for a very long time.

Efforts were made to synthesize metal derivatives of unsaturated alcohols via three routes: (a) by the use of readily available and thermodynamically stable alcohols, (b) by the generation of thermodynamically less stable alcohols in situ followed by coordination to the metal centers (Eq. 192), and (c) dehydrogenation of a metal alkoxide derivative.

Route (a) may be subdivided into two main groups: the alcohol exchange reactions (Eq. 193), and reactions of metal(loid) chlorides with alcohols in the presence of a suitable chloride acceptor (Eqs. 195–197). Some of these types of reactions are illustrated by the Eqs. 193–197:

$$M(O\text{-}i\text{-}Pr)_n + nROH \xrightarrow{\text{benzene}} M(OR)_n + n\ i\text{-}Pr\text{-}OH$$

$$R = CH_2{=}CHCH_2 \text{ (Ref. 466)}, CH_3CH{=}CHCH_3 \text{ (Ref. 467)},$$

$$CH_2{=}CHC(CH_3)_2 \text{ (Ref. 468)}, CH_2{=}(CH_3)CH_2 \text{ (Ref. 469)}, \quad (193)$$

$$CH_2{=}CHCH_2CH_2 \text{ (Ref. 470)},$$

$$\text{when } M = B, Al\ (n = 3); M = Ti \text{ or } Ge\ (n = 4);$$

$$M = Nb \text{ or } Ta\ (n = 5)$$

$$CH_2{=}CHOH + LiBu \xrightarrow[(471)]{\text{THF}} LiOCH{=}CH_2 + C_4H_{10} \quad (194)$$

$$[(C_5H_4R)_2LnCl]_2 + 2LiOCH{=}CH_2 \xrightarrow[-2LiCl]{\text{THF}}$$

$$\text{(Ref. 472)} \quad [(C_5H_4R)_2Ln(\mu\text{-}OCH{=}CH_2)]_2 \quad (195)$$

$$Ln = Y, Yb, \text{ or } Lu; R = H$$

$$Ln = Y; R = Me$$

$$Cp_2MCl_2 + 2LiOCH{=}CR_2 \xrightarrow[(471)]{} Cp_2M(OCH{=}CR_2)_2 + 2LiCl \quad (196)$$

$$M = Ti; R = H \text{ or } Me$$

$$M = Zr; R = Me$$

$$Cp_2M(Me)Cl + LiOCH{=}CR_2 \xrightarrow[(471)]{} Cp_2M(Me)OCH{=}CR_2 + LiCl \quad (197)$$

M = Ti; R = H or Me

M = Zr; R = Me

$$Zn(OCH_2CH_2NMeCH_2CH_2NMe_2)_2 \xrightarrow[-2H_2]{170°C(10^{-4}atm)}$$

$$Zn(1,4,7{-}\eta^3{-}OCH{=}CHNMeCH_2CH_2NMe_2)_2 \text{ (Ref. 106)} \quad (198)$$

### 2.  Properties and Structures

The IR spectra of these derivatives exhibit a strong absorption in the $1600{-}1620\text{-cm}^{-1}$ region, which is attributable to a $C{=}C$ stretch (471, 472). Each spectrum also contains a strong band between 1160 and 1120 cm$^{-1}$ in the $C{-}O$ stretching region. Absorptions in the 1260–1310 and 1320–1390-cm$^{-1}$ regions may be attributed to in-place $={}$CH and $={}$CH$_2$ deformation vibrations, respectively (471, 472).

The $^1$H NMR spectra (471, 472) of yttrium, lutetium, and titanium derivatives are consistent with structures in which the oxygen atom of the enolate bonds to the metal center. Resonances due to the $={}$CH group appear as doublets of doublets. The signals due to the $={}$CH$_2$ protons appears as two doublets of doublets.

X-ray crystal structure of $[(MeC_5H_4)_2Y(\mu{-}OCH{=}CH_2)]_2$ is shown in Fig. 69; the complex consists of a dimer in which the two methylcyclopentadienyl

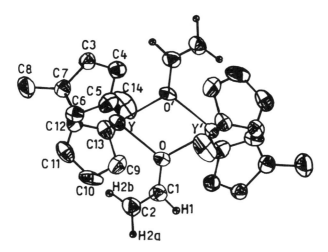

Figure 69.   The crystal structure of $[(MeC_5H_4)_2Y(\mu{-}OCH{=}CH_2)]_2$. [Adapted from (472).]

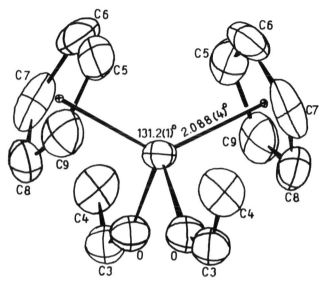

Figure 70. The crystal structure (an ORTEP plot) of $(C_5H_5)_2Ti(OCH=CH_2)_2$. [Adapted from (471).]

rings and the two bridging enolate oxygen atoms coordinate to each yttrium atom in a roughly tetrahedral geometry. The structure shown in Fig. 69 is typical of bent metallocene derivatives that contain two additional ligands.

X-ray crystallography established the structure of $(C_5H_5)_2Ti(OCH=CH_2)_2$ shown in Fig. 70 (471). The molecule possesses imposed $C_2$ symmetry. The staggering of the Cp rings is similar to that found in $Cp_2TiCl_2$ (473).

Figure 71 shows X- ray crystallographically determined structure of the unsolvated lithium enolate of *tert*-butyl methyl ketone. This structure is novel. The crystallographic asymmetric unit consists of two independent half-hexamers (474).

## V.  CHEMISTRY OF OXO-ALKOXIDES

### A.  Introduction

Because of the extremely high susceptibility of metal alkoxides to even traces of moisture, stringent precautions for maintenance of anhydrous conditions (solvents, reagents, equipment, etc.) as well as careful protection from atmospheric moisture were generally regarded as essential conditions for the synthesis of simple binary or heterometallic alkoxides in a pure state.

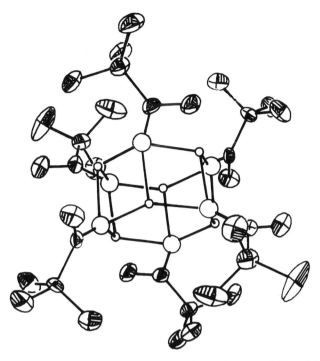

Figure 71.   The crystal structure of an unsolvated lithium enolate anion. [Adapted from (474).]

This facile hydrolyzability of alkoxides is now being utilized extensively in the preparation of ceramic materials by the sol–gel process (Section VII). Interest in the mechanism of such hydrolysis reactions [cf. Bradley et al. (6)] and the characterization of oxo-alkoxides has therefore grown. The structures of only a few oxo-metal alkoxides, $[Ti_7O_4(OC_2H_5)_{20}]$ (475a), $[Nb_8O_{10}(OC_2H_5)_{20}]$ (476a), $[Zr_{13}O_8(OCH_3)_{36}]$ (477a) and $[Sn_6O_4(OCH_3)_4]$ (478a) were studied up to 1978, with a reconfirmation of the titanium (475b) and niobium (476b) species done in 1991. However, since 1989 there were a growing number of examples of the isolation and characterization by sophisticated techniques (specially X-ray crystallography) of the oxo-alkoxide species of several metals that were obtained (for reasons still not fully understood) by synthetic routes (see Sections II.A.1, II.A.2, II.C.1, II.C.2, and II.F.4) by which normal homo- and heterometal alkoxides were expected.

Initially, there was a natural skepticism that the formation of these oxo-alkoxides could have resulted from some stray hydrolytic side reaction(s). For example, while reporting such a species, $Y_5O(O-i-Pr)_{13}$, for the first time in 1989 with the reaction of yttrium metal and isopropanol (478), the Poncelet et al. (478) reported that these isolated oxo-isopropoxide crystals did not form

$Y_5\{Al(O-i-Pr)_4\}_3$ on condensation with 3 mol of $Al(O-i-Pr)_3$, whereas the mother liquor of the reaction mixture could yield the expected $Y\{Al(O-i-Pr)_4\}_3$ in fairly high ($\sim 70\%$) yield. Although similar comments were made initially by Mehrotra (479), many of these oxo-alkoxides were formed by a variety of conventional routes, which were expected to yield the normal alkoxides. This number has been increasing so fast in spite of the most stringent anhydrous conditions employed by the investigators, that concerted efforts are being made to suggest novel pathways by which such oxo-alkoxide species could result under nonhydrolytic conditions as well.

For example, the research group of Turova et al. (480) found evidence for the formation of such oxide species through the elimination of ether from homo- or heterometal alkoxides:

$$>M<\genfrac{}{}{0pt}{}{\overset{\delta-}{O}\overset{\delta+}{R}}{O\underset{\delta-}{R'}} \quad \xrightarrow[-R_2O]{} \quad >M=O \qquad (199)$$

Ether elimination appears to be involved in the formation of $Pb_4Ti_4O_3(O-i-Pr)_8$ (481) during the condensation of isopropoxides of lead and titanium.

$$\frac{4}{\infty}[Pb(O-i-Pr)_2]_\infty + 4Ti(O-i-Pr)_4 \xrightarrow[-3i-Pr_2O]{toluene, isopropanol} Pb_4Ti_4O_3(O-i-Pr)_{18} \qquad (200)$$

An alternative route of conversion of rather sterically hindered ditertiary and disecondary alkoxides of barium [e.g., $Ba(OCEt_3)_2$, $Ba(OCMeEt-i-Pr)_2$, $Ba(OCMe_3)_2$, and $Ba\{OCH(CMe_3)_2\}_2$] into oxo-alkoxides was shown (346) (by mass spectroscopy) to occur by elimination of small amounts of alkenes (with parent alcohols) or by the ketone, [e.g., $(Me_3C)_2C=O$]. Although the exact mechanism has not been investigated to date, the conversion of $KZr_2(O-i-Pr)_9$ into the oxo-species $K_4Zr_2O(O-i-Pr)_{10}$ was proposed (117) to involve the fission of C—O bonds of the isopropoxide groups $Me_2CHO$, yielding the hydrocarbon $(Me_2CH)_2$ and the oxozirconium species, $K_4Zr_2O(O-i-Pr)_{10}$.

The latter two mechanisms required detailed mechanistic studies for understanding the involved rationale. The mechanism of the conversion of condensation with acetone of hydrocarbon soluble $Zn(OCEt_3)_2$ to a rigid gel (Eq. 201), followed by conversion into ZnO (Eq. 202) by an aldol condensation was elucidated (482) by isolation of intermediate products.

$$4 \underset{\text{Acetone}}{\overset{O}{\|}} + Zn(OCEt_3)_2 \xrightarrow{C_6H_6} Gel \qquad (201)$$

$$(202)$$

The role of atmospheric oxygen in the crystallization of $[Ca_6(\mu_4\text{-}O)_2(\mu_3\text{-}OEt)_4(OEt)_4]\cdot 14EtOH$ from a refluxing solution of $Ca(OEt)_2$ in ethanol was emphasized by Turova et al. (483) as this oxo-ethoxide species does not crystallize out under complete isolation from oxygen (deaerated alcohol under an helium atmosphere). It was suggested that the oxo groups originate from the oxidation of $Ca(OEt)_2$ with the formation of peroxide derivatives and their subsequent decomposition.

The interconversion (164) between oxo and alkoxo complexes through the removal of an alkyl group as carbocation ($R^+$), a radical (R), or a carbanion ($R^-$) was already mentioned in Section III.A.2.

## B.  Methods of Formation

### 1.  General

As mentioned in the above introduction, the formation of a number of hydroxo- and oxo-alkoxide products of different metals was reported (till 1978) (1, 6) during the extensive hydrolytic studies carried out on metal alkoxides and even crystal structures of a few oxide-alkoxides of Ti (475a), Nb (476a), Zr (477a), and Sn(II) (477b).

Since 1989, however, various oxo-alkoxides (both homo as well as hetero) were isolated by a variety of methods that could be expected to yield the normal alkoxides under nonhydrolytic conditions. These are summarized in Section V.B.2 by a few illustrative equations. This is followed by Section V.B.3, which indicates a few special types of reactions that yield oxo-alkoxide derivatives. Finally, salient crystal data of some oxo-alkoxides are summarized in Table VI.

### 2.  Formation During Routine Procedures

Although most of the preparations appear to have been so far unplanned, the following equations represent the isolation and characterization of oxo-alkoxide species by conventional routes (Sections II.A.1, II.A.2, II.C.1, II.C.2, and II.F.4), which are expected to yield the normal homo- or heterometal alkoxides.

TABLE VI

Crystal Data for Some Oxo-Alkoxides of Metals and Bridging Modes of Oxo Ligands

| Compound (Bridging Mode of Oxo Ligand) | Crystal Class | Space Group | Z | $a$ (Å) $b$ (Å) $c$ (Å) | $\alpha$ (°) $\beta$ (°) $\gamma$ (°) | References |
|---|---|---|---|---|---|---|
| *Homometallic Derivatives* | | | | | | |
| [Ca$_6$O$_2$(OEt)$_8$]·14EtOH ($\mu_4$) | Monoclinic | $P2_{1/c}$ | 2 | 12.697(3) 22.501(4) 13.835(3) | 90.0 112.88(2) 90.0 | 483 |
| [H$_3$Ba$_6$(O)(O-$t$-Bu)$_{11}$(OCEt$_2$CH$_2$O)(thf)$_3$] ($\mu_5$) | Triclinic | $P\bar{1}$ | 2 | 14.537(5) 22.707(7) 14.524(5) | 91.35(1) 102.64(1) 102.67(1) | 486 |
| [H$_4$Ba$_6$(O)(OCH$_2$CH$_2$OMe)$_{14}$] ($\mu_6$) | Triclinic | $P\bar{1}$ | 1 | 12.980(2) 13.059(2) 12.044(2) | 96.57(1) 106.51(1) 114.95(1) | 372 |
| [Y$_5$O(O-$i$-Pr)$_{13}$] ($\mu_5$) (Fig. 77) | Orthorhombic | $Pbca$ | 8 | 25.512(14) 21.549(12) 20.716(13) | 90.0 90.0 90.0 | 478 |
| [Ce$_4$O(O-$i$-Pr)$_{13}$(HO-$i$-Pr)] ($\mu_4$) (Fig. 75) | Monoclinic | $C2/c$ | 4 | 21.405(6) 14.077(3) 20.622(6) | 90.0 103.97(1) 90.0 | 500 |
| [Nd$_5$O(O-$i$-Pr)$_{13}$(HO-$i$-Pr)$_2$] ($\mu_5$) (Fig. 78) | Orthorhombic | $Aba2$ | 4 | 21.343(6) 21.109(4) 14.338(3) | 90.0 90.0 90.0 | 302 |
| [Yb$_5$O(O-$i$-Pr)$_{13}$] ($\mu_5$) | Monoclinic | $P2_{1/n}$ | 4 | 12.878(4) 21.621(4) 21.377(3) | 90.0 91.38(2) 90.0 | 484 |
| [U$_3$O(O-$t$-Bu)$_{10}$] ($\mu_3$) (Fig. 74) | Hexagonal | $P6_3mc$ | 2 | 18.256(4) 18.256(4) 11.013(2) | 90.0 90.0 120.0 | 487 |

## TABLE VI
### Crystal Data for Some Oxo-Alkoxides of Metals and Bridging Modes of Oxo Ligands

| Compound (Bridging Mode of Oxo Ligand) | Crystal Class | Space Group | Z | a (Å) b (Å) c (Å) | α (°) β (°) γ (°) | References |
|---|---|---|---|---|---|---|
| | | | | *Homometallic Derivatives* (Continued) | | |
| [UO$_2$(O-t-Bu)$_2$(OPPh$_3$)$_2$] (terminally bonded) | Monoclinic | P2$_{1/n}$ | 4 | 10.801(1) 21.758(5) 18.386(5) | 90.0 97.98(1) 90.0 | 494 |
| [Ti$_7$O$_4$(OEt)$_{20}$] (μ$_3$/μ$_4$) | Triclinic | P$\bar{1}$ | 2 | 13.91(5) 20.212(9) 12.162(5) | 90.49(3) 108.20(4) 74.65(2) | 475, 475b |
| [VO(OCH$_2$CH$_2$Cl)$_3$] (terminally bonded) (Fig. 73) | Triclinic | P$\bar{1}$ | 2 | 7.662(2) 9.075(4) 9.570(2) | 82.08(3) 66.90(2) 79.42(3) | 496 |
| [Nb$_8$O$_{10}$(OEt)$_{20}$] (μ$_2$/μ$_3$) | Monoclinic | P2$_{1/n}$ | 2 | 14.9169(9) 14.2541(7) 16.8726(8) | 90.0 91.754(4) 90.0 | 476a, 476b |
| [Mo$_3$O(O-i-Pr)$_{10}$] (μ$_3$) | Orthorhombic | Pbcn | 8 | 35.56(2) 18.97(1) 19.34(1) | 90.0 90.0 90.0 | 506 |
| [Mo$_6$O$_{10}$(O-i-Pr)$_{12}$] (μ$_2$/terminally bonded) | Triclinic | P$\bar{1}$ | 1 | 13.082(3) 11.478(2) 9.760(2) | 106.40(1) 91.85(1) 99.81(1) | 507 |
| [W$_3$O$_2$(O-t-Bu)$_8$] (μ$_2$) | Monoclinic | P2$_{1/n}$ | 4 | 17.886(2) 12.374(1) | 90.0 99.80(1) 90.0 | 498 |
| [W$_4$O$_2$(O-i-Pr)$_{12}$] (μ$_2$) | Triclinic | P$\bar{1}$ | 2 | 13.386(7) 19.426(15) 10.250(6) | 99.28(4) 104.20(3) 94.52(3) | 497 |
| [Al$_4$O(OCH$_2$CF$_3$)$_{11}$]$^-$ (μ$_4$) | Orthorhombic | Pna2$_1$ | 4 | 16.875(8) 13.575(5) 19.423(8) | 90.0 90.0 90.0 | 489 |

*Heterometallic Derivatives*

| Compound | Crystal system | Space group | Z | a, b, c | α, β, γ | Ref. |
|---|---|---|---|---|---|---|
| [Ba$_4$Ti$_4$O$_4$(OR)$_{16}$(ROH)$_4$][Ba$_4$Ti$_4$O$_4$(OR)$_{16}$(ROH)$_3$] (μ$_4$) R = *i*-Pr | Triclinic | $P\bar{1}$ | 2 | 13.947(4)<br>14.742(4)<br>50.035(10) | 82.81(2)<br>84.59(2)<br>62.09(2) | 486 |
| [La$_2$Mo$_4$O$_8$(O-*i*-Pr)$_{14}$] (μ$_4$/terminally bonded) | Tetragonal | *I4/m* | 2 | 11.871(6)<br>11.871(6)<br>21.205(7) | 90.0<br>90.0<br>90.0 | 491 |
| [Sm$_4$TiO(O-*i*-Pr)$_{14}$] (μ$_5$) | Tetragonal | *I4$_1$cd* | 8 | 21.510(6)<br>21.510(6)<br>25.823(5) | 90.0<br>90.0<br>90.0 | 490 |
| [K$_4$Zr$_2$O(O-*i*-Pr)$_{10}$] (μ$_6$) | Tetragonal | *P4$_{2/n}$* | 4 | 18.899(2)<br>18.899(2)<br>12.931(1) | 90.0<br>90.0<br>90.0 | 117 |
| [Pb$_2$Ti$_2$O(O-*i*-Pr)$_{10}$] (μ$_4$) (Fig. 76) | Monoclinic | *P2$_{1/n}$* | 4 | 10.286(14)<br>22.564(4)<br>18.709(5) | 90.0<br>97.93(5)<br>90.0 | 36b |
| [Mn$_8$Sb$_4$O$_4$(OEt)$_{20}$] (μ$_5$) | Monoclinic | *C2/c* | 4 | 17.565(6)<br>17.578(7)<br>21.870(5) | 90.0<br>93.37(5)<br>90.0 | 78 |
| Na$_2$[Fe$_6$O(OMe)$_{18}$] · 6MeOH (μ$_6$) (Fig. 79) | Tetragonal | *P4$_1$2$_1$2* | 4 | 14.367(3)<br>14.367(3)<br>24.39(1) | 90.0<br>90.0<br>90.0 | 508 |
| [Ni$_5$Sb$_3$O$_2$(OEt)$_{15}$(HOEt)$_4$] | Orthorhombic | *Pca2$_1$* | 4 | 17.909(2)<br>20.608(3)<br>17.221(2) | 90.0<br>90.0<br>90.0 | 77 |
| [Pb$_6$Nb$_4$O$_4$(OEt)$_{24}$] (μ$_4$) | Monoclinic | *P2$_{1/n}$* | 4 | 15.444(5)<br>23.403(5)<br>23.572(8) | 90.0<br>92.630(3)<br>90.0 | 492 |
| [Na$_4$Sb$_2$O(O-*t*-Bu)$_8$] (μ$_6$) | Trigonal | *R$\bar{3}$c* | 6 | 12.404(6)<br>12.404(6)<br>50.81(3) | 90.0<br>90.0<br>120.0 | 58 |
| [Na$_4$Bi$_2$O(O-*t*-Bu)$_8$] (μ$_6$) | Trigonal | *R$\bar{3}$c* | 6 | 12.449(6)<br>12.449(6)<br>51.08(3) | 90.0<br>90.0<br>120.0 | 58 |

**(II.A.1)**[*]

$$5M + 15i\text{-Pr-OH} \xrightarrow{\text{HgCl}_2,\ \text{I}_2}$$

$$M_5O(O\text{-}i\text{-Pr})_{13} + \tfrac{15}{2}H_2 \uparrow (+i\text{-Pr}_2\text{-O}) \tag{203}$$

[M = Y (Ref. 478), Sc,In,Yb (Ref. 484), and Nd (Ref. 302)]

**(II.A.1)**

$$Ba + t\text{-BuOH} \xrightarrow[\substack{70°C \\ (1\ h)}]{\text{THF}} \text{Product}$$

crystallized from
toluene–pentane

(Ref. 485)   $[H_3Ba_6(O)(O\text{-}t\text{-Bu})_{11}(OCEt_2CH_2O)(thf)_3]$     (204)

*Note:* The formation of a diolate moiety $(OCEt_2CH_2O)$ in the oxo-alkoxide produced in Eq. 204 appears to arise in a side reaction with the solvent THF, as the same is not formed in the absence of THF.)

**(II.A.1)**

$$6Ba + 16HOCH_2CH_2OMe \xrightarrow{\text{toluene}}$$

$$[H_4Ba_6(\mu_6\text{-O})(\mu_3\text{-}\eta^2\text{-OC}_2H_4OMe)_8(\eta^2\text{-OC}_2H_4OMe)_4$$

(Ref. 372)  $(OC_2H_4OMe)_2] + MeOCH_2CH_2OCH_2CH_2OMe$     (205)

**(II.A.1)**

$$6Ca + \underset{(\text{excess})}{12EtOH} \xrightarrow{\text{reflux}}$$

(Ref. 483)  $[Ca_6(\mu_4\text{-O})_2(\mu_3\text{-OEt})_4(OEt)_4.14EtOH + 2Et_2O$     (206)

**(II.A.2)**

$$8Ba + 23i\text{-Pr-OH} + 8Ti(O\text{-}i\text{-Pr})_4 \xrightarrow{i\text{-Pr-OH}}$$

(Ref. 486)  $[Ba_4Ti_4(\mu_4\text{-O})_4(\mu_3\text{-O-}i\text{-Pr})_2(\mu\text{-O-}i\text{-Pr})_8(O\text{-}i\text{-Pr})_8(i\text{-Pr-OH}_4] +$

$$[Ba_4Ti_4(\mu_4\text{-O})_4(\mu_3\text{-O-}i\text{-Pr})_2(\mu\text{-O-}i\text{-Pr})_9(O\text{-}i\text{-Pr})_5(i\text{-Pr-OH}_3]$$

$$(+8i\text{-Pr}_2\text{O}) + 8H_2 \tag{207}$$

**(II.C.1)**[*]

$$UCl_4 + K(O\text{-}t\text{-}Bu) \longrightarrow U_3O(O\text{-}t\text{-}Bu)_{10} + (Ref.\ 487) \cdots \quad (208)$$

**(II.C.1)**

$$YCl_3 + 2NaO\text{-}t\text{-}Bu \xrightarrow{\text{THF}}$$

$$Y_3(\mu_3\text{-}O\text{-}t\text{-}Bu)(\mu_3\text{-}Cl)(\mu\text{-}O\text{-}t\text{-}Bu)_3(O\text{-}t\text{-}Bu)_3(thf)_2(Cl)$$

standing in toluene over
N$_2$ for 2 weeks

$$Y_{14}(O\text{-}t\text{-}Bu)_{28}Cl_{10}O_2(thf)_4 \quad (209)$$

*Note:* On standing, the above conversion of the initial product in toluene was suggested as an indication of the manner in which alkoxides could change into oxo-alkoxides, even during a sol–gel process.)

$$4(Me)_3Al + 12CF_3CH_2OH \xrightarrow{\text{hexane}}$$

$$(Ref.\ 489)\quad [Al_4O(H)(OCH_2CF_3)_{11}] + 11CH_4 + EtCF_3 \quad (210)$$

**(II.C.2)**

$$SmI_2 + 2NaTi(O\text{-}i\text{-}Pr)_5 \xrightarrow[\text{reflux}]{\text{THF}}$$

$$(Ref.\ 490)\quad Sm_4Ti(\mu_5\text{-}O)(\mu_3\text{-}O\text{-}i\text{-}Pr)_2(\mu\text{-}O\text{-}i\text{-}Pr)_6(O\text{-}i\text{-}Pr)_6 \quad (211)$$

---

[*]On a personal note, it may be relevant to place an earlier observation of Mehrotra (488) on record. He found that repeated efforts to purify the simple $Zr(OEt)_4$ [prepared by the reaction: $ZrCl_4$ + $4EtOH + 4NH_3 \rightarrow Zr(OEt)_4 + 4NH_4Cl \downarrow$] finally yielded a derivative with the approximate composition: $Zr_3O(OEt)_{10}$. In fact, efforts at sublimation of the initial product **a** with apparently correct analysis for Zr and OEt also yielded a similar product, $Zr_3O(OEt)_{10}$. Efforts to sublime an insoluble product with the approximate analysis of $ZrO(OEt)_2$ also gave a small amount of $Zr_3O(OEt)_{10}$ sublimate. The investigations were not pursued further as crystalline $Zr(O\text{-}i\text{-}Pr)_4 \cdot i\text{-}Pr\text{-}OH$ could be obtained by a very facile procedure. We are not aware of any publication dealing with the synthesis of a simple compound such as $Zr(OC_2H_5)_4$. In fact, Turova and co-workers (197) found evidence for the partial formation of species such as $Zr_3O(O\text{-}i\text{-}Pr)_{10}$ when the crystalline adduct $Zr(O\text{-}i\text{-}Pr)_4 \cdot i\text{-}OH$ is distilled under reduced pressure.

(II.C.2)

$$NiCl_2 + Sb(OEt)_3 + 2Na \xrightarrow[\text{reflux}]{\text{EtOH}} Ni_5Sb_3O_2(OEt)_{15}(EtOH)_4 + 2\ NaCl \downarrow$$

(Ref. 77)                                                                    (212)

(II.C.2)

$$MnCl_2 + NaSb(OEt)_4 \xrightarrow[\text{toluene}]{\text{EtOH}} Mn_8Sb_4(\mu_5\text{-O})_4(\mu_3\text{-OEt})_4(\mu\text{-OEt})_{16}$$

(Ref. 78)                                                                    (213)

(II.F.4)

"La(O-$i$-Pr)$_3$" + 2MoO(O-$i$-Pr)$_4$ $\longrightarrow$ "monooxo complex" $\xrightarrow{4(i\text{-Pr})_2O}$

(Ref. 491)   La$_2$Mo$_4$O$_8$(O-$i$-Pr)$_{14}$                              (214)

(II.F.4)

$$Pb_4O(OEt)_6 + Nb_2(OEt)_{10} \xrightarrow[\text{r.t.}]{\text{EtOH}}$$

(Ref. 492)   [Pb$_6$Nb$_4$($\mu_4$-O)$_4$($\mu_3$-OEt)$_4$($\mu$-OEt)$_{12}$(OEt)$_8$      (215)

(II.F.4)

$$2Pb(O\text{-}i\text{-Pr})_2 + 2Ti(O\text{-}i\text{-Pr})_4 \xrightarrow{\text{toluene}}$$

(Ref. 36b)   [Pb$_2$Ti$_2$($\mu_4$-O)($\mu_3$-O-$i$-Pr)$_2$($\mu$-O-$i$-Pr)$_4$(O-$i$-Pr)$_4$]    (216)

(II.F.4)

$$Sr(OEt)_2 + 5Sb(OEt)_5 \xrightarrow[\substack{\text{EtOH} \\ \text{excess}}]{\text{toluene}} \frac{1}{n}[Sr_2Sb_4O(OEt)_{14}]_n$$    (217)

(Ref. 493)

(II.F.4)

$$nE\ (O\text{-}t\text{-Bu})_3 + mNa(O\text{-}t\text{-Bu}) \longrightarrow Na_4E_2O(O\text{-}t\text{-Bu})_8$$

(Ref. 58) (E = Sb or Bi)                                                      (218)

In addition to the formation of unexpected oxo-alkoxides by routine proce-
dures, the following equations represent the synthesis of $UO_2(O\text{-}t\text{-Bu})_2(Ph_3PO)_2$

(494) and $VO(OR)_3$ (495) by the reactions of $UO_2Cl_2$ and $VOCl_3$ with the corresponding alkali alkoxides. These reactions can be cited as normal extensions of the procedure in Section II.C.1.

$$UO_2Cl_2(Ph_3PO)_2 + 2KO\text{-}t\text{-}Bu \xrightarrow{\text{THF}}$$

$$\text{(Ref. 494)} \quad UO_2(O\text{-}t\text{-}Bu)_2(Ph_3PO)_2 + 2KCl\downarrow \qquad (219)$$

$$VOCl_3 + 3NaOR \longrightarrow VO(OR)_3 + 3NaCl \qquad (220)$$

$$\text{(Ref. 495)} \quad R = Me, Et, i\text{-}Pr, t\text{-}Bu, \text{etc.}$$

The ovovanadium alkoxides were also prepared by the reactions of $V_2O_5$ with alcohols.

(II.B)

$$V_2O_5 + 10ROH \xrightarrow{\text{benzene}} 2VO(OR)_3 + 5H_2O + 2R_2O \qquad (221)$$

(Ref. 496)

$R = Et, n\text{-}Pr, i\text{-}Pr, s\text{-}Pr, t\text{-}Bu, CH_2CH_2Cl, CH_2CH_2F, \text{and } CH_2CCl_3$

### 3. Formation by Novel Procedures

In addition to these routine procedures, the following equations illustrate rather unusual (novel) routes by which the formation of oxo-alkoxides of metals were reported.

$$2W_2(O\text{-}i\text{-}Pr)_6(Py)_2 + 2Me_2C{=}O \longrightarrow$$

$$\text{(Ref. 497)} \quad W_4O_2(O\text{-}i\text{-}Pr)_{12} + Me_2C{=}CMe_2 + 4py \qquad (222)$$

$$W_2(O\text{-}t\text{-}Bu)_6 + 3N_2O \xrightarrow[-15°C]{\text{pentane}} \text{``}W_3O_2(O\text{-}t\text{-}Bu)_8\text{''} + 3N_2 \qquad (223)$$

(Ref. 498)

$$[W_2(O\text{-}t\text{-}Bu)_7]^- \longrightarrow [W_2(O\text{-}t\text{-}Bu)_6(\mu\text{-}O)(\mu\text{-}H)] + H_2C{=}CMe_2 \qquad (224)$$

(Ref. 499)

$$Ce_2(O\text{-}i\text{-}Pr)_8(Pr\text{-}i\text{-}OH)_2 \xrightarrow[\text{visible light}]{\text{irradiation}}$$

$$\text{(Ref. 500)} \quad Ce_4(\mu_4\text{-}O)(\mu_3\text{-}O\text{-}i\text{-}Pr)_2(\mu\text{-}O\text{-}i\text{-}Pr)_4(O\text{-}i\text{-}Pr)_7(Pr\text{-}i\text{-}OH) \qquad (225)$$

## 4.  Ester Elimination Procedures

Ester elimination between a metal alkoxide and the carboxylate of another metal is the most convenient method for the preparation of heterobimetallic oxo-alkoxides, with the general formula, $(RO)_{n-1}-M'-O-M''-O-M'(OR)_{n-1}$, where $M' = Al^{III}$ ($n = 3$) or $Ti^{IV}$ ($n = 4$), and so on, and $M'' = Cr^{II}$, $Mn^{II}$, $Fe^{II}$, $Co^{II}$, $Ni^{II}$, $Zn^{II}$, $Mo^{II}$, $Ca^{II}$, or $R_2Sn^{II}$, and so on (501, 502). For example,

$$M''(OAc)_2 + 2Al(O\text{-}i\text{-}Pr)_3 \xrightarrow{\text{decalin}} M''\{OAl(O\text{-}i\text{-}Pr)_2\}_2 + 2i\text{-}PrOAc \uparrow$$

$$(226)$$

$$M''(OAc)_2 + 2Ti(O\text{-}i\text{-}Pr)_4 \xrightarrow{\text{decalin}} M''\{OTi(O\text{-}i\text{-}Pr)_3\}_2 + 2i\text{-}PrOAc \uparrow$$

$$(227)$$

Ester elimination reactions involving different metal alkoxides in stepwise manner may lead (503) to ter and even higher heterometal $\mu$-oxo-alkoxides.

$$Al(O\text{-}i\text{-}Pr)_3 + Ca(OAc)_2 \longrightarrow (O\text{-}i\text{-}Pr)_2Al-O-Ca(OAc) + i\text{-}PrOAc \uparrow$$

$$(228)$$

$$(i\text{-}PrO)_2Al-O-Ca(OAc) + Ti(O\text{-}i\text{-}Pr)_4 \longrightarrow$$

$$(i\text{-}PrO)_2Al-O-Ca-O-Ti(O\text{-}i\text{-}Pr)_3 + i\text{-}PrOAc \uparrow \qquad (229)$$

The important role played by the solvent in ester elimination reactions of the previous type can be illustrated by the following example. In two recent publications (120, 121), it was shown that the reactions between $Sn(O\text{-}t\text{-}Bu)_4$ with $Sn(OAc)_4$, as well as with $Me_3Si(OAc)$ in a refluxing hydrocarbon (e.g., toluene) solvent occurred with elimination of *tert*-butyl acetate and formation of oxo-alkoxides.

$$3Sn(O\text{-}t\text{-}Bu)_4 + 3Sn(OAc)_4 \xrightarrow{\text{toluene}} Sn_6O_6(O\text{-}t\text{-}Bu)_6(OAc)_6 + 6t\text{-}BuOAc$$

(Ref. 120)                                                                        $$(230)$$

$$Sn(O\text{-}t\text{-}Bu)_4 + Me_3SiOAc \xrightarrow{\text{toluene}} (Bu\text{-}t\text{-}O)_3SiOSiMe_3 + t\text{-}BuOAc$$

(Ref. 121)                                                                        $$(231)$$

However, only ligand exchanges of the following types were found to occur when the above reactions were carried out in a coordinating solvent such as pyridine.

$$3Sn(O\text{-}t\text{-}Bu)_4 + 3Sn(OAc)_4 \xrightarrow{\text{pyridine}}$$

$$\text{(Ref. 120)} \quad 2Sn(O\text{-}t\text{-}Bu)_3(OAc) + 2Sn(O\text{-}t\text{-}Bu)_2(OAc)_2$$

$$+ 2Sn(O\text{-}t\text{-}Bu)(OAc)_3 \tag{232}$$

$$Sn(O\text{-}t\text{-}Bu)_4 + Me_3Si(OAc) \xrightarrow{\text{pyridine}}$$

$$\text{(Ref. 121)} \quad Sn(O\text{-}t\text{-}Bu)_3(OAc)\cdot py + Me_3SiO\text{-}t\text{-}Bu \tag{233}$$

In keeping with our general theme of presenting similarities between hetero- and homometal alkoxides, it may be pointed out that ester elimination was found to be a side reaction in the preparation of a metal carboxylate from the reaction between the alkoxide of a metal and a carboxylic acid leading to an oxo-carboxylate and sometimes to an oxo-alkoxide at the intermediate stage(s) [cf. (6, 145)]:

$$Al(O\text{-}i\text{-}Pr)_3 + RCOOH \longrightarrow Al(O\text{-}i\text{-}Pr)_2(OOCR) + Pr\text{-}i\text{-}OH$$
$$\downarrow \Delta$$

$$Al(O)(O\text{-}i\text{-}Pr) + RCOO\text{-}i\text{-}Pr \tag{234}$$

$$Al(O\text{-}i\text{-}Pr)_3 + 2RCOOH \longrightarrow Al(O\text{-}i\text{-}Pr)(OOCR)_2 + 2Pr\text{-}i\text{-}OH$$
$$\downarrow \Delta$$

$$Al(O)(OOCR) + RCCO\text{-}i\text{-}Pr \tag{235}$$

The reaction between aluminum isopropoxide and carboxylic acid could, however, be pushed to completion, if the isopropanol liberated was continuously fractionated out azeotropically with benzene. Then aluminium tricarboxylates could be successfully isolated for the first time.

$$Al(O\text{-}i\text{-}Pr)_3 + 3RCOOH \longrightarrow Al(OOCR)_3 + 3Pr\text{-}i\text{-}OH \tag{236}$$

In the case of the reaction between titanium and zirconium alkoxides and excess carboxylic acids, however, ester elimination could not be avoided at the later stages and the final product was oxo-acetate $(AcO)_3TiOTi(OAc)_3$ or $(AcO)_3Zr\text{-}O\text{-}Zr(OAc)_3$.

$$Zr(OAc)_3(O\text{-}i\text{-}Pr) + Zr(OAc)_4 \longrightarrow (AcO)_3ZrOZr(OAc)_3 + i\text{-}PrAc \uparrow$$

$$\tag{237}$$

## C. Structural Features

The last decade has witnessed mushrooming growth (4, 23, 42, 504) in X-ray structural determinations (Table VI) of metal oxo-alkoxide complexes.

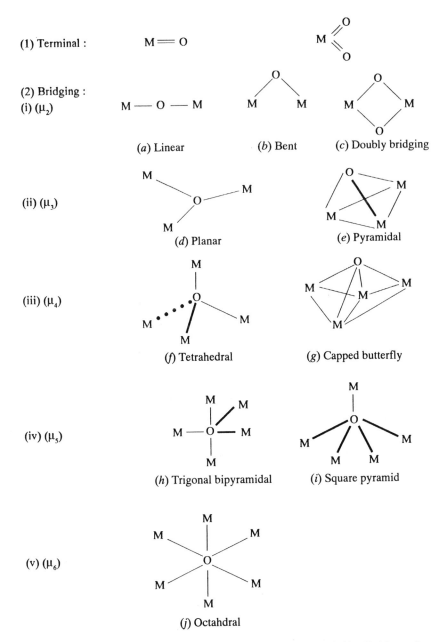

Figure 72.    The common X-ray established structural types of oxo ligands identified in metal oxo-alkoxides.

These complexes exhibited an extraordinary variety of structures that revealed interesting, and in many cases, new types of metal–ligand bonding. The *oxo ligand* functions as a simple terminal ligand, in some cases, for example, $[UO_2(O\text{-}t\text{-}Bu)_2(Ph_3PO)_2]$ (494) and $[VO(OCH_2CH_2Cl)_3]$ (496). In most of the other cases (504), for example, $[Pb_2Ti_2O(O\text{-}i\text{-}Pr)_{10}]$ (36b), $[Nd_5O(O\text{-}i\text{-}Pr)_{13}(HO\text{-}i\text{-}Pr)_2]$ (302), $[Yb_5O(O\text{-}i\text{-}Pr)_{13}]$ (484), $[H_4Ba_6O(OCH_2CH_2OMe)_{14}]$ (372), $[Sn_6O_4(OMe)_4]$ (477b), $[U_3O(O\text{-}t\text{-}Bu)_{10}]$ (487), $[Al_4O(OCH_2CF_3)_{11}]^-$ (489), $[W_3O_2(O\text{-}t\text{-}Bu)_8]$ (498), $[Ce_4O(O\text{-}i\text{-}Pr)_{13}(HO\text{-}i\text{-}Pr)]$ (500), $[LiTiO(O\text{-}i\text{-}Pr)_3]_4$ (505), $[Mo_3O(O\text{-}i\text{-}Pr)_{10}]$ (506), $[Mo_6O_{10}(O\text{-}i\text{-}Pr)_{12}]$ (507), $Na_2[Fe_6O(OMe)_{18}]\cdot 6MeOH$ (508), and $[ZnTaI_2O_2(O\text{-}i\text{-}Pr)_7]_2$ (509), the oxo ligand functions as a bridging ligand. General examples of different types of oxo-metal bonding are illustrated diagrammatically (Fig. 72) by representing only the metal–oxygen cores (excluding the alkoxide groups).

An interesting variety of structures has been found for homo- and heterometal oxo-alkoxide complexes (4a, 23, 42, 504) involving the above depicted (Fig. 72) bonding modes of oxo ligands. Many of these are included in Table VI. A few typical structures are shown in Figs. 73–79 along with a brief comment on their salient structural features.

The X-ray crystal structures of a terminally oxo-bonded vanadium alkoxide $[VO(OCH_2CH_2Cl)_3]_2$ (496), shown in Fig. 73, reveals dimeric association of molecules belonging to adjacent unit cells via long $V—OCH_2CH_2Cl$ bonds [2.261(2) Å], and a trigonal bipyramidal geometry for each vanadium atom in the dimeric molecule.

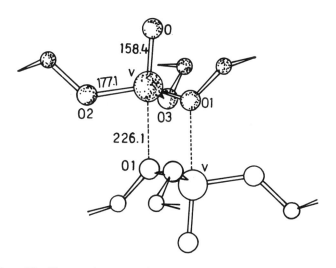

Figure 73.   The crystal structure of $VO(OCH_2CH_2Cl)_2$. [Adapted from (496).]

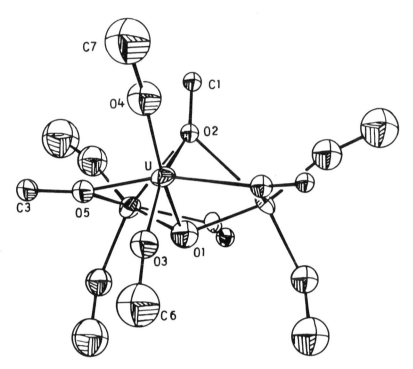

Figure 74.  The crystal structure of [U₃O(O-*t*-Bu)₁₀]. [Adapted from (487).]

The discrete trinuclear structure of [U₃O(O-*t*-Bu)₁₀] consists of three distorted mutually confacial octahedra sharing an edge (Fig. 74) (487) with $C_{3v}$ crystallographic symmetry.

The compound [Ce₄O(O-*i*-Pr)₁₃(HO-*i*-Pr)] (Fig. 75) (500) consists of a butterfly (rather than common tetrahedral) arrangement of four cerium atoms around the oxo ligand. The molecule possesses a crystallographic $C_2$ axis that passes through the oxo ligand and the center of a symmetric hydrogen bond between the coordinated isopropanol and one terminal isopropoxide.

An interesting structure was found for [Pb₂Ti₂O(O-*i*-Pr)₁₀] (Fig. 76) (36b). The metals are displayed around a $\mu_4$-oxo ligand O(1) and form a regular tetrahedron. The titanium atoms display distorted octahedral geometry while both lead atoms are five coordinate with a distorted tetragonal stereochemistry and a stereochemically active lone pair. A number of other homo- and heterometallic oxo-alkoxides such as [Al₄O(OCH₂CF₃)₁₁]⁻ (489), [Ca₆O₂(OEt)₈(EtOH)₁₄] (483), [La₂Mo₄O₈(O-*i*-Pr)₁₄] (491), and [Pb₆Nb₄O₄(OEt)₂₄] (492) also contain $\mu_4$-oxo ligands in tetrahedral environments.

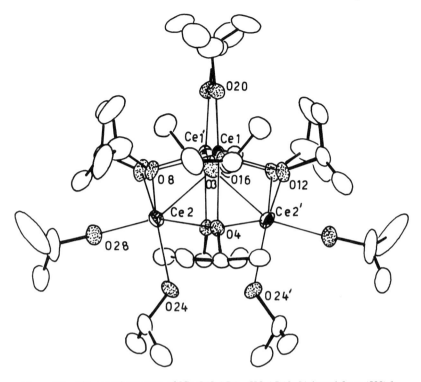

Figure 75. The crystal structure of [Ce$_4$O(O-$i$-Pr)$_{13}$(HO-$i$-Pr)]. [Adapted from (500).]

The architecture of the core in metal oxo-alkoxide complexes is determined to a large extent by the nature of both the metal and the ligands bound to it, and by its preference for (a) the square pyramidal, as in [Y$_5$O(O-$i$-Pr)$_{13}$] (Fig. 77) (478) and [Mn$_8$Sb$_4$O$_4$(O-$i$-Pr)$_{20}$] (78) and (b) the trigonal bipyramidal, as in [Nd$_5$O(O-$i$-Pr)$_{13}$(HO-$i$-Pr)$_2$] (Fig. 78) (302) as well as in [Sm$_4$TiO(O-$i$-Pr)$_{14}$] (490), coordination symmetries of the oxo ligand(s) in the core. The trigonal bipyramidal environment around a oxo ligand in the last two complexes is unprecedented in oxo-alkoxide chemistry.

The complex Na$_2$Fe$_6$O(OCH$_3$)$_{18}$·6MeOH (Fig. 79) (508) consists of an oxo ligand in the center of an octahedron formed by the six iron(III) ions, which are themselves linked to each other by 12 $\mu_2$-methoxide bridges. A distorted octahedral coordination geometry of the ferric ions is completed by six terminal methoxide ligands (508). Each of the two Na$^+$ ions, is linked to three methoxide bridges of the [Fe$_6$O(OMe)$_{18}$]$^{2-}$ complex ion and to three additional methanol molecules attaining an idealized $C_3$-coordination geometry (CN$_{Na}$ = 6).

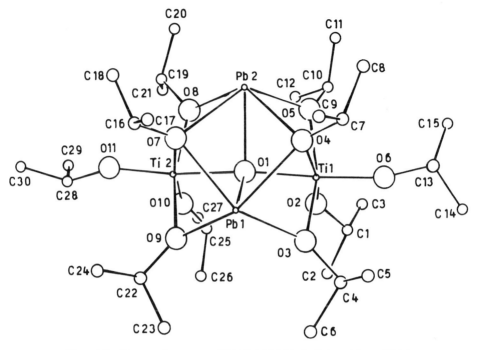

Figure 76. The crystal structure of $[Pb_2Ti_2O(O\text{-}i\text{-}Pr)_{10}]$. [Adapted from (36b).]

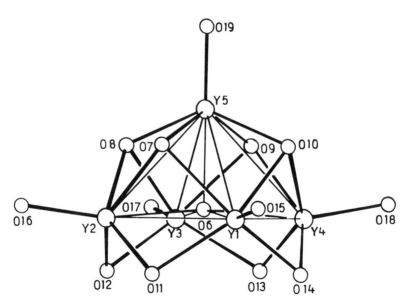

Figure 77. The crystal structure (stereoscopic ORTEP drawing of the $Y_5O_{14}$ core) of $[Y_5O(O\text{-}i\text{-}Pr)_{13}]$. [Adapted from (478).]

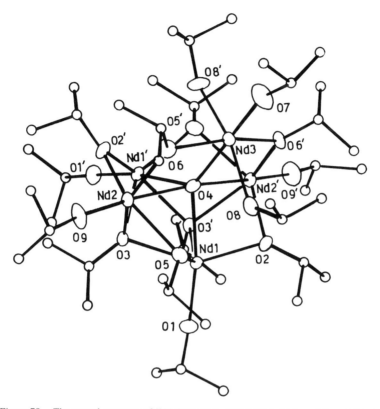

Figure 78. The crystal structure of $[Nd_5O(O\text{-}i\text{-}Pr)_{13}(HO\text{-}i\text{-}Pr)_2]$. [Adapted from (302).]

Such $\mu_6$-oxo ligand bonding was also crystallographically established in an interesting homometallic complex $[H_4Ba_6O(OCH_2CH_2OMe)_{14}]$ (372).

## VI. METAL ALKOXIDES AS UNIQUE SYNTHONS FOR A VARIETY OF ORGANOMETALLIC DERIVATIVES

The initial development of $\sigma$-bonded organotransition metal chemistry has mainly centered around the proper choice of soft, $\pi$-acceptor ancillary ligands (510–512), such as CO, $PR_3$, Cp, and cod. Alkoxide ligands, $RO^-$, share some features with the above. Noteworthy among these are (a) the capability of $RO^-$ ligands to act as terminal, $\mu_2$- and $\mu_3$-bridging groups (4a, 6, 28, 34, 38, 39) and their fluxional behavior similar to that of CO, (b) sterically demanding alkoxide groups (6, 28, 34, 38, 39) favor low-coordination states for the central metal atom in a manner similar to bulky phosphines or substituted cyclopenta-

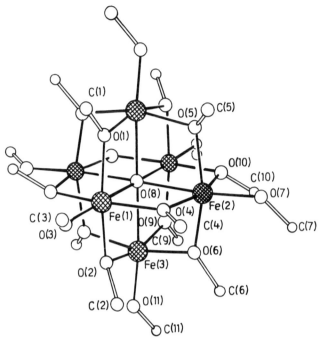

Figure 79.   The crystal structure of the $[Fe_6O(OMe)_{18}]^{2-}$ complex. [Adapted from (508).]

dienyls. Further alkoxide ligands are electronically flexible (28, 39) and the degree of $\pi$ donation can respond to the needs of a metal atom. In general, the electron-releasing strength of $^-$OR ligands (both $\sigma$ and $\pi$) follows the order $^-$O-$t$-Bu > O-$i$-Pr > OCH$_2$-$t$-Bu > OEt.

Besides other similarities already mentioned, a number of isolobal relationships are becoming apparent. The $d^2$M(OR)$_4$ and $d^3$M(OR)$_3$ groups (where M = Mo or W) are isolobal with say Fe(CO)$_4$ (16e) and Co(CO)$_3$ (15e), respectively.

In this section, we will highlight the development in the use of metal alkoxides for the synthesis of new and interesting organometallic compounds, many of these are either inaccessible or difficult to synthesize by common synthetic procedures. We will not discuss (a) the chemistry of organometallic compounds containing alkoxides as supporting ligands, for which excellent reviews by Chisholm and co-workers (154, 513, 514) are available; and (b) intramolecular cyclometalation (i.e., C—H bond activation) reactions of metal aryloxides due to the availability of an excellent account of this topic in a review article by Rothwell (515). Furthermore, a brief mention of the use of a related metal derivative (i.e., metal aryloxide) will be made merely for comparison.

Since the first preparation of a $\sigma$-bonded transition metal organometallic derivative PhTl(O-*i*-Pr)$_3$ in the 1950s by the reaction of Ti(O-*i*-Pr)$_4$ with LiPh in ether (516), a number of interesting homo- and heteroleptic organometallic compounds were prepared (see Scheme 9) by a similar procedure. The yields of these compounds were higher than by the reaction of the corresponding metal chloride or its complex.

Although metal alkoxides were successfully used as a synthon for the synthesis (see Scheme 9) of more interesting homo- and heteroleptic metal alkyls, they have not attained the same importance as their sterically hindered aryloxide analogues. This finding might be due to the general solubility of both the products [i.e., desired metal alkyls and alkali metal (generally lithium) alkoxides] in hydrocarbon solvents. This limitation has made a cleaner separation of the products more difficult.

It may not be out of place to note that metal aryloxides are finding continuous growing importance as a synthon for the synthesis of new and interesting metal alkyls, which are often difficult to synthesize by other routes.

$$M(OAr)_n + nLiR' \xrightarrow[25°C]{\text{pentane}} MR'_n + nLiOAr \downarrow$$

$$M = Mg \text{ (Ref. 524)}, n = 2, R' = CH(SiMe_3)_2,$$

$$Ar = C_6H_2t\text{-Bu}_2\text{-2,6-Me-4} \tag{238}$$

$$M = Y, La, Sm, \text{ or } Lu \text{ (Ref. 525)}, n = 3, R' = CH(SiMe_3)_2,$$

$$Ar = C_6H_2t\text{-Bu}_3\text{-2,4,6}$$

$$M = U \text{ (Ref. 525d)}, n = 3, R' = CH(SiMe_3)_2, Ar = C_6H_3Bu_2\text{-2,6}$$

For this type of transformation, the insolubility of one of the products (i.e., lithium aryloxide) in pentane appears to be the driving force. Another distinct feature of metal aryloxides is their susceptibility to undergo intramolecular C—H bond activation (i.e., cyclometalation) reactions to afford new organometallic systems supported by aryloxide ligands (515).

The dinuclear $d^3$-$d^3$ (RO)$_3$M≡M(OR)$_3$ alkoxide system has provided a fertile inorganic redox active and coordinatively unsaturated template for the development of organometallic chemistry via substrate activation. Some of these interesting reactions are depicted in Scheme 10 and Eqs. 239, 241, and 242.

Although alkynes may be cleaved by dimetal alkoxides to yield terminal alkylidyne species (see Scheme 10), bridging alkylidynes are sometimes also accessible (530) from alkynes.

$$W_2(O\text{-}t\text{-Bu})_6 + PhC≡CPh \longrightarrow W_2(O\text{-}t\text{-Bu})_4(\mu\text{-CPh})_2 \tag{239}$$

$$\underrightarrow{\text{4 LiR'}} \quad \text{Ti(adme)}_4 + \text{4 LiO-}i\text{-Pr}$$

-60°C, light petroleum;
R' = 1-adamantylmethyl (adme);
R = i-Pr; M = Ti (Ref.517); $n = 4$; $x = 0$

$$\underrightarrow{\text{4 LiR'}} \quad \text{CrR'}_4 + \text{4 LiO-}t\text{-Bu}$$

-78°C, light petroleum,
M = Cr; R = t-Bu;
R' = t-Bu (Ref.518), 1-adamantylmethyl (Ref.517)
$n = 4$; $x = 0$

$$\underrightarrow{\text{4 LiCH}_2\text{SiMe}_3} \quad \text{Ta(OMe)(CH}_2\text{SiMe}_3)_4 + \text{4 LiOMe}$$

M = Ta (Ref.519); R = Me;
$n = 5$; $x = 0$

$$\underrightarrow{\text{4 Li-}t\text{-Bu}} \quad [\text{Li(tmeda)}_2][\text{M(}t\text{-Bu)}_4] + \text{3 LiO-}t\text{-Bu}$$

TMEDA; M = Er, Lu (Ref.520);
R = t-Bu; $n = 3$; $x = 0$

$$\text{M(OR)}_n\text{L}_x \xrightarrow{\text{3 LiR'}} \text{M(OR)}_2\text{R'}_3 + \text{3 LiOR}$$

M = Nb, Ta (Ref.519);
R = Me, i-Pr, C$_6$H$_3$Me$_2$-2,6;
R' = Me, CH$_2$-t-Bu;
$n = 5$; $x = 0$

$$\underrightarrow{\text{16 LiR' (excess)}} \quad \text{2 Li}_3\text{UR'}_8\text{.dioxane} + \text{10 LiOEt}$$

dioxane, M(OR)$_n$L$_x$ = U$_2$(OEt)$_{10}$;
R' = Me, CH$_2$-t-Bu, CH$_2$SiMe$_3$ (Ref.521)

$$\underrightarrow{\text{5 AlEt}_3/1.5\ \text{C}_8\text{H}_8(\text{cot})} \quad \tfrac{1}{2}\text{Ce}_2(\text{cot})_3 + 5/2\ \text{AlEt}_2(\text{O-}i\text{-Pr}) + \text{C}_2\text{H}_6 + \text{4 [Et]}$$

C$_7$H$_8$, 140°C; M = Ce (Ref.522a), L$_x$ = Pr-i-OH,
R = i-Pr, $n = 4$

$$\underrightarrow{\text{5 AlEt}_3/\text{C}_8\text{H}_8(\text{cot})} \quad \text{Ce(cot)}_2 + 5\ \text{AlEt}_2(\text{O-}i\text{-Pr}) + \text{C}_2\text{H}_6 + \text{4 [Et]}$$

R = i-Pr, $n = 4$, L$_x$ = Pr-i-OH
100°C; M = Ce (Ref.522a),

$$\underrightarrow{\text{4 AlEt}_3/\text{C}_8\text{H}_8(\text{cot})} \quad \text{Ce(cot)(O-}i\text{-Pr)}_2\text{AlEt}_2$$

toluene, 100°C
M = Ce (Ref.522b); R = i-Pr; $n = 4$, L$_x$ = Pr-i-OH

$$\underrightarrow{n\ \text{Ph}_2\text{C=C=O}} \quad \text{(OR)}_{4-n}\text{M(CPh}_2\text{C(O)-OR)}_n$$

M = Ti, Zr (Ref.523), $n = 4$; L$_x$ = O, R = i-Pr

Scheme 9.   Synthesis of organometallic derivatives from metal alkoxides.

$$HC \equiv CH/n \text{ Py}$$

$$\longrightarrow \quad M_2(OR)_6(\mu\text{-}C_2H_2)(Py)_n$$

$R = i\text{-Pr};$
$M = \text{Mo, W(Ref.526)}; n = 2$

$$XC \equiv CX \text{ or}/XC \equiv CR$$

$$\longrightarrow \quad 2(Bu\text{-}t\text{-}O)_3W \equiv CX$$

$M = \text{W (Ref.527)}; R = t\text{-Bu};$
$R = \text{Me, Et}; X = \text{Me, Et, } t\text{-Bu, Ph}$

$$R'C \equiv N$$

$$\longrightarrow \quad (Bu\text{-}t\text{-}O)_3W \equiv CR' + (Bu\text{-}t\text{-}O)_3W \equiv N$$

$M = \text{W(Ref.527)};$
$R = t\text{-Bu}; R' = \text{Me, Ph, CH}_2\text{Ph}$

$$3 \text{ CO}$$

$$(RO)_3M \equiv M(OR)_3 \quad \longrightarrow \quad \tfrac{1}{2} \text{ Mo(CO)}_6 + 3/2 \text{ Mo(O-}t\text{-Bu})_4$$

$M = \text{W(Ref.528)};$
$R = t\text{-Bu}$

$$2 \text{ C}_2\text{H}_4; 22^\circ\text{C}$$

$$\longrightarrow \quad W_2(OCH_2\text{-}t\text{-Bu})_6(\eta^2\text{-}C_2H_4)_2$$

$M = \text{W (Ref.529)};$
$R = \text{CH}_2\text{-}t\text{-Bu}$

$$3 \text{ C}_2\text{H}_4, 0^\circ\text{C}$$

$$\longrightarrow \quad W_2(O\text{-}i\text{-Pr})_6(CH_2)_4(\eta^2\text{-}C_2H_4)$$

$R = i\text{-Pr}; M = \text{W (Ref.529)}$

$$3 \text{ C}_2\text{H}_4$$

$$\longrightarrow \quad W_2(OR)_6(\mu\text{-CCH}_2\text{CH}_2\text{CH}_2) + C_2H_2$$

$22^\circ\text{C, hydrocarbon}$
$M = \text{W (Ref.529)};$
$R = c\text{-Hex, } c\text{-pent}; i\text{-Pr, CH}_2\text{-}t\text{-Bu}$

Scheme 10.   Synthesis of some novel organometallics from multiply bonded metal alkoxides.

Some other reactions of $(RO)_3M \equiv M(OR)_3$ and $W_4(OR)_{12}$ compounds leading to the formation of exciting products are shown below.

*Reactions Involving Carbon–Oxygen Double Bonds*

$$W_2(OCH_2\text{-}t\text{-Bu})_6(py)_2 + 2R_2C{=}O \xrightarrow[22^\circ C]{\text{hexane}}$$

$$(Ref. (531) \quad W_2O_2(OCH_2\text{-}t\text{-Bu})_6 + R_2C{=}CR_2 + 2py \qquad (240)$$

$$2W_2(O\text{-}i\text{-}Pr)_6(py)_2 + 2Me_2C{=}O \xrightarrow[22°C]{\text{hexane}}$$

$$\text{(Ref. 497)} \quad W_4(O)_2(O\text{-}i\text{-}Pr)_{12} + Me_2C{=}CMe_2 + 2py$$

$$\text{(cf. Eq. 222)} \tag{240a}$$

*Activation of Carbon Monoxide*

The $Mo_2(OR)_6$ compounds fail to exhibit comparable reactivity with $C{-}C$ and $C{-}O$ double bonds. This difference between $M{\equiv}Mo$ and $W$ reflects the relative orbital energetics of the $M{\equiv}M$ bond and emphasizes the greater reducing ability of the $W{\equiv}W$ bond.

$$M_2(O\text{-}t\text{-}Bu)_6 + Co \longrightarrow M_2(O\text{-}t\text{-}Bu)_6(\mu_2\text{-}CO) \tag{241}$$

(Refs. 528, 532)

The second approach involves the use of $W_4(OR)_{12}$.

$$W_4(OR)_{12} + CO \xrightarrow[22°C]{\text{hexane}} W_4(C)(O)(OR)_{12} \tag{242}$$

(Ref. 169)

## VII. METAL ALKOXIDES AS PRECURSORS FOR CERAMIC MATERIALS

### A. Introduction

In this current age of advanced ceramics, a continuous search is being made for novel oxide materials (533) for use in electronics and optical devices as well as in superconductivity. In addition, these materials have other well-known applications. A variety of new processes were investigated for the preparation of these materials. These materials were obtained for thousands of years by heating together mixtures of finely ground solid oxides (e.g., $SiO_2$, $Al_2O_3$, $CaO$, $TiO_2$, and $BaO$) or their decomposible compounds (such as carbonates), at rather high temperatures (say 1000–2000°C) for different durations (hours to days) of time. In fact, the word ceramics itself was probably derived from the Greek word *keramikos* (meaning pottery) or the Sanskrit root *shrapak* (meaning to heat on fire). Out of the new procedures evolved during the past three to four decades, the solution–sol–gel (or just SG) (23, 31, 534) and MOCVD (24, 27, 308) techniques assumed special significance.

In view of the solubility of metal alkoxides (binary or homometallic as well as heterometallic) in organic solvents and their volatility, these materials are strongly preferred as precursors in both the SG as well as MOCVD processes. Interestingly, the structural relationships (243) among metal oxides (243), oxo-alkoxides (535), and alkoxides, coupled with their versatility led to their increasing use as starting compounds (precursors) of various types. A critical review of the sol–gel chemistry of transition metal oxides by Livage et al. (536) appeared in 1988. The chemistry of the alkoxides of lanthanons was reviewed (28) in 1991, followed by review articles on the alkoxide chemistry of Groups 1 (IA), 2 (IIA), and 12 (IIB) metals (38) as well as the *p* block elements (39) in 1994.

## 1.  The Sol–Gel Procedure

The SG procedure (Fig. 80) consists of the following steps: (a) preparation of a homogeneous solution of easily purifiable precursor(s) in an organic solvent that is miscible with water or the reagent used in the next step; (b) converting the homogeneous solution to a sol (colloidal form) by treatment with a suitable reagent (e.g., water with $HCl/NaOH/NH_4OH$), (c) letting the sol change into a gel by self-polymerization; (d) shaping the gel (or highly viscous sol) to the finally desired form/shape (537) such as thin films (538), fibers (539), or spheres (540); and (e) finally converting (sintering) the shaped gel to the desired ceramic material (541) generally at temperatures much ($\sim 500°C$) lower than those required in the conventional process of melting the oxides together.

One of the highly attractive features of the SG process is the possibility [Fig. 81] of obtaining the final ceramic material in the different forms desired (e.g., bulk, fibre, coating, or aerogel) by a control of the conditions as depicted for the most typical oxide, $SiO_2$.

Shrinkage of the gel is the result of the surface tension effect at the interface during the removal of the liquid. If this shrinkage is avoided by removal of the

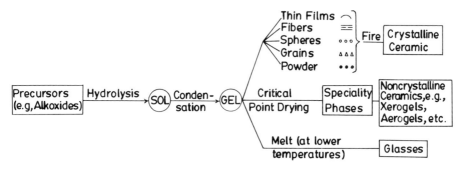

Figure 80.   Steps in the sol–gel process for ceramic materials.

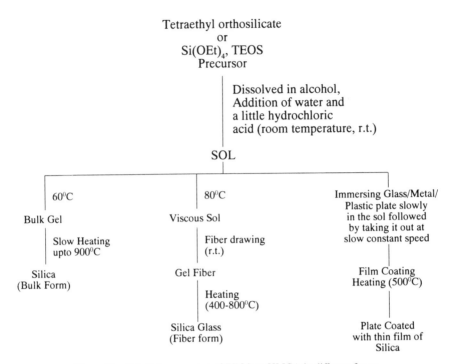

Figure 81.  Sol–Gel conversion of TEOS to SILICA in different forms.

liquid under supercritical conditions (542), an extremely porous form of light material, called aerogel (543), with densities as low as 0.01 g cm$^{-1}$ and a surface area as high as 1500 m$^2$ g$^{-1}$ with up to 99% air in the framework can be obtained. Although the nature of the liquid that was to be removed from the sol was changed from water to alcohol (542) and later to liquid carbon dioxide, the supercritical removal of a liquid during the sol–gel conversion has remained a key step in aerogel technology. Teichner (544) introduced the use of $Si(OMe)_4$ in place of $Si(OEt)_4$ for silica aerogels. Sometimes aerogels are called the "fourth state of matter" to signify their exceptional characteristics (545). Aerogels of oxides of other metals (e.g., Al, W, Fe, Th, and Sn) were used in important applications such as superinsulating spacers in windows, acoustic impedance matching in piezoelectric devices, and more effective catalysts. Mehrotra (11b) has emphasized the advantages of the use of homo- and heterometal alkoxides as precursors for aerogels.

Although noticed (9) as early as 1846 during the first synthesis of $Si(OEt)_4$, it is only in the last three to four decades that the sol–gel process has received increasing attention (10a, 31, 533, 534, 546, 547). A new *Journal of Sol–Gel*

*Science and Technology* has been launched since 1993, in which the significance of this simple technique has been reemphasized by Sakka (548).

From many possible precursors [e.g., metal nitrates or acetates (12), monodispersed metal hydrous oxides (549), oxides dissolved in alcohols (12), oxo-alkoxides (550), and alkoxides (12)], metal alkoxides were considered as specially suitable precursors (12, 31, 551) for the preparation of oxide ceramics since the 1950s, mainly due to the ease of their purification. This purification was generally distillation and in some cases crystallization, for example, $Zr(O$-$i$-$Pr)_4(i$-$Pr$-$OH)$, solubility in organic solvents such as the parent alcohols that are miscible with water, and their extremely facile hydrolyzability, which can be modulated effectively by substitution of some of the alkoxide groups with chelating ligands such as $\beta$-diketonates (35, 536). The extraordinary future potential and possibilities of the SG process were reemphasized recently by Sakka (534) and Roy (552) in two review articles.

The formation of oxides from metal alkoxides generally involves the following steps, which tend to overlap.

1. Hydrolysis of precurors

$$M(OR)_n + HOH \longrightarrow M(OR)_{n-1}(OH) + ROH \uparrow \qquad (243)$$

followed by

2. Dealcoholation reactions

$$M(OR)_{n-1}(OH) + (RO)_nM \longrightarrow$$

$$(RO)_{n-1}M-O-M(OR)_{n-1} + ROH \uparrow \qquad (244)$$

and

3. Dehydration reactions

$$2M(OR)_{n-1}(OH) \longrightarrow (RO)_{n-1}M-O-M(OR)_{n-1} + H_2O \uparrow \quad (245)$$

and so on

Using alkoxides of more than one metal (including that of silicon, whenever required) generally involves complexation (13) or/and transesterification (553) among the precursors at the initial stage itself. These processes enable the real precursor to become either a heterometallic alkoxide (resulting from complexation; eq. 246) or a mixture of the alkoxides of the metal and say, silicon as exemplified below (eq. 249).

$$M(OR)_n + M'(OR)_m \xrightarrow{\text{complexation}} MM'(OR)_{n+m} \qquad (246)$$

$$M(OR)_n + Si(OR')_4 \xrightarrow[\text{transesterification}]{} M(OR)_{n-x}(OR')_x$$

$$+ Si(OR')_{4-x}(OR)_x \qquad\qquad (247)$$

The different steps occurring during the SG process (e.g., hydrolysis, de-alcoholation, dehydration, as well as those at the initial stage), were followed by NMR techniques involving different metals such as $^{29}Si$ (554, 555) or more recently $^{17}O$ (556).

A unique advantage of the SG process is the possibility of proceeding all the way from the molecular precursor to the final ceramic procedure, allowing better control of the intermediate steps to a large extent. Thus ultrahomogeneous materials can be obtained through "complexation" reactions envisaged (13) in the starting solution of different metal alkoxides, or even better by the use of specifically synthesized heterometallic complexes (3, 32, 33, 35), which opens up the possibilities of using "single source precursors" (21, 22), for tailor made materials.

Mixed-metal oxides constitute a significant proportion of electroceramics (e.g., ferroelectrics or superconductors). In addition, electrooptical ceramics such as $Pb(LaZrTi)O_3(PLZT)$, $PbNb_{2/3}Mg_{1/3}O_3(PNM)$, and $Bi_4Ti_3O_{12}$ received considerable attention. It may be pointed out that the low-temperature SG route appears to be more suitable for lead containing materials in view of the comparatively more volatile characteristic of lead oxide, which tends to disturb the desired stoichiometry of the multimetal oxide material involving lead, prepared by the MOCVD procedure.

In the hydrolysis of $Mg\{Al(O\text{-}i\text{-}Pr)_4\}_2$, carried out in the presence of ethanolamine during the preparation of the polycrystalline $MgAl_2O_4$ (spinel) by the SG procedure, a careful $^{27}Al$ NMR study (557) showed that the basic framework of the bimetallic alkoxide, $Mg\{\mu\text{-}O\text{-}i\text{-}Pr)_2Al(OR)_2\}_2$, does not break down, at least in the initial stages of the hydrolysis. This result confirmed that the conversion took place at the molecular level in the process. Similar indications were obtained in the conversion of the bimetallic alkoxide $Mg\{Nb(OC_2H_4OMe)_6\}_2$ to the perovskite phase of PNM (558).

The preparation of new suitable precursors of yttrium, barium, and copper received special attention since the discovery of the superconducting properties of $YBa_2Cu_3O_{7-x}$ (YBCO). The reaction of $[Y(OCH_2CH_2OMe)_3]_{10}$ with $Cu(acac)_2$ in toluene yielded (384) two products, that is, $[Y_3(\mu_3, \eta^2\text{-}OCH_2CH_2OMe)_2(\mu_2, \eta^2\text{-}OCH_2CH_2OMe)_2(\mu_2, \eta^1\text{-}OCH_2CH_2OMe)(acac)_4]$ and $[Cu(OCH_2CH_2OMe)(acac)]$. In addition to these products, the volatile trinuclear tertiary butoxide of yttrium, $[Y_3(\mu_3\text{-}O\text{-}t\text{-}Bu)_2(\mu_2\text{-}O\text{-}t\text{-}Bu)_3(O\text{-}t\text{-}Bu)_4(Bu\text{-}t\text{-}OH)_2]$ and even more volatile $tert$-heptyl and $tert$-hexyl derivatives were described by Bradley et al. (103). The volatile mononuclear adducts $[Y(hftb)_3(thf)_3]$ and $[Y(hftb)_3(diglyme)]$ were obtained by using the hftb ligand (337).

Both of the adducts were shown (by X-ray crystallography) to be octahedrally coordinated. The latter complex sublimed readily without the loss of the neutral diglyme ligand (337). The work has since been extended (340) to include hexafluoro *tert*-butoxide derivatives of lanthanum, praseodymium, and europium with similar results.

In the case of barium (an important constituent in oxide–ceramics of wide-ranging properties and utilities, e.g., $BaTiO_3$, $Ba_{0.6}K_{0.4}BiO_3$, $YBa_2Cu_3O_{7-x}$, $HgBa_2CuO_4$, and $HgBa_2Ca_2Cu_3O_8$), even the bulkiest alkoxide ligands failed to yield volatile monomeric derivatives. Instead, there appears to be a strong tendency for formation of oxoderivatives, for example, $[H_3Ba_6(\mu_5\text{-}O)(O\text{-}t\text{-}Bu)_{11}(OCEt_2CH_2O)(thf)_3]$ (485) and $[H_4Ba(\mu_6\text{-}O)(OCH_2CH_2OMe)_{14}$ (372). A soluble tetrameric alkoxide, $[Ba(O\text{-}t\text{-}Bu)_2(HO\text{-}t\text{-}Bu)_2]$ was recently described (106b). For monomeric derivatives, recource had to be taken to chelate polyfunctional alkoxide ligands, as illustrated by the preparation of derivatives such as $[Ba\{OC_2H_4)(HOC_2H_4)_2N\} \cdot 2EtOH]$ (559), and $[Ba\{OC\text{-}t\text{-}Bu(CH_2OEt)_2\}_2]$ (288, 560).

The search for soluble/volatile alkoxides of copper also yielded many interesting derivatives, for example, $[Cu\{OC_2H_4NEt_2\}_2]$ (404), which sublimes at $\sim 60°C$ under reduced pressure. It would be pertinent to mention also a few heterometal alkoxides containing Ba/Cu/Y. These compounds were also synthesized with the same objective, for example, $[Ba\{Cu(O\text{-}t\text{-}Bu)_3\}_2]$ and $[Ba\{Cu(OCH(CF_3)_2)_3\}]_2$ (561). Among a number of precursors suggested for the superconducting materials, such as $YBa_2Cu_3O_{7-\delta}$, mention may be made of the heterotrimetallic complex $[YBa_2Cu_3O_3(OCH_2CH_2OMe)_7]$ (562). Similar possibilities for the use of single-source heterometallic alkoxide precursors of a variety of other materials were also emphasized from our laboratories (563).

In addition to the use of heterometal alkoxides, metal alkoxides are often associated with more easily available precursors such as acetates for the SG route to multicomponent oxides. A number of such alkoxide acetate precursors [e.g., $MNb_2(\mu\text{-}OAc)_2(\mu\text{-}OR)_4(OR)_6$ (M = Cd or Mg), $PbZr_3(\mu_4\text{-}O)(\mu\text{-}OAc)_2(\mu\text{-}OR)_5(OR)_5$, and $Gd_2Zr_6(\mu_4\text{-}O)_2(\mu\text{-}OAc)_6(\mu\text{-}OR)_{10}(OR)_{10}$ (with R = *i*-Pr)] were characterized (564) by X-ray crystallography. Their hydrolytic studies indicate their potential use as precursors for the synthesis of electrooptical materials, for example, $Pb(ScNb)O_3$ (PSN), and dielectric ceramics, for example, $[PbMg_{1/3}Nb_{2/3}O_3]$ (PNM).

Mention is also made of soluble 2-(2-methoxy) ethoxyethoxides of Y, Ba, and Cu in a $1:2:3$ molar ratio, respectively, for the sol–gel synthesis of $YBa_2Cu_3O_{7-\delta}$ (565).

The applications of the SG methods for the preparation of catalysts, particularly in the form of aerogels (566), was recently reviewed (11a, 567). Attention may be drawn to other recent reviews (175) on the applications of the alkoxide derivatives of tin (175), as well as *p* block (39) and other main group metals (568) for the preparation of ceramic materials by the SG process.

A rather distinctively advantageous feature of the SG technique is that at the low temperature(s) of operation, it opens up the possibilities of synthesizing Organically Modified Ceramics (ORMOCERS) or Organically Modified Silicates (ORMOSILS). Schmidt (569) was successful in introducing organics into inorganic components by a number of procedures such as

1. Penetrating porous glasses/gel by organic monomers, followed by their polymerization (570).
2. Linking organic groups to an inorganic backbone through chemical bonds of the types.

a. Covalent
$$-O\!\!\!\diagdown \atop -O\!\!\!\diagup \!\!\Rightarrow Si\!-\!C \diagup\!\!\diagdown\!\!\diagup\!\!\diagdown R \qquad\qquad R = \diagup\!\!\diagdown\!\!\diagup\!\!\diagdown NMe$$

b. Ionic
$$-O\!\!\!\diagdown \atop -O\!\!\!\diagup \!\!\Rightarrow M^+\!\!-\!O\!-\!\underset{\underset{O}{\|}}{C}\diagup\!\!\diagdown\!\!\diagup\!\!\diagdown R \qquad\qquad M = Ti, Zn, Sn,\ \text{and so on}$$

c. Coordinating
$$\begin{array}{c} \!\!\!\Rightarrow Si \diagup\!\!\diagdown\!\!\diagup\!\!\diagdown L \diagdown \\ \qquad\qquad\qquad M'\!\!<^{X}_{X} \qquad\qquad M' = Zn, Sn,\ \text{and so on} \\ \!\!\!\Rightarrow Si \diagup\!\!\diagdown\!\!\diagup\!\!\diagdown L \diagup \end{array}$$

Mehrotra (31) as well as Sakka (534) recently reviewed the current literature on a number of gels with microstructures [e.g., porous gels (546)], optically useful gels containing dyes such as rhodamines (571), and inorganic–organic composites (572).

Hybrid siloxane–oxide materials were prepared (573) by the SG techniques by using either monomeric species such as $Me_2Si(OEt)_2$ or by OH-terminated PDMS cross-linked by metallic alkoxides for example, $M(OR)_4$ with M = Si, Ti, Zr, and so on.

A detailed account of the hydrolysis and condensation reactions of alkyl orthosilicates (554) and alkoxides of other metals (555) was presented by Brinker and Scherer (546).

Special mention may be made of a building "block" approach that was initiated by Klemperer et al. (574). During their studies of the hydrolysis of $Si(OMe)_4$, these workers synthesized a new species, $[Si_8O_{12}]$ $(OMe)_8$, in addition to the expected lower building blocks such as $(Si_2O)(OMe)_6$, and $[Si_3O_2](OMe)_8$. It was shown that the silica xerogels obtained by the SG processing of the $(Si_8O_{12})(OMe)_8$ [Fig. 82(b)] building block have microstructural features like surface areas and porosites different from those obtained by similar SG processing of $Si(OMe)_4$ [Fig. 82(a)]: This difference was explained (575)

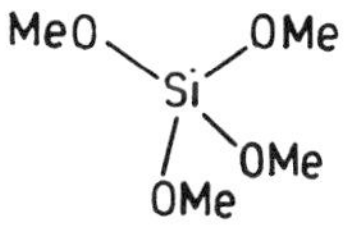

**Si(OMe)₄**

(a)

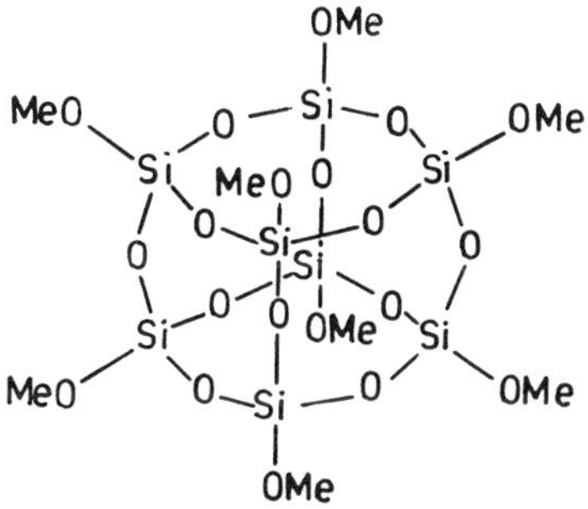

**[Si₈O₁₂](OMe)₈**

(b)

Figure 82. Sol–gel building blocks.

on the basis of rigidity and the large size of the cubic ($Si_8O_{12}$) core structure, inhibiting the extensive cross-linking possible in the Si(OMe$_4$) derived gels.

Compared to alkyl orthosilicates, alkoxides of transition and other main group metals are generally much more susceptible to hydrolysis. Extensive work on the identification of products formed during the controlled hydrolysis of alkoxides of a number of metals [e.g., Ti$^{IV}$, Zr$^{IV}$, Ta$^{V}$, Sn$^{IV}$, Ce$^{IV}$, and U$^{V}$] was initiated by Bradley et al. (6). They estimated ebulliometrically the average molecular complexities of the hydrolysis products formed by adding calculated quantities of water in solutions of metal alkoxides in the parent alcohols. In spite of inherent difficulties (72) in the structural elucidation of alkoxide derivatives of metals by X-ray crystallography, the actual structures of many oxide-alkoxides (cf. Section V on oxide-alkoxides) were determined. However, a novel technique employing $^{17}$O NMR spectroscopy, which was employed by Klemperer and co-workers (556, 576, 577) yielded interesting information on the study of possible intermediate hydrolyzed alkoxide products (building blocks technique) during the use of titanium alkoxides in the SG process.

Oxygen-17 occurs naturally at an abundance of 0.037% only (578). When water was enriched with $^{17}$O to about 10%, reacts with a titanium alkoxide sample, which was not enriched in $^{17}$O, the kinetically inert C—O bonds are hardly affected in contrast to the Ti—O bonds. The resulting hydrolysis reaction therefore can be represented as follows:

$$x\text{Ti(OR)}_4 + y\text{H}_2\text{O}^* \longrightarrow (\text{Ti}_x\text{O}_y^*)(\text{OR})_{4x-2y} + 2 \text{ ROH} \uparrow \cdots \quad (248)$$

In view of the above selective enrichment, $^{17}$O NMR spectra (556) of the resultant species depict strong resonances for the "oxide" oxygen atoms only, and not for alkoxide oxygen atoms. The consequent spectral simplification allows for identification of individual oxo-species present in the complex mixtures of polyoxoalkoxides (e.g., ethoxides, isopropoxides, and benzyloxides), the structures of many of which were elucidated (579) by solid state $^{17}O$ NMR spectroscopy as well as X-ray crystallography.

A significant contribution of these investigations was the revelation of the roles played by different species in the mechanism of the sol–gel process. For example, it was shown (577) that in the titanium ethoxide system during the sol–gel process, the well-known $(Ti_7O_4](OEt)_{20}$ species (475b, 556) is highly reactive toward ethanol and its $[Ti_7O_4]$ core is extensively decomposed within 10 min, whereas the $[Ti_{16}O_{16}]$ core of the $[Ti_{16}O_{16}](OEt)_{32}$ species (475b, 556) is relatively stable and is preserved in good yield during sol–gel polymerization.

## 2. Metalloorganic Chemical Vapor Deposition (MOCVD) Technique

Although MBE continues to be the best technique for controlled deposition of thin layers (10–100 Å) of materials, the MOCVD process does offer the advantage for rapid deposition over large areas of substrates. The basic principles of the MOCVD technique is the thermal decomposition of volatile molecular precursors to the desired combined form (or even the constituent metal itself) on a selected substrate at not too high a temperature. In addition to the volatility of the precursor and its facile decomposition to the desired combined form for deposition, the whole operation should not, from the practical point of view, involve any toxic/hazardous byproducts, that might entail any environmental problems.

Different steps involved in this technique can be represented diagrammatically (580–582), as shown in Fig. 83.

Although metal alkoxides proved to be excellent precursors for the preparation of ceramics by the sol–gel process, their applications in the MOCVD technique are not as widespread, in spite of the attractive features of their volatility and tendency to decompose to the metal oxides. Both of these desirable prerequisites appear to be enhanced, as discussed earlier, by choosing more ramified alkoxide groups. Despite these attractive features, the MOCVD applications of metal alkoxides appear to be limited by (a) their commercial nonavailability, (b) handling difficulties arising from their ready hydrolyzability, and (c) the lack of a clear understanding of their decomposition pathways, which could lead to improvement(s) in the purity of the deposited material.

As early as 1959, Bradley and Factor (583) suggested the use of zirconium tertiary butoxide $[Zr(O\text{-}t\text{-}Bu)_4$, bp, $\sim 50°C/1 \text{ mm}^{-1}]$ as a suitable precursor. This use is based on its decomposition to $ZrO_2$ at around 200–250°C through a chain reaction involving the hydrolysis of the alkoxide by water, formed in a

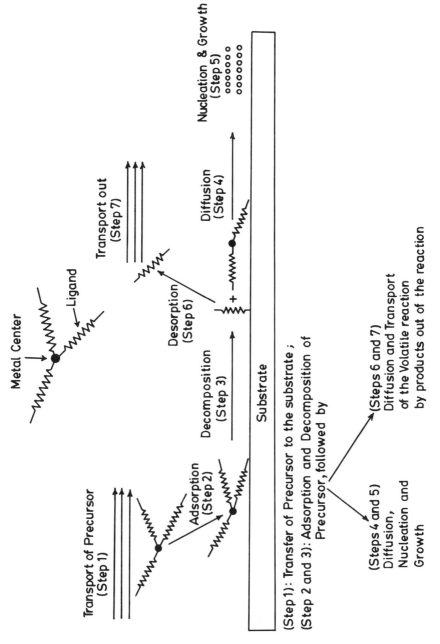

(Step 1): Transfer of Precursor to the substrate ;
(Step 2 and 3): Adsorption and Decomposition of
Precursor, followed by

(Steps 4 and 5)
Diffusion,
Nucleation and
Growth

(Steps 6 and 7)
Diffusion and Transport
of the Volatile reaction
by products out of the reaction

Figure 83.    Different steps involved in MOCVD technique.

side dehydration reaction of the tertiary butanol as indicated by the following plausible steps:

$$2Zr(O\text{-}t\text{-}Bu)_4 \longrightarrow 2ZrO(O\text{-}t\text{-}Bu)_2 + H_2O + 2H_2C{=}CMe_2$$

$$\downarrow$$

$$2ZrO_2 + 2H_2C{=}CMe_2 \qquad\qquad (249)$$

and so on . . .

Mazdiyasni et al. (584) were successful in depositing $ZrO_2/HfO_2$ by a similar route as well as the oxides of yttrium, dysprosium, and ytterbium from their isopropoxides at about 250°C under an atmosphere of nitrogen. More recently, deposition of $Al_2O_3$ on InP was reported (585) by the thermolysis of $Al(O\text{-}i\text{-}Pr)_3$ and also of $TiO_2$ and $Ta_2O_5$, respectively, from their ethoxides (586).

Rather limited information is available on the mode(s) of pyrolysis of metal alkoxides in general. For example, a study of flash vacuum pyrolysis of different titanium alkoxides, $Ti(OR)_4$ (R = $C_2H_5$, $C_3H_7$, $t\text{-}C_4H_9$, $H_2C\cdot C(Me)_3$, and so on) revealed (587) that dehydration processes occur mainly in the case of the neopentyloxide, which appears to undergo decomposition especially on the $\gamma$ hydrogen. In other cases, alcohols, ethers, alkenes, and carbonyl derivatives are obtained depending on the nature of R in the $Ti(OR)_4$ starting material; for example, the ratio of alcohol to ether appears to increase from primary to secondary to tertiary alkoxides, indicating the influence of crowding on the $\alpha$-carbon atom. However, the carbonyl compounds appear to emanate from a $\beta$-hydrogen elimination step mainly in the cases of ethoxide and isopropoxide.

It may be emphasized that metal alkoxides sometimes yield deposits other than those of the metal oxides by the MOCVD technique. For example, although the tertiary butoxides of aluminum, molybdenum, and tungsten all deposit the corresponding oxides, the cyclohexyloxides of molybdenum and tungsten depict a different behavior (588). The cyclohexyloxides of Mo and W eliminate a mixture of cyclohexanol, cyclohexanone, and cyclohexane, leaving $Mo_2C_4O_4$ and $W_2C_4O_4$, which respectively, decompose at higher temperatures to $Mo_2C$ and metallic W.

Similarly, the copper(I) *tert*-butoxide, $[Cu(O\text{-}t\text{-}Bu)]_4$, sublimes at $10^{-5}$ torr and its chemical vapor is deposited at about 400°C, yielding (589) thin layers of metallic copper on substrates such as silicon, quartz, graphite, aluminum, and glass. Intuitively, its more volatile (80°C/$10^{-3}$ torr) adduct $Cu(O\text{-}t\text{-}Bu)(PMe_3)$ also leads (590) to a deposit of metallic copper.

Obviously, the lower oligomerization tendencies of the branched alkoxides of metals in valence states 5, 4, and 3 endow upon them sufficient volatility to make them useful in the MOCVD technique. Generally however, the lower volatility of alkoxides of di- and monovalent metals, which arise from their higher association coupled with the dominant electrovalent nature of the metal–oxygen bond(s), renders them less suitable for this purpose. Efforts are continuing to overcome these difficulties by exploring other possibilities, for example, (a) by using their heterometallic alkoxides, which are generally less associated and more volatile; (b) by using alkoxides of alkanols with chelating-type characteristics as in alkoxy- and aminoalkanols; (c) by enhancing the volatility of alkoxides by using fluoroalkanols in place of simple alkanols; (d) by employing oxide-alkoxide derivatives that can be handled more easily and sometimes tend to be decomposed more neatly.

A few illustrative examples of these are described below.

1. Bimetallic alkoxides such as $LiNb(OR)_6$ and $LiTa(OR)_6$ are not only soluble and volatile, but such heterometallic alkoxides tend to yield the desired polyheterooxides (e.g., $LiNbO_3$ and $LiTaO_3$). However, although such heterometallic alkoxide precursors have proved to be highly valuable in the sol–gel processes (563), even the slightest tendencies of their dissociation at the higher temperatures required in the MOCVD process has to be carefully guarded against. In spite of such difficulties, attempts are being made for the above and other useful ceramic materials such as $KSr_2Nb_5O_{15}$, $Sr_xBa_{1-x}Nb_2O_6$, $KTa_xNb_{1-x}O_3$, and $PbMg_{0.33}Nb_{0.67}O$ and superconductors such as $YBa_2Cu_3O_{7-\delta}$ by the use of suitable presynthesized heterometallic alkoxide precursors or their molecular mixtures in the desired proportion(s). The use of a more volatile lithium derivative of, the $\beta$-diketone 2,2,6,6-tetramethylheptan-3,5-dione (308), for example, with $Nb(OMe)_5$, tends to overcome this deficiency due to the much lower volatility of LiOMe.

2. In addition to the use of chelating $\beta$-diketonate ligands including their fluoranated derivatives (308), the synthesis of copper(II) bis(dimethylaminoethoxide), $Cu(OCH_2CH_2NMe_2)_2$, which tends to volatilize at about 60°C under reduced pressure, shows promise as a MOCVD precursor (404). Similarly, the first structurally characterized monomeric barium aminoalkoxide derivative, $[Ba\{(OC_2H_4)(HOC_2H_4)N\}_2 \cdot 2EtOH]$, with an eight-coordinated barium (559) atom appears to be useful for this purpose; even more attractive is the barium derivative of a chelating ramified alkanol, $[BaOC-t-Bu(CH_2OEt)_2]_2$, with which the MOCVD deposit could be neater, as the precursor does not involve either fluorine or nitrogen contamination (288, 560).

The above rather brief selective account clearly shows the need of further concerted investigations for the choice of optimally suited precursor(s) and their decomposition pathways to achieve the full potentialities of the use of homo-as well as heterometallic alkoxides for the preparation of novel ceramic materials via the MOCVD technique.

## VIII.  CONCLUSION

An upsurge of interest, created by the elucidation of steric and inductive effects in the early 1950s, resulted in an exhaustive study (4, 6, 7, 28, 34, 38, 39, 42) during the following two-to-three decades on the synthesis and properties of the alkoxides of almost all the elements in the periodic table. Similarly, the apparently covalent behavior of double (bimetallic) alkoxides of even strongly electropositive metals led in the early 1970s to novel routes of their isolation and bridged structural formulation (2, 4, 14). These routes were suggested on the basis of the differences in the reactivity and the NMR spectra of the terminal and bridging alkoxy groups of a metal alkoxide species under study. The high reactivity of the alkoxy derivatives with protic agents (including water) and a variety of organic substrates, opens up a new direction of future research for alkoxides as unique synthons for a variety of novel metalloorganic (6, 7) and organometallic derivatives (513, 514, 524, 525), which in many cases are not easily accessible by any other route. Finally, the potentiality of their applications as precursors for ceramic materials has provided a new impetus to (a) their structural elucidation by X-ray crystallography (4a, 28, 34, 38, 39, 42, 504) and other sophisticated physicochemical techniques, which have reemphasized the role of alkoxide bridges between similar and dissimilar elements in the chemistry of homo- as well as heterometal alkoxides, respectively, and also to (b) the extension of active interest in alkoxide derivatives of substituted (mainly fluoro) and chelating alcohols particularly of metals such as Ba, Cu, Ln, Bi, and Pb, the mixed oxide-ceramics of which have begun to find applications in superconducting materials since 1986.

A new dimension has been opened during the last decade in the field of heterometal coordination chemistry because of the synthesis of a variety of thermally stable ter and higher heterometallic alkoxides. This area is expected to receive increasing attention because of the first X-ray structural elucidation done in 1996 (39a) of a termetallic isopropoxide of Ba, Zr, and Cd. This derivative also revealed an even more fascinating behavior of the shifting of the alkoxometallate ligands to suit the nature of the three metals involved in the species.

The inadvertent isolation of a large number and variety of oxo-alkoxide derivatives (42) during the last decade has added a fascinating dimension to metal alkoxide chemistry as a whole. This dimension will direct it into a new phase of unprecedented novelty in the chemistry of these compounds and reveal their possible role(s) between the metal alkoxides and oxide-ceramic materials (243).

This unprecedented interest, which emerged in the chemistry of alkoxides, has not only broadened our insight into the field but has also clearly pointed to the need for more penetrating physicochemical investigations on their formation and a much clearer understanding of the nature of bond(s) involved therein. A unique feature of this chapter is juxtaposed treatment of homo- as well as heterometallic alkoxides. This juxtaposition emphasizes the close similarity in their fascinating chemistry, which has unfolded, particularly during the past decade.

## ABBREVIATIONS

| | |
|---|---|
| a, b, c, $\alpha$, $\beta$, $\gamma$, | Crystallographic lattice parameters |
| acac | Acetylacetonate |
| adme | 1-Adamantylmethyl |
| An | An actinide |
| AOH | Aminopropanols |
| Ar | Aryl (substituted phenyl) |
| bpy | 2,2'-Bipyridine |
| $t$-Bu | $tert$-Butyl |
| $c$ | Cyclo |
| CAN | Ceric ammonium nitrate |
| Cat | Catecholate dianion |
| cod | Cyclooctadiene |
| cot | Cyclooctatetraene |
| Cp | Cyclopentadienyl |
| CVD | Chemical vapor deposition |
| Cy | Cyclohexyl |
| $\beta$-dik | $\beta$-Diketonate |
| $\beta$-dikH | $\beta$-Diketone |
| diglyme | Diethylene glycol dimethylether |
| dma | $N,N'$-Dimethylacetamide |
| dme | 1,2-Dimethoxyethane (ligand) |
| dmf | $N,N'$-Dimethylformamide (ligand) |
| dmso | Dimethyl sulfoxide (ligand) |
| DMSO | Dimethyl sulfoxide (solvent) |
| ESR | Electron spin resonance |
| Et | Ethyl |
| EXAFS | Extended X-ray absorption fine structure |
| HDBM | Dibenzoylmethane |
| HDPM | Dipivaloylmethane |
| hfd | 1,1,1,5,5,5-Hexafluoroacetylacetonato |
| hfip | Hexafluoroisopropoxide, $^-OCH(CF_3)_2$ |
| hftb | Hexafluoro-$tert$-butoxide, $^-OCMe(CF_3)_2$ |

| | |
|---|---|
| HMPA | Hexamethylphosphoramide |
| *i* | Iso |
| *i*-Pr | Isopropyl |
| IR | Infrared |
| *J* | Coupling constant |
| L | Generally a neutral monodentate ligand (unless otherwise indicated) |
| $L_{Al}$ | Tetraisopropoxoaluminate, $\{Al(O\text{-}i\text{-}Pr)_4\}^-$ |
| $L_{Nb}$ | Hexaisopropoxoniobate, $\{Nb(O\text{-}i\text{-}Pr)_6\}^-$ |
| $L_{Ta}$ | Hexaisopropoxotantalate, $[Ta(O\text{-}i\text{-}Pr)_6]^-$ |
| $L_{Zr}$ | Nonaisopropoxodizirconate, $\{Zr_2(O\text{-}i\text{-}Pr)_9\}^-$ |
| Ln | A lanthanide |
| MAS | Magic angle spinning |
| MBE | Molecular beam epitaxy |
| Me | Methyl |
| Mes | Mesityl |
| MOCVD | Metal–organic chemical vapor deposition or Metalloorganic chemical vapor deposition |
| *n* | Normal |
| NMR | Nuclear magnetic resonance |
| OAc | Acetate |
| $OB_z$ | Benzyloxide |
| ORMOCERS | Organically modified ceramics |
| ORMOSILS | Organically modified silicates |
| PDMS | Poly(dimethylsiloxanes) |
| Ph | Phenyl |
| PhCN | Benzonitrile |
| Phen | 1,10-Phenanthroline |
| PhMe | Toluene |
| $\alpha$-pic | $\alpha$-Picoline |
| $\beta$-pic | $\beta$-Picoline |
| $\gamma$-pic | $\gamma$-Picoline |
| PLZT | Lead lanthanum zirconium titanate |
| PNM | Lead niobium magnesate |
| PSN | Lead scandium niobate |
| py | Pyridine |
| R | An alkyl group |
| $R_F$ | Fluorinated alkyl groups such as $CH(CF_3)_2$, $CMe(CF_3)_2$, $CMe_2(CF_3)$ |
| *sec* | Secondary |
| SG | Sol–gel |
| TBP | Trigonal bipyramidal |

| teed | $N,N,N',N'$-Tetraethylethylenediamine |
| TEOS | Tetraethyl orthosilicate |
| thf | Tetrahydrofuran (ligand) |
| THF | Tetrahydrofuran (solvent) |
| thd | 2,2,6,6-Tetramethyl-3,5-heptanedionato |
| thdH | 2,2,6,6-Tetramethyl-3,5-heptanedione |
| THME-$H_3$ | Tris(hydroxymethyl)ethane |
| THMP-$H_3$ | Tris(hydroxymethyl)propane |
| tmeda | $N,N,N',N'$-Tetramethylethylenediamine (ligand) |
| tmu | Tetramethylurea (ligand) |
| TMU | Tetramethylurea (solvent) |
| X | Generally a monodentate anionic ligand like alkoxide, amide, halide (unless otherwise indicated) |
| XANES | X-ray absorption near edge spectroscopy |
| YBCO | Yttrium barium copper oxide |

## ACKNOWLEDGMENTS

The continued financial support of our research on this topic by the Department of Science and Technology, New Delhi is gratefully acknowledged. We wish to acknowledge the invaluable collaboration of the students whose names appear in the cited references.

## REFERENCES

1. D. C. Bradley, *Progress in Inorganic Chemistry*, Wiley-Interscience, New York, 1960, vol. 2, p. 303.

2. R. C. Mehrotra, *Inorg. Chim. Act, Rev.*, *1*, 99 (1967).

3. R. C. Mehrotra, *Chemtracts*, *2*, 389 (1990).

4. (a) K. G. Caulton and L. G. Hubert-Pfalzgraf, *Chem. Rev.*, *90*, 969 (1990). (b) H. Meerwein and T. Bersin, *Annalen*, *476*, 113 (1929).

5. R. C. Mehrotra and A. Mehrotra, *Inorg. Chim. Acta Rev.*, *5*, 127 (1971).

6. D. C. Bradley, R. C. Mehrotra, and D. P. Gaur, *Metal Alkoxides*, Academic, London, UK, 1978.

7. R. C. Mehrotra, *Adv. Inorg. Radiochem.*, *26*, 289 (1983).

8. R. C. Mehrotra and M. Aggarwal, *Polyhedron*, *4*, 845, 1141 (1985).

9. J. J. Ebelman, *Annalen*, *57*, 331 (1846).

10. R. Roy, *J. Am. Ceram. Soc.*, *49*, 145 (1956); *Science*, *238*, 1664 (1987).

11. R. C. Mehrotra, *J. Non-Cryst. Solids*, *100*, 1 (1988); *145*, 1 (1992).

12. R. C. Mehrotra, in *Sol–Gel Science and Technology*, M. A. Aegerter, M. Jaffel-icci Jr., D. F. Souza, and E. D. Zanotto, Eds., World Scientific, Singapore, 1989, pp. 1–16, 17–39, 40–60, and 421–431.

13. H. Dislich, *Angew. Chem. Int. Ed. (Engl)*, *10*, 367 (1971).

14. R. C. Mehrotra and P. N. Kapoor, *Coord. Chem. Rev.*, *14*, 1 (1974).

15. R. C. Mehrotra, *Proc. Ind. Natl. Sci. Acad.*, *42*, 1 (1976).

16. R. C. Mehrotra, *Kemiai Kozlemenyek*, *45*, 197 (1976).

17. R. C. Mehrotra and J. V. Singh, *J. Ind. Chem. Soc.*, *54*, 109 (1977).

18. R. C. Mehrotra, J. M. Batwara, and P. N. Kapoor (Sessional Lecture at the International Conference on Coordination Chemistry, San Paulo, 1977) *Coord. Chem. Rev.*, *31*, 67 (1980).

19. R. C. Mehrotra (Sessional lecture at the International Conference on Coordination Chemistry, Toulouse, 1980) *Coord. Chem. (IUPAC)*, *21*, 113 (1981).

20. R. C. Mehrotra (Invited Lecture at the Golden Jubilee of Indian Association for the Cultivation of Science, Calcutta, 1985), *Proc. Ind. Natl. Sci. Acad.*, *52A*, 954 (1986).

21. S. Hirano, T. Hayashi, K. Nosaki, and K. Kato, *J. Am. Ceram. Soc.*, *72*, 707 (1989).

22. A. Nazeri-Eshghi, A. X. Kuang, and J. D. Mackenzie, *J. Mater. Sci.*, *25*, 3333 (1990).

23. L. G. Hubert-Pfalzgraf, *Polyhedron*, *13*, 1181 (1994).

24. D. C. Bradley, *Polyhedron*, *13*, 1111 (1994).

25. R. K. Dubey, A. Singh, and R. C. Mehrotra, *J. Organomet. Chem.* (Eaborn Issue), *341*, 869 (1988).

26. R. K. Dubey, A. Shah, A. Singh, and R. C. Mehrotra, *Recl. Trav. Chim. Pays-Bas* (Van der Kirk Issue), *107*, 237 (1988).

27. D. C. Bradley, *Chem. Rev.*, *89*, 1317 (1989).

28. R. C. Mehrotra, A. Singh, and U. M. Tripathi, *Chem. Rev.*, *91*, 1287 (1991).

29. R. C. Mehrotra, *Natl. Acad. Sci. Lett.*, *14*, 153 (1991).

30. R. C. Mehrotra and A. K. Rai, *Polyhedron*, *10*, 1967 (1991).

31. R. C. Mehrotra, Present Status and Future Potential of the Sol–Gel process, In Chemistry, Spectroscopy and Applications of Sol–Gel Process, R. Relsfeld and C. K. Jorgensen, Eds., *Structure and Bonding*, Vol. LXXVII, Springer-Verlag, 1992, pp. 1–36.

32. R. C. Mehrotra, *Natl. Acad. Sci. Lett.*, *16*, 41 (1993).

33. R. C. Mehrotra, J. Sol–Gel Sci. Tech., *2*, 1 (1994).

34. M. H. Chisholm and I. P. Rothwell, in *Comprehensive Coordination Chemistry*, G. Wilkinson, R. D. Gillard, and J. A. McCleverty, Eds., Vol. 2, Pergamon, London, UK, 1987, pp. 335–364.

35. R. C. Mehrotra, *Mater. Res. Soc. Symp.*, *121*, 81 (1988).

36. (a) R. C. Mehrotra, Heterometal Alkoxides: A Novel Series of Stable Polymetallic Coordination Compounds; in *Facets of Coordination Chemistry*, B. V. Agarwala

and K. N. Munshi, Eds., World Scientific, Singapore, 1993, pp. 46–55. (b) S. Daniele, R. Papiernik, and L. G. Hubert-Pfalzgraf, S. Jagner, and M. Hakansson, *Inorg. Chem.*, *34*, 628 (1995).

37. R. C. Mehrotra, *Ind. J. Chem.*, *31A*, 492 (1992).

38. R. C. Mehrotra, A. Singh, and S. Sogani, *Chem. Soc. Rev.*, *23*, 215 (1994).

39. R. C. Mehrotra, A. Singh, and S. Sogani, *Chem. Rev.*, *94*, 1643 (1993). (a) M. Veith, S. Mathur, and V. Huch, *J. Am. Chem. Soc.*, *96*, (1996). (a) M. Veith, S. Mathur, and V. Huch, *J. Am. Chem. Soc.*, *118*, 903 (1996).

40. R. C. Mehrotra, S. K. Agarwal, and Y. P. Singh, *Coord. Chem. Rev.*, *68*, 100 (1985).

41. R. G. Bergman, *Polyhedron*, *14*, 3227 (1995).

42. R. C. Mehrotra and A. Singh, *Chem. Soc. Rev.*, *26*, 1 (1996).

43. (a) M. H. Chisholm, *Chem. Soc. Rev.*, *24*, 79 (1995); (b) M. I. Khan and J. Zubieta, *Progress in Inorganic Chemistry*, Wiley-Interscience New York, 1995, vol. 43, pp. 1–149.

44. W. G. Van Der Sluys, J. C. Huffman, D. S. Ehler, and N. N. Sauer, *Inorg. Chem.*, *31*, 1316 (1992).

45. B. A. Vaartstra, J. C. Huffman, W. E. Streib, and K. G. Caulton, *J. Chem. Soc. Chem. Commun.*, 1750 (1990).

46. J. A. Meese-Marktscheffel, R. E. Cramer, and J. W. Gilje, *Polyhedron*, *13*, 1045 (1994).

47. J. A. Meese-Marktscheffel, R. Fukuchi, M. K. Kido, G. Tachibanna, C. M. Jensen, and J. W. Gilje, *Chem. Mater.*, *5*, 755 (1993).

48. M. Bhagat, A. Singh, and R. C. Mehrotra, unpublished results.

49. H. Lehmkuhl and M. Eisenbach, *Annalen*, 672 (1975).

50. V. A. Schreider, E. P. Turevskaya, N. I. Koslova, and N. Ya Turova, *Inorg. Chim. Acta*, *53*, L73 (1981); N. Ya Turova, A. V. Korolea, D. E. Tchebukov, A. L. Yanovsky, and Yu. T. Struchkov, *Polyhedron*, *15*, 3869 (1996).

51. J. S. Banait and P. K. Pahil, *Polyhedron*, *5*, 1965 (1986).

52. J. S. Banait, S. K. Deol, and B. Singh, *Synth. React. Inorg. Met-Org. Chem.*, *20*, 1331 (1990).

53. N. Ya Turova, private communication.

54. W. G. Power and G. A. Ozin, *Adv. Inorg. Chem. Radiochem.*, *23*, 140 (1982).

55. G. B. Goodwin and M. E. Kenney, *Inorg. Chem.*, *29*, 1216 (1990).

56. C. W. Dekock and L. V. McAfee, *Inorg. Chem.*, *24*, 4293 (1985).

57. R. C. Mehrotra, K. N. Mahendra, and M. C. Aggarwal, *Proc. Ind. Acad. Sci.* (Chem. Soc.), *93*, 719 (1984).

58. M. Veith, E-Chul. Yu, and V. Huch, *Chem. Eur. J.*, *1*, 25 (1995).

59. K. F. Tesh, T. P. Hanusa, J. C. Huffman, and C. J. Huffman, *Inorg. Chem.*, *31*, 5572 (1992).

60. R. C. Mehrotra, A. K. Rai, and A. Jain, *Polyhedron*, *10*, 1030 (1990).

61. W. J. Evans, M. S. Sollberger, and T. P. Hanusa, *J. Am. Chem. Soc.*, *110*, 1841 (1988).

62. W. J. Evans and M. S. Sollberger, *Inorg. Chem.*, *27*, 4417 (1988).

63. P. S. Gradeff, F. G. Schreiber, K. C. Brooks, and R. E. Sievers, *Inorg. Chem.*, *24*, 1110 (1985).

64. W. J. Evans, T. J. Deming, J. M. Olofson, and J. W. Ziller, *Inorg. Chem.*, *28*, 4027 (1989). (a) H. C. Aspinall and M. W. Williams, *Inorg. Chem.*, *35*, 255 (1996).

65. E. J. Thaler, K. Rypdal, A. Haaland, and K. G. Caulton, *Inorg. Chem.*, *28*, 2431 (1989).

66. S. Sogani, A. Singh, and R. C. Mehrotra, *Main Group Metal Chem.*, *13*, 375 (1990).

67. R. C. Mehrotra, M. Aggarwal, and C. K. Sharma, *Synth. React. Inorg. Met.-Org. Chem.*, *13*, 571 (1984).

68. R. K. Dubey, A. Singh, and R. C. Mehrotra, *Inorg. Chim. Acta*, *143*, 169 (1988).

69. A. Shah, A. Singh, and R. C. Mehrotra, *Ind. J. Chem.*, *27A*, 372 (1988).

70. R. C. Chhipa, A. Singh, and R. C. Mehrotra, *Synth. React. Inorg. Met.-Org. Chem.*, *20*, 989 (1990).

71. U. M. Tripathi, A. Singh, and R. C. Mehrotra, *Polyhedron*, *10*, 949 (1991).

72. D. C. Bradley, *Philos. Trans. Soc. London*, *A330*, 167 (1990).

73. S. Sogani, R. Bohra, M. Noltemeyer, A. Singh, and R. C. Mehrotra, *J. Chem. Soc. Chem. Commun.*, 738 (1991).

74. U. M. Tripathi, A. Singh, R. C. Mehrotra, S. C. Goel, M. Y. Chiang, and W. E. Buhro, *J. Chem. Soc. Chem. Commun.*, 152 (1992).

75. A. Edelmann, J. W. Gilje, and F. T. Edelman, *Polyhedron*, *11*, 2421 (1992).

76. A. P. Purdy and C. F. George, *Polyhedron*, *13*, 709 (1994).

77. U. Bemm, R. Norrestam, M. Nygren, and G. Westin, *Inorg. Chem.*, *31*, 2050 (1992).

78. U. Bemm, R. Norrestam, M. Nygren, and G. Westin, *Inorg. Chem.*, *34*, 2367 (1995).

79. U. M. Tripathi, A. Singh, and R. C. Mehrotra, unpublished results.

80. S. Sogani, A. Singh, and R. C. Mehrotra, *Main Group Metal Chem.*, *15*, 197 (1992).

81. S. Sogani, A. Singh, and R. C. Mehrotra, *Indian J. Chem.*, *32A*, 345 (1993).

82. S. Mathur, A. Singh, and R. C. Mehrotra, *Polyhedron*, *11*, 341 (1992).

83. S. Mathur, Ph.D. Thesis, "Studies on Heterometallic Alkoxides of Tin and Lead" University of Rajasthan, Jaipur, 1992.

84. R. K. Dubey, A. Singh, and R. C. Mehrotra, *Indian J. Chem.*, *31A*, 156 (1992).

85. R. Gupta, A. Singh, and R. C. Mehrotra, *Indian J. Chem.*, *30A*, 592 (1991).

86. R. Gupta, A. Singh, and R. C. Mehrotra, *New J. Chem.*, *15*, 665 (1991).

87. R. Gupta, A. Singh, and R. C. Mehrotra, *Indian J. Chem.*, *32A*, 310 (1993).

88. G. Garg, A. Singh, and R. C. Mehrotra, *Indian J. Chem.*, *30A*, 688, 866 (1991).

89. G. Garg, A. Singh, and R. C. Mehrotra, *Synth. React. Inorg. Met.-Org. Chem.*, *21*, 1047 (1991).

90. G. Garg, A. Singh, and R. C. Mehrotra, *Polyhedron*, *10*, 1733 (1991).

91. R. C. Chhipa, A. Singh, and R. C. Mehrotra, *Synth. React. Inorg. Met.-Org. Chem.*, *20*, 989 (1990).

92. R. C. Chhipa, A. Singh, and R. C. Mehrotra, *Indian J. Chem.*, *30A*, 1024 (1991).

93. S. Sogani, A. Singh, and R. C. Mehrotra, *Indian J. Chem.*, *34A*, 449 (1995).

94. C. K. Narula, *Mater. Res. Soc. Symp. Proc.*, *271*, 181 (1992).

95. J. Rai and R. C. Mehrotra, *Main Group Metal Chemistry*, *15*, 209 (1992).

96. J. Rai and R. C. Mehrotra, *J. Non-Cryst. Solids*, *152*, 118 (1993).

97. R. G. Jones, G. Karmas, G. A. Martin, Jr., and H. Gilman, *J. Am. Chem. Soc.*, *78*, 4285 (1956).

98. I. M. Thomas, *Can J. Chem.*, *39*, 1386 (1961).

99. B. Horvath, R. Moseler, and E. G. Horvarth, *Z. Anorg. Allg. Chem.*, *449*, 41 (1979).

100. W. G. Van der Sluys, A. P. Sattelberger, and M. McElfresh, *Polyhedron*, *9*, 1843 (1990).

101. M. A. Matchett, M. Y. Chiang, and W. E. Buhro, *Inorg. Chem.* *29*, 358 (1990).

102. H. A. Stecher, A. Sen, and A. L. Rheingold, *Inorg. Chem.*, *28*, 3280 (1989).

103. D. C. Bradley, H. Chudzynska, M. B. Hursthouse, and M. Motevalli, *Polyhedron*, *10*, 1049 (1991).

104. H. A. Stecker, A. Sen, and A. Rheingold, *Inorg. Chem.*, *27*, 1130 (1988).

105. W. Zwick, L. R. Avens, and A. P. Sattelberger, unpublished results.

106. (a) S. C. Goel, M. Y. Chiang, and W. E. Buhro, *Inorg. Chem.*, *29*, 4646 (1990). (b) B. Borup, J. A. Samuels, W. E. Streib, and K. G. Caulton, *Inorg. Chem.*, *33*, 994 (1994).

107. M. H. Chisholm, V. F. Distase, and W. E. Streib, *Polyhedron*, *9*, 253 (1990).

108. R. G. Jones, E. Bindschadler, D. Blume, G. Karmas, G. A. Martin, Jr., J. R. Thirtle, F. A. Yoeman, and H. Gilman, *J. Am. Chem. Soc.*, *78*, 6030 (1956).

109. D. G. Berg, R. A. Anderson, and A. Zalkin, *Organometallics*, *7*, 1858 (1988).

110. A. Sen, H. A. Stecher, and A. L. Rheingold, *Inorg. Chem.*, *32*, 473 (1992).

111. J. L. Stewart and R. A. Andersen, *J. Chem. Soc. Chem. Commun.*, 1846 (1987).

112. K. G. Moloy and T. J. Marks, *J. Am. Chem. Soc.*, *106*, 7051 (1984).

113. E. E. Van Tamelen, *Acc. Chem. Res.*, *3*, 361 (1970).

114. H. Nöth and H. Suchy, *Z. Anorg. Allg. Chem.*, *358*, 44 (1968).

115. W. J. Evans and M. S. Sollberger, *J. Am. Chem. Soc.*, *108*, 6095 (1986).

116. W. J. Evans, M. S. Sollberger, S. I. Khan, and R. Bau, *J. Am. Chem. Soc.*, *110*, 439 (1988).

117. B. A. Vaartstra, W. E. Streib, and K. G. Caulton, *J. Am. Chem. Soc.*, *112*, 8593 (1990).

118. Y. J. Kim, K. Osakada, K. Sugita, T. Yamamoto, and A. Yamamoto, *Organometallics*, 7, 2182 (1988); Y. J. Kim, K. Osakada, A. Takenaka, and A. Yamamoto, *J. Am. Chem. Soc.*, *112*, 1096 (1990).

119. P. T. Matsunga, J. C. Mavropoulos, and G. L. Hillhouse, *Polyhedron*, *14*, 175 (1995).

120. J. Caruso, M. J. Hampden-Smith, A. L. Rheingold, and G. Yap, *J. Chem. Soc. Chem. Commun.*, 157 (1995).

121. J. Caruso, C. Roger, F. Schwertfeger, M. J. Hampden-Smith, A. L. Rheingold, and G. Yap, *Inorg. Chem.*, *34*, 449 (1995).

122. M. Arora and R. C. Mehrotra, *Indian J. Chem.*, *7A*, 399 (1969).

123. R. C. Mehrotra and M. Arora, *Z. Anorg. Allg. Chem.*, *370*, 300 (1969).

124. S. S. Al-Juaid, C. Eaborn, M. N. A. El-Kheli, P. B. Hitchcock, P. D. Lickiss, H. E. Molla, J. D. Smith, and J. A. Zora, *J. Chem. Soc. Dalton*, 447, 1989.

125. M. M. Agrawal, Ph.D. Thesis, "Double alkoxides of some ter-, quadri- and quinque-valent metals," University of Rajasthan, Jaipur, India, 1968.

126. J. A. Meese-Marktscheffel, R. Weimann, H. Schumann, and J. W. Gilje, *Inorg. Chem.*, *32*, 5894 (1993).

127. W. A. Herrmann, N. W. Huber, and O. Runte, *Angew. Chem. Int. Ed. Engl. 34*, 2187 (1995).

128. M. J. Hampden-Smith, D. E. Smith, and E. G. Duesler, *Inorg. Chem.*, *29*, 3399 (1989).

129. M. Veith and M. Reimers, *Chem. Ber.*, *123*, 1941 (1990).

130. T. Athar, R. Bohra, and R. C. Mehrotra, *Main Group Metal Chem.*, *10*, 399 (1987).

131. M. J. Hampden Smith, D. S. Williams, and A. L. Rheingold, *Inorg. Chem.*, *29*, 4076 (1990).

132. R. C. Mehrotra and M. M. Agrawal, *J. Chem. Soc. (A)*, 1026 (1967).

133. C. K. Sharma, S. Goel, and R. C. Mehrotra, *Indian J. Chem.*, *14A*, 878 (1976).

134. D. J. Teff, J. C. Huffman, and K. G. Caulton, *Inorg. Chem.*, *33*, 6289 (1994).

135. R. C. Mehrotra, M. M. Agrawal, and P. N. Kapoor, *J. Chem. Soc. (A)*, 2673 (1968).

136. A. P. Purdy and C. F. George, *Polyhedron*, *14*, 761 (1995).

137. A. Mehrotra and R. C. Mehrotra, *Inorg. Chem.*, *11*, 2170 (1972).

138. L. G. Hubert-Pfalzgraf and J. G. Riess, *Inorg. Chem.*, *14*, 2854 (1975).

139. R. C. Mehrotra and M. M. Agrawal, unpublished work.

140. S. Govil, P. N. Kapoor, and R. C. Mehrotra, *Inorg. Chim. Acta*, *15*, 43 (1975).

141. M. Veith, *Chem. Rev.*, *90*, 3 (1990).

142. M. Veith, D. Kafer, and V. Huch, *Angew. Chem. Int. Ed. Engl.*, *25*, 375 (1986).

143. J. A. Samuels, J. W. Zwanzinger, E. B. Lobkovsky, and K. G. Caulton, *Inorg. Chem.*, *31*, 4046 (1992).

144. G. Westin and M. Nygren, *J. Mater. Soc.*, *27*, 1617 (1992).

145. R. C. Mehrotra and R. Bohra, *Metal Carboxylates*, Academic, London, UK, 1983.

146. R. C. Mehrotra, R. Bohra, and D. P. Gaur, *Metal β-Diketonates and Allied Derivatives*, Academic, London, UK, 1978.

147. D. C. Bradley, *Nature (London)*, *182*, 1211 (1958).

148. D. J. Teff, J. C. Huffman, and K. G. Caulton, *Inorg. Chem.*, *34*, 2491 (1995).

149. B. A. Vaarstra, J. C. Huffman, W. E. Streib, and K. G. Caulton, *Inorg. Chem.*, *30*, 3068 (1991).

150. N. Ya Turova, V. A. Kuzunov, A. I. Yanovskii, N. G. Borkii, Yu T. Struchkov, and B. L. Tarnopolskii, *J. Inorg. Nucl. Chem.*, *41*, 5 (1979).

151. R. C. Mehrotra, *J. Ind. Chem. Soc.*, *30*, 585 (1953).

152. F. A. Cotton, D. O. Marler, and W. Schwotzer, *Inorg. Chem.*, *23*, 4211 (1984).

153. R. H. Cayton, M. H. Chisholm, E. R. Davidson, V. F. DiStasi, P. Du, and J. C. Huffman, *Inorg. Chem.*, *30*, 1020 (1991).

154. M. H. Chisholm, *Polyhedron*, *2*, 681 (1983).

155. R. R. Schrock, *Polyhedron*, *14*, 3177 (1995).

156. A. R. Barron, *Polyhedron*, *14*, 3197 (1995).

157. A. Haaland, in Coordination Chemistry of Aluminum, G. H. Robinson, Ed., VCH, New York, 1993, Chapter I.

158. D. M. Lunder, E. B. Lobkovsky, W. E. Streib, and K. G. Caulton, *J. Am. Chem. Soc.*, *113*, 1187 (1991).

159. T. J. Johnson, J. C. Huffman, and K. G. Caulton, *J. Am. Chem. Soc.*, *114*, 2725 (1992).

160. J. T. Poulton, K. Folting, W. E. Steib, and K. G. Caulton, *Inorg. Chem.*, *31*, 3190 (1992).

161. H. Rothfuss, J. C. Huffman, and K. G. Caulton, *Inorg. Chem.*, *33*, 187 (1994).

162. I. M. Atagi and J. M. Mayer, *Angew. Chem. Int. Ed. Engl.*, *32*, 439 (1993).

163. J. L. Kerschner, P. E. Fanwick, I. P. Rothwell, and J. C. Huffman, *Inorg. Chem.*, *28*, 780 (1989).

164. J. M. Mayer, *Polyhedron*, *14*, 3273 (1995) and references cited therein.

165. D. C. Bradley, R. C. Mehrotra, and W. Wardlaw, *J. Chem. Soc.*, 2027, 4204, and 5020 (1952).

166. J. H. Rogers, A. W. Apblett, W. M. Cheaver, A. N. Tyler, and A. R. Barron, *J. Chem. Soc. Dalton Trans.*, 3179 (1992).

167. R. J. Butcher, D. L. Clark, S. K. Grumbine, R. L. Vincent-Hollis, B. L. Scott, and J. G. Watkin, *Inorg. Chem.*, *34*, 5468 (1995).

168. B. A. Vaarstra, J. C. Huffman, P. S. Gradeff, L. G. Hubert-Pfalzgraf, J. C. Daran, S. Parraud, K. Yunlu, and K. G. Caulton, *Inorg. Chem.*, *29*, 3126 (1990) and references cited therein.

169. M. H. Chisholm, K. Folting, C. E. Hammond, M. J. Hampden-Smith, and K. G. Moodley, *J. Am. Chem. Soc.*, *111*, 5300 (1989).

170. F. A. Cotton, M. P. Diebold, and W. J. Roth, *Inorg. Chem.*, *24*, 3509 (1985); *26*, 3323 (1987).

171. K. Osakada, Y. J. Kim, M. Tanaka, S. Ishiguro, and A. Yamamoto, *Inorg. Chem.*, *30*, 197 (1991).

172. D. M. Barnhart, D. L. Clark, J. C. Gordon, J. C. Huffman, J. G. Watkin, and B. D. Zurich, *Inorg. Chem.*, *34*, 5416 (1995).

173. F. Labrize, L. G. Hubert-Pfalzgraf, J. Vaissermann, and C. B. Knobler, *Polyhedron*, *15*, 577 (1996).

174. C. D. Chandler, J. Caruso, M. J. Hampden-Smith, and A. L. Rheingold, *Polyhedron*, *14*, 2491 (1995).

175. M. J. Hampden-Smith, T. A. Wark, and C. J. Brinker, *Coord. Chem. Rev.*, *112*, 81 (1992).

176. F. Baboneau, J. Maquet, and J. Livage, *Chem. Mater.*, *7*, 1050 (1995).

177. V. W. Day, T. A. Eberspacher, W. G. Klemperer, C. W. Park, and F. S. Rosenberg, *J. Am. Chem. Soc.*, *113*, 8190 (1991).

178. R. K. Harris and B. E. Mann, Eds., NMR and the Periodic Table, Academic, London, UK, 1978.

179. J. Mason, *Polyhedron*, *8*, 1657 (1989).

180. D. C. Crans, R. A. Felty, and M. M. Miller, *J. Am. Chem. Soc.*, *113*, 265 (1991).

181. U. M. Tripathi, A. Singh, and R. C. Mehrotra, *Polyhedron*, *12*, 1947 (1993).

182. S. Mathur, A. Singh, and R. C. Mehrotra, *12*, 1073 (1993); *Indian J. Chem.*, *32A*, 585 (1993); *34A*, 454 (1995).

183. R. Papiermik, L. G. Hubert-Pfalzgraf, and M. C. Massiani, *Polyhedron*, *10*, 1657 (1991).

184. T. J. Boyle, D. C. Bradley, M. J. Hampden-Smith, A. Patel, and J. W. Ziller, *Inorg. Chem.*, *34*, 5893 (1995).

185. G. R. Hatfield and K. R. Gardiner, *J. Mater. Sci.*, *24*, 4209 (1989).

186. B. K. Teo, *Acc. Chem. Res.*, *13*, 412 (1980).

187. B. K. Teo, in *EXAFS: Basic Principles and Data Analysis*, Springer-Verlag, Berlin, 1986.

188. D. E. Sayers and B. A. Bunker, in *X-ray Absorption: Principles Applications, Techniques of EXAFS, SEXAFS and SANES*, D. C. Konigsberger and R. Prins, Eds., Wiley-Interscience, New York, 1988, pp. 211–253.

189. F. Babonneau, S. Doeuff, A. Leaustic, C. Sanchez, C. Cartier, and M. Verdaguer, *Inorg. Chem.*, *27*, 3166 (1988).

190. J. Choy, J. Yoon, D. Kim, and S. Huang, *Inorg. Chem.*, *34*, 6524 (1995).

191. E. A. Cuellar, S. S. Miller, T. J. Marks, and E. Weitz, *J. Am. Chem. Soc.*, *105*, 4580 (1983).

192. A. P. Purdy, C. F. George, and G. A. Brewer, *Inorg. Chem.*, *31*, 2633 (1992).

193. R. H. Holm, S. Ciurli, and J. Weigel, *Progress in Inorganic Chemistry*, Wiley-Interscience, New York, 1990, vol. 38, p. 1.

194. D. N. Hendrickson, G. Christou, E. A. Schmitt, E. Libby, J. S. Bashkin, S. Wang, H. L. Tsai, J. B. Vincent, P. D. W. Boyd, J. C. Huffman, K. Folting, Q. Li, and W. E. Streib, *J. Am. Chem. Soc.*, *114*, 2455 (1992) and references cited therein.

195. K. L. Taft, A. Caneschi, L. E. Pence, C. D. Delfs, G. C. Papaefthymiou, and S. J. Lippard, *J. Am. Chem. Soc.*, *115*, 11753 (1993).

196. R. C. Mehrotra and A. Mehrotra, unpublished results. A. Mehrotra, Ph.D. Thesis, "Alkoxides and Double Alkoxides of Some Rare Earth and other Metallic Elements," University of Rajasthan, Jaipur, India, 1971.

197. E. P. Turevskaya, N. I. Kozlova, N. Ya. Turova, A. I. Belokon, D. V. Berdyev, V. G. Kessler, and Yu. K. Grishin, *Russ. Chem. Bull.*, *44*, 734 (1995).

198. K. Folting, W. E. Streib, K. G. Caulton, O. Poncelet, and L. G. Hubert-Pfalzgraf, *Polyhedron*, *10*, 1639 (1991).

199. M. Veith, S. Weidner, K. Kunze, D. Käfer, J. Hans, and V. Huch, *Coord. Chem. Rev.*, *137*, 297 (1994).

200. B. Cetinkaya, I. Gümrükcü, M. F. Lappert, J. L. Atwood, R. D. Rogers, and M. J. Zaworotko, *J. Am. Chem. Soc.*, *102*, 2088 (1980).

201. M. F. Lappert, A. Singh, and R. G. Smith, *Inorg. Synth.*, *27*, 164 (1990).

202. W. G. Van der Sluys and A. P. Sattelberger, *Chem. Rev.*, *90*, 1027 (1990).

203. P. B. Hitchcock, M. F. Lappert, G. A. Lawless, and B. Royo, *J. Chem. Soc. Chem. Commun.*, 1141 (1990).

204. S. R. Drake, D. J. Otway, M. B. Hursthouse, and K. M. A. Malik, *Polyhedron*, *11*, 1995 (1992).

205. K. G. Caulton, M. H. Chisholm, S. R. Drake, and K. Folting, *Inorg. Chem.*, *30*, 1500 (1991).

206. R. A. Andersen and G. E. Coates, *J. Chem. Soc. Dalton Trans.*, 2153 (1972).

207. K. Ruhlandt-Senge, R. A. Bartlett, M. M. Olmstead, and P. P. Power, *Inorg. Chem.*, *32*, 1724 (1993).

208. D. L. Clark and J. G. Watkin, *Inorg. Chem.*, *32*, 1766 (1993).

209. M. J. Hampden-Smith, T. A. Wark, A. L. Rheingold, and J. C. Huffman, *Can. J. Chem.*, *69*, 121 (1991).

210. B. Cetinkaya, I. Gümrükcü, M. F. Lappert, J. L. Atwood, R. Shakir, *J. Am. Chem. Soc.*, *102*, 2086 (1980).

211. S. R. Drake, W. E. Streib, K. Folting, M. H. Chisholm, and K. G. Caulton, *Inorg. Chem.*, *31*, 3205 (1992).

212. W. J. Evans, R. E. Golden, and J. W. Ziller, *Inorg. Chem.*, *30*, 4963 (1991).

213. D. C. Bradley, R. C. Mehrotra, J. D. Swanwick, and W. Wardlaw, *J. Chem. Soc.*, 2025 (1953).

214. C. D. Chandler, G. D. Fallon, A. J. Koplick, and B. O. West, *Aust. J. Chem.*, *40*, 1427 (1987).

215. J. A. Dean, *Lange's Handbook of Chemistry*, McGraw-Hill, New York, 13th Ed., 1985, Section 3-120.

216. G. W. Svetich and A. A. Voge, *Acta Crystallogr.*, *B28*, 1760 (1972).

217. (a) W. S. Harwood, D. DeMarco, and R. A. Walton, *Inorg. Chem.*, *23*, 3077 (1984). (b) L. B. Anderson, F. A. Cotton, D. DeMarco, A. Fang, W. H. Ilsley, B. W. S. Kolthammer, and R. A. Walton, *J. Am. Chem. Soc.*, *103*, 5078 (1981).

(c) F. A. Cotton, L. R. Falvello, M. F. Fredrich, D. DeMarco, and R. A. Walton, *J. Am. Chem. Soc.*, *105*, 3088 (1983).

218. M. H. Chisholm, K. Folting, J. C. Huffman, and R. Tatz, *J. Am. Chem. Soc*, *106*, 1153 (1984).

219. G. M. Kapteijn, D. M. Grove, G. V. Koten, W. J. J. Smeets, and A. L. Spek, *Inorg. Chim Acta*, *207*, 131 (1993).

220. (a) K. Osakada, Y. J. Kim, and A. Yamamoto, *J. Organomet. Chem.*, *382*, 303 (1990). (b) K. Osakada, Y. J. Kim, M. Tanaka, S. Ishiguro, and A. Yamamoto, *Inorg. Chem.*, *30*, 197 (1991).

221. T. Fjeldberg, P. B. Hitchcock, M. F. Lappert, S. J. Smith, and A. J. Thorne, *J. Chem. Soc. Chem. Commun.*, 939 (1985).

222. M. Veith, P. Hobein, and R. Rösler, *Z. Naturforsch.*, *44b*, 1067 (1989).

223. E. P. Turevskaya, N. Ya. Turova, A. V. Korolev, A. I. Yanovsky, and Y. T. Struchkov, *Polyhedron*, *14*, 1531 (1995).

224. A. A. Pinkerton, D. Schwarzebach, L. G. Hubert-Pfalzgraf, and J. G. Riess, *Inorg. Chem.*, *15*, 1196 (1976).

225. W. J. Evans, T. J. Boyel, and J. W. Ziller, *Polyhedron*, *11*, 1093 (1992).

226. M. Veith, J. Hans, L. Stahl, P. May, V. Huch, and A. Sebald, *Z. Naturforsch.*, *Teil B*, *46*, 403 (1991).

227. M. Veith and R. Rösler, *Angew. Chem.*, *94*, 867 (1982); *Angew. Chem. Int. Ed. Engl.*, *21*, 858 (1982).

228. S. C. Goel, M. Y. Chiang, and W. E. Buhro, *Inorg. Chem.*, *29*, 4640 (1990).

229. J. Sassmannshausen, R. Riedel, K. B. Pflanz, and H. Chmiel, *Z. Naturforsch*, *48b*, 7 (1993).

230. K. W. Kirby, *Mater. Res. Bull.*, *23*, 881 (1988).

231. B. A. Vaartstra, J. A. Samuels, E. H. Barash, J. D. Martin, W. E. Streib, C. Gasser, and K. G. Caulton, *J. Organomet. Chem.*, *449*, 191 (1993).

232. R. K. Dubey, A. Singh, and R. C. Mehrotra, *Inorg. Chim. Acta*, *118*, 151 (1986).

233. R. K. Dubey, A. Singh, and R. C. Mehrotra, *Polyhedron*, *6*, 427 (1987).

234. S. Sogani, A. Singh, and R. C. Mehrotra, unpublished results (S. Sogani, Ph.D. Thesis, "Studies on Simple and Polymetallic Alkoxides of Some Group 2 and 12 Metals" University of Rajasthan, Jaipur, India, 1992).

235. M. Veith, *Angew. Chem. Int. Ed. Engl.*, *26*, 1 (1987).

236. H. M. M. Shearer and C. B. Spencer, *Acta Crystallogr. Sect. B*, *36*, 2046 (1980).

237. (a) M. S. Bains, *Can. J. Chem.*, *42*, 945 (1964). (b) G. E. Hartwell and T. L. Brown, *Inorg. Chem.*, *5*, 1257 (1966).

238. M. H. Chisholm, S. R. Drake, A. A. Naiini, and W. E. Streib, *Polyhedron*, *10*, 337 (1991).

239. M. H. Chisholm, S. R. Drake, A. A. Naiini, and W. E. Streib, *Polyhedron*, *10*, 805 (1991).

240. (a) T. Greiser and E. Weiss, *Chem. Ber.*, *110*, 3388 (1977). (b) J. E. Davies, J. Kopf, and E. Weiss, *Acta Crystallogr.*, *Part B.*, *38*, 2251 (1982). (c) E. Weiss, H. Alsdorf, H. Kühr, and H. F. Grützmacher, *Chem. Ber.*, *101*, 3777 (1968).

241. T. Greiser and E. Weiss, *Chem. Ber.*, *109*, 3142 (1976).

242. J. A. Ibers, *Nature (London)*, *197*, 686 (1963).

243. M. H. Chisholm, *ACS Symp. Ser.*, *211*, 243 (1983).

244. D. A. Wright and D. A. Williams, *Acta Crystallogr. Sect. B.*, *24*, 1107 (1968).

245. M. H. Chisholm, J. C. Huffman, C. C. Kirkpatrick, and J. Leonelli, *J. Am. Chem. Soc.*, *103*, 6093 (1981).

246. S. R. Drake, W. E. Streib, M. H. Chisholm, and K. G. Caulton, *Inorg. Chem.*, *29*, 2708 (1990).

247. M. Veith and R. Rösler, *Z. Naturforsch.*, *B41*, 1071 (1986).

248. S. Liu, S. N. Shaikh, and J. Zubieta, *Inorg. Chem.*, *26*, 4303 (1987).

249. D. M. Barnhart, D. L. Clark, J. C. Gordon, J. C. Huffman, J. G. Watkin, and B. D. Zwick, *J. Am. Chem. Soc.*, *115*, 8461 (1993).

250. D. M. Barnhart, D. L. Clark, J. C. Gordon, J. C. Huffman, and J. G. Watkin, *Inorg. Chem.*, *33*, 3939 (1994).

251. M. Wijk, R. Norrestam, M. Nygren, and G. Westin, *Inorg. Chem.*, *35*, 1077 (1996).

252. R. C. Mehrotra and A. Singh, *J. Ind. Chem. Soc.*, *70*, 885 (1993).

253. W. J. Evans, R. E. Golden, and J. W. Ziller, *Inorg. Chem.*, *32*, 3041 (1993).

254. P. J. Wheatley, *J. Chem. Soc.*, 4270 (1960).

255. L. Malpezzi, U. Zucchini, and T. Dall'Occo, *Inorg. Chim. Acta*, *180*, 245 (1991).

256. R. A. Andersen, D. H. Templeton, and A. Zalkin, *Inorg. Chem.*, *17*, 1962 (1978).

257. J. J. H. Edema, S. Gambarotta, F. Bolhuis, W. J. J. Smeets, and A. L. Spek, *Inorg. Chem.*, *28*, 1407 (1989).

258. H. Schumann, G. Kociok-Koehn, and J. Loebel, *Z. Anorg. Chem.*, *581*, 69 (1990).

259. P. G. Edwards, G. Wilkinson, M. B. Hursthouse, and K. M. A. Malik, *J. Chem. Soc. Dalton Trans.*, 2467 (1980).

260. K. L. Taft, C. D. Delfs, G. C. Papaeffthymiou, S. Foner, D. Gatteschi, and S. J. Lippard, *J. Am. Chem. Soc.*, *116*, 823 (1994).

261. R. C. Mehrotra and A. Singh, *Chemtracts*, *5*, 346 (1993).

262. J. H. Wengrovius, M. F. Garbauskas, E. A. Williams, R. C. Going, P. E. Donahue, and J. E. Smith, *J. Am. Chem. Soc.*, *108*, 982 (1986).

263. P. Maleki and M. J. Schwing-Weill, *J. Inorg. Nucl. Chem.*, *37*, 435 (1975).

264. A. Dhamani, R. Bohra, and R. C. Mehrotra, *Polyhedron*, *14*, 733 (1995) and *15*, . . . (1996); *Main Group Metal Chem.*, *18*, 687 (1995).

265. B. Marcinec and P. Krzyzanowski, *J. Organomet. Chem.*, *493*, 261 (1995).

266. P. Krzyzanowski, M. Kubicki, and R. Marcinec, *Polyhedron*, *15*, 1 (1996).

267. W. J. Evans, J. M. Olofson, and J. W. Ziller, *J. Am. Chem. Soc.*, *112*, 2308 (1990).

268. W. J. Evans, M. S. Sollberger, J. L. Shreeve, J. M. Olofson, J. H. Hain, Jr., and J. W. Ziller, *Inorg. Chem.*, *31*, 2492 (1992).

269. W. J. Evans, T. J. Deming, and J. W. Ziller, *Organometallics*, *8*, 1581 (1989).

270. M. H. Chisholm, F. A. Cotton, C. A. Murillo, and W. W. Reichert, *Inorg. Chem.*, *16*, 1801 (1977).

271. M. H. Chisholm, K. Folting, J. C. Huffman, E. F. Putilina, W. E. Streib, and R. J. Tatz, *Inorg. Chem.*, *32*, 3771 (1993).

272. D. M. Hoffman, D. Lappas, and D. A. Wierda, *J. Am. Chem. Soc.*, *111*, 1531 (1989).

273. D. M. Hoffman, D. Lappas, and D. A. Wierda, *J. Am. Chem. Soc.*, *115*, 10538 (1993).

274. T. H. Lemmen, G. V. Goeden, J. C. Huffman, R. L. Geerts, and K. G. Caulton, *Inorg. Chem.*, *29*, 3680 (1990).

275. W. M. Cleaver and A. R. Barron, *Organometallics*, *12*, 1001 (1993).

276. H. Schumann, K. K. Gabriele, A. Dietrich, and F. H. Goerlitz, *Z. Naturforsch.*, *46b*, 896 (1991).

277. D. J. Eichorst, D. A. Payne, S. R. Wilson, and K. E. Howard, *Inorg. Chem.*, *29*, 1458 (1990).

278. M. Veith, D. Käfer, J. Koch, P. May, L. Stahl, and V. Huch, *Chem. Ber.*, *125*, 1033 (1992).

279. S. Boulmaaz, R. Papiernik, L. G. Hubert-Pfalzgraf, and J. C. Daran, *Eur. J. Solid State Inorg. Chem.*, *30*, 583 (1993).

280. P. Biagini, G. Lugli, L. Abis, and R. Millini, *J. Organomet. Chem.*, *474*, C16–C18 (1994).

281. J. Hvoslef, H. Hope, B. D. Murray, and P. P. Power, *J. Chem. Soc. Chem. Commun.*, 1438 (1983).

282. D. M. Barnhart, D. L. Clark, J. C. Huffman, R. L. Vincent, and J. G. Watkin, *Inorg. Chem.*, *32*, 4077 (1993).

283. W. A. Hermann, R. Anwander, M. Kleine, and W. Scherer, *Chem. Ber.*, *125*, 1971 (1992).

284. M. Bochmann, G. Wilkinson, G. B. Young, M. B. Hursthouse, and K. M. A. Malik, *J. Chem. Soc. Dalton Trans.*, 901 (1980).

285. M. Bochmann, G. Wilkinson, G. B. Young, M. B. Hursthouse, and K. M. A. Malik, *J. Chem. Soc. Dalton Trans.*, 1863 (1980).

286. B. D. Murray, H. Hope, and P. P. Power, *J. Am. Chem. Soc.*, *107*, 169 (1985).

287. B. D. Murray and P. P. Power, *J. Am. Chem. Soc.*, *106*, 7011 (1984).

288. W. A. Hermann, N. W. Huber, and T. Priermeir, *Angew. Chem. Int. Ed. Engl.*, *33*, 105 (1994).

289. M. M. Olmstead, P. P. Power, and G. Sigel, *Inorg. Chem.*, *25*, 1027 (1986).

290. M. Wedler, J. W. Gilje, U. Pieper, D. Stalke, M. Noltemeyer, and F. T. Edelmann, *Chem. Ber.*, *124*, 1163 (1991).

291. F. T. Edelmann, A. Steiner, D. Stalke, J. W. Gilje, S. Jagner, and M. Hakansson, *Polyhedron*, *13*, 539 (1994).

292. J. R. vanden Hende, P. B. Hitchcock, and M. F. Lappert, *J. Chem. Soc. Chem. Commun.*, *1413 (1994).*

293. K. C. Malhotra and R. L. Martin, *J. Organomet. Chem.*, *239*, 159 (1982).

294. A. W. Duff, R. A. Kamarudin, M. F. Lappert, and R. J. Norton, *J. Chem. Soc. Dalton Trans.*, 489 (1986).

295. T. V. Lubben, P. T. Wolczanski, and G. D. Van Duyne, *Organometallics*, *3*, 977 (1984).

296. G. Beck, P. B. Hitchcock, M. F. Lappert, and I. A. Mackinnon, *J. Chem. Soc. Chem. Commun.*, 1312 (1989).

297. J. L. Stewart and R. A. Andersen, *J. Chem. Soc. Chem. Commun.*, 1846 (1987).

298. C. Baudin and M. Ephritikhine, *J. Organomet. Chem.*, *364*, C1–C2 (1989).

299. D. L. Clark, J. C. Huffman, and J. G. Watkin, *J. Chem. Soc. Chem. Commun.*, 266 (1992).

300. G. A. Sigel, R. A. Bartlett, D. Decker, M. M. Olmstead, and P. P. Power, *Inorg. Chem.*, *26*, 1773 (1987).

301. P. B. Hitchcock, M. F. Lappert, and I. A. Mackinnon, *J. Chem. Soc. Chem. Commun.*, 1557 (1988).

302. G. Helgesson, S. Jagner, O. Poncelet, and L. G. Hubert-Pfalzgraf, *Polyhedron*, *10*, 1559 (1991).

303. C. J. Willis, *Coord. Chem. Rev.*, *88*, 133 (1988).

304. J. A. Samuels, K. Folting, J. C. Huffman, and K. G. Caulton, *Chem. Mater.*, *7*, 929 (1975).

305. J. L. Kiplinger, T. G. Richmond, and C. E. Osterberg, *Chem. Rev.*, *94*, 373 (1994).

306. R. E. Sievers, J. J. Brooks, J. A. Cunningham, and W. E. Rhine, in *Inorganic Compounds with Unusual Properties*, R. B. King, Ed., Advances in Chemistry 150, ACS, Washington, DC, 1976, pp. 222–231.

307. A. R. Siedle, in *Comprehensive Coordination Chemistry*, G. Wilkinson, R. D. Gillard, and J. A. McCleverty, Eds., Vol. 2, Pergamon, New York, 1987, pp. 365–412.

308. L. G. Hubert-Pfalzgraf, *Appl. Organomet. Chem.*, *6*, 627 (1992).

309. A. P. Purdy, A. D. Berry, R. T. Holm, M. Fatemi, and D. K. Gasskill, *Inorg. Chem.*, *28*, 2799 (1989).

310. P. N. Kapoor and R. C. Mehrotra, *Chem. Ind. (London)*, 1034 (1966).

311. P. N. Kapoor, R. N. Kapoor, and R. C. Mehrotra, *Chem. Ind.*, 1314 (1968).

312. A. Merbach and J. P. Carrard, *Helv. Chim. Acta*, *54*, 2771 (1971).

313. K. S. Mazdiyasni and B. J. Schaper, *J. Less-Commun. Met.*, *30*, 105 (1973).

314. K. S. Mazdiyasni, B. J. Schaper, and L. M. Brown, *Inorg. Chem.*, *10*, 889 (1971).

315. D. K. Huggins and W. B. Fox, U.S. Patent 3631081 (1971).

316. S. L. Chadha and L. Mohini, *Indian J. Chem.*, *21A*, 317 (1981).

317. R. A. Andersen, *Inorg. Nucl. Chem. Lett.*, *15*, 57 (1979).

318. R. A. Andersen and G. E. Coates, *J. Chem. Soc. Dalton Trans.*, 1244 (1975).

319. J. M. Canich, G. L. Gard, and J. M. Shreeve, *Inorg. Chem.*, *23*, 441 (1984).

320. (a) R. C. Paul, P. Sharma, P. K. Gupta, and S. L. Chadha, *Inorg. Chim. Acta*, *20*, 7 (1976); (b) R. C. Paul, P. K. Gupta, M. Gulati, and S. L. Chadha, *Inorg. Nucl. Chem. Lett.*, *13*, 665 (1977).

321. M. Basso-Bert and D. Gervais, *Inorg. Chem. Acta*, *34*, 191 (1979).

322. R. C. Paul, P. Sharma, L. Subbiah, H. Singh, and S. L. Chadha, *J. Inorg. Nucl. Chem.*, *38*, 169 (1976).

323. D. C. Bradley and C. E. Holloway, *J. Chem. Soc. A*, 282 (1969).

324. R. Choukron and D. Gervais, *Inorg. Chim. Acta*, *27*, 163 (1978).

325. R. Choukron, A. Dia, and D. Gervais, *Inorg. Chim Acta*, *34*, 187 (1978); *Inorg. Chim Acta*, *34*, 211 (1979).

326. L. B. Handy, *J. Fluorine Chem.*, *7*, 641 (1976).

327. D. A. Johnson, J. C. Taylor, and A. B. Waugh, *J. Inorg. Nucl. Chem.*, *42*, 1271 (1980).

328. F. E. Brinckman, R. B. Johannesen, R. B. Hammerschmidt, and L. B. Brandy, *J. Fluorine Chem.*, *6*, 427 (1975).

329. S. L. Chadha and K. Uppal, *Bull. Soc. Chim. Fr.*, *3*, 431 (1987).

330. S. L. Chadha, V. Sharma, and A. Sharma, *J. Chem. Soc. Dalton Trans.*, 1253 (1987).

331. S. L. Chadha, *Inorg. Chim Acta*, *156*, 173 (1989).

332. T. M. Gilbert, A. M. Landes, and R. D. Rogers, *Inorg. Chem.*, *31*, 3438 (1992).

333. H. Rothfuss, J. C. Huffman, and K. G. Caulton, *Inorg. Chem.*, *33*, 187 (1994).

334. P. M. Jeffries, S. R. Wilson, and G. S. Girolami, *Inorg. Chem.*, *31*, 4503 (1992).

335. C. M. Jones, M. D. Burkart, and K. H. Whitmire, *Angew. Chem. Int. Ed. Engl.*, *31*, 451 (1992).

336. C. M. Jones, M. D. Burkart, R. E. Bachman, D. L. Serra, S. J. Hwu, and K. H. Whitmire, *Inorg. Chem.*, *32*, 5136 (1993).

337. D. C. Bradley, H. Chudzynska, M. B. Hursthouse, and M. Motevalli, *Polyhedron*, *12*, 1907 (1993).

338. D. C. Bradley, H. Chudzynska, M. B. Hursthouse, M. Motevalli, and R. Wu, *Polyhedron*, *13*, 1 (1994).

339. D. C. Bradley, H. Chudzynska, M. E. Hammond, M. B. Hursthouse, M. Motevalli, and W. Ruowen, *Polyhedron*, *11*, 375 (1992).

340. D. C. Bradley, H. Chudzynska, M. B. Hursthouse, and M. Motevalli, *Polyhedron*, *13*, 7 (1994).

341. M. E. Gross, *J. Electrochem. Soc.*, *138*, 2422 (1991).

342. J. A. Samuels, E. B. Lobkovsky, W. E. Streib, K. Folting, J. C. Huffman, J. W. Zwanziger, and K. G. Caulton, *J. Am. Chem. Soc.*, *115*, 5093 (1993).

343. D. J. Darensbourg, B. L. Mueller, J. H. Reibenspies, and C. J. Bischoff, *Inorg. Chem.*, *29*, 1789 (1990).

344. P. C. Wales and H. Weigold, *J. Organomet. Chem.*, *24*, 413 (1970).

345. R. Choukron and D. Gervais, *Synth. React. Inorg. Met.-Org. Chem.*, *8*, 137 (1978).

346. A. P. Purdy, C. F. George, and J. H. Callahan, *Inorg. Chem.*, *30*, 2812 (1991).

347. A. P. Purdy and C. F. George, *Inorg. Chem.*, *30*, 1969 (1991).

348. J. Zhao, K. Dahmen, H. O. Marcy, L. M. Tonge, T. J. Marks, B. W. Wessels, and C. R. Kannewurf, *Appl. Phys. Lett.*, *53*, 1750 (1988).

349. D. J. Larkin and L. V. Interrante, *J. Mater. Res.*, *5*, 2706 (1990).

350. C. Campbell, S. G. Bott, R. Larsen, and W. G. Van Der Sluys, *Inorg. Chem.*, *33*, 4950 (1994).

351. C. H. Winter, P. H. Sheridan, and M. J. Heeg, *Inorg. Chem.*, *30*, 1962 (1991).

352. R. H. King, M. H. Chisholm, D. L. Clark, and C. E. Hammond, *J. Am. Chem. Soc.*, *111*, 2751 (1989).

353. S. L. Chadha and V. Sharma, *Z. Anorg. Allg. Chem.*, *545*, 227 (1987).

354. S. L. Chadha, V. Sharma, C. Mohini, S. P. Taneja, and D. Raj, *Z. Anorg. Allg. Chem.*, *536*, 164 (1986).

355. S. L. Chadha, V. Sharma, and K. Uppal, *Indian J. Chem.*, *25A*, 625 (1986).

356. S. L. Chadha, V. Sharma, and K. Uppal, *Synth. React. Inorg. Met.-Org. Chem.*, 409 (1987).

357. S. L. Chadha and V. Sharma, *Inorg. Chim. Acta*, *132*, 237 (1987).

358. S. L. Chadha and V. Sharma, *Monatsh. Chem.*, *119*, 553 (1988).

359. S. L. Chadha and V. Sharma, *Transition Metal Chem. 13*, 219 (1988).

360. S. L. Chadha, T. Singh, K. Uppal, and C. M. Jaswal, *Indian J. Chem.*, *24A*, 781 (1985).

361. S. L. Chadha and V. Sharma, *Inorg. Chim. Acta*, *118*, L43 (1986).

362. S. L. Chadha and V. Sharma, *Research Bulletin (Science) of the Panjab University*, *36*, 141 (1985).

363. T. Saegusa and T. Ueshima, *Inorg. Chem.*, *6*, 1679 (1967).

364. S. L. Chadha and V. Sharma, *Inorg. Chim. Acta*, *131*, 101 (1987).

365. S. L. Chadha, V. Sharma, S. P. Taneja, and D. Raj, *Transition Metal Chem.*, *11*, 369 (1986).

366. (a) J. G. Bednorz and A. Müller, *Z. Phys. B. Condens. Matter*, *64*, 189 (1986). (b) A. Müller and J. G. Bednorz, *Science*, *217*, 1133 (1987).

367. S. C. Goel, M. A. Matchett, M. Y. Chiang, and W. E. Buhro, *J. Am. Chem. Soc.*, *113*, 1844 (1991).

368. W. S. Res, Jr., and D. A. Moreno, *J. Chem. Soc. Chem. Commun.*, 1759 (1991).

369. O. Poncelet, L. G. Hubert-Pfalzgraf, J. C. Daran and R. Astier, *J. Chem. Soc. Chem. Commun.*, 1846 (1989).

370. N. Ya. Turova, E. P. Turevskaya, V. G. Kessler, N. I. Kozlova, and A. I. Belocon, *Zh. Neorg. Khim.*, *37*, 50 (1992).

371. F. Schmidt and A. Feltz, *Z. Chem.*, *30*, 109 (1990).

372. K. G. Caulton, M. H. Chisholm, S. R. Drake, and J. C. Huffman, *J. Chem. Soc. Chem. Commun.*, 1498 (1990).

373. F. Schmidt and A. Feltz, *Z. Chem.*, *30*, 228 (1990).

374. F. Schmidt and A. Feltz, *Z. Chem.*, *30*, 229 (1990).

375. (a) M. C. Massiani, R. Papiernik, L. G. Hubert-Pfalzgraf, and J. C. Daran, *J. Chem. Soc. Chem. Commun.*, 301 (1990). (b) M. C. Massiani, R. Papiernik, L. G. Hubert-Pfalzgraf, and J. C. Daran, *Polyhedron*, *10*, 437 (1991).

376. S. D. Ramamurthi and D. A. Payne, *J. Am. Ceram. Soc.*, *73*, 2547 (1990).

377. S. C. Goel, K. S. Kramer, P. C. Gibbons, and W. E. Buhro, *Inorg. Chem.*, *28*, 3619 (1989).

378. V. G. Kessler, N. Ya. Turova, A. V. Korolev, A. I. Yanovskü, and Yu. T. Struchkov, *Mendeleev Commun.*, *3*, 89 (1991).

379. W. Bidell, V. Shklover, and H. Berke, *Inorg. Chem.*, *31*, 5561 (1992).

380. S. Sogani, A. Singh, and R. C. Mehrotra, unpublished results.

381. S. Sogani, A. Singh, and R. C. Mehrotra, *Indian J. Chem.*, *32A*, 345 (1993).

382. S. Boulmaaz, R. Papiernik, L. G. Hubert-Pfalzgraf, J. Vaissermann, and J. C. Daran, *Polyhedron*, *11*, 1331 (1992).

383. N. W. Huber, Ph.D. Thesis, "Flüchtige Metall-Alkoxide für die Gasphasenabscheidung," Technische Universität München, 1994.

384. O. Poncelet, L. G. Hubert-Pfalzgraf, and J. C. Daran, *Inorg. Chem.*, *29*, 2883 (1990).

385. J. F. Campion, D. A. Payne, H. K. Chae, J. K. Maurin, and S. R. Wilson, *Inorg. Chem.*, *30*, 3244 (1991).

386. N. N. Sauer, E. Garcia, K. V. Salazar, R. R. Ryan, and J. A. Martin, *J. Am. Chem. Soc.*, *112*, 1524 (1990).

387. W. Bidell, J. Döring, H. W. Bosch, H. U. Hund, E. Plappert, and H. Berke, *Inorg. Chem.*, *32*, 502 (1993).

388. A. F. Wells, in *Structural Inorganic Chemistry*, 5th ed., Oxford Science Publications, Clarendon Press, Oxford, UK, 1986.

389. W. J. Evans, J. H. Hain, Jr., and J. W. Ziller, *J. Chem. Soc. Chem. Commun.*, 1628 (1989).

390. J. A. Bertrand and P. G. Eller, *Progress in Inorganic Chemistry*, Wiley-Interscience, New York, 1976, Vol. 21, p. 29.

391. J. A. Bertrand, E. Fujita, and D. G. Vanderveer, *Inorg. Chem.*, *19*, 2022 (1980).

392. W. A. Nugent and R. L. Harlow, *J. Am. Chem. Soc.*, *116*, 6142 (1994).

393. M. G. Voronkov, *Chem. Br.*, *9*, 411 (1973); *Top. Curr. Chem.*, *84*, 77 (1979).

394. M. G. Voronkov, *Pure Appl. Chem.*, *13*, 35 (1966).

395. J. W. Turley and F. P. Boer, *J. Am. Chem. Soc.*, *90*, 4026 (1968).

396. L. Párkányi, K. Simon, and J. Nagy, *Acta Crystallogr.*, *B30*, 2338 (1974).

397. L. Párkányi, J. Nagy, and K. Simon, *J. Organomet. Chem.*, *101*, 11 (1975).

398. J. J. Daly and F. Sanz, *J. Chem. Soc. Dalton Trans.*, 2051 (1974).

399. (a) A. A. Naiini, V. Young, and J. G. Verkade, *Polyhedron*, *14*, 393 (1995). (b) V. G. Kessler, L. G. Hubert-Pfalzgraf, S. Halut, and J. C. Daran, *J. Chem. Soc. Chem. Commun.*, 705 (1994).

400. (a) S. Katayama and M. Sekine, *J. Mater. Res.*, *5*, 683 (1990). (b) S. Katayama and M. Sekine, *J. Mater. Res.*, *6*, 36 (1991). (c) S. Katayama and M. Sekine, *Nippon Seramikkusu Kyokai Gakujutsu Ronbunshi*, *99*, 345 (1991).

401. L. G. Hubert-Pfalzgraf, F. Labrize, C. Bois, and J. Vaissermann, *Polyhedron*, *13*, 2163 (1994).

402. (a) T. Lindgren, R. Sillanpää, T. Nortia, and K. Pihlaja, *Inorg. Chim. Acta*, *73*, 153 (1983); *Inorg. Chim. Acta*, *82*, 1 (1984). (b) R. Sillanpaa, T. Lindgren, and L. Hiltunen, *Inorg. Chim. Acta*, *131*, 85 (1987).

403. W. M. P. B. Menze and J. G. Verkade, *Inorg. Chem.*, *30*, 4628 (1991).

404. S. C. Goel, K. S. Kramer, M. Y. Chiang, and W. E. Buhro, *Polyhedron*, *9*, 611 (1990).

405. L. G. Hubert-Pfalzgraf, N. El Khokh, and J. C. Daran, *Polyhedron*, *11*, 59 (1992).

406. J. Pinkas and J. G. Verkade, *Inorg. Chem.*, *32*, 2711 (1993).

407. K. S. Chong, S. J. Rettig, A. Storr, and J. Trotter, *Can. J. Chem.*, *57*, 586 (1979).

408. M. G. Voronkov and V. P. Baryshok, *J. Organomet. Chem.*, *239*, 199 (1982).

409. R. Contreras, C. Garcia, T. Mancilla, and B. Wrackmeyer, *J. Organomet. Chem.*, *246*, 213 (1983).

410. N. W. Alcock, A. W. G. Platt, and P. Pringle, *J. Chem. Soc. Dalton Trans.*, 2273 (1987) and references cited therein.

411. N. W. Alcock, A. W. G. Platt, and P. G. Pringle, *Inorg. Chim. Acta*, *128*, 215 (1987).

412. H. E. Bryndza, S. A. Kretchmor, and T. H. Tulip, *J. Am. Chem. Soc.*, *108*, 4805 (1986) and references cited therein.

413. A. Davis and F. R. Hartley, *Chem. Rev.*, *81*, 79 (1981) and references cited therein.

414. H. E. Bryndza, L. K. Fong, R. A. Paciello, W. Tam, and J. E. Bercaw, *J. Am. Chem. Soc.*, *109*, 1444 (1987).

415. C. D. Montgomery, N. C. Payne, and C. J. Willis, *Inorg. Chem.*, *26*, 519 (1987).

416. P. B. Hitchcock, M. F. Lappert, and I. A. Mackinnon, *J. Chem. Soc. Chem. Commun.*, 1557 (1988).

417. P. B. Hitchcock, M. F. Lappert, and I. A. Mackinnon, *J. Chem. Soc. Chem. Commun.*, 1015 (1993).

418. L. M. Engelhardt, J. M. Harrowfield, M. F. Lappert, I. A. Mackinnon, B. H. Newton, C. L. Raston, B. W. Skelton, and A. H. White, *J. Chem. Soc. Chem. Commun.*, 846 (1986).

419. N. W. Alcock, A. W. G. Platt, and P. G. Pringle, *J. Chem. Soc. Dalton Trans.*, 139 (1989).

420. A. A. G. Tomlinson and B. J. Hathaway, *J. Chem. Soc. A*, 1685 (1968).

421. T. Distler and P. A. Vaughan, *Inorg. Chem.*, *6*, 126 (1967).

422. T. Greiser and E. Weiss, *Chem. Ber.*, *109*, 3142 (1976).

423. C. H. Yoder and J. J. Zuckerman, *Heterocyclic Compounds of the Group IV Elements*, Vol. 6, W. L. Jolly, Ed., Wiley-Interscience, New York, 1971, pp. 81–155.

424. For a recent review on the coordination chemistry of the catecholate dianion see: C. G. Pierpont and C. W. Lange, *Progress in Inorganic Chemistry*, Wiley-Interscience, New York, 1994, vol. 41, p. 331.

425. S. R. Sofen, K. Abu-Dari, D. P. Freyberg, and K. N. Raymond, *J. Am. Chem. Soc.*, *100*, 7882 (1978).

426. S. R. Sofen, S. R. Cooper, and K. N. Raymond, *Inorg. Chem.*, *18*, 161 (1979).

427. G. E. Freeman and K. N. Raymond, *Inorg. Chem.*, *24*, 1410 (1985).

428. J. R. Hartman, B. M. Foxman, and S. R. Cooper, *Inorg. Chem.*, *23*, 1381 (1984).

429. R. R. Holmes, S. Shafieezad, V. Chandrasekhar, A. C. Sau, J. M. Holmes, and R. O. Day, *J. Am. Chem. Soc.*, *110*, 1168 (1988).

430. J. J. Flynn and F. P. Boer, *J. Am. Chem. Soc.*, *91*, 5756 (1969).

431. K. N. Raymond, S. S. Isied, L. D. Brown, F. R. Fronczek, and J. H. Nibert, *J. Am. Chem. Soc.*, *98*, 1767 (1976).

432. S. R. Sofen, D. C. Ware, S. R. Cooper, and K. N. Raymond, *Inorg. Chem.*, *18*, 234 (1979).

433. S. R. Cooper, Y. B. Koh, and K. N. Raymond, *Inorg. Chem.*, *104*, 5092 (1982).

434. L. A. deLearie and C. G. Pierpont, *Inorg. Chem.*, *27*, 3842 (1988).

435. C. G. Pierpont and H. H. Downs, *J. Am. Chem. Soc.*, *97*, 2123 (1975).

436. L. A. deLearie, R. C. Haltiwanger, and C. G. Pierpont, *Inorg. Chem.*, *26*, 817 (1987).

437. L. A. deLearie, R. C. Haltiwanger, and C. G. Pierpoint, *J. Am. Chem. Soc.*, *111*, 4324 (1989).

438. M. B. Hursthouse, T. Fram, L. New, W. P. Griffith, and A. J. Nielson, *Transition Metal Chem.*, *3*, 255 (1978).

439. Q. Chen and J. Zubieta, *Coord. Chem. Rev.*, *114*, 107 (1992).

440. Q. Chen and J. Zubieta, *Inorg. Chim. Acta*, *198–200*, 95 (1992).

441. T. J. Boyle, R. W. Schwartz, R. J. Doedens, and J. W. Ziller, *Inorg. Chem.*, *34*, 1110 (1995).

442. C. P. Rao, S. P. Kaiwar, and M. S. S. Raghavan, *Polyhedron*, *13*, 1895 (1994).

443. D. C. Crans, R. A. Felty, H. Chen, H. Eckert, and N. Das, *Inorg. Chem.*, *33*, 2427 (1994).

444. D. C. Crans, R. A. Felty, and M. M. Miller, *J. Am. Chem. Soc.*, *113*, 265 (1991).

445. G. J. Gainsford, T. Kemmitt, C. Lensink, and N. B. Milestone, *Inorg. Chem.*, *34*, 746 (1995).

446. G. J. Gainsford, T. Kemmitt, and N. B. Milestone, *Inorg. Chem.*, *34*, 5244 (1995).

447. C. P. Love, C. C. Torardi, and C. J. Page, *Inorg. Chem.*, *31*, 1784 (1992).

448. M. H. Chisholm, I. P. Parkin, W. E. Streib, and O. Elsenstein, *Inorg. Chem.*, *33*, 812 (1994).

449. S. Pasynkiewicz and W. Ziemkowska, *J. Organomet. Chem.*, *423*, 1 (1992).

450. Z. Zhu, A. M. Al-Ajlouni, and J. H. Espenson, *Inorg. Chem.*, *35*, 1408 (1996).

451. C. D. Chandler, M. J. Hampden-Smith, and E. N. Duesler, *Inorg. Chem.*, *31*, 4891 (1992).

452. A. K. S. Gupta, R. Bohra, and R. C. Mehrotra, *Synth. React. Inorg. Met.-Org. Chem.*, *21*, 445 (1991).

453. D. C. Crans, R. W. Marshman, M. S. Gottlieb, O. P. Anderson, and M. M. Miller, *Inorg. Chem.*, *31*, 4939 (1992).

454. A. A. El-Hadad, J. E. Kickham, S. J. Loeb, L. Taricani, and D. G. Tuck, *Inorg. Chem.*, *34*, 120 (1995).

455. V. J. Scherle and F. A. Schröder, *Acta Crystallogr.*, *B20*, 2772 (1974).

456. J. M. Manriquez, D. R. McAlister, R. D. Sanner, and J. E. Bercaw, *J. Am. Chem. Soc.*, *100*, 2716 (1978).

457. T. J. Marks and R. D. Ernst, *Comprehensive Organometallic Chemistry*, Vol. 3, G. Wilkinson, F. G. A. Stone, and E. W. Abel, Eds., Pergamon, New York, 1982, Chapter 21, pp 249–254, and references cited therein.

458. P. J. Fagan, K. G. Moloy, and T. J. Marks, *J. Am. Chem. Soc.*, *103*, 6959 (1981).

459. W. J. Evans, J. W. Grate, and R. J. Doedens, *J. Am. Chem. Soc.*, *107*, 1671 (1985).

460. J. M. Manriquez, P. J. Fagan, T. J. Marks, C. S. Day, and V. W. Day, *J. Am. Chem. Soc.*, *100*, 7112 (1978).

461. W. J. Evans, A. L. Wayda, W. E. Hunter, and J. L. Atwood, *J. Chem. Soc. Chem. Commun.*, 706 (1981).

462. (a) R. P. A. Sneeden, *Comprehensive Organometallic Chemistry*, Vol. 8, G. Wilkinson, F. G. A. Stone, and E. W. Abel, Eds., Pergamon, New York, 1982, Chapter 50.2. (b) I. Tkatchenko, *Comprehensive Organometallic Chemistry*, Vol. 8, G. Wilkinson, F. G. A. Stone, and E. W. Abel, Eds., Pergamon, New York, 1982, Chapter 50.3.

463. P. T. Wolczanski and J. E. Bercaw, *Acc. Chem. Res.*, *13*, 121 (1980).

464. J. A. Marsella and K. G. Caulton, *J. Am. Chem. Soc.*, *102*, 1747 (1980).

465. J. H. Haslam, *U.S. Patent*, *2*, 719, 869 (October 4, 1955).

466. P. N. Kapoor, S. K. Mehrotra, R. C. Mehrotra, R. B. King, and K. C. Nainan, *Inorg. Chim. Acta*, *12*, 273 (1975).

467. S. C. Goel and R. C. Mehrotra, *Z. Anorg. Allg. Chem.*, *440*, 281 (1978).

468. S. C. Goel, V. K. Singh, and R. C. Mehrotra, *Z. Anorg. Allg. Chem.*, *447*, 253 (1978).

469. S. C. Goel and R. C. Mehrotra, *Indian J. Chem.*, *20A*, 440 (1981).

470. S. C. Goel and R. C. Mehrotra, *Indian J. Chem.*, *20A*, 1054 (1981).

471. M. D. Curtis, S. Thanedar, and W. M. Butler, *Organometallics*, *3*, 1855 (1984).

472. W. J. Evans, R. Dominguez, and T. P. Hanusa, *Organometallics*, *5*, 1291 (1986).

473. M. D. Curtis, J. J. D'Errico, D. N. Duffy, P. S. Epstein, and L. G. Bell, *Organometallics*, *2*, 1808 (1983).

474. P. G. Williard and G. B. Carpenter, *J. Am. Chem. Soc.*, *107*, 3345 (1985).

475. (a) K. Watenpaugh and C. N. Caughlan, *J. Chem. Soc. Chem. Commun.*, 76 (1967). (b) R. Schmid, A. Mosset, and J. Galy, *J. Chem. Soc. Dalton Trans.*, 1999 (1991).

476. (a) D. C. Bradley, M. B. Hursthouse, and P. F. Rodesilar, *J. Chem. Soc. Chem. Commun.*, 1112 (1968). (b) V. G. Kessler, N. Ya. Turova, A. I. Yanovskii, A. I. Belokon, and Yu. T. Struchkov, *Russ. J. Inorg. Chem.*, 36, 938 (1991).

477. (a) B. Morrison, *Acta Crystallogr.*, *B33*, 303 (1977). (b) P. G. Harrison, B. J. Haylett, and T. J. King, *J. Chem. Soc. Chem. Commun.*, 112 (1978).

478. O. Poncelet, W. J. Sartain, L. G. Hubert-Pfalzgraf, K. Folting, and K. G. Caulton, *Inorg. Chem.*, 28, 263 (1989).

479. R. C. Mehrotra, *Chemtracts*, 2, 14 (1990).

480. N. Ya Turova, V. G. Kessler, and S. I. Kuchciko, *Polyhedron*, 10, 2617 (1991).

481. R. Papiernik, L. G. Hubert-Pfalzgraf, and F. Chaput, *J. Non-Cryst. Solids*, 147, 36 (1992).

482. S. C. Goel, M. Y. Chiang, F. C. Gibbons, and W. E. Buhro, *Mater. Res. Soc. Symp. Proc.*, 271, 3 (1992).

483. N. Ya Turova, E. P. Turevskaya, V. G. Kessler, A. I. Yanovsky, and Yu T. Struchkov, *J. Chem. Soc. Chem. Commun.*, 21 (1993).

484. D. C. Bradley, H. Chudzynska, D. M. Frigo, M. E. Hammond, M. B. Hursthouse, and M. A. Mazid, *Polyhedron*, 9, 719 (1990).

485. K. G. Caulton, M. H. Chisholm, S. R. Drake, and K. Folting, *J. Chem. Soc. Chem. Commun.*, 1349 (1990).

486. A. I. Yanovsky, M. I. Yanovskaya, V. K. Limar, V. G. Kessler, N. Ya Turova, and Y. T. Struchkov, *J. Chem. Soc. Chem. Commun.*, 1605 (1993).

487. F. A. Cotton, D. O. Marler, and W. Schwotzer, *Inorg. Chim. Acta*, 95, 207 (1984).

488. R. C. Mehrotra, Ph.D. Thesis, 'Group IV Metal Alkoxides,' London University, 1952.

489. S. Sangokoya, W. T. Pennington, J. Byers-Hill, G. H. Robinson, and R. D. Rogers, *Organometallics*, 12, 2429 (1993).

490. S. Danielle, L. G. Hubert-Pfalzgraf, J. C. Daran, and S. Halut, *Polyhedron*, 13, 927 (1994).

491. V. G. Kessler, N. Ya. Turova, and A. N. Panov, A. I. Yanovsky, A. D. Pisarevsky, and Y. T. Struchkov, *Polyhedron*, 15, 335 (1996).

492. R. Papiernik, L. G. Hubert-Pfalzgraf, J. C. Daran, and Y. Jeannin, *J. Chem. Soc. Chem. Commun.*, 695 (1990).

493. U. Bemmn, K. Lashgari, R. Norrestam, M. Nygren, and G. Westin, *J. Solid State Chem.*, *108*, 243 (1994).

494. C. J. Burns, D. C. Smith, A. P. Sattelberger, and H. B. Gray, *Inorg. Chem.*, *31*, 3724 (1992).

495. D. Rehder, *Angew. Chem. Int. Ed. Engl.*, *30*, 148 (1991).

496. W. Priebsch and D. Rehder, *Inorg. Chem.*, 29, 3013 (1990).

497. T. P. Blatchford, M. H. Chisholm, K. Folting, and J. C. Huffman, *J. Chem. Soc. Chem. Commun.*, 1295 (1994).

498. M. H. Chisholm, C. M. Cook, and K. Folting, *J. Am. Chem. Soc.*, *114*, 2721 (1992).

499. T. A. Budzichowski, M. H. Chisholm, and W. E. Streib, *J. Am. Chem. Soc.*, *116*, 387 (1994).

500. K. Yunlu, P. S. Gradeff, N. Edelstein, W. Kot, G. Shalimoff, W. E. Streib, B. A. Vaartstra, and K. G. Caulton, *Inorg. Chem.*, *30*, 2317 (1991).

501. Sonika, A. K. Narula, O. P. Vermani, and H. K. Sharma, *J. Organomet. Chem.*, *470*, 67 (1994) and references cited therein.

502. J. Rai and R. C. Mehrotra, *J. Non-Cryst. Solids*, *134*, 23 (1991).

503. R. C. Mehrotra and J. Rai, unpublished results.

504. C. D. Chandler, C. Roger, and M. J. Hampden-Smith, *Chem. Rev.*, *93*, 1205 (1993).

505. R. Kuhlman, B. A. Vaartstra, W. E. Streib, J. C. Huffman, and K. G. Caulton, *Inorg. Chem.*, *32*, 1272 (1993).

506. M. H. Chisholm, K. Folting, J. C. Huffman, and C. C. Kirkpatrick, *J. Am. Chem. Soc.*, *103*, 5967 (1981).

507. M. H. Chisholm, K. Folting, J. C. Huffman, and C. C. Kirkpatrick, *Inorg. Chem.*, *23*, 1021 (1984).

508. K. Hegetschweiler, H. W. Schmalle, H. M. Streit, V. Gramlich, H. U. Hund, and I. Erni, *Inorg. Chem.*, *31*, 1299 (1992).

509. S. Boulmaaz, L. G. Hubert-Pfalzgraf, S. Halut, and J. C. Daran, *J. Chem. Soc. Chem. Commun.*, 601 (1994).

510. For general comment on transition metal alkyls see R. C. Mehrotra and A. Singh, *Organometallic Chemistry*, Wiley/Wiley Eastern, New York/New Delhi, 1992, pp. 558–565.

511. F. A. Cotton and G. Wilkinson, *Advanced Inorganic Chemistry*, 4th ed. Wiley-Interscience, New York, 1988, p. 1122.

512. J. P. Collman, L. S. Hegedus, J. R. Norton, and R. G. Finke, *Principles and Applications of Organotransitions Metal Chemistry*, University Science Books, Mill Valley, CA, 1987, pp. 94–104.

513. M. H. Chisholm, D. M. Hoffman, and J. C. Huffman, *Chem. Soc. Rev.*, *14*, 69 (1985).

514. (a) M. H. Chisholm, *Angew. Chem. Int. Ed. Engl.*, *25*, 21 (1986). (b) M. H. Chisholm, *J. Organomet. Chem.*, *400*, 235 (1990). (c) M. H. Chisholm, *Acc. Chem. Res.*, *23*, 419 (1990).

515. I. P. Rothwell, *Acc. Chem. Res.*, *21*, 153 (1988).

516. D. F. Herman and W. K. Nelson, *J. Am. Chem. Soc.*, *75*, 3877 (1953).

517. M. Bochmann, G. Wilkinson, and G. B. Young, *J. Chem. Soc. Dalton Trans.*, 1879 (1980).

518. W. Mowat, A. J. Shortland, N. J. Hill, and G. Wilkinson, *J. Chem. Soc. Dalton Trans.*, 770 (1973).

519. L. Chamberlain, J. Keddington, and I. P. Rothwell, *Organometallics*, *1*, 1098 (1982).

520. H. Schumann, in *Organometallics of the f-Elements*, T. J. Marks and R. D. Fischer, Eds., D. Reidel, Dordrecht, The Netherlands, 1979, p. 103.

521. E. R. Sigurdson and G. Wilkinson, *J. Chem. Soc. Dalton Trans.*, 812 (1977).

522. (a) A. Greco, S. Cesca, and G. Bertolini, *J. Organomet. Chem.*, *113*, 321 (1976). (b) A. Greco, G. Bertolini, and S. Cesca, *Inorg. Chim. Acta*, *21*, 245 (1977).

523. C. Blandy and D. Gervais, *Inorg. Chim. Acta*, *47*, 197 (1981).

524. P. B. Hitchcock, J. A. K. Howard, M. F. Lappert, W.-P. Leung, and S. A. Mason, *J. Chem. Soc. Chem. Commun.*, 847 (1990).

525. (a) P. B. Hitchcock, M. F. Lappert, R. G. Smith, R. A. Bartlett, and P. P. Power, *J. Chem. Soc. Chem. Commun.*, 1007 (1988). (b) J. L. Atwood, M. F. Lapport, R. G. Smith, and H. Zhang, *J. Chem. Soc. Chem. Commun.*, 1308 (1988). (c) C. J. Schaverien and A. G. Orpen, *Inorg. Chem.*, *30*, 4968 (1991). (d) W. G. van der Sluys, C. J. Burns, and A. P. Sattleberger, *Organometallics*, 8, 855 (1989).

526. (a) M. H. Chisholm, K. Folting, J. C. Huffman, and I. P. Rothwell, *J. Am. Chem. Soc.*, *104*, 4389 (1982). (b) M. H. Chisholm, K. Folting, D. M. Hoffman, and J. C. Huffman, *J. Am. Chem. Soc.*, *106*, 6794 (1984).

527. R. R. Schrock, M. L. Listemann, and L. G. Sturgeoff, *J. Am. Chem. Soc.*, *104*, 4291 (1982).

528. M. H. Chisholm, F. A. Cotton, M. W. Extine, and R. L. Kelly, *J. Am. Chem. Soc.*, *101*, 7645 (1979).

529. (a) M. H. Chisholm, J. C. Huffman, and M. J. Hampden-Smith, *J. Am. Chem. Soc.*, *111*, 5284 (1989). (b) M. H. Chisholm and M. J. Hampden-Smith, *J. Am. Chem. Soc.*, *109*, 5871 (1987). (c) M. H. Chisholm and M. J. Hampden-Smith, *Angew. Chem. Int. Ed. Engl.*, *26*, 903 (1987).

530. F. A. Cotton, W. Schwotzer, and E. S. Shamshoum, *Organometallics*, *2*, 1167 (1983).

531. (a) M. H. Chisholm and J. A. Klang, *J. Am. Chem. Soc.*, *111*, 2324 (1989). (b) M. H. Chisholm, K. Folting, and J. A. Klang, *Organometallics*, *9*, 602 (1990). (c) M. H. Chisholm, K. Folting, and J. A. Klang, *Organometallics*, *9*, 607 (1990).

532. (a) M. H. Chisholm, F. A. Cotton, M. W. Extine, and R. L. Kelly, *J. Am. Chem. Soc.*, *100*, 2256 (1978). (b) M. H. Chisholm, D. M. Hoffman, and J. C. Huffman, *Organometallics*, *4*, 986 (1985).

533. D. Segal, *Chemistry Synthesis of Advanced Ceramic Materials*, Cambridge University Press, Cambridge, UK, 1989.

534. S. Sakka, *J. Sol-Gel Sci. Tech.*, *3*, 69 (1994).

535. N. Ya. Turova, E. P. Turevskaya, V. G. Kessler, and M. I. Yanovskaya, *J. Sol-Gel Sci. Tech.*, *2*, 17 (1994).

536. J. Livage, M. Henry, and C. Sanchez, *Prog. Solid State Chem.*, *18*, 259 (1988).

537. L. C. Klein, Ed., *Sol-Gel Technology for Thin Films, Fibres, Preforms Electronic and Speciality Shapes*, Noyes, NJ, 1988.

538. C. J. Brinker, A. J. Hurd, G. C. Frye, K. J. Ward, and C. S. Ashley, *J. Non-Cryst. Solids*, *121*, 294 (1990).

539. S. Sakka, *J. Non-Cryst. Solids*, *121*, 417 (1990).

540. L. D. Raymond, A. B. Matthias, and J. M. Wayne, *Sol–Gel Technology for Thin Films, Fibres, Preforms Electronic and Speciality Shapes*, Noyes, NJ, 1988, pp. 330–381.

541. G. W. Scherer, in *Sol–Gel Science and Technology*, M. A. Aegerter, M. Jaffelicci, Jr., D. R. Sauza, and E. D. Zanotts, Eds., World Scientific, Singapore, 1989, pp. 221–256.

542. S. S. Kistler, *J. Phys. Chem.*, *46*, 19 (1942).

543. J. Fricke, *Sol–Gel Technology for Thin Films, Fibres, Preforms Electronic and Speciality Shapes*, pp. 226–246, Aerogels, North-Holland, 1992.

544. S. Teichner, *Chem. Tech.*, *21*, 372 (1991).

545. R. C. Mehrotra, *J. Ind. Chem. Soc.*, *70*, 4 (1993).

546. C. J. Brinker and G. W. Scherer, *Sol–Gel Science: The Physics and Chemistry of Sol–Gel Processing*, Academic, New York, 1990.

547. L. L. Hench and J. K. West, *Chem. Rev.*, *90*, 33 (1990).

548. S. Sakka, *J. Sol–Gel Sci. Tech.*, *4*, 5 (1995).

549. E. Matijevic, *Acc. Chem. Res.*, *14*, 22 (1981).

550. K. Kezuke, Y. Hiyashi and T. Yamaguchi, *J. Am. Ceram. Soc.*, *72*, 1660 (1989).

551. R. Roy and E. F. Osborne, *Am. Mineral*, *39*, 583 (1956).

552. R. Roy, *J. Solid State Chem.*, *111*, 11 (1994).

553. S. Dire and F. Babonneau, *J. Non-Cryst. Solids*, *167*, 29 (1994).

554. C. J. Brinker and G. W. Scherer, *Sol–Gel Science: The Physics and Chemistry of Sol–Gel Processing*, Academic, New York, 1990, pp. 97–234.

555. C. J. Brinker and G. W. Scherer, *Sol–Gel Science: The Physics and Chemistry of Sol–Gel Processing*, Academic, New York, 1990, pp. 21–96.

556. V. W. Day, T. A. Eberspachev, W. G. Klemperer, C. W. Park, and F. S. Rosenberg, in *Chemical Processing of Advanced Materials*, L. L. Hench and J. K. West, Eds., Wiley, New York, 1992, pp. 257–263.

557. K. Jones, T. J. Davies, M. G. Emblem and P. Parkes, *Mater. Res. Soc. Symp. Proc.*, *73*, 111 (1986).

558. L. F. Francis, V. J. Oh, and D. A. Payne, *J. Mater. Soc.*, *25*, 5007 (1990).

559. O. Poncelet, L. G. Hübert-Pfalzgraf, L. Toupet, and J. C. Daran, *Polyhedron*, *10*, 2045 (1991).

560. R. C. Mehrotra and A. Singh, *Chemtracts*, *6*, 27 (1994).

561. O. K. Liu, R. J. Chin, and A. L. Lai, *Chem. Mater.*, *3*, 13 (1991).

562. M. W. Rupich, B. Lagos, and J. P. Hackney, *Appl. Phys. Lett.*, *55*, 2447 (1989).

563. R. C. Mehrotra, in *Sol–Gel Processing and Applications*, Y. A. Attia, Ed., Plenum Publications, New York, 1994, pp. 41–60.

564. L. G. Hübert-Pfalzgraf, *Mater. Res. Soc. Symp. Proc.*, *271*, 15 (1992).

565. C. J. Page, C. S. Houk, and G. A. Burgoine, *Mater. Res. Soc. Symp. Proc.*, *271*, 155 (1992).

566. G. M. Pajonk, *Appl. Catal.*, *72*, 217 (1991).

567. M. A. Cauqui, J. M. Rodriguez-Izquierdo, *J. Non-Cryst. Solids*, *147–148*, 724 (1992).

568. R. C. Mehrotra, *Proc. Ind. Natl. Sci. Acad.*, India *61A*, 253 (1995).

569. H. Schmidt, *Preparation, Application and Potential of Ormocers*, in *Sol–Gel Science and Technology*, M. A. Aegerter, M. Jaffelicci Jr., D. F. Souza, and E. D. Zanotto, Eds., World Scientific, Singapore, 1989, pp. 432–469.

570. H. Schmidt, H. Scholze, and G. Tinker, *J. Non-Cryst. Solids*, *80*, 557 (1986).

571. D. Anvir, D. Levy, and R. Reisfeld, *J. Phys. Chem.*, *88*, 5956 (1984).

572. H. Schmidt, *J. Sol–Gel Sci. Tech.*, *1*, 217 (1994).

573. F. Babonneau, *Polyhedron*, *13*, 1123 (1994).

574. V. W. Day, W. G. Klemperer, V. V. Mainz, and D. M. Millar, *J. Am. Chem. Soc.*, *107*, 8262 (1985).

575. P. C. Cagle, W. G. Klemperer, and C. A. Semmons, *Mater. Res. Soc. Symp. Proc.*, *180*, 29 (1990).

576. V. W. Day, T. A. Eberspacher, W. G. Klemperer, C. W. Park, and F. S. Rosenberg, *J. Am. Chem. Soc.*, *113*, 8190 (1991).

577. Y. W. Chen, W. G. Klemperer, and C. W. Park, *Mater. Res. Soc. Symp. Proc.*, *271*, 57 (1992).

578. C. Brevard and P. Granger, *Handbook of High Resolution Multinuclear NMR*, Wiley, New York, 1981, pp. 92, 93.

579. Y. Chen, J. Hao, W. G. Klemperer, C. W. Park, and F. S. Rosenberg, *Polym. Prepr. Am. Chem. Soc. Div. Polym. Chem.*, *34*, 250 (1993).

580. M. J. Hampden-Smith and T. T. Kodas, *Polyhedron*, *14*, 699 (1994); *The Chemistry of Metal CVD*, VCH, Weinheim, Germany, 1994.

581. M. L. Hitchman and K. F. Jensen, *Chemical Vapour Deposition, Principles and Applications*, Academic, San Diego, CA, 1993.

582. J. T. Spencer, *Progress in Inorganic Chemistry*, Wiley-Interscience, New York, 1994, vol. 41, p. 145.

583. D. C. Bradley and M. M. Factor, *Trans. Faraday Soc.*, *55*, 2117 (1959).

584. K. S. Mazdiyasni, C. T. Lynch, and J. S. Smith, *Inorg. Chem.*, *5*, 342 (1966).

585. D. C. Cameron, L. D. Irving, G. R. Jones, and J. Woodward, *J. Thin Solid Films*, *91*, 339 (1982).

586. C. C. Wang, K. H. Zaninger, and M. T. Duffy, *R.C.A. Rev.*, *31*, 728 (1970).

587. T. Nandi, M. Rhubright and A. Sen, *Inorg. Chem.*, *29*, 3065 (1990).

588. D. V. Baxter, M. H. Chisholm, V. F. Distasi, and A. J. Klang, *Chem. Mater.*, *3*, 221 (1991).

589. P. M. Jeffries and G. S. Girolami, *Chem. Mater.*, *1*, 8 (1989).

590. M. J. Hampden Smith, T. T. Kordas, M. F. Pfaffet, J. D. Farr, and H. K. Shin, *Chem. Mater.*, *2*, 636 (1990).

# Subject Index

π-Acceptor ancillary ligands, alkoxide ligands and, 411
Acetonitrile, hydrogen bond formation, 55
[(Acetylacetonate)$_2$Al($\mu$-O-$i$-Pr)$_2$Al(O-$i$-Pr)$_2$], 326
Acetylene metathesis reaction, 269–270
Actanides, alkoxide complexes, 283
Adenosine triphosphate (ATP), biological role, 3
Aerogels, 418
[Ag(diphenylpyrazole)]$_3$, crystal structure, 157
[Ag(4-NO$_2$dimethylpyrazole)$_6$](Et$_3$NH)$_2$, 160
[Ag$_{14}$(S-$t$-Bu)$_{14}$(PPh$_3$)$_4$], 121
[Ag$_6$(SC$_6$H$_4$Cl-4)$_6$(PPh$_3$)$_5$], 119
[Ag$_6$(SC$_5$H$_3$N(SiMe$_3$)-6)$_6$], 119
[Ag$_4$(SC$_6$H$_4$SiMe$_3$-2)$_4$], 116
[Ag$_4$(SCH(SiMe$_3$)$_2$)$_4$], 116
[Ag$_8$(SCMe$_2$Et)$_8$], 120
[Ag$_8$(SCMeEt$_2$)$_8$(PPh$_3$)$_2$], 120
[Ag$_4$(SC(SiMe$_3$)$_3$)$_4$], 116
[Ag$_3${SC(SiMe$_2$Ph)$_3$}$_3$], 115
Al{Al(O-$i$-Pr)$_4$}$_3$, properties, 266–267
Alcohol exchange reactions, 345–346, 391
Alcoholysis
  alkoxyalcohol synthesis, 359
  chloroalkoxide synthesis, 355
  in fluoroalkoxide synthesis, 344–345
  metal alkoxide reactions, 255–257
  S$_N$2 mechanism, 256
  sterically congested tri- and tetravalent metal alkoxides, 332
  sterically hindered metal alkoxide synthesis, 331
Aliphatic diols, metalloid complexes of, 383
Alkali alkoxometalate synthesis, 263
Alkali tri-$tert$-butoxyzincates, 253
Alkaline earth metals, dinuclear alkoxides of, 283
Alkenes, alkoxide ligands and, 269–270
Alkoxide ligands
  π-acceptor ancillary ligands and, 411
  electronic spectra and EPR studies, 275
  electronically flexible, 412
  sterically hindered, 411–412

Alkoxide to metal π bonding, 269
Alkoxometalate chlorides
  conversion to alkoxometalate or $\beta$-diketonate, 253
  partially substituted, 252–253
Alkoxyalkoxides
  as bridging or bridging-chelating ligands, 362
  synthesis, 358–361
    anodic dissolution, 358–359
    tetrameric cubane structure, 373
    x-ray crystallography, 361–374
Alkoxyethoxides, coordination modes, 361–362
2-Alkoxyethoxides, polymeric structures, 370
Alkoxysilanes, 248
Alkyl orthosilicates, hydrolysis and condensation reactions, 422
Alkyltin halides, exchange reactions, 262
Alkyltin isopropoxides, exchange reactions, 262
Alkynes, yielding terminal alkylidyne species, 413, 415
[($\eta^3$-Allyl)Pd($\mu$-pz)($\mu$-Cl)Pd($\mu$-pz)($\mu$-Cl)Pd($\eta^3$-allyl)], 191
Al$_2$(NMe$_2$)$_6$, with $t$-BuOH, 259
Al{(O-$t$-Bu)$_3$}$_2$, 259
Al$_2$(O-$t$-Bu)$_6$, M—O bonding, 269
[Al(O-$i$-Pr)$_4$]$^-$, bidentate ligational behavior, 278
[Al{($\mu$-O-$i$-Pr)$_2$Al(O-$i$-Pr)$_2$}$_3$], structural preferences, 277–280
Aluminium alkoxides, alkali alkoxometalates preparation, 263–264
Aluminium trihydride, reaction with alcohols, 261
Aluminum alkoxides, with $\beta$-diketones/$\beta$-ketoesters, 326
Aluminum aryloxides, Al—O bond interaction, 270
Aluminum isopropoxides
  aging phenomenon, 267
  reaction with carboxylic acid, 405
Amide hydrogen bonding, 73–74

# Cumulative Index, Volumes 1–46